Applied Dairy Microbiology

FOOD SCIENCE AND TECHNOLOGY

A Series of Monographs, Textbooks, and Reference Books

1. Flavor Research: Principles and Techniques, *R. Teranishi, I. Hornstein, P. Issenberg, and E. L. Wick*
2. Principles of Enzymology for the Food Sciences, *John R. Whitaker*
3. Low-Temperature Preservation of Foods and Living Matter, *Owen R. Fennema, William D. Powrie, and Elmer H. Marth*
4. Principles of Food Science
   Part I: Food Chemistry, *edited by Owen R. Fennema*
   Part II: Physical Methods of Food Preservation, *Marcus Karel, Owen R. Fennema, and Daryl B. Lund*
5. Food Emulsions, *edited by Stig E. Friberg*
6. Nutritional and Safety Aspects of Food Processing, *edited by Steven R. Tannenbaum*
7. Flavor Research: Recent Advances, *edited by R. Teranishi, Robert A. Flath, and Hiroshi Sugisawa*
8. Computer-Aided Techniques in Food Technology, *edited by Israel Saguy*
9. Handbook of Tropical Foods, *edited by Harvey T. Chan*
10. Antimicrobials in Foods, *edited by Alfred Larry Branen and P. Michael Davidson*
11. Food Constituents and Food Residues: Their Chromatographic Determination, *edited by James F. Lawrence*
12. Aspartame: Physiology and Biochemistry, *edited by Lewis D. Stegink and L. J. Filer, Jr.*
13. Handbook of Vitamins: Nutritional, Biochemical, and Clinical Aspects, *edited by Lawrence J. Machlin*
14. Starch Conversion Technology, *edited by G. M. A. van Beynum and J. A. Roels*
15. Food Chemistry: Second Edition, Revised and Expanded, *edited by Owen R. Fennema*

Applied Dairy Microbiology

edited by

Elmer H. Marth

James L. Steele
University of Wisconsin—Madison
Madison, Wisconsin

MARCEL DEKKER, INC.   NEW YORK · BASEL · HONG KONG

ISBN: 0-8247-0116-X

The publisher offers discounts on this book when ordered in bulk quantities. For more information, write to Special Sales/Professional Marketing at the address below.

This book is printed on acid-free paper.

MARCEL DEKKER, INC.
270 Madison Avenue, New York, New York 10016
http://www.dekker.com

Current printing (last digit):
10 9 8 7 6 5 4 3 2 1

PRINTED IN THE UNITED STATES OF AMERICA

Preface

Two books on dairy microbiology were published in 1957: *Dairy Bacteriology, 4th Edition* (B. W. Hammer and F. J. Babel, John Wiley and Sons, New York) and *Dairy Microbiology* (E. M. Foster, F. E. Nelson, M. L. Speck, R. N. Doetsch, and J. C. Olson, Jr., Prentice-Hall, Englewood Cliffs, New Jersey). Since then, no book on this subject has been published in the United States (although a two-volume work on dairy microbiology appeared in Europe).

When the two aforementioned books were published, there were numerous small dairy farms and dairy factories, and they produced a limited number of products. As time went on, dairy farms evolved into fewer but larger units with cows that produced more milk than in earlier years. Factories, too, decreased in number and increased in size and complexity. Furthermore, these factories began producing a far greater array of products than in the 1950s. All these changes have had an impact on dairy microbiology as it is currently understood and practiced.

Much of the information in the dairy microbiology books of the 1950s resulted from research done in dairy industry or closely related departments of most land grant universities. These departments also trained many of the workers in the dairy industry. As time went on, when problems occurred in other segments of the food industry, faculty in dairy industry departments were often consulted. In some instances, existing faculty responded to the new challenges; in others, faculty were added to work in various non-dairy segments of the food industry. Eventually, most dairy industry departments evolved into food science departments. This led to publication of several books on food microbiology—these books usually contain a chapter or two on dairy microbiology but offer no thorough discussion of the subject.

Although food science departments have replaced most dairy industry departments in land grant universities, research on dairy microbiology has not stopped. In the 1980s, six centers for dairy research were established at various U.S. universities—the availability of funds through these centers and through national and several state promotional organizations served to stimulate research on dairy foods in general and on dairy microbiology in particular. Industrial research in this field has also expanded, but often the resulting information is proprietary.

This book updates and extends information available in earlier texts on dairy microbiology. In a manner unique to this book, it begins with a discussion of the microbiology of the milk-producing animal and how this relates to biosynthesis and quality of raw milk. This is followed by a series of chapters dealing with the microbiology of unfermented (except in a few instances) dairy foods: raw milk, fluid milk products, dried and concentrated milks and whey, frozen dairy desserts, and butter and related products. The book then considers fermented dairy foods by devoting two chapters to microorganisms used to manufacture these foods. The first of these describes starter cultures and how they are used. The second deals with genetics and metabolism of starter bacteria. Fermented dairy foods are discussed in the succeeding two chapters: cultured milks and cream in one, cheese products in the other. Another unique feature of this book is the discussion of probiotics in the chapter on cultured milks and creams. Probiotics refers to the purposeful ingestion of certain bacteria, usually dairy-related lactic acid bacteria, to improve the health and well-being of humans. Use of various microorganisms to produce valuable products through fermentation of whey, the principal by-product of the dairy industry, concludes this part of the book.

During the last four decades of the twentieth century there have been major and minor outbreaks of foodborne illness associated with dairy foods. Some of the outbreaks have been salmonellosis (nonfat dried milk, pasteurized milk, cheese, ice cream), staphylococcal food poisoning (butter, cheese, chocolate milk), and listeriosis (pasteurized milk, cheese, chocolate milk). In addition, pathogens responsible for these and other diseases have occasionally been found in dairy foods that did not cause illness. These developments have prompted concerns about public health in the food industry in general and the dairy industry in particular. Consequently, the largest chapter in this book deals with this important subject. The next chapter discusses control of pathogenic and spoilage microorganisms in processing dairy foods in which the concept of Hazard Analysis and Critical Control Points (HACCP) is emphasized.

Various microbiological tests are done to ensure the quality and safety of dairy foods. Sampling and testing are discussed in the penultimate chapter of the book. Another unique feature of this book is the last chapter which provides information on treatment of dairy wastes, processes that are microbiological in nature.

There is some overlap among chapters in this book. For example, *Listeria monocytogenes, Salmonella,* psychrotrophic bacteria, lactic acid bacteria, milk composition, and bacterial standards for milk and some products are mentioned in more than one chapter. We could have exercised our prerogative as editors and eliminated the duplication, but we elected not to do so because: (a) many persons who use this book will not read it from cover to cover but instead will read one or two chapters of immediate interest and so the information in each chapter should be as complete as possible, (b) removing repetitive material, in most instances, would be detrimental to the flow of thought within a chapter and hence its readability, and (c) repetition enhances the educational value of the book—it's been said that the "three Rs" of learning are repetition, repetition, and repetition.

This book is intended for use by advanced undergraduate and graduate students in food/dairy science and food/dairy microbiology. It will also be useful to persons in the dairy industry—both those producing products and those doing research. In addition, it should be beneficial to students in veterinary medicine and to veterinarians whose practice includes dairy animals. Finally, the book will be helpful to many persons in local, state, and federal regulatory agencies.

Elmer H. Marth
James L. Steele

Contents

Contributors

J. Russell Bishop   Center for Dairy Research, University of Wisconsin—Madison, Madison, Wisconsin

Robert D. Byrne   International Dairy Foods Association, Washington, D.C.

Warren S. Clark, Jr.   American Dairy Products Institute, Chicago, Illinois

Russell S. Flowers   Silliker Laboratories Group, Inc., Homewood, Illinois

Joseph F. Frank   Department of Food Science and Technology, University of Georgia, Athens, Georgia

Stanley E. Gilliland   Department of Animal Science, Oklahoma State University, Stillwater, Oklahoma

Ashraf N. Hassan   Department of Dairy Science, Minia University, Minia, Egypt

David R. Henning   Department of Dairy Science, South Dakota State University, Brookings, South Dakota

Mark E. Johnson   Wisconsin Center for Dairy Research, University of Wisconsin—Madison, Madison, Wisconsin

Jeffrey L. Kornacki Silliker Laboratories of Wisconsin, Inc., Madison, Wisconsin

Richard A. Ledford Department of Food Science, Cornell University, Ithaca, New York

Robert T. Marshall Department of Food Science and Human Nutrition, University of Missouri, Columbia, Missouri

Elliot T. Ryser Department of Animal and Food Sciences, University of Vermont, Burlington, Vermont

James L. Steele Department of Food Science, University of Wisconsin—Madison, Madison, Wisconsin

Paul J. Weimer U.S. Dairy Forage Research Center, Agricultural Research Service, U.S. Department of Agriculture, and Department of Bacteriology, University of Wisconsin—Madison, Madison, Wisconsin

William L. Wendorff Department of Food Science, University of Wisconsin—Madison, Madison, Wisconsin

Charles H. White Department of Food Science and Technology, Mississippi State University, Mississippi State, Mississippi

1
Microbiology of the Dairy Animal

PAUL J. WEIMER

Agricultural Research Service, U.S. Department of Agriculture, and
University of Wisconsin—Madison, Madison, Wisconsin

I. INTRODUCTION

Domestication of ruminant animals and their use to produce milk, meat, wool, and hides represents one of the cornerstone achievements in the history of agriculture. The essential feature of the ruminant animal that has fostered its utility as a dairy animal is the presence of a large pregastric chamber where microbial digestion of feed (particularly fibrous feeds not directly digestible by humans) provides various fermentation products that serve as precursors for efficient and voluminous synthesis of milk. Without this symbiosis between animal and microbe, the dairy industry would not have developed and, indeed, human culture would be vastly different in its food-gathering methods.

The dairy animal is a host to a wide variety of microorganisms. Most of these are the intestinal microbes that are essential for fermentative digestion of the animal's feed. However, a number of other bacteria, fungi, and viruses can induce a pathogenic state in various organ systems, resulting in fatal or nonfatal diseases. This chapter focuses first on the microbiology of digestion by the normal flora and its occasional alteration by opportunistic microbes. Second, this chapter gives a brief overview of the major infectious diseases and their effects on the animal and on the quantity and quality of milk produced. Most of the information pertains to cows, but much of it applies to sheep and goats as well.

II. THE DAIRY ANIMAL

A. Populations and Production

There are nearly 3 billion domestic ruminants in the world, the most numerous and economically important of which are cattle, sheep, and goats (Table 1). Lactating dairy cattle (not including replacement heifers and dry cows) represent nearly one-fifth of the world's domestic cattle population and provide most of the world's milk supply. The numbers of sheep and goats actually used for milk production are difficult to estimate, but they are of major importance in providing protein and energy to the human populations of developing countries, and they fill niche markets for specialty foods in developed countries. Both sheep and goats are regarded as superior to cattle in poor-quality grazing and browsing environments, in part because of their more efficient retention of water and nitrogen (Devendra and Coop, 1982). Several other species (water buffalo, yak, camel, reindeer, and even the nonruminant horse) normally used in some cultures as sources of meat, hides, hair, or draft power also provide milk for human consumption.

Because of their large size and abundant milk production, the Holstein is the predominant breed of dairy cow in use. Improvements in animal breeding and genetics have yielded substantially larger animals over the years, with corresponding increases in feed intake. This factor, combined with a gradual shift

TABLE 1 Worldwide Population of Domestic Ruminants and Worldwide Milk Production, 1992

Species	Population (10^6 head)	Milk production (10^6 metric tons)
Cattle	1,284	
Dairy cattle	225	455.0
Sheep	1,138	7.8
Goats	574	9.6
Buffalo	147	0.5

Source: Food and Agricultural Organization (1993).

to diets having higher energy contents (i.e., higher proportions of grain), has resulted in a progressive increase in average milk production per cow, which, in well-bred and well-managed herds, may approach 13,600 kg (approximately 30,000 lb) per lactation.

Dairy cows are usually maintained on a 305-day lactation schedule, after which the cow is "dried" (by reducing feed and by not milking) for 2 months before calving to permit full development of the calf and to allow the buildup of the body reserves necessary for the next lactatation. After calving, milk production steadily increases over a 6- to 8-week period, then slowly decreases for the rest of the lactation. Normally, the cow is bred again 11 to 12 weeks after calving, and delivers her next calf approximately 40 weeks later. Thus, the cow is pregnant through most of her lactation.

B. Organization of the Digestive Tract

The rumen is the first of the four preintestinal digestive chambers in ruminant animals and is physically proximate to the second chamber, the reticulum (Figure 1). Because of the location and similar function of the rumen and reticulum, their physiology and microbiology are usually considered together. At birth, the ruminant animal is essentially a monogastric animal having a functional abomasum that digests a liquid diet (colostrum and milk) high in protein (Van Soest, 1994). As solids and fiber are gradually introduced into the diet, the other three preintestinal chambers develop over a period of approximately 7 weeks. The rumen is a large organ (approximately 10 L in sheep and goats but up to 150 L in high-producing dairy cows). The reticulum and rumen together constitute approximately 85% of the stomach capacity and contain digesta having 10% to 20% of the animal's weight (Bryant, 1970). The rumen functions as the site of an active microbial fermentation in which feed components are converted to a mixture of volatile fatty acids (VFAs)—acetate, propionate, and butyrate)[*]— which are absorbed through the rumen wall for use by the animal as sources of energy and biosynthetic precursors. Thus, the ruminant animal cannot directly use carbohydrates for energy, and it is absolutely dependent upon its microflora to, in effect, predigest its food.

By virtue of its large size, the rumen slows the rate of passage of feed through the organ, which permits microbial digestion of essentially all of the nonstructural carbohydrate of the feed (starches and sugars) as well as more than

---

[*] In this chapter, these and other organic acids are referred to as their anionic forms, although they are normally metabolized and transported across the cell membrane in their protonated (uncharged) form. An exception is made in the discussion of lactic acidosis (see Sec. IV.D.1), wherein the acid itself is responsible for the pathological condition.

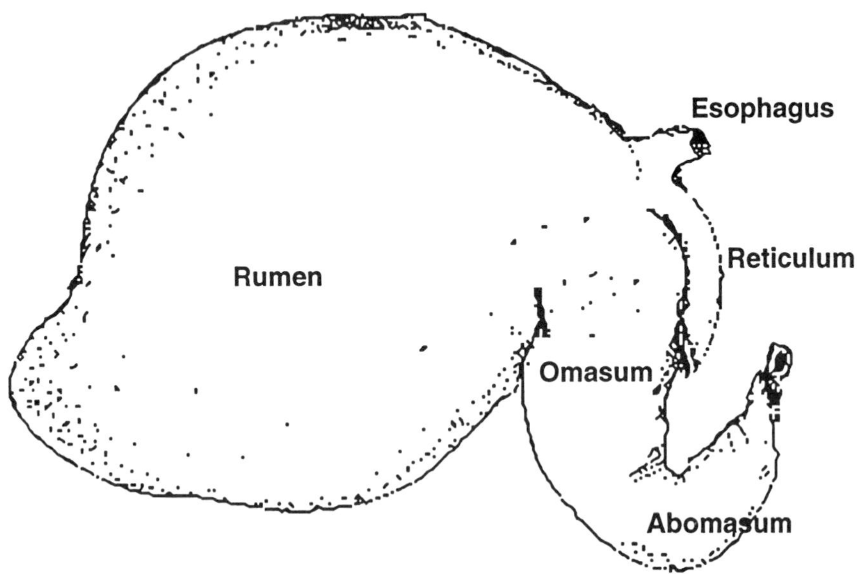

FIGURE 1 Schematic representation of the arrangement of the four preintestinal digestive chambers in the ruminant, illustrating the dominant size of the rumen.

half of the more recalcitrant feed fiber (cellulose and hemicelluloses) (Van Soest, 1994). Rumen contents, which contains 6% to 18% dry matter, are mixed by strong muscular movement and are periodically returned via the esophagus to the mouth for additional chewing (rumination). Even so, the solids have a tendency to stratify, with some maintaining a suspension in the rumen liquor, some settling to the bottom of the rumen, and some being borne up by gas bubbles to form a floating mat at the liquid surface. Passage rates vary with intake, with the rates for solids averaging approximately twice that for liquids. From several published experiments, mean retention times for the rumen liquid range from 8 to 24 hours, whereas that of the particulate phase range from 14 to 52 hours (Broderick et al., 1991). The consequence of these long retention times for solids is that ruminant animals can use fibrous feeds (forages and certain agricultural by-products) that are not used by humans and other monogastric animals, with the ultimate conversion of these feedstuffs to useful products.

In addition to VFAs, other products of the fermentation include microbial cells and fermentation gases. The microbial cells eventually pass through the omasum and into the abomasum (the acidic "true stomach"), where the microbial cell protein is hydrolyzed to amino acids that are available for subsequent intestinal absorption. This microbial protein is a major contributor to the protein requirements of the animal and acts to counterbalance somewhat the considerable loss of feed protein that occurs as a result of microbial proteolysis and amino acid fermentation in the rumen (see Sec. IV.C.5).

Fermentation gases include primarily carbon dioxide (50%–70%) and methane (30%–40%). Rates of gas production immediately after a meal can exceed 30 L/h, and a typical cow may release 500 L of methane per day (Wolin, 1990). Although some gas is absorbed across the rumen wall and carried by the blood to the lungs for exhalation, most is eructated through the mouth.

III. MILK

A. Milk Composition

In the United States, milk has a strict legal definition as "the lacteal secretion, practically free of colostrum, obtained by complete milking of one or more healthy cows" (Office of the Federal Register, 1995); parallel definitions are provided for milk from goats and sheep (United States Public Health Service, 1993). Because of the central role of milk in the food supply and its ease of microbial contamination, the production and processing of milk used for consumption is subject to tight regulation in most developed countries. In the United States, most milk production is regulated according to the "Grade A Pasteurized Milk Ordinance" (U. S. Public Health Service, 1993), a document that sets the standards for all aspects of milk production and processing. From a microbiological standpoint, the "Grade A Pasteurized Milk Ordinance" is important primarily in its setting of the standards for acceptable numbers of viable microorganisms in milk before and after pasteurization. The ordinance sets limits for microbial counts in raw milk for pasteurization at 1×10^5 for milk from an individual producer and 3×10^5 for commingled milk from multiple producers. The ordinance also establishes the permissible levels of antibiotic residues in milk, which affects the selection and implementation of antibiotic therapies to control infectious diseases in dairy animals.

In addition to the direct contamination of milk with pathogens, many microorganisms that are not pathogenic can alter the composition of milk after its synthesis. For example, mycotoxins can produce a deleterious effect on milk. Mycotoxins are secondary metabolites of fungi that can produce various toxic effects ranging from acute poisoning to carcinogenesis. The most widely known mycotoxins are the aflatoxins, produced by *Aspergillus flavus*, *Aspergillus parasiticus*, and *Aspergillus nomius*. Numerous structurally distinct aflatoxins have been identified (Figure 2). The most notorious of these is aflatoxin B_1,

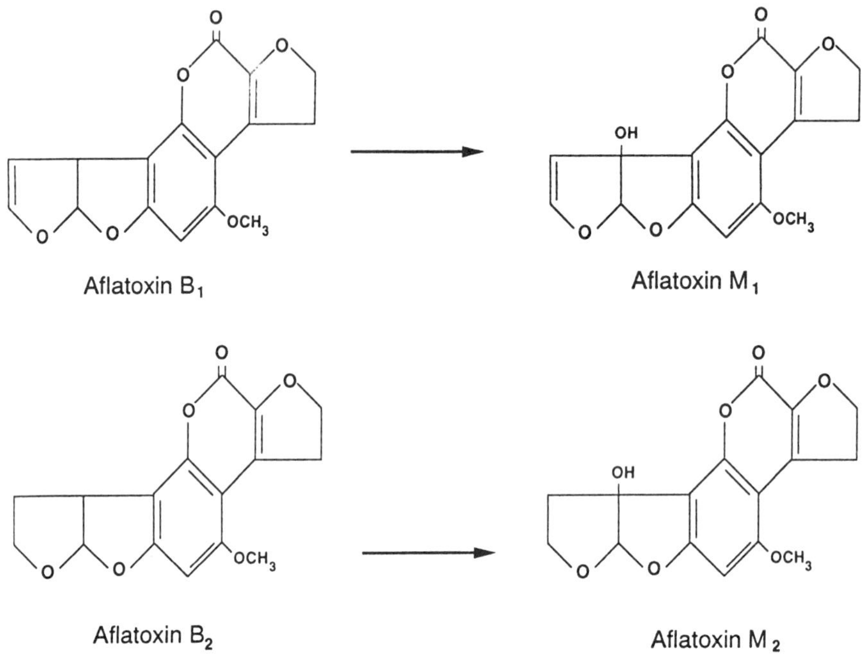

Aflatoxin B_1

Aflatoxin M_1

Aflatoxin B_2

Aflatoxin M_2

FIGURE 2 Ruminal bioconversion of aflatoxins B_1 and B_2 to M_1 and M_2, respectively.

one of the most potent carcinogens known. Milk and dairy products may be contaminated by mycotoxins either directly (by contamination of milk or other dairy products with fungi) or indirectly (by contamination of animal feed with subsequent passage of the mycotoxin to milk) (van Egmond, 1989). In either event, contamination is largely dependent upon environmental conditions that determine the ability of the fungi to grow and product toxins.

Two of the more potent aflatoxins, B_1 and B_2, can be converted in the rumen to their respective 4-hydroxy derivatives, the somewhat less toxic M_1 and M_2. The extent of this conversion varies greatly among cows. For example, Patterson et al. (1980) reported that the M_1 concentration in the milk of six cows fed approximately 10 µg aflatoxin B_1/kg feed varied from 0.01 to 0.33 µg/L milk; on average, approximately 2.2% of the ingested B_1 was converted to M_1. Several other researchers have noted substantial differences in M_1 concentration among cows at similar or different stages of milk production and milk yield, and between milkings of the same cow (Kiermeier et al., 1977; Lafont et al., 1980).

B. Milk Biosynthesis

In evaluating the microbial role in providing the animal with milk precursors, it is useful to briefly describe the biosynthesis of milk. A more detailed treatment of the process is provided by Bondi (1983).

Although the mammary gland comprises only 5% to 7% of the dairy cow's body weight, it represents perhaps the animal's highest concentration of metabolic activity. Careful breeding and advances in nutrition over the years have resulted in the annual production of milk nutrients from a single cow sufficient to provide the nutrients required by 50 calves.

Milk is produced in secretory cells clustered in groups known as alveoli. These cells feed milk through an arborescent duct system that collects milk into the udder. Production of milk is strongly controlled by endocrine hormones. Following parturition, the cells secrete antibody-rich colostrum for several days until milk secretion begins. Continued production of milk is stimulated by suckling or by milking, through the stimulation of several hormones, particularly prolactin.

Nutrients for milk synthesis are provided to the udder through the blood via a pair of major arteries. The ability of the mammary gland to effectively capture milk precursors from the arterial blood supply—expressed as a "percent extraction" calculated from the difference of precursor concentrations in arterial and venous blood—is impressive (Table 2), especially considering that the rapid flow of arterial blood through the udder can approach 20 L/min in dairy cows. Production of one liter of milk requires ~500 L of arterial blood flow through the udder.

TABLE 2 Arterial Concentrations of Milk Precursors and the Efficiency of Their Extraction in the Udder of Goats

Precursor	Arterial concentration (mg/L)	Extraction efficiency (%)
Blood		
O_2	119	45
Glucose	445	33
Acetate	89	63
Lactate	67	30
Plasma		
3-Hydroxybutyrate	58	57
Triglycerides	219	40

Source: Bondi (1983).

Milk is predominantly water (approximately 80%–87%). The major components of milk solids are lactose, protein, and fat. The composition of milk varies with feeding regimens, individual animal, and breed. Marked differences are noted among different ruminant species as well, with sheep's milk having substantially greater content of protein and fat than the milk of cows or goats (Table 3). Much of the energy required for biosynthesis of milk in the udder is produced by oxidation of glucose (30%–50%) or acetate (20%–30%). In the ruminant animal, glucose is not derived directly from dietary carbohydrate, but is instead produced by gluconeogenic pathways, primarily using propionate, a major product of the ruminal fermentation.

Lactose, a disaccharide of D-glucose and D-galactose linked by an α-1,4-glycosidic bond, is synthesized by a series of reactions using D-glucose as the starting substrate. Approximately 60% of the glucose consumed in the mammary gland is used for lactose synthesis. Lactose concentration in milk is relatively invariant with diet and stage of lactation, although its concentration declines substantially in mastitic cows (see Sec. V.A).

Milk fat is a heterogeneous combination of triglycerides, with very few (<2%) phospholipids or sterols. Triglycerides are composed of glycerol esterified to three molecules of fatty acids having 4 to 20 carbon atoms (almost exclusively even numbered). In all mammalian species, the fatty acids are derived in part from circulatory lipoproteins produced from dietary or body fat. These lipoproteins are hydrolyzed at the endothelial capillary wall and are subsequently recombined to produce milk triglycerides. In ruminant animals, almost half of the fatty acids are synthesized from acetate produced in the ruminal fermentation and from 3-hydroxybutyrate produced in the rumen wall from butyrate, another ruminal fermentation product. Milk fat content is subject to variations in diet. Because milk fat is an important determinant of selling price, diets that depress

TABLE 3 Mean Composition of Milk from Domestic Ruminants

	Percent by weight in milk of		
Component	Cow	Goat	Sheep
Fat	3.5	4.5	7.4
Protein	2.9	2.9	5.5
Lactose	4.9	4.1	4.8
Ca	0.12	0.13	0.20
P	0.10	0.11	0.16

Source: Bondi (1983).

milk fat yield are avoided even if they provide good milk yields. The pasteurized milk ordinance stipulates that whole milk in its final packaged form for beverage use shall contain greater than or equal to 8.25% "milk solids not fat" and greater than or equal to 3.25% fat (U. S. Public Health Service, 1993).

Protein in milk is predominantly (82%–86%) casein, with smaller amounts of globulins. Milk proteins are synthesized from amino acids extracted from the arterial blood supply. These amino acids, in turn, are derived from several sources: synthesis by the animal, dietary protein that escapes the rumen, and microbial protein produced in the rumen and hydrolyzed to amino acids and peptides by passage through the abomasum (see Sec. IV.C.5).

IV. MICROBIOLOGY OF THE RUMEN

A. Methods

Rumen microbiology is of historical importance in that the rumen was the first anaerobic habitat whose microbiology was systematically investigated. Many of the techniques for study of strictly anaerobic microbes were developed in these research programs, beginning with the pioneering studies of the research groups of Robert Hungate and Marvin Bryant in the 1940s. Despite the difficulties inherent in studying a habitat of limited accessibility and the requirements for experimental work under strictly anaerobic conditions, the rumen has come to be regarded as one of the best understood of all microbial habitats.

Most studies of ruminal microbes have been conducted in batch culture, usually at fairly high substrate concentrations. This growth mode has been useful in examining the products and kinetics of digestion by mixed ruminal microflora (so called in vitro digestion experiments), in isolating and characterizing pure cultures, and in examining interactions among microorganisms at different trophic levels (e.g., interspecies H_2 transfer reactions; see Sec. IV.C.4). Studies have also been carried out in continuous culture, in which substrates are fed either continuously or at defined (e.g., hourly) intervals. This mode of growth is more useful for some types of studies because, under proper conditions, it can simulate the feeding schedule of the animal.

One type of continuous culture, the chemostat, has been widely used in growth studies. In this mode of culture, one substrate in the feed medium is present at a concentration that limits microbial growth. Feeding of the culture vessel at different volumetric flow rates results in the achievement of a steady-state in which the rate of microbial growth is equal to the dilution rate, that is, (volumetric flow rate)/(working volume of the culture vessel). The chemostat allows the experimenter to examine the microbial response to growth at sub-optimal rates, an important consideration because microbes in nature normally grow at rates well below their maxima (Slater, 1988). Appropriate fitting of data to theoretically derived equations permits determination of fundamental growth parameters such as affinity constants, true growth yields, and maintenance coefficients (see Sec. IV.C.5.b). Until recently, chemostat studies were limited to using soluble substrates, but several new configurations have permitted growth in a continuous mode on insoluble substrates such as cellulose (Kistner and Kornelius, 1990; Weimer et al., 1991). Culture systems have also been constructed that allow differential flow rates for solids and liquids, further approximating conditions in the rumen (Hoover et al., 1983). However, no laboratory culture method can fully simulate the complexities of digestion within the rumen itself, because in vivo digestion involves not only microbial activity but also rumination and mastication, salivary secretions, and recycling of some nutrients.

B. The Ruminal Environment

Much of current understanding of the physiology and microbiology of the rumen has come from in vitro studies of rumen contents. Early studies with rumen contents were obtained by recovering samples from animals at the slaughterhouse, but the microbiology of the rumen under such conditions does not represent that of the living ruminant animal because food is withheld from the animal for at least 24 hours before slaughter. More realistic studies of rumen microbiology were stimulated by development of procedures to sample the rumen via a stomach tube or a surgically implanted fistula (Figure 3), which allows recovery of a more representative grab sample containing both solids and liquor and provides a port for periodic insertion or removal of test materials (e.g., feedstuffs placed in nylon mesh bags) for measurement of digestion in situ.

Ruminal studies have revealed that the physical and chemical conditions within the rumen are fairly constant. Rumen temperature remains within a few degrees of 39°C as a result of heat production by both animal tissues and the intestinal microflora. Despite the continuous influx of oxygen into the rumen through the swallowing of feed and water and through diffusion from the bloodstream via the capillaries feeding the gut epithelial cells, the rumen remains highly anaerobic, with oxygen concentrations ranging from 0.25 to 3 μM (Ellis et al., 1989). Maintenance of these low concentrations of oxygen appears to result from the combined effects of facultative anaerobes and strict anaerobes (protozoa and bacteria). The strict anaerobes can apparently consume substantial amounts of oxygen in reactions involving H_2 oxidation, as long as concentrations of oxygen remain below 7 μM (Ellis et al., 1989). The rumen is not only anaerobic but also highly reducing, with an oxidation-reduction potential near –400 mV.

Ruminal pH varies within the range of approximately 5 to 7 because of opposing forces of microbial fermentation to produce acids on the one hand and

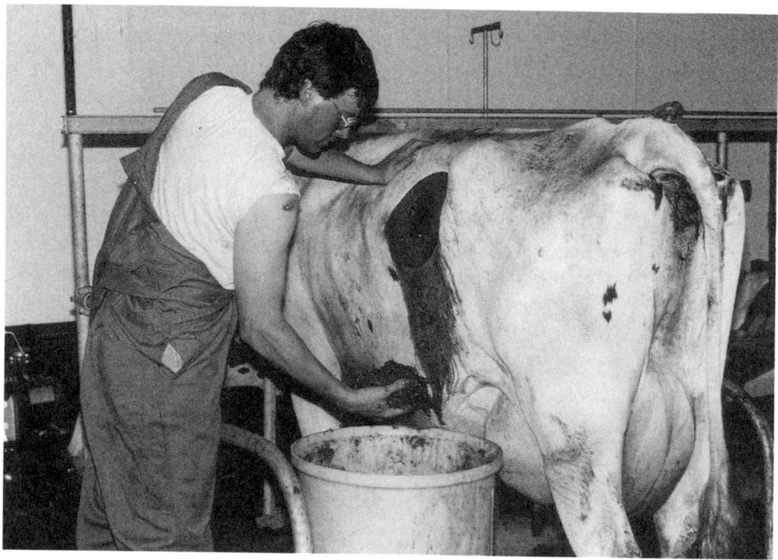

FIGURE 3 A researcher removing a sample of digesta from a ruminally fistulated cow.

their absorptive removal on the other (Table 4). Buffering is provided by secretion of bicarbonate-rich saliva, which, in high-producing dairy cows, may approximate 150 L/d (Church, 1988). Normally, pH is highest immediately before feeding. pH values below 5 are associated with certain undesirable conditions (e.g., lactic acidosis; see Sec. V.E.1).

Total concentrations of ruminal VFAs and their molar proportions vary with diet, but total VFAs are generally near 100 mM, with the molar proportions of acetate, propionate, and butyrate approximately 68%, 20%, and 10% respectively (Mackie and Bryant, 1994); small amounts of isobutyrate, isovalerate, valerate, and caproate are also usually present.

C. The Ruminal Microbial Population

The microbial population in the rumen includes numerous species of bacteria, protozoa, and fungi. There appear to be few differences among cattle, goats, and sheep with regard to either the digestibility of feeds or the species of microbes inhabiting the rumen (Baumgardt et al., 1964; Jones et al., 1972). In terms of sheer numbers of cells, the bacteria far outstrip the eukaryotes, but the latter group, because of their large cell size, contribute considerably to ruminal microbial biomass.

TABLE 4 Factors Controlling Ruminal pH

Factor	Determinants and remarks
pH of feed	Near neutrality for fresh herbage hay and grains Acidic (pH <5) for silages
Acid production	Diet composition (maximum rate and extent of digestion) Feeding schedule (pH highest just before feeding) Microbial populations (species composition and fermentation pathways)
Acid absorption across rumen wall	Fermentation product ratios (volatile fatty acids absorbed faster than lactate)
Salivation	Amount of saliva Buffer capacity of saliva (concentrations of bicarbonate and phosphate)

1. Bacteria

More than 200 different bacterial species have been isolated from rumen contents and their properties determined, but only about 2 dozen species are thought to be of major importance in ruminal metabolism (Table 5). Like any natural environment, the rumen probably contains many other species that have resisted isolation. Moreover, the recent use of phylogenetic criteria (i.e., sequences of evolutionarily conserved macromolecules such as 16S ribosomal RNAs) in taxonomy has altered microbiologists' concepts of what constitutes a microbial species. As a result, new species continue to be described, although the major functional groups of bacteria have probably been identified. The bacterial population can carry out essentially all of the enzymatic reactions that occur in the rumen with regard to digestion of feed materials and bacteria are probably the main agents of ruminal digestion of carbohydrate and protein in feed. The pathways for conversion of carbohydrate (the ruminant animal's major energy source) to different fermentation end products are shown in Figure 4.

Total populations of bacteria in the rumen are hard to measure with accuracy, because a large fraction (perhaps up to 70%) of the cells are attached to solid surfaces, mostly to feed particles (Hobson and Wallace, 1982; Costerton et al., 1987), but, to a certain degree, to the rumen wall as well (Mead and Jones, 1981). Thus, bacterial cell counts of 10^7 to 10^9 cells/mL, normally determined by counting unattached cells under the microscope or by plating onto nonselective culture media, must be regarded as considerable underestimates of the total population. The same must be said for the many studies on quantitating individual species or physiological groups by traditional culture methods. Recent use of nucleotide probes directed toward 16S rRNAs of specific phylogenetic units (e.g., kingdom, species, or strain) has shown great promise for in situ studies of ruminal microbial ecology (Stahl et al., 1988) and has been applied successfully to in vitro studies of ruminal contents (Krause and Russell, 1996) and defined cocultures of ruminal bacteria (Odenyo et al., 1994).

In general, ruminal bacteria are adapted to grow within a fairly narrow range of environmental conditions, which is expected given the relative constancy of environmental conditions in the rumen. Ruminal bacteria are mesophilic, but are highly stenothermal (i.e., they grow within a narrow temperature range). Most have growth optima near the mean ruminal temperature of 39°C and many exhibit poor or no growth at room temperature. Most ruminal bacteria also have some requirements for vitamins and amino acids that are present in low concentrations in the ruminal liquor (Bryant, 1970). Many species also require carbon dioxide for growth (Dehority, 1971). Because environmental conditions are fairly constant and organic growth substrates are continuously available, few ruminal bacteria have developed the capability to form resistant morphological forms, such as cysts or spores. Even though various endospore-forming *Clostridium* species have been isolated from the rumen, they are rarely abundant and, in some instances, may simply be transients that have little involvement in ruminal metabolism (Varel et al., 1995).

2. Protozoa

Because of their large size (100 μm or more in length), protozoa were readily observed microscopically and thus were first described in 1843. Many species of ruminal protozoa have been identified, primarily based on morphological criteria (Hungate, 1966). These can be classified into flagellates and ciliates. Flagellates dominate the ruminal protozoan population of young animals, but are gradually displaced by the ciliates with ageing. The ciliates contain two main groups: the relatively simple holotrichs (e.g., *Isotricha* or *Dasytricha*) or the structurally more complex oligotrichs (e.g., *Entodinia* and *Diplodinia*). The populations of protozoa in the rumen vary widely, but are usually in the range of 10^2 to 10^6/mL. These densities are much lower than those of the bacteria; however, because of their large size, the protozoa may in fact represent up to half of the microbial biomass in the rumen (Van Soest, 1994).

All of the ruminal protozoa appear to have a strictly fermentative metabolism. Relative to the bacteria, much less is known regarding the physiology and biochemistry of the protozoa, for two reasons. First, the protozoa are rather difficult to cultivate in the laboratory (Coleman et al., 1963); ruminal

TABLE 5 Physiological Properties of Ruminal Bacteria

Nutritional type	Gram	Substrates utilized[d]	Products formed[e]	Additional characteristics
Fibrolytic				
Fibrobacter succinogenes	–	C, Cd	Suc, Ac, For	
Ruminococcus albus	+	C, Cd, Xn, Xd	Ac, EtOH, H_2, For, CO_2	
Ruminococcus flavefaciens	+	C, Cd, Xn, Xd, P	Ac, Suc, H_2, For, CO_2	
Butyrivibrio fibrisolvens	–[c]	C, Cd, Xn, Xd, P	For, But, Ac, Lac, EtOH, CO_2	
Clostridium sp.[a]	+	C, G_2, Hx	For, Ac, But, CO_2	Form endospores
Lachnospira multiparus	–	P, G_2	For, Ac, EtOH, Lac, H_2, CO_2	
Succinivibrio dextrinosolvens	–	P, Hx	Suc, Ac, For	
Starch and sugar digesters				
Streptococcus bovis	+	S, G_2, Hx, Prot	Lac, EtOH, Form, Ac, CO_2	Hydrolyzes pectin
Succinomonas amylolytica	–	S, G, Mt	Suc, Ac, Pro, H_2	
Ruminobacter amylophilus	–	S, Mt	Suc, For, Ac	Hydrolyzes pectin
Prevotella ruminicola	–	S, Cd, Hx, Xd, L, F, P, Prot	For, Ac, Pro, Suc	
Selenomonas ruminantium	–[c]	S, Cd, Hx, X, A, Gol, P, Prot, Xd, Suc	Pro, Ac, But, For, Suc, Lac, H_2	Chief agent of Suc→Pro and Lac→Pro
Eubacterium ruminantium	+	Hx, G_2, F, MeOH, P, Xd	For, Ac, But, Lac	
Megasphaera elsdenii	–[c]	S, Mt, Sc, Gol, Pep	For, Ac, Pro, But, H_2, CO_2	Converts Lac→Pro
Treponema bryantii	–	Hx, X, A, G_2, L, Mt	For, Ac, Suc	
Proteolytic/amino acid fermenting				
Peptostreptococcus anaerobius	+	Pep, AA	Ac, But, BCVFAs, NH_3, CO_2	
Clostridium sticklandii	+	Pep, AA	Ac, But, BCVFAs, NH_3, CO_2	
Clostridium aminophilum	+	Pep, AA	Ac, But, BCVFAs, NH_3, CO_2	

		Substrate[d]	Product[e]	
Hydrogen consumers				
Methanobrevibacter bryantii	+	$H_2 + CO_2$	CH_4	Autotrophic
Methanosarcina barkeri	+	$H_2 + CO_2$, MeOH	CH_4	Autotrophic, methylotrophic
Desulfovibrio ruminis	−	$H_2 + SO_4^=$, EtOH, Lac	H_2S; $Ac + H_2$	Produces H_2 from EtOH and Lac in presence of methanogens
Acetitomaculum ruminis	+	$H_2 + CO_2$, Hx	Ac	Probably not an important H_2 consumer in the rumen
Wolinella succinogenes	−	Fumarate + H_2, For, or H_2S	Suc	Can also reduce inorganic nitro compounds (e.g., NO_3^-)
Other nutritional specialists				
Acidaminococcus fermentans	−[c]	Glu, Cit, TAA	Ac, But, H_2	Detoxifies TAA
Anaerovibrio lipolytica	−	TG, Gol, F, Rib	Pro, Ac, But, Suc, H_2, CO_2	
Oxalobacter formigenes	−	Oxalate	For, CO_2	Detoxifies oxalate
Succiniclasticum ruminis	−	Suc	Pro, CO_2	
Synergistes jonesii	+	Arg, His, DHP	Ac, Pro, H_2	
Veillonella parvula[b]	−[c]	Lac, Gol	Ac, Pro, H_2, CO_2	

[a]Includes *Clostridium cellobioparum, Clostridium chartatabidium, Clostridium lochheadii, Clostridium longisporum, Clostridium polysaccharolyticum*. A few of these species also produce ethanol.
[b]Abundant in ovine rumen but not bovine rumen.
[c]Stain gram-negative, but phylogenetically related to gram-positive eubacteria.
[d]Substrates: AA, amino acids; Arg, arginine; Cd, cellodextrins (except where indicated, glucose also fermented); Cit, citrate; DHP, 2,3- and 3,4-dihydroxypyridinediols; EtOH, ethanol; F, fructose; For, formate; G, glucose; G₂, cellobiose; Glu, glutamate; Gol, glycerol; His, histidine; Hx, most common hexose sugars; L, lactose; Lac, lactate; MeOH, methanol; P, pectin; Pep, peptides; Prot, protein; TAA, trans-aconitate; TG, triglycerides; X, xylose, Xd, xylodextrins; Xn, xylan.
[e]Products: Ac, acetate; BCVFAs, branch-chain volatile fatty acids (isobutyrate, isovalerate, 2-methylbutyrate); But, butyrate; EtOH, ethanol; For, formate; Lac, lactate; Pro, propionate; Suc, succinate.

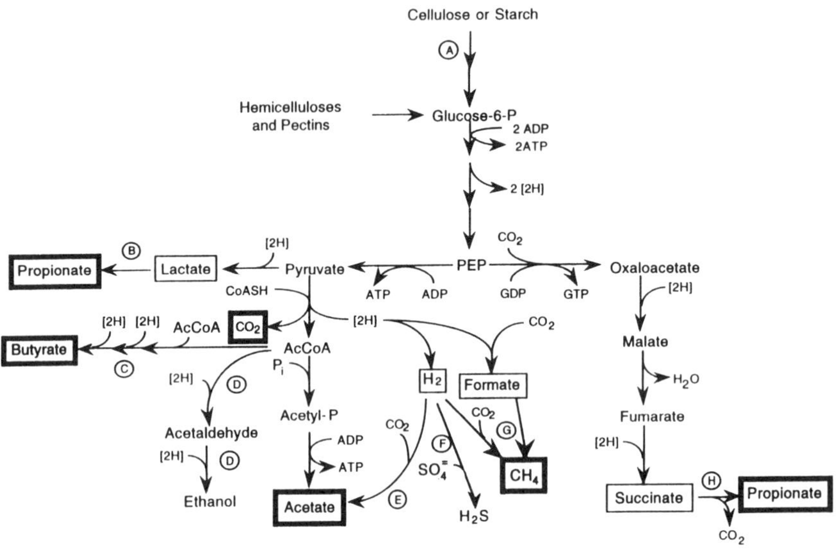

FIGURE 4 Generalized pathway of carbohydrate fermentations in the rumen. Fermentation products in dark-bordered boxes are maintained in substantial concentrations in the normal rumen. Fermentation products in light-bordered boxes are produced and excreted by some organisms but do not accumulate under normal conditions. Abbreviations: [2H], pairs of reducing equivalents; ADP and ATP, adenosine diphosphate or triphosphate; GDP and GTP, guanosine diphosphate and triphosphate; PEP, phosphoenolpyruvate; AcCoA, acetyl coenzyme A. Reactions coded by a circled letter are restricted to a few species, as follows: A, fibrolytic or amylolytic microbes; B, lactate utilizers, particularly *Selenomonas ruminantium* and *Megasphaera elsdenii*; C, *Butyrivibrio fibrisolvens*; D, *Ruminococcus albus, S. ruminantium, Streptococcus bovis*; E, homoacetogenic bacteria (e.g., *Acetitomaculum ruminis*); F, sulfate-reducing bacteria; G, methanogenic bacteria; H, *S. ruminantium* and *Succiniclasticum ruminis*.

protozoa generally die within hours of transferring mixed rumen microflora into most laboratory culture environments. Second, many protozoa in a variety of habitats contain intracellular or surface-attached bacterial symbionts that engage in syntrophic interactions with their hosts (Fenchel et al., 1977; Vogels et al., 1980). Thus, even when "pure" cultures of protozoa (i.e., single protozoal species in the absence of free-living bacteria) are established and maintained, it is difficult to evaluate the potential contribution of the associated bacteria to the metabolic activities of the protozoa. Some continuous culture systems have successfully maintained protozoa by including a floating-mat matrix that allows the protozoa to resist washout from the vessel at fluid dilution rates similar to those operating in the rumen (Abe and Kurihara, 1984), and it is likely that ruminal protozoa associate in vivo with the ruminal mat or the ruminal wall in a similar manner.

Because the relatively large size of protozoa permits microscopic identification of species and behavioral examination, much of the current knowledge of these organisms has come from study of the rumen itself, particularly from comparisons of *faunated* animals (i.e., those having a natural protozoal population) and *defaunated* animals (i.e., those whose protozoal populations have been nearly or completely removed, usually by treatment with chemical agents such as 1,2-dimethyl-5-nitroimidazole or dioctyl sodium sulfosuccinate).

The holotrichs appear to be adapted to growth purely on soluble carbohydrates. On the other hand, microscopic observations have revealed that the entodiniomorphs can engulf plant particles or can attach to the cut ends of plant fiber and can obtain their nutrition from starches and perhaps some structural polysaccharides as well. Despite the observed associations of protozoa and particulate feeds, it is widely believed that the primary ecological role of the entodiniomorph protozoa is the grazing of bacteria (Clarke, 1977; Hobson and Wallace, 1982). Under phase-contrast microscopy, these protozoa can be observed to rapidly ingest bacteria, and bacterial cell concentrations are approximately 10-fold higher in rumen samples from defaunated than in faunated animals. Despite numerous studies (reviewed by Hobson and Wallace, 1982), no specific predatory relationships between particular species of protozoa and bacteria have been identified. Protozoal grazing of bacteria can reduce the availability of microbial protein to ruminants, a notion reflected by lower weight gain in faunated than in defaunated cattle and lambs when tests were conducted with protein-deficient diets—an effect that disappears at higher levels of feed protein.

Protozoa are not the only agents that control bacterial numbers; the rumen maintains substantial populations of bacteriophages (viruses that infect bacteria). Characterization of phage DNAs from rumen contents by pulsed-field electrophoresis (Swain et al., 1996) has revealed that individual animals harbor their own unique populations of phages. Regardless of these differences among host animals, phage populations (as measured by total phage DNA) follow diurnal population cycles related to the populations of the bacterial hosts, with minima and maxima at approximately 2 hours and 10 to 12 hours after feeding, respectively.

3. Fungi

Orpin (1975) demonstrated that several microorganisms originally thought to be flagellated protozoa were actually the zoospore stage of anaerobic fungi. These fungi alternate between a freely motile zoospore stage and a particle-associated thallus. Fungal populations in rumen contents range from 10^4 to 10^5 thallus-forming units per gram of ruminal fluid (Theodorou et al., 1990). Approximately 2 dozen species of these fungi have been identified on the basis of morphology and 16S rRNA sequences (Trinci et al., 1994). Much of current understanding of the metabolic capabilities of the ruminal fungi has been derived from a single species, *Neocallimastix frontalis*.

Ruminal fungi are strictly anaerobic and have a catabolism based on fermentation of carbohydrate. Several species can digest cellulose via extracellular enzymes that are produced in low titer, but have very high specific activities (Wood et al., 1986). The major products of carbohydrate fermentation are acetate, formate, and H_2, with lesser amounts of lactate (primarily the D-isomer), carbon dioxide, and traces of succinate (Borneman et al., 1989). H_2 production occurs via hydrogenosomes, intracellular organelles containing high levels of the enzyme hydrogenase. In pure culture, the amounts of soluble and gaseous fermentation products essentially equal the amount of carbohydrate consumed (Borneman et al., 1989), suggesting that the yield of fungal mycelia is very small. As a result, the ruminal fungi probably contribute little to the total microbial biomass in the rumen. However, the ruminal fungi appear to have specific roles not readily duplicated by bacteria. For example, there is considerable evidence that fungi can attach to and physically disrupt plant tissue during growth by penetration through cell walls and expansion into the pit fields between cells (Akin et al., 1989). This physical disruption is thought to make the plant material more easily broken apart during rumination and thus more available to bacteria, which are more efficient at digesting the individual plant cell components such as cellulose.

D. Microbial Fermentations in the Rumen

1. Structural Carbohydrates

Plant cell walls (the fibrous component of most forages) are composed primarily of cellulose, hemicellulose, pectin, and lignin. These polymers are differentially localized into the different layers of the cell wall (Figure 5). The architecture of the plant cell wall varies greatly with cell type (Harris, 1990). Some cell types such as mesophyll and collenchyma are thin-walled and essentially unlignified, and are thus easily digested. Other cell types such as sclerenchyma and xylem tracheary elements display more complex architectures, with clearly distinct structures. Groups of these cell types are separated from one another by a middle lamella, a highly lignified region that is also rich in pectin. Interior to the middle lamella is the primary wall, the region where wall growth initiates; it is composed primarily of xyloglucans and other hemicelluloses, as well as various wall-associated proteins. The secondary wall is laid down later in development and is

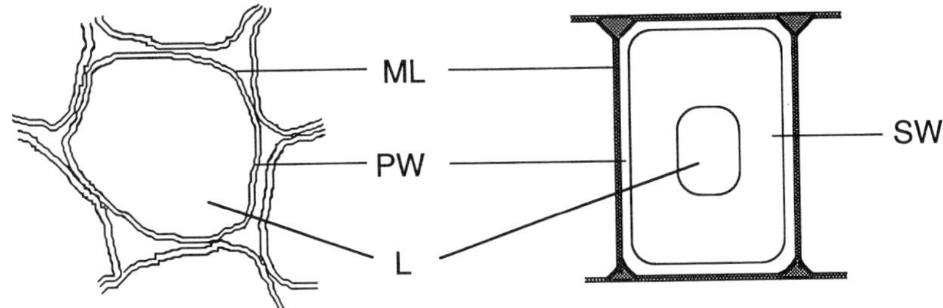

FIGURE 5 Schematic cross-sectional view of the cell wall of two plant cell types. Left panel: Mesophyll cell, characterized by a thin, essentially unlignified primary cell wall that is digested rapidly from both the outer and inner (luminal) surface. The middle lamella is thin and unlignified, and is usually separated from the middle lamellae of adjacent cells by air spaces. Right panel: Sclerenchyma cell, characterized by a thin primary wall and thick secondary walls consisting primarily of cellulose but also containing moderate amounts of hemicelluloses and lignin. Adjacent cells are separated by middle lamellae having a high lignin content. As a result, sclerenchyma cell walls are digested only from the luminal surface outward, and at a relatively slow rate and incomplete extent. Abbreviations: ML, middle lamella; PW, primary wall; SW, secondary wall; L, lumen, which, in the living cell, contains the cytoplasm but is replaced with ruminal fluid during ruminal digestion.

very thick in mature plants. This region, which contains mostly cellulose with smaller amounts of hemicelluloses and lignin, can be further differentiated into layers (S1, S2, S3) based on the orientation of the cellulose microfibrils.

Cellulose

Cellulose is the major component of forage fiber, comprising 35% to 50% of dry weight. Individual cellulose molecules are linear polymers of β-1,4-linked glucose molecules. These chainlike molecules are assembled via extensive intrachain and interchain hydrogen bonds to form crystalline microfibrils that, in turn, are bundled into larger cellulose fibers. The packing of the cellulose chains within the microfibrils is so tight that even water cannot penetrate. Cellulose fibers thus have a fairly low ratio of exposed surface to volume. Ruminal cellulose digestion appears to follow first-order kinetics with respect to cellulose concentration (i.e., the rate of cellulose digestion is limited by the availability of cellulose, rather than by any inherent property of the cellulolytic microbes themselves) (Waldo et al., 1972; Van Soest, 1973).

A number of species of bacteria, fungi, and protozoa have been reported to digest cellulose in vitro, but only three species of bacteria—*Fibrobacter* (formerly *Bacteriodes*) *succinogenes*, *Ruminococcus flavefaciens*, and *Ruminococcus albus*—are thought to be of major importance in cellulose digestion in the rumen (Weimer, 1992; Dehority, 1993). In pure culture, these three species digest crystalline cellulose as a first-order process with rate constants of 0.05 to 0.10 h^{-1}, higher than those of any cellulolytic microbes that grow at a similar temperature in nonruminal habitats (Weimer, 1996). These relatively rapid rates of cellulose digestion are derived, in part, from the ability of these species to attach directly to the cellulosic substrate (Figure 6) and digest the cellulose via cell-bound enzymes; this adherence appears to be a prerequisite to rapid cellulose digestion (Latham et al., 1978; Costerton et al., 1987; Kudo et al., 1987). The cell-associated cellulolytic enzymes are apparently organized into supramolecular complexes resembling the cellulosome, an organelle that has been well characterized in the nonruminal thermophilic bacterium *Clostridium thermocellum* (Felix and Ljungdahl, 1993). Although cellulose digestion in the rumen is more rapid than in nonruminal environments, the process is slow relative to the digestion of nonstructural carbohydrates and proteins. Because of this, forages, with

FIGURE 6 Stereo-optic view of the adherence of the ruminal cellulolytic bacterium *Fibrobacter succinogenes* onto a particle of cellulose. Proper focusing of the eyes or use of a stereo-optic viewer permits a three-dimensional view of the subject. Bar represents 10 µm.

their high rumen fill and slow digestion, must be supplemented with more rapidly digested cereal grains to adequately balance energy and protein requirements for high-producing dairy animals (Van Soest, 1994).

The products of cellulose hydrolysis are cellodextrins (short water-soluble β-1,4-glucosides of two to eight glucose units) that are subject to fermentation by both cellulolytic and noncellulolytic species (Russell, 1985). Whereas the individual cellulolytic species can compete directly for cellulose in vitro, they show differential ability to adhere to different plant cell types (Latham et al., 1978), which may indicate separate but overlapping niches in the rumen. Moreover, it appears that degradations of some plant cell types is delayed by the slow diffusion of nonmotile fibrolytic bacteria into the cell plant lumen (Wilson and Mertens, 1995). These cell types may provide a niche for motile cellulolytic species such as *Butyrivibrio fibrisolvens.*

The three major cellulolytic species form different fermentation end products (Hungate, 1966). *F. succinogenes* produces primarily succinate (an important precursor of propionate) with lesser amounts of acetate. *R. flavefaciens* produces the same acids, but with acetate predominating. *R. albus* produces primarily acetate and ethanol in pure culture, but in the rumen produces mostly acetate and H_2. Unlike other ruminal bacteria, the ratio of fermentation end products formed by each of the predominant cellulolytic species changes little with growth conditions (pH or growth rate). As a result, the relative populations of each species may have a major role in determining the proportions of acetate and propionate in the rumen, particularly in forage-fed animals (Weimer, 1996).

Hemicelluloses

Hemicelluloses are polymers that may comprise up to one-third of plant cell wall material. They are a diffuse class of structural carbohydrates that may contain any of a number of monomeric units (Stephen, 1983). Most hemicelluloses contain a main backbone, usually having β-1, 4- or β-1,3-glycosyl linkages; various types and degrees of branching from the main chain are frequently observed. Because of the multiplicity of hemicellulose structures present in each plant species, it is extremely difficult to isolate pure substrates of known structure, a fact that has severely limited the laboratory study of hemicellulose digestion. Among the most abundant of the hemicelluloses are the xylans (unbranched β-1,4-linked polymers of xylose) and the arabinoxylans (xylans containing pendant arabinose side chains). The latter are particularly important because they are thought to be covalently linked to lignin via cinnamic acid derivatives such as ferulic acid and *p*-coumaric acid (Hatfield, 1993).

Hemicelluloses are hydrolyzed by enzymes that may be extracellular or cell associated, depending on the species (Hespell and Whithead, 1990).

The most active hemicellulose digesters among the ruminal bacterial isolates include *B. fibrisolvens* and the cellulolytic species *R. flavefaciens, R. albus,* and *F. succinogenes*; the last of these can hydrolyze hemicelluloses in vitro, but cannot use the hydrolytic products for growth (Dehority, 1973).

Pectic Materials

Pectins are polymers of galacturonic acids, some of which also contain substantial amounts of neutral sugars (e.g., arabinose, rhamnose, and galactose). Pectins are more abundant in leaf tissue than in stems and are major components of some by-product feeds (citrus pulp and fruit processing waste). Although purified pectins from forages are fairly water soluble, they can be considered to be structural carbohydrates because they are localized in the plant cell wall, particularly in the middle lamellae between cells.

Pectins are, in many respects, an ideal substrate for ruminal fermentation. They are rapidly digested out of both alfalfa leaves and stems (rate constants of approximately 0.3 h^{-1}), but, unlike starch, do not yield lactic acid as a fermentation product (Hatfield and Weimer, 1995). The acetate–propionate ratio resulting from fermentation of pectins is in the range of 6 to 12, well above that of most substrates and useful in maintaining milk fat levels in lactating dairy cows. Production of these acids is accompanied by consumption of the galacturonic acid moieties of the pectin, thus assisting in the maintenance of rumen pH. Several bacterial species actively degrade pectin, including *Lachnospira multiparus, B. fibrisolvens, Prevotella* (formerly *Bacteroides*) *ruminicola,* some strains of the genus *Ruminococcus* (Gradel and Dehority, 1972) and some spirochetes (Ziolecki, 1979).

Lignin

Lignin, the third major component of the forage cell wall, is a polymer of phenylpropanoid units assembled by a random free-radical condensation mechanism during cell wall biosynthesis. Lignin is indigestible under anaerobic conditions and constitutes the bulk of the indigestible material leaving the digestive tract. Moreover, the covalent linkages between lignin (or phenolic acids) and hemicelluloses reduce the digestibility of these forage components (Hatfield, 1993). Electron microscopic studies clearly reveal the recalcitrance of lignified tissues to ruminal digestion (Akin, 1979).

2. Nonstructural Carbohydrates

Nonstructural carbohydrates are those carbohydrates in plant cells that are contained in the cytoplasm or in storage vacuoles. The most abundant of these are the starches (the linear amylose and the branched amylopectin), which are major components of cereal grains (e.g., corn) that comprise much of the diet of high-producing dairy cows.

Starch

Starches are depolymerized fairly rapidly by extracellular enzymes (amylases and pullulanases) that produce maltodextrins (α-1,4-oligomers of glucose), which are easily converted by other α-glucosidases to glucose and maltose—substrates usable by almost all of the carbohydrate-fermenting microbes in the rumen (Hungate, 1966). Consequently, starches have the potential to be completely digestible, although the form of the starch is an important determinant of the rate of digestion. Wheat and barley starch are digested more rapidly than is that of high moisture corn, which, in turn, is digested more rapidly than dried corn or dried sorghum. The more rapidly digesting starches have first-order rate constants of digestion of approximately 0.25 h^{-1}.

Several bacterial species are important in starch digestion, including *Ruminobacter* (formerly *Bacteroides*) *amylophilus, B. fibrisolvens, P. ruminicola, Succinomonas amylolytica, Succinivibrio dextrinosolvens*, and *Streptococcus bovis*. The latter species can grow extremely rapidly, particularly on glucose (minimum doubling time is 13 minutes) and is the causative agent of lactic acidosis (see Sec. V.D.1). Some protozoa actively engulf starch granules but do not appear to produce lactate, thus sequestering these granules from serving as substrates for bacterial lactate production.

Even though diets high in grain content are usually preferred for high-producing cows because of their greater energy density, presence of an adequate level of fiber in the diet is important for several reasons (Van Soest, 1994). Fiber promotes the long-term health of the ruminant animal by providing a modest rate of carbohydrate digestion and by stimulating rumination and salivation, all of which aid in maintaining rumen pH within a range desirable for balanced microbial activity. Moreover, fiber in the diet helps the animal avoid milkfat depression, a syndrome resulting primarily from a relative deficiency in acetic acid (a precursor of short-chain fatty acids in milk triglycerides) and a relative excess of propionate, which inhibits mobilization of body fat (a precursor of long-chain fatty acids in milk triglycerides).

Soluble Sugars and Oligomers

Many ruminal carbohydrate-fermenting bacteria can use most of the different monosaccharides that comprise the various plant polysaccharides (Hungate, 1966): D-glucose, D-xylose, D-galactose, L-arabinose, and D- or L-rhamnose. Many can also use at least some oligosaccharides that are released from the plant cytoplasm by cell wall breakage or that are produced by enzymatic hydrolysis of plant polysaccharides. The latter include cellodextrins (Russell, 1985) and xylooligosaccharides (Cotta, 1993) having seven or fewer glycosyl residues. Concentrations of soluble sugars and their oligomers are maintained at very low levels in the rumen, indicating that biopolymer hydrolysis is the rate-limiting step in digestion and that competition for soluble carbohydrates is probably an important determinant of species composition in the rumen (Russell and Baldwin, 1979a).

In the few cases that have been systematically examined, sugar fermenters have shown dramatic changes in fermentation product ratios with changes in growth rate. Both *S. bovis* (Russell and Hino, 1985) and *Selenomonas ruminantium* (Melville et al., 1988) carry out mixed acid fermentations at low growth rates but nearly homolactic fermentations at growth rates near their maxima.

3. Conversion of Fermentation Intermediate Compounds to Volatile Fatty Acids

Microbial fermentation of both structural and nonstructural polysaccharides produces a mixture of VFAs (usually acetic acid with some butyric acid) and other fermentation acids (succinic, lactic, and formic) that are further metabolized by other ruminal microbes. Most of these bacteria require additional growth factors such as amino acids, peptides, and vitamins. Succinate is decarboxylated to propionate (see Figure 4) by several ruminal species, including the metabolically versatile *S. ruminantium* and the metabolically specialized *Succiniclasticum ruminis* (van Gylswyk, 1995). Lactate is converted to propionate by several bacterial species, particularly *S. ruminantium, Megasphaera elsdenii, Veillonella parvula, Anaerovibrio lipolytica,* and some *Propionibacterium* spp. (Mackie and Heath, 1979). Formate is produced in abundance in the rumen, both from carbohydrate fermentation and from reduction of carbon dioxide. Formate is rapidly turned over to methane and rarely accumulates (Hungate et al., 1970).

4. H_2 Consumption and Interspecies Hydrogen Transfer

Anaerobic metabolism requires that electrons (reducing equivalents) generated from biological oxidations be transferred to terminal electron acceptors other than oxygen. Most anaerobes that ferment carbohydrates dispose of these electrons by transfer to one or more organic intermediate compounds in the catabolic pathways such as pyruvate (producing lactate), acetyl-coenzyme A and acetaldehyde (producing ethanol), and carbon dioxide (producing formate) (see Figure 4). An alternative electron acceptor is the protons present in all aqueous environments, resulting in production of hydrogen gas (H_2). Disposal of electrons as H_2 is particularly advantageous in that it does not consume carbon-containing intermediate compounds that may be used as biosynthetic precursors. However, production of H_2 is thermodynamically unfavorable unless its production is coupled to its continuous removal by H_2-consuming reactions. This spatial and temporal coupling of H_2 production with H_2 use, referred to as interspecies H_2 transfer, is one of the most important processes in the ecology of anaerobic habitats (Oremland, 1988; Wolin, 1990). Interspecies H_2 transfer benefits both the H_2 consumer, which directly receives its energy source, and the H_2 producer, which can channel more of its substrate into the ATP-yielding production of acetate as a fermentation end product (Table 6).

The dominant H_2-consuming reaction in the rumen is the reproduction of carbon dioxide to methane gas:

$$4 H_2 + CO_2 \rightarrow CH_4 + 2 H_2O$$

This reaction is carried out by a specialized group of organisms, the methanogenic bacteria. These organisms are classified with the Archaea, a phylogenetically distinct group that represents an early evolutionary lineage distinct from both eubacteria (true bacteria) and eukaryotes (Woese and Olsen, 1986). Methanogens are highly specialized metabolically. Most are restricted in their catabolism to reduction of carbon dioxide to methane, using H_2 as an electron donor, whereas a few have the ability to convert one or more simple organic compounds (methanol, methylamine, formate, or acetate) to methane (Oremland, 1988). Methanol may be periodically available in the rumen from deesterification of pectins. Formate, although not a major ruminal fermentation product, is probably

TABLE 6 Fermentation Products From Cellulose in *Ruminococcus albus* Monocultures and *R. albus/Methanobrevibacter smithii* Cocultures, Illustrating Changes caused by Interspecies Transfer of H_2 to the Methanogen

Product	mmol/100 mmol glucose equivalents consumed[a]	
	R. albus alone	*R. albus + M. smithii*
Ethanol	81	22
Formate	14	0
Acetate	89	151
CO_2	156	98
H_2	140	0
CH_4	0	75[b]

Source: Pavlostathis et al. (1990).
[a]Mean values from continuous culture runs conducted at five different dilution rates.
[b]Equivalent to 300 mmol H_2 consumed (at a stoichiometry of 4 mol H_2 consumed per mol CH_4 formed).

produced by carbon dioxide reduction in amounts sufficient to contribute slightly to ruminal methanogenesis. Acetate, which is in the rumen, does not support the growth of "aceticlastic" methanogens, whose growth rates, even under ideal conditions, are well below the dilution rates of both liquids and solids in the rumen. Most methanogens are also autotrophs, that is, they can obtain all of their cell carbon from carbon dioxide. Thus, they can produce microbial protein for the ruminant host without consumption of otherwise useful organic matter.

The energy associated with reduction of the abundant ruminal carbon dioxide to methane is sufficient to permit both growth of the methanogens and thermodynamic displacement or "pulling" of the reduction of protons to H_2. As a result, the concentration of H_2 in ruminal fluid is very low—normally approximately 1 μM, with only occasional excursions to approximately 20 μM for a few minutes after feeding (Smolenski and Robinson, 1988). Thus, ruminal methanogenesis, which is viewed unfavorably by nutritionists as a loss of approximately 8% of the metabolizable energy of the feed, in fact has an important thermodynamic function that permits an adequate rate and extent of carbohydrate fermentation.

Representatives of another group of bacteria, the carbon dioxide–reducing homoacetogens, have been isolated from the rumen and appear to be present at low cell densities. Like the methanogens, these eubacteria can reduce carbon dioxide with H_2, but according to the stoichiometry:

$$4\,H_2 + 2\,CO_2 \rightarrow CH_3COOH + 2\,H_2O$$

The homoacetogens have attracted interest as potential competitors of the methanogens, in that they could, in principle, remove fermentatively produced H_2 while at the same time producing acetic acid, an energy source and biosynthetic precursor that the ruminant animal is well equipped to use (Mackie and Bryant, 1994). Numerous in vitro studies have shown that the acetogens are ineffective competitors of the methanogens because of the latter's superior affinity for low concentrations of H_2. The actual role of the acetogens in the rumen is presently unclear; because this metabolically diverse group is capable of sugar fermentation and removal of methoxyl groups from some feed constituents, its members may fill several niches.

A third group of H_2 utilizers, the sulfate-reducing bacteria, can couple the oxidation of H_2 or certain organic compounds such as lactate to reduction of sulfate (Odom and Singleton, 1993):

$$4\,H_2 + 2\,H^+ + SO_4^= \rightarrow H_2S + 4\,H_2O$$

Sulfate-reducing bacteria have an affinity for H_2 that even surpasses that of the methanogens; sulfate reduction is the dominant means of disposal of excess electrons in sulfate-rich environments (e.g., ocean sediments). Sulfate-reducing bacteria have the unusual capacity to act as H_2 consumers when sulfate is abundant or as H_2 producers (from lactate) when sulfate is absent (Bryant et al., 1977). In the latter situation, the sulfate reducers are probably maintained in the rumen by a symbiotic interaction with methanogens, wherein the sulfate reducers oxidize lactate to H_2, whose concentration is kept low by methanogenic activity.

5. Nitrogen Metabolism in the Rumen

Protein Degradation

Availability of protein to the ruminant animal is determined by the amount of protein in the feed, its loss in the rumen from microbial fermentation, and the efficiency of microbial protein synthesis that occurs in the rumen. It is estimated that approximately 35% to 80% of the protein of most forages and grains is degraded by ruminal fermentation and is thus not directly available for intestinal absorption (National Research Council, 1985). Hydrolysis of protein is dependent upon a number of factors, particularly solubility, which determines both its availability to ruminal microbes and its rate of escape from the rumen. The generalized scheme of protein degradation (Figure 7) suggests some similarities to polysaccharide degradations. Proteins are hydrolyzed extracellularly or at the microbial cell surface to produce soluble oligomers that serve as the actual growth substrates. Major proteolytic species in the rumen are *B. fibrisolvens,*

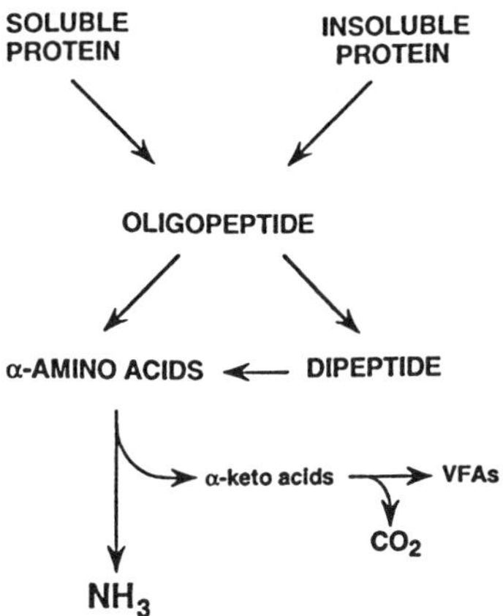

FIGURE 7 Generalized scheme of protein degradation in the rumen. Both bacteria and protozoa participate in the process. α-Keto acids may be used intracellularly as anabolic intermediates or decarboxylated to volatile fatty acids (VFAs), which are then exported.

S. bovis, and *P. ruminicola*. These species also have important roles in carbohydrate fermentation.

The fermentation of amino acids and peptides released from protein hydrolysis is carried out by a number of ruminal species. The most active appear to be *Clostridium aminophilum, Clostridium sticklandii*, and *Peptostreptococcus anaerobius*. Classical proteolytic species such as *P. ruminicola* may be important in protein hydrolysis, but are probably less important in amino acid fermentations, because their rates of ammonia production from amino acids in vitro are one or two orders of magnitude lower. Both *C. sticklandii* and *P. anaerobius* are monensin sensitive, which may explain the protein-sparing effect observed upon inclusion of monensin in ruminant diets (Krause and Russell, 1996). Because the concentrations of peptide and free amino acids in the rumen are very low, competition for these substrates among both proteolytic and nonproteolytic microbes is probably intense.

Protein Synthesis

Whereas the ruminal microflora is responsible for this extensive loss of feed protein, they also contribute up to half of the nitrogen requirements of the animal through synthesis of microbial cell protein, which is hydrolyzed in the abomasum and is subsequently available to the animal (Ørskov, 1982). Protein synthesis by ruminal bacteria occurs primarily from ammonia and organic acids. Most ruminal bacteria will grow in vitro on ammonia as the sole nitrogen source, and many species cannot incorporate significant amounts of amino acids or peptides. Ruminal ammonia is supplied either as a direct product of the ruminal degradation of feed proteins or from urea recycled back into the rumen by the animal. The organic acids used for protein synthesis are derived from both protein and carbohydrate fermentation. Availability of these organic acids is important for adequate carbohydrate nutrition. For example, the predominant ruminal cellulolytic bacteria require isobutyrate, isovalerate, and/or 2-methylbutyrate as precursors for intracellular synthesis of the branch-chain amino acids valine, leucine, and isoleucine, respectively (Bryant, 1970). This is an excellent example of both the interactions among different physiological groups of ruminal bacteria, and the interaction between energy and protein metabolism in ruminant nutrition.

Because of their impact on production of microbial protein, the quantitative aspects of microbial cell yield have received considerable attention. The efficiency of microbial growth (growth yield) varies among species and with growth conditions. Important determinants of growth yield include (a) efficiency of energy conservation (ATP production per unit substrate consumed), (b) the ability to import and incorporate preformed organic compounds (e.g., amino acids) into cell material, (c) maintenance energy (the amount of energy that must be expended to maintain cellular constituents and function), and (d) the extent to which cells carry out other nongrowth related functions such as polysaccharide storage or wasteful "energy spilling" (Russell and Cook, 1995). The growth rate of a microbe also has an impact on cell yield. At low growth rates, yields are depressed somewhat because a larger portion of the total energy expenditure is devoted to maintenance.

Carbohydrate-fermenting ruminal bacteria have true growth yields (cell yields not corrected for maintenance) within the range of 0.1 to 0.6 g cells/g carbohydrate; in some instances, these yields may be artificially high if the organisms synthesize storage polysaccharides (Table 7). Cell yields of ruminal bacteria decline when the pH of the environment decreases below 6 (Russell and Dombrowski, 1980). Nevertheless, the growth yields of ruminal bacteria are generally higher than those of anaerobic bacteria native to other anaerobic environments (Hespell, 1979).

Microbial growth yield is affected by growth rate–induced metabolic shifts that alter the ATP yield. For example, increased growth rate on sugars in some species is accomplished by a shift in fermentation products from acetate to lactate and a reduced ATP yield (because conversion of pyruvate to acetate results in formation of one unit of ATP, whereas conversion of pyruvate to lactate does not) (see Figure 4). In this instance, the organisms have increased growth rate by selecting a pathway with an inherently high substrate flux (rate of substrate

TABLE 7 Growth Yields and Maintenance Coefficients for Several Species of Ruminal Bacteria Grown in Continuous Culture

Bacterium	Substrate	Y_g[a]	m[b]	Reference
Butyrivibrio fibrisolvens	Glucose	0.40	0.049	c
Megasphaera elsdenii	Glucose	0.46	0.187	c
Prevotella ruminicola	Glucose	0.50	0.135	c
Selenomonas ruminantium	Glucose	0.58	0.022	c
Streptococcus bovis	Glucose	0.40	0.150	c
Fibrobacter succinogenes	Cellulose	0.24	0.05	d
Ruminococcus albus	Cellulose	0.11	0.10	e
	Cellobiose	0.28	0.04	f
Ruminococcus flavefaciens	Cellulose	0.24	0.07	g

[a]True growth yield (g cells/g substrate consumed), calculated in the absence of maintenance.
[b]Maintenance coefficient (g substrate consumed/g cells/h).
[c]Russell and Baldwin (1979b).
[d]Weimer (1993).
[e]Pavlostathis et al. (1988).
[f]Thurston et al. (1993).
[g]Weimer et al. (1991).

consumed per unit time) at the sacrifice of some ATP yield. By contrast, inter-species H_2 transfer reactions increase the ATP yield of the H_2 producers by allowing more of the organic substrate to be converted to acetate and less to other compounds (e.g., ethanol or lactate) (Wolin, 1990).

E. Microbial Contributions to Rumen Dysfunction

Under some conditions, the normal ruminal microflora contribute through their activities to certain metabolic diseases (i.e., diseases that are neither infectious nor degenerative and that are preventable by proper feeding and management).

1. Lactic Acidosis

Lactic acidosis is an acute acidification of the rumen resulting from the microbial overproduction of lactic acid (Slyter, 1976). The condition is normally observed in ruminant animals fed diets high in grains, particularly after a switch from diets higher in fiber content. These concentrates are rich in starches and have a relatively poor buffering capacity. The starches are fermented rapidly to lactic acid, primarily by *S. bovis*, a normal rumen inhabitant. At near-neutral pH, *S. bovis* produces primarily formic and acetic acids and only small amounts of lactic acid, but, under conditions of rapid growth, it carries out a homolactic fermentation, producing the D-isomer. The explosive growth of *S. bovis* outpaces the activities of ruminal lactate consumers (e.g., *S. ruminantium* and *Megasphera elsdenii*). As a result, lactic acid levels may increase from normal values of less than or equal to 1 mM to 20 to 300 mM. Because the acidity of lactic acid is 10-fold greater (pK_a = 3.8) than for the VFAs—acetic, propionic and butyric acids—(pK_a = 4.7–4.8), rumen pH may decrease to 4.5 or below. At high lactic acid concentrations, blood and body tissues attempt to restore proper osmolality to the rumen, leading to a systemic dehydration that may be fatal.

Once acidosis has begun, a number of factors combine to further exacerbate the problem (Russell and Hino, 1985). When pH decreased sufficiently, *S. bovis* maintains its homolactic metabolism even as its growth rate decreases. Reduced pH also inhibits degradation of lactate by *S. ruminantium* and *M. elsdenii*, and establishes a ruminal niche for other homolactic fermenters such as the facultatively anaerobic lactobacilli.

Even in nonfatal cases, animal health is severely affected. D-Lactic acid is absorbed into the bloodstream, where it is metabolized more slowly than is the L-isomer. As a result, blood pH decreases and pathological condition of other tissues becomes important (e.g., ulceration of the rumen wall, liver abscess, and foot disorders). Low rumen pH also negatively affects milk production and live weight gain, because fiber digestion is inhibited and feed intake is reduced (Van Soest, 1994).

2. Foamy Bloat

Foamy (or frothy) bloat is an acute condition resulting from formation of a rigid, persistent foam mat at the rumen liquor surface that prevents normal eruptive release of fermentation gases (Clarke and Reid, 1974). Foamy bloat is particularly common in pastured dairy cattle grazing certain lush feeds, especially some legumes (clovers and alfalfa). Gas accumulation results in substantial distention of the reticulorumen. In severe cases, this distention can interfere with respiratory function and produce death within an hour of feeding unless strong remedial action (i.e., puncture of the rumen wall) is taken. Even in cases of mild bloat, dairy production and animal weight gain may be affected substantially because of reduced feed intake.

Plant factors that have been suggested as contributing to induction of bloat include (a) a high content of certain constituents that may contribute to the structure of the foam mat (soluble proteins, pectin, saponins, or certain classes of lipids) and (b) a high rate of fermentation (usually related to high concentration of soluble sugars and an easily digested cell wall). The amount and characteristics of the plant protein appear to be particularly important. Forages containing high levels of condensed tannins (e.g., birdsfoot trefoil) do not cause bloat, and feeding of condensed tannins usually prevents bloat, apparently because of their capacity to precipitate proteins (Tanner et al., 1995). Animal factors are also involved in bloating; there is a clear genetic predisposition toward bloat resistance and bloat sensitivity (Clarke and Reid, 1974).

The involvement of microbes in bloat is controversial (Clarke and Reid, 1974). Microbes certainly are involved to the extent that the ruminal fermentation produces CH_4 and carbon dioxide gas and the acids reduce the ruminal pH and cause release of carbon dioxide from the ruminal bicarbonate pool. More direct roles of individual species of bacteria and protozoa have been difficult to establish. However, microbial involvement is suggested by two lines of evidence: (a) bloat is routinely and effectively inhibited by controlled release of monensin into the rumen (Cameron and Malmo, 1993) and (b) complete switching of rumen contents between fistulated cattle having a high or low susceptibility to bloat results in a change of susceptibility that is maintained for approximately 24 hours before the animal's natural susceptibility or resistance reasserts itself (Clarke and Reid, 1974).

3. Polioencephalomalacia

Polioencephalomalacia (PEM) also known as cerebrocorticol necrosis, is an acute toxicosis that causes destruction of tissues of the central nervous system. It manifests as lethargy and sometimes blindness that progresses to muscular tremors and coma, with death following within a few days. PEM was attributed to a thiamin deficiency that may result from elevated levels of thiaminases. More recent data indicate that, in many cases, the disease results from conversion of ingested sulfates to highly toxic hydrogen sulfide (H_2S) by sulfate-reducing bacteria (Cummings et al., 1995) (see Sec. IV.D). Sulfate is not normally a component of dairy rations, but it can be present in high concentrations in some groundwaters and surface waters used for watering stock, particularly in the western United States where the disease was first described and is especially common.

F. Microbes in the Causation and Mitigation of Plant Toxicoses

Many wild forages (and a few cultivated ones) contain compounds that have the potential to poison ruminant animals (James et al., 1988). In some instances, the toxicosis occurs as a result of microbial conversion of a nontoxic plant constituent to a more toxic form. Alternatively, microbes may be involved in detoxifying a poisonous agent in the ingested plant. Specific microorganisms have been identified in three different toxicoses: grass tetany, oxalate poisoning, and mimosine poisoning.

1. Grass Tetany

Grass tetany is a type of hypomagnesemia observed in ruminant animals grazing lush pastures, most commonly during periods of cool, cloudy weather in the spring and autumn. Several clinical forms of the disease have been reported (Littledike et al., 1983). Symptoms of the most common type include nervous and excited behavior followed within hours or days by strong convulsions that may lead to coma and death. Several causes of magnesium deficiency have been put forward, including inhibition of magnesium uptake by potassium and formation of $MgNH_4PO_4$ precipitates. Alternative, more feasible explanations revolve around *trans*-aconitate (TAA; Russell and Forsberg, 1986). This compound, an isomer of the tricarboxylic acid cycle (TCA) intermediate *cis*-aconitate, represents up to 7% of the dry weight of some grasses. Itself a potent chelator of Mg^{2+} in vitro, *trans*-aconitate is also reduced by some ruminal microbes (particularly *S. ruminantium*) to tricarballylate. This compound is readily absorbed into the bloodstream and acts as both a strong chelator of Mg^{2+} and as a structural analog of citrate that inhibits the enzymatic conversion of citrate to isocitrate, a key reaction sequence of the oxidative TCA cycle (Figure 8). At least one ruminal bacterium, *Acidaminococcus fermentans*, can detoxify *trans*-aconitate by stoichimetric conversion to acetate (Cook et al., 1994).

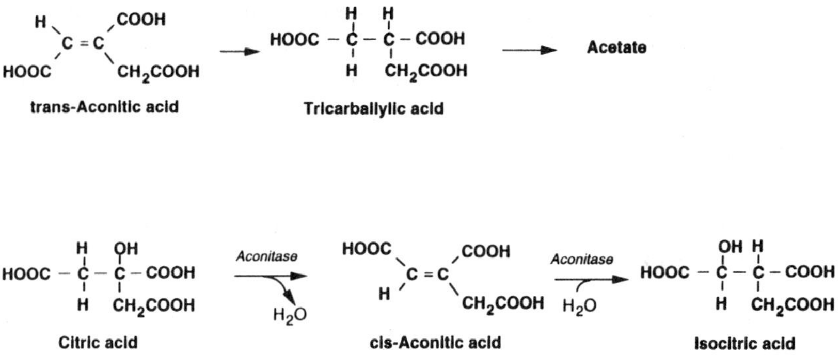

FIGURE 8 Ruminal metabolism of *trans*-aconitate, a common component of some forages that is thought to be involved in eliciting grass tetany. The reduced intermediate tricarballylic acid can chelate magnesium and is a potent inhibitor of the enzymatic conversion of the tricarboxylic acid cycle enzyme *cis*-aconitase. Some ruminal bacteria can degrade tricarballylate to acetate, but only slowly. (From Russell and Forsberg, 1986.)

2. Oxalate Poisoning

Oxalate is widely distributed in plants and in some wild forages (e.g., halogeton) and may comprise several percent of dry weight. Because oxalate is a potent chelator of calcium (and to a lesser extent magnesium), ingestion of these forages can cause hypocalcemia. Oxalate can be metabolized by a dismutation reaction:

$$HOOC\text{-}COOH \rightarrow HCOOH + CO_2$$

carried out by *Oxalobacter formigenes*, a gram-negative bacterial specialist unable to use other substrates as energy sources (Allison et al., 1985).

3. Mimosine Poisoning

Mimosine, a nonprotein amino acid, is present in some tropical forages, particularly the shrub *Leucaena leucocephale*. In the rumen, the pyridone group is released and metabolized to the toxic goiterogen 3,4-dihydroxypyridine. Resistance to mimosine poisoning is dependent on the ruminal bacterium *Synergistes jonesii* (Allison et al., 1992). This species has been found in goats from Hawaii and Indonesia, and has been successfully transferred to ruminant animals in Australia (Jones and Megarrity, 1986) and the United States (Hammond et al.,

1989), where it also confers resistance to mimosine poisoning. In the latter case, the bacterium was maintained in the rumen over a winter during which *Leucaena* was not fed in the diet of the host cattle; maintenance probably resulted from the ability of the bacterium to successfully compete with the native microflora for other amino acids. *S. jonesii* is unique among ruminal bacteria in that it exhibits a specific geographic distribution.

G. Potential for Altering the Rumen Fermentation and the Composition of Milk

The ruminal symbiosis has developed over eons in response to selective pressures on both the animal and the ruminal microflora (Van Soest, 1994). The high levels of production achieved in the animal industry have come, in part, by use of feeding and management strategies that have placed new challenges on the ruminal microflora (e.g., feeding of starches that induce lactic acidosis). Numerous proposals have been put forward to "improve" the rumen fermentation. These proposals have aimed at one or more objectives: (a) increase the rate and extent of digestion of fiber, (b) improve nitrogen availability, either by decreasing the rate and extent of degradation of feed protein or by improving microbial protein synthesis, (c) redirect the microbial fermentation to enhance the amounts or ratios of products that serve as precursors for milk or meat, and (d) detoxify feed or forage components.

Increasing the rate and extent of fiber digestion is complicated by the nature of the plant cell wall (see Sec. IV.D.1). Introduction of enhanced fibrolytic capabilities by genetic engineering has been touted as a means to improve fiber digestion (Russell and Wilson, 1988). Under normal conditions, cellulose digestion in the rumen appears to be limited by cellulose accessibility and not by properties of the microflora (Waldo et al., 1972; Van Soest, 1973). However, under conditions of low pH, most fibrolytic species—particularly the cellulolytics—have limited activity. Introduction of fibrolytic activities into acid-tolerant but nonfibrolytic species may be a viable route to improve fiber digestion, as long as the introduced organism can maintain itself in the rumen both at low pH (when competition for fiber may be minimal) and at more normal pH (when competition for fiber is more intense). A second approach to enhancing the ruminal digestion of fiber involves improvements in plant breeding to produce plant varieties having cell wall structures of improved digestibility (Buxton and Casler, 1993).

Reducing the ruminal degradation of feed protein can be accomplished by a variety of means, including chemical (formaldehyde) or physical (heat) treatment, or incorporation of tannins into the diet (Broderick et al., 1991).

Alternative means of controlling the microbes—either reducing their proteolytic activity or increasing microbial growth yield—have shown little promise.

Controlling the ratios of fermentation end products is already exploited in the beef industry through use of monensin and other ionophores. These compounds are more effective against gram-positive than gram-negative bacteria. Because these groups contain some of the more notable producers of acetate and propionate, respectively, treatment with monensin has several effects, including increasing ruminal propionate and decreasing ruminal acetate and the acetate–propionate ratio. This effect, along with an increase in intake, leads to improved gluconeogenesis, feed efficiency, and body weight gain in beef animals (summarized by Goodrich et al., 1984). Effects in heifers have been more equivocal, although monensin does significantly decrease the age at breeding and at calving (Meinert et al., 1992). The opposite strategy to shift the fermentation balance toward acetate production may be useful for dairy animals, because the undesired reduction in ruminal acetate–propionate ratio that occurs in some diets is associated with a reduced level of milk fat (Shaver et al., 1986; Woodford and Murphy, 1988; Klusmeyer et al., 1990).

There is considerable interest in redirecting ruminal H_2 away from production of methane and toward acetate (Mackie and Bryant, 1994). Although this has not been accomplished practically, recent evidence suggests that yeast may enhance the competitiveness of acetogenic bacteria for H_2, although this effect has only been demonstrated in vitro at H_2 concentrations well above those found in the rumen (Chaucheyras et al., 1995). Yeasts are an example of a direct-fed microbial agent (or probiotic, a natural strain of microbe that improves digestive function). Incorporation of some yeasts and fungi into ruminant diets improves fiber digestion and milk production (Williams et al., 1991; Wohlt et al., 1991), although the mechanism remains unclear (Martin and Nisbet, 1992). Bacteria may also be useful as probiotics. For example, it has been shown recently that lactic acidosis can be avoided in sheep abruptly switched to a grain diet if the lactate-using bacteria *S. ruminantium* and *M. elsdenii* are fed as a probiotic (Wiryawan and Brooker, 1995). Use of probiotics in the dairy industry is expanding, although they have not assumed the same status as in the poultry industry, wherein bacterial probiotics are widely used to prevent colonization of young chicks with *Salmonella*.

As discussed (see Sec. IV.F.3), implantation of mimosine-degrading bacteria has been proven to confer resistance of ruminants to mimosine toxicity. Once established in an animal, these bacteria apparently can be readily transferred to other herd members through normal close contact (Quirk et al., 1988). The probiotic use of other detoxifying organisms holds promise for more productive utilization of toxigenic forages in ruminants diets.

Several milk fat components have been implicated in having the ability to prevent or reduce the incidence of cancer. Two of these components, butyric acid and conjugated linoleic acid, are produced primarily by ruminal bacteria. Butyric acid is produced by many common ruminal bacteria (see Table 5). It is maintained at concentrations of several millimolar in the rumen and is efficiently absorbed across the rumen wall. Among its various metabolic fates is its incorporation into milk fat, where it accounts for 7.5 to 13 mol % of the fatty acids (Parodi, 1996). Butyric acid has been demonstrated to have a variety of anticarcinogenic activities (Parodi, 1996) and its production in the colon of humans on high-fiber diets has been implicated in reducing colon cancer (McIntyre et al., 1993).

Conjugated linoleic acids (CLA) are a class of isomers of linoleic acid having conjugated double bonds. CLA, of which milk fat is the richest natural source, has been reported to have anticarcinogenic, antiatherogenic, and immuno-modulating activities (reviewed by Parodi, 1996). The most abundant CLA isomer, *cis*-9, *trans*-11-octadecadienoic acid, is produced as an intermediate compound in the hydrogenation of linoleic acid by the ruminal fibrolytic bacterium *B. fibrisolvens* (Kepler et al., 1966). This synthetic activity is in accord with the higher levels of milk CLA observed in pastured cows whose diets are particularly rich in fiber (Dhiman et al., 1996). It appears that CLA can also be produced by the gut microflora of monogastric animals, because normal rats have been shown to have higher amounts of CLA in their tissues than do germ-free rats (Chin et al., 1994). The higher levels of the linoleic acid substrate that are present in the rumen, purportedly resulting from hydrolysis of the ruminal bacteria themselves, are thought to explain the unusually high production of CLA by ruminants (Chin et al., 1994).

H. Fermentations in the Hindgut

Hindgut fermentations received very little attention until development of intestinal cannulae permitted quantitative studies. It was long assumed that the extent of digestion that occurs in the hindgut is only a small fraction of that of the total tract. However, the fraction of total tract digestibility that occurs in the hindgut varies with several factors, particularly feed intake (Tamminga, 1993). In cattle fed at high intakes, up to 37% of the total energy digestion can occur in the cecum and large intestine (Zinn and Owens, 1981). Digestion in the hindgut should be of greater importance in high-producing ruminant animals, which, in general, have both high levels of feed intake and ruminal pH values sufficiently low to depress fiber digestion and some other microbial activities in the rumen. The microbiology of the hindgut fermentation in ruminant animals has not been extensively explored, but, in many respects, probably resembles that of monogastric animals.

V. INFECTIOUS DISEASES OF DAIRY ANIMALS

Dairy animals are subject to numerous infections by different species of pathogenic microorganisms. All groups of microbes—bacteria, fungi, viruses, protozoa, and even algae—contain species that are pathogenic to dairy animals. The diseases caused by these organisms are tremendously costly to the dairy producer. Even if animals survive infection, the producer can suffer severe economic hardship in treatment costs, lost production of milk or calves, and disposal of infected milk or milk tainted by antibiotic residues. Quantitative data on the effects of bacterial infections on milk yield and milk composition are lacking for most infectious diseases, mastitis being a notable exception.

A listing of the more common bacterial diseases is provided in Table 8. For more detail, refer to veterinary texts, particularly the recent two-volume treatise of Coetzer et al. (1994).

A. Mastitis

Mastitis is an inflammation of the mammary gland that can affect virtually any mammalian species, but is especially important in dairy animals because of their large udder sizes, high milk production rates, and extensive handling of teats. Mastitis remains the most costly disease of the dairy animal (DeGraves and

TABLE 8 Major Bacterial Diseases of Cattle

Disease	Causative agent
Anthrax	*Bacillus anthracis*
Bovine tuberculosis	*Mycobacterium bovis*
Botulism	*Clostridium botulinum*
Brucellosis	*Brucella abortus*
Clostridial enterotoxemia	*Clostridium perfringens* types B, C, and D
Fusobacterium infections	*Fusobacterium necrophorum*
Gas gangrene	*Clostridium chauvoei, Clostridium novyi, Clostridium septicum*
Genital campylobacteriosis	*Campylobacter* sp.
Haemophilus somnus complex	*Haemophilus somnus*
Leptospirosis	*Leptospira pomona*
Listeriosis	*Listeria monocytogenes*
Mastitis	Many agents (see Table 9)
Paratuberculosis	*Mycobacterium paratuberculosis*
Salmonellosis	*Salmonella* serovars
Tetanus	*Clostridium tetani*

Fetrow, 1993). Economic losses are well over $2 billion annually in the United States alone. Most of the economic losses associated with the disease result from the decrease in milk output and in the discard of milk from infected animals. When the costs associated with additional labor, veterinary fees, and therapeutic agents are added, the total represents 10% to 11% of the productive capacity of the dairy cattle industry.

Mastitis is classified as clinical or subclinical based on its severity, cause, and the characteristics of the exudate fluid; additional subclassifications can also be made (duPreez and Giesecke, 1994). Clinical mastitis is accompanied by macroscopic signs of disease in the animal (e.g., fever, swelling) and in the milk. Clinical mastitis appears to cause similar reductions in yield in high- and low-yielding herds (Firat, 1993).

Subclinical mastitis can only be detected by laboratory methods, and is most commonly revealed by routine microscopic counts of somatic cells ($>5 \times 10^5$ cells/mL, usually leukocytes) in the milk; if mastitis is caused by infection, the causative agent can be observed and often identified at the same time. Even subclinical mastitis is usually associated with a decrease in milk volume.

Mastitis may have any of several causes, chief among which are bacterial infections. Although the udder is constantly exposed to potential pathogens, development of mastitis requires both that the agent be sufficiently numerous and virulent and that the host be susceptible to infection. Susceptibility is a complex function of the animal and of management practices, including milking technique. From an epidemiological standpoint, mastitis is regarded as *contagious* if it is transmitted from infected animals (i.e., almost exclusively by the milking process) or *environmental* if the reservoir of the pathogen and the source of infection is the animal's environment. Numerous species of bacteria have been implicated in causing mastitis (Table 9), but the importance of individual species has changed with changes in dairy practice (Fox and Gay, 1993). *Streptococcus agalactiae* was once the most common causative agent, but it has been displaced over the past few decades by *Staphylococcus aureus*. Several genera of the family Mollicutes (bacteria having very simple genomes and lacking a cell wall), including *Mycoplasma* sp., appear to have a growing involvement as causative agents of mastitis, as does *Listeria monocytogenes*.

Mastitic infection can occur via the blood or by trauma to the udder, but far more commonly occurs via the streak canal of the teat. Although the arrangement of cells and folding of tissues within the teat provide considerable defense against invading pathogens, this defense weakens in cows with age or under conditions of high production. Infection, regardless of route, results in a suite of host responses. Among these are phagocytosis by polymorphonuclear neutrophils (Craven and Williams, 1985), production of antibodies that resist bacterial adherence to epithelial cells, and neutralization of toxins.

TABLE 9 Causative Agents of Bovine Mastitis

Common agents
  Staphylococcus aureus
  Streptococcus spp. (especially *Streptococcus agalactiae, Streptococcus dysgalactiae, Streptococcus uberis*)
  Coliform bacteria (especially *Escherichia coli, Citrobacter freundii, Enterobacter* sp., and *Klebsiella* sp.)
  Actinomyces pyogenes
Less common agents
  Listeria monocytogenes
  Pseudomonas aeruginosa
  Mycoplasma bovis
  Corynebacterium bovis and *Corynebacterium diphtheriae*
  Nocardia sp. (especially *Nocardia asteroides*)
  Coagulase-negative *Staphylococcus* sp. (many species)
  Bacillus cereus
  Brucella abortus
  Clostridium perfringens
  Coxiella burnetii
  Leptospira sp.
  Mycobacterium bovis
  Serratia marcesens
  Prototheca zopfii (alga)

Source: duPreez and Giesecke (1994).

Infectious mastitis results in changes—often dramatic—in milk composition (duPreez and Giesecke, 1994). Fat content is reduced to less than 3%, chloride is increased about 1.5-fold, lactose decreases substantially (often by 5-fold or more) because the pathogen uses this substrate for growth. Total protein content may show only slight changes, but the amount of casein may be reduced at the expense of protein from antibodies, somatic cells, and bacterial cells. In addition to its nutritional inferiority, mastitic milk is visually and organoleptically unappealing because of the presence of microbial polymers, release of free fatty acids as a result of lipase activity, and reduced lactose and increased chloride content.

S. aureus, the most common agent of clinical mastitis, is a gram-positive, nonmotile coccus that grows in characteristic aggregates resembling bunches of grapes. The virulence of *S. aureus* appears to result from a variety of characteristics, including production of extracellular polysaccharide (EPS) capsule, ability to involute into the epithelial cells, production of exotoxins (e.g., leukocidin and coagulase), and causation of tissue necrosis. Chronic mastitic infections are often characterized by bacterial growth in the form of adherent colonies embedded within a large EPS matrix (Brown et al., 1988). Most *S. aureus* isolates that have been recovered from mastitic milk show a characteristic "diffuse colony morphology" resulting from the constitutive or inducible production of an EPS capsule (Baselga et al., 1994). The specific EPS is normally determined by direct serotyping of capsular antigens. Although the EPS is apparently involved in adhesion of bacterial cells to ducts and alveoli in the mammary gland, it is not yet clear whether the EPS is involved in the initial adhesion event or more firmly attaches the bacteria in place following initial adhesion of the cells to the mammary tissue. Regardless, these matrices provide the bacteria with resistance to antibiotic treatment (because of inaccessibility) and phagocytosis (because of the substantial size of the cellular complex).

Much has been written regarding the potential increase in mastitis that may arise from treatment of bovine somatotropin (BST) in cows. Whereas BST treatments undoubtedly increase the prevalence of mastitis, there is considerable evidence (reviewed by Burton et al., 1994) that this effect is not the result of a reduced immunological capacity to resist infection, but instead is caused by extra stress placed on udders from increased milk volume. Thus, the enhanced levels of mastitis are similar to those observed in cows geared to high production by any of a number of feeding and management strategies, regardless of exogenous BST supplementation.

B. Tuberculosis

Tuberculosis is a contagious, chronic disease resulting from infection by species of the genus *Mycobacterium*. Tuberculosis has been one of the most pervasive and destructive diseases of both humans and animals throughout recorded history, and Robert Koch's isolation in 1882 of *Mycobacterium tuberculosis* (the main causative agent in humans) is one of the greatest achievements of clinical microbiology.

Bovine tuberculosis is caused by *Mycobacterium bovis*, an organism with an unusually wide host range that includes not only cattle but humans and other primates, along with many domestic animals (e.g., dogs, cats, pigs, and goats) (O'Reilly and Daborn, 1995). Reservoirs of tuberculosis are also maintained in many wild animals, including bison (*Bison bison*) and elk (*Cervus elaphus*) in North America, badgers (*Meles meles*) in England, and opposum (*Trichosurus velpecula*) in New Zealand. These wild species represent a potential source of infection of domesticated ruminant animals or more commonly provide sufficient exposure to elicit positive tuberculin tests that complicate the undertaking of prophylactic measures to control the disease. In most nonbovine species, the infection is not self- maintaining; even in sheep and goats, the disease is rare.

M. bovis infections of humans through the drinking of milk from infected dairy cows was a serious public health problem early in the 20th century, creating the impetus for compulsory disinfection of the U. S. public milk supply by pasteurization (Myers and Steele, 1969). These and other advances in sanitation, along with aggressive culling of infected animals, has largely controlled bovine tuberculosis in many parts of the world, but it remains an impending threat to the dairymen.

Bovine tuberculosis is normally spread among herds as a result of the introduction of infected cattle into noninfected herds. Infections are generally spread among animals by inhalation of aerosol microdroplets (2–5 μm diameter, small enough to reach the lung alveoli) released by infected animals upon sneezing and coughing; however, transmission is also thought to be possible via feces and various body fluids that may contain the bacilli. The spread of the disease within a herd is largely governed by the susceptibility of its cows, which, in turn, depends on management conditions (e.g., stock density, the overall health of the herd, and control measures adopted by the producer) and by the relative number of young stock. Control measures are complicated by the generally chronic, subclinical nature of the disease. In most cases, the lesions are small in size and number, and clinical signs are often not readily apparent. In clinical forms of the disease, the lymph nodes are the most common targets, with the lungs less often affected. Other organs are affected only rarely, usually as a result of spread through the bloodstream; included among these are infections of the udder (discussed previously as a form of mastitis). The pathogenesis of the disease has been recently reviewed (Neill et al., 1994).

As a genus, the mycobacteria are straight or slightly curved rods that lack motility or the ability to form resistant endospores. Because of their high content of lipids, the cells do not stain readily by the Gram-staining method, although electron microscopy reveals that the cell walls are clearly gram-positive. The lipids are responsible for the characteristic property of acid-fastness (i.e., resistance to decolorization by an acid–alcohol mixture following initial staining by heated carbol fuchsin), a characteristic sufficiently rare among bacteria as to constitute strong preliminary evidence for a mycobacterial infection. The lipids are also responsible for the considerable resistance of the mycobacteria to chemical agents, and this property is used to advantage in the isolation of mycobacterium from clinical samples. Tissues are ground in a saline solution and pretreated for 30 minutes or less with 1 M of NaOH or 2% HCl before neutralization, centrifugation (to concentrate the cells), and plating onto solid media.

The mycobacteria are notoriously slow growers in culture media, including the preferred rich diagnostic media such as Löwenstein-Jensen, Ogawa, Dubos, or Middlebrook 7H10 medium. Even in these media, growth is often not detected before 3 or 4 weeks of incubation at 37°C. Clinical and veterinary microbiologists should recognize that, in addition to host specificity, *M. bovis* and *M. tuberculosis* display several physiological differences (Table 10).

Elimination of tuberculosis in infected herds is usually accomplished by either immediate slaughter of infected animals, or by gradual isolation of infected animals until all of the remaining cattle are free of tuberculosis.

C. Paratuberculosis

Paratuberculosis (Johne's disease) is a chronic and infectious disease of the intestinal tract caused by *Mycobacterium paratuberculosis* (Huchzermeyer et al., 1994). The disease affects both domestic and wild ruminant animals, and causes a severe diarrhea and debilitating weight loss. Infection normally occurs either congenitally or via ingestion by young animals of feces from infected animals. Older animals may largely resist infection because mycobacteria do not survive well in the rumen. In infected animals, the incubation period varies enormously, but clinical signs of the disease apparently require multiple exposures and are not normally manifested for 3 to 5 years. Even in totally infected herds, however, only a few percent of the herd may display clinical signs while the remaining subclinically infected animals may or may not be actively shedding the agent in their feces. As a result of the low percentage of clinical cases in infected herds, the mortality rate within the herd is fairly low (Blood et al., 1989). The long incubation period and subclinical nature of the disease make antibiotic therapy difficult and fairly ineffective in clinical cases. Vaccination is effective only in conjunction with a strong emphasis on animal hygiene and must be used only in tuberculosis-free herds because the vaccine interferes with serological or allergic tests. In humans, *M. paratuberculosis* is thought to cause Crohn's disease.

TABLE 10 Phenotypic Characteristics Differentiating *Mycobacterium bovis* from *Mycobacterium tuberculosis*

Characteristic	*M. bovis*	*M. tuberculosis*
Primary host	Cattle	Human
Colony morphology	Moist, smooth, flat	Dry, wrinkled
Colony development	≥ 3 weeks	10–14 days
Nitrate reduction	Negative	Positive
Niacin production	Negative	Positive
Glycerol	Inhibits growth	Stimulates growth
Pyrazinamide	Resistant	Sensitive
Thiophene-2-carboxylic acid hydrazide	Sensitive	Resistant

M. paratuberculosis is a short, thin, gram-positive, acid-fast rod connected by intercellular filaments that give the organism an aggregated appearance under microscopic observation. Like the mycobacterial agent of bovine tuberculosis, *M. paratuberculosis* grows extremely slowly, even in the preferred Herrold's egg yolk medium, and requires exogenous mycobactin (a class of lipid-soluble cell wall components) for growth. Because of this slow growth, successful isolation of the bacterium requires that feces or intestinal tissue be macerated and exposed briefly to chemical agents (e.g., NaOH or various disinfectants) to eliminate other bacterial contaminants.

D. *Brucella* Infections

Bacteria of the genus *Brucella* include several infectious disease agents, including *Brucella abortus*, which causes bovine brucellosis (contagious abortion) in cattle, bison, and other bovines; *Brucella ovis*, which causes epididimytis and orchitis in sheep; and *Brucella melitensis*, which causes abortion and orchitis in sheep and goats. *B. abortus* can also be transmitted to humans, where it causes undulant fever; this debilitating and often misdiagnosed disease (Latter, 1984) most often afflicts workers having extensive contact with cattle, but it has been reported in some cases to result from contamination of unpasteurized dairy products from infected animals (Bishop et al., 1994).

Members of the genus *Brucella* are gram-negative, nonmotile, nonsporulating cells having a coccus or coccobacillus morphology. They are fairly fastidious in their growth requirements, most requiring complex media containing serum and an atmosphere enriched to 5% to 10% CO_2 for growth. One distinguishing feature of *B. abortus* is its use of erythritol, a four-carbon sugar alcohol, as an energy source. This substrate is abundant in the pregnant uterus, stimulating the localization of the organism at that site.

Because the disease is often subclinical in nature, an extensive battery of tests is often used to detect *Brucella* infections (Bishop et al., 1994). These include direct culture of the agent, detection of specific antibodies, and detection of allergic responses to the agent. Various inocula are used for direct culture, particularly of uterine discharge, colostrum, or milk (from live animals); supramammary lymph nodes (from slaughtered animals); and lung, stomach, and liver (from aborted fetuses and full-term calves). The complement fixation test is regarded as the most definitive of the antibody tests. Other tests involve the reaction of serum antibodies with antigens stained with Rose Bengal or the reaction of milk fat antibodies with stained *B. abortus* cells.

Vaccination with avirulent strains of *B. abortus* is somewhat effective in controlling infection, particularly in heifers. Such vaccination enhances resistance to the disease but does not provide absolute immunity.

E. Enteropathogenic *Escherichia coli*

Several serotypes of *Escherichia coli*, including O157:H7, are highly pathogenic to humans as a result of the ability of the agent to both colonize the intestine and produce verotoxins, a class of Shiga-like toxins. These strains cause severe (and sometimes fatal) intestinal illnesses, including bloody diarrhea and hemolytic uremic syndrome. *E. coli* O157:H7 toxemia in humans is usually the result of eating contaminated meat, sometimes at infectious doses as low as approximately 10^2 cells.

Cattle have been identified as a major reservoir of some of these *E. coli* strains (reviewed by Bettelheim, 1996). The bacteria proliferate primarily in the hindgut and are shed in the feces, where they may remain viable for months (Wang et al., 1996). A survey of 1131 dairy cattle and 659 calves in Ontario, Canada, for Shiga-like toxin-producing strains of *E. coli* (Wilson et al., 1992) revealed that approximately 10% of all cows and 25% of all calves were infected; in some herds, the infection rates were 60% and 100%, respectively. However, few of the 206 verotoxin-producing strains were serovars that had been isolated from humans, and none were serovar O157:H7. In contrast, five of 60 dairy herds in Washington State had cows with fecal O157:H7 present, although overall prevalence (only 10 of 3570 cows) was low (Hancock et al., 1994).

E. coli O157:H7 has also been identified in the bovine rumen. Under normal conditions of adequate feeding, these bacteria and other Enterobacteriaceae are suppressed by low ruminal pH, high concentrations of VFAs, and competition from well-adapted microbial species for nutrients. In fasted animals, however, growth of these transient species is facilitated by elevated pH, reduced VFA concentrations, and the poor survival of native ruminal microbes during starvation (Rasmussen et al., 1993). The effect of dietary stress on increasing the populations of toxigenic enterobacteria has considerable implications for the meat industry, owing to the practice of fasting animals before slaughter. In well-managed dairies, enteropathogenic *E. coli* is probably not a major public health concern.

F. Viral Diseases

Most of the major classes of viruses contain strains that are pathogenic to dairy animals (Table 11). The bovine leukemia virus is the most serious in the United States, and 10% to 30% of dairy herds may be infected. In tropical countries, rinderpest and hoof-and-mouth disease are probably the most serious viral infections of cattle. Unlike many bacterial infections of ruminant animals that can also be transmitted to humans, most viruses that infect ruminant animals have narrower host specificities. As a result, most viruses in a given class that infect ruminant animals do not normally infect humans. Exceptions include the

TABLE 11   Viral Agents of Disease in Cattle

Viral family	Disease
Adenoviridae	Adenovirus infection[a]
Bunyaviridae	Crimean-Congo hemorrhagic fever Rift Valley fever
Coronaviridae	Coronavirus infection
Flaviviridae	Weselbron disease Louping-ill
Herpesviridae	Bovine herpes mammilitis Malignant catarrhal fever Pseudorabies
Paramyxoviridae	Bovine respiratory syncytial virus[a] Parainfluenza type 3 (shipping fever)[a] Rinderpest
Parvoviridae	Bovine parvovirus infection
Picornaviridae	Bovine rhinovirus infection Foot-and-mouth disease
Retroviridae	Bovine leucosis

[a]Also affects goats, as do caprine arthritis-encephalitis and peste de petits ruminants.
Source: Adapted from Coetzer et al. (1994).

following: some of the Orthomyxoviridae (influenza viruses), Flaviviridae, which cause mild influenza-like diseases, and the parainfluenza type 3 virus, which causes a pneumonia-like condition. The more serious exceptions include the Bunyaviridae, causative agents of Rift Valley fever and Crimean-Congo hemor-rhagic fever. The former is, in humans, a mild influenza with various and occasionally fatal complications; the latter is a serious disease with a mortality rate in humans of approximately 30% (Swanepol, 1994).

The lack of response of viruses to antibiotics makes treatment of viral diseases particularly problematic, and stresses the importance of both animal hygiene and good management techniques to ward off viral infections.

Viral infections have variable effects on milk production. Bovine diarrhea virus has been reported to have severe economic impact in dairy herds, both through lower milk yield and more severe disease in calves (Moerman et al., 1994). Bovine respiratory syncytial virus has no significant effect on milk production (Van der Poel et al., 1993). Bovine leukemia virus has been reported in one case to decrease milk yield and in another to increase yield (Rulka et al., 1993). Dairy cattle having genetic potential for high milk production have a greater tendency toward infection with bovine leukemia virus, which probably explains why cows having subclinical infections with this virus sometimes produce more milk (albeit with lower milk fat content) than do uninfected animals in the same herd (Wu et al., 1989).

REFERENCES

Abe M, Kurihara Y. Long-term cultivation of certain rumen protozoa in a continuous fermentation system supplemented with sponge materials. J Appl Bacteriol 56: 201–213, 1984.

Akin, DE. Microscopic evaluation of forage digestion by rumen microorganisms—a review. J Anim Sci 148:701–710, 1979.

Akin DE, Lyon CE, Windham WR, Rigsby LL. Physical degradation of lignified stem tissues by ruminal fungi. Appl Environ Microbiol 55:611–616, 1989.

Allison MJ. Dawson KA, Mayberry WR, Foss JG. *Oxalobacter formigenes* gen. nov., sp. nov.: oxalate-degrading anaerobes that inhabit the gastrointestinal tract. Arch Microbiol 141:1–7, 1985.

Allison MJ, Mayberry WR, McSweeney CS, Stahl DA. *Synergistes jonesii*, gen. nov., sp. nov.: a rumen bacterium that degrades toxic pyridinediols. Syst Appl Microbiol 15:522–529,1992.

Baselga R, Albizu I, Amorena B. *Staphylococcus aureus* and slime as virulence factors in ruminant mastitis. A review. Vet Microbiol 39:195–205, 1994.

Baumgardt BR, Byer WJ, Jumah HF, Krueger CR. Digestibility in the steer, goat, and artificial rumen as measures of forage nutritive value. J Dairy Sci 47:160–164, 1964.

Bettleheim KA. Enterohaemorrhagic *Escherichia coli*: a new problem, an old group of organisms. Aust Vet J 73:20–26, 1996.

Bishop, GC, Bosman PP, Herr S. Bovine brucellosis. In: Coetzer JAW, Thomson GR, Tustin RC, eds. Infectious Diseases of Livestock, with Special Reference to Southern Africa. Cape Town, South Africa: Oxford University Press, 1994, pp 1056–1066.

Blood DC, Radostits OM, Henderson JA. Diseases caused by bacteria. Vol. IV. Veterinary Medicine. 7th ed. London: Bailliére, Tindall, and Cox, 1989.

Bondi A. Animal Nutrition. London: John Wiley and Sons, 1983, pp. 437–475.

Borneman WS, Akin DE, Ljungdahl LG. Fermentation products and plant cell wall degrading enzymes produced by monocentric and polycentric anaerobic rumen fungi. Appl Environ Microbiol 55:1066–1073, 1989.

Broderick GA, Wallace RJ, Ørskov ER. Control of the rate and extent of protein degradation. In: Physiological Aspects of Digestion and Metabolism in Ruminants: Proceedings of the Seventh International Symposium on Ruminant Physiology. New York: Academic Press, 1991, pp 541–592.

Brown, MRW, Allison DG, Gilbert P. Resistance of bacterial biofilms to antibiotics: a growth-rate related effect? J Antimicrob Chemother 22:777–783, 1988.

Bryant MP. Normal flora—rumen bacteria. Am J Clin Nutr 23:1440–1450, 1970.

Bryant MP, Campbell LL, Reddy CA, Crabill MR. Growth of *Desulfovibrio* in lactate and ethanol media low in sulfate in association with H_2-utilizing methanogenic bacteria. Appl Environ Microbiol 33:1162–1169, 1977.

Burton JL, McBride BW, Block E, Glimm DR, Kennelly JJ. A review of bovine growth hormone. Can J Anim Sci 74: 167–201, 1994.

Buxton DR, Casler MD. In: Jung HG, Buxton DR, Hatfield RD, Ralph J. eds. Forage Cell Wall Structure and Digestibility. Madison, WI: American Society for Agronomy—Crop Science Society of America—Soil Science Society of America, 1993, pp 685–714.

Cameron AR, Malmo J. A survey of the efficacy of sustained-release monensin capsules in the control of bloat in dairy cattle. Aust Vet J 70:1–4, 1993.

Chaucheyras F, Fonty G, Bertin G, Gouet P. In vitro H_2 utilization by a ruminal acetogenic bacterium alone or in association with an archaea methanogen is stimulated by a probiotic strain of *Saccharomyces cerevesiae*. Appl Environ Microbiol 61:3466–3467, 1995.

Chin SF, Storkson JM, Liu W, Albright KJ, Pariza MW. Conjugated linoleic acid (9,11- and 10,12-octadecadienoic acid) is produced in conventional but not germ-free rats fed linoleic acid. J Nutr 124:694–701, 1994.

Church DC, ed. The Ruminant Animal. Englewood Cliffs, NJ: Prentice Hall, 1988, p 564.

Clarke RTJ. Protozoa in the rumen ecosystem. In: Clarke RTG, Bauchop T, eds. Microbial Ecology of the Gut. New York: Academic Press, 1977, pp 251–275.

Clarke RTJ, Reid CSW. Foamy bloat of cattle: a review. J Dairy Sci 57:753–785, 1974.

Coetzer JAW, Thomson GR, Tustin RC, eds. Infectious Diseases of Livestock, with Special Reference to Southern Africa. Cape Town, South Africa: Oxford University Press, 1994.

Coleman GS. The growth and metabolism of rumen ciliate protozoa. In: Dubos R, Kessler A, eds. Symbiotic Associations. 13th Symposium of the Soc Gen Microbiol. Cambridge, England: Cambridge University Press, 1963, pp 298–324.

Cook GM, Rainey FA, Chen G, Stackebrandt E, Russell JB. Emendation of the description of *Acidaminococcus fermentans*, a *trans*-aconitate and citrate-oxidizing bacterium. Int J Syst Bacteriol 44:576–578, 1994.

Costerton JW, Cheng KJ, Geesey GG, Ladd TI, Nickel DC, Dasgupta M, Marrie TJ. Bacterial biofilms in nature and disease. Ann Rev Microbiol 41:435–464, 1987.

Cotta MA. Utilization of xylooligosaccharides by selected ruminal bacteria. Appl Environ Microbiol 59:3557–3563, 1993.

Craven N, Williams MR. Defences of the bovine mammary gland against infection and prospects for their enhancement. Vet Immunol Immunopathol 10:71–127, 1985.

Cummings BA, Gould DH, Caldwell DR, Hamar DW. Ruminal microbial alterations associated with sulfide generation in steers with dietary-induced polioencephalomalacia. Am J Vet Res 56:1390–1395, 1995.

DeGraves FJ, Fetrow J. Economics of mastitis and mastitis control. In: Anderson KL, ed. The Veterinary Clinics of North America Food Animal Practice. Vol. 9. Philadelphia, Pa.: W. B. Saunders, 1993, pp 421–434.

Dehority BA. Carbon dioxide requirements of various species of rumen bacteria. J Bacteriol 105:70–76, 1971.

Dehority BA. Hemicellulose degradation by rumen bacteria. Fed Proc 32:1819–1825, 1973.

Dehority BA. Microbial ecology of cell wall degradation. In: Jung HG, Buxton DR, Hatfield RD, Ralph J, eds. Forage Cell Wall Structure and Digestibility. Madison, WI: American Society for Agronomy—Crop Science Society of America—Soil Science Society of America, 1993, pp 425–454.

Devendra C, Coop IE. Ecology and distribution. In: Coop IE, ed. World Animal Science. C. Production-System Approach, Vol. 1. Sheep and Goat Production. Amsterdam: Elsevier 1982, p 8.

Dhiman TR, Anand GR, Satter LD, Pariza M. Conjugated linoleic acid content of milk from cows fed different diets. J Dairy Sci 79(supp 1):137, 1996.

duPreez JH, Giesecke WH. Mastitis. In: Coetzer JAW, Thomson GR, Tustin RC, eds. Infectious Diseases of Livestock, with Special Reference to Southern Africa. South Africa: Cape Town, Oxford University Press, 1994, pp 1564–1595.

Ellis JE, Williams AG, Lloyd D. Oxygen consumption by ruminal microorganisms: protozoal and bacterial contributions. Appl Environ Microbiol 55:2583–2587, 1989.

Felix CR, Ljungdahl LG. The cellulosome: the exocellular organelle of *Clostridium*. Ann Rev Microbiol 47:791–819, 1993.

Fenchel T, Perry T, Thane A. Anaerobiosis and symbiosis with bacteria in free-living ciliates. J Protozool 24:154–163, 1977.

Firat MZ. An investigation into the effects of clinical mastitis on milk yield in dairy cows. Livestock Prod Sci 36:311–321, 1993.

Food and Agricultural Organization. Production Yearbook. Vol 46 (1992). FAO Statistics Series No. 112. United Nations, Rome, Italy: Food and Agricultural Organization, 1993.

Fox LK, Gay JM. Contagious mastitis. In: Anderson KL, ed. The Veterinary Clinics of North America Food Animal Practice. Vol. 9. Philadelphia, Pa.: W. B. Saunders, 1993, pp 475–488.

Goodrich RD, Garrett JE, Gast DR, Kirick MA, Larson DA, Meiscke JC. Influence of monensin on the performance of cattle. J Anim Sci 58:1484–1498, 1984.

Gradel CM, Dehority BA. Fermentation of isolated pectin and pectin from intact forages by pure cultures of rumen bacteria. Appl Microbiol 23:332–340, 1972.

Hammond AC, Allison MJ, Williams MJ, Prine GM, Bates DB. Prevention of *Leucaena* toxicosis of cattle in Florida by ruminal inoculation with 3-hydroxy-4-(1H)-pyridone-degrading bacteria. Am J Vet Res 50:2176–2180, 1989.

Hancock DD, Besser TE, Kinsel ML, Tarr PI, Rice DH, Paros G. The prevalence of *Escherichia coli* O157:H7 in dairy and beef cattle in Washington State. Epidemiol Infect 113:199–207, 1994.

Harris PJ. Plant cell wall structure and development. In: Akin DE, Ljungdahl LG, Wilson JR, Harris PJ, eds. Microbiol and Plant Opportunities to Improve Lignocellulose Utilization in Ruminants. New York: Elsevier, 1990, pp 71–90.

Hatfield RD. Cell wall polysaccharide interactions and degradability. In: Jung HG, Buxton DR, Hatfield RD, Ralph J, eds. Forage Cell Wall Structure and Digestibility. Madison WI: American Society for Agronomy—Crop Science Society of America—Soil Science Society of America. 1993, pp 285–314.

Hatfield RD, Weimer PJ. Degradation characteristics of isolated and in situ cell wall lucerne pectic polysaccharides by mixed ruminal microbes. J Sci Food Agric 69: 185–196, 1995.

Hespell RB. Efficiency of growth by ruminal bacteria. Fed Proc 38:2707–2712, 1979.

Hespell RB, Whitehead TR. Physiology and genetics of xylan degradation by gastro-intestinal tract bacteria. J Dairy Sci 73:3013–3022, 1990.

Hobson PN, Wallace RJ. Microbial ecology and activities in the rumen: part I. CRC Crit Rev Microbiol 9:165–225, 1982.

Hoover WH, Kincaid CR, Varga GA, Thayne WV, Junkins LL Jr. Effects of solids and liquid flows on fermentation in continuous cultures. IV. pH and dilution rate. J Anim Sci 58:692–699, 1983.

Huchzermeyer HFAK Brückner GK, Bastianello SS. Paratuberculosis. In: Coetzer JAW, Thomson GR, Tustin RC, eds. Infectious Diseases of Livestock, with Special Reference to Southern Africa. Cape Town, South Africa: Oxford University Press, 1994, pp 1445–1457.

Hungate, RE. The Rumen and Its Microbes. New York: Academic Press, 1966.

Hungate RE, Smith W, Bauchop T, Yu I, Rabinowitz JC. Formate as an intermediate in the rumen fermentation. J Bacteriol 102:389–397, 1970.

James LF, Ralphs MH, Nielsen DB. The Ecology and Economic Impact of Poisonous Plants on Livestock Production. Boulder, CO: Westview Press, 1988.

Jones GM, Larsen RE, Javed AH, Donefer E, Gaudreau JM. Voluntary intake and nutrient digestion of forages by goats and sheep. J Anim Sci. 34:830–838, 1972.

Jones RJ, Megarrity RG. Successful transfer of DHP-degrading bacteria from Hawaiian goats to Australian ruminants to overcome the toxicity of *Leucaena*. Aust Vet J. 63:259–262, 1986.

Kepler CR, Hirons KP, McNeill JJ, Tove SB. Intermediates and products in the biohydrogenation of linoleic acid by *Butyrvibrio fibrisolvens.* J Biol Chem 241: 1350–1354, 1966.

Kiermeier F, Weiss G, Behringer G, Miller M, Ranfft K. Presence and content of aflatoxin M_1 in milk supplied to a dairy. Z Lebens Unters Forsch 163:71–74, 1977.

Kistner A, Kornelius JH. A small-scale, three-vessel, continuous culture system for quantitative studies of plant fibre degradation by anaerobic bacteria. J Microbiol Meth 12:173–182, 1990.

Klusmeyer TH, Cameron MR, McCoy GC Clark JH. Effects of feed processing and frequency of feeding on ruminal fermentation, milk production, and milk composition. J Dairy Sci 73:3538–3543, 1990.

Krause DO, Russell JB. An rRNA approach for assessing the role of obligate amino acid–fermenting bacteria in ruminal amino acid deamination. Appl Environ Microbiol 62:815–821, 1996.

Kudo H, Cheng KJ, Costerton JW. Electron microscopic study of the methylcellulose-mediated detachment of cellulolytic rumen bacteria from cellulose fibers. Can J Microbiol 33:267–272.

Lafont P, Lafont J, Mousset J, Frayssinet C. Etude de la contamination du lait de vache lors de l'ingestion de faibles quantités d'aflatoxine. Ann Nutr Alim 34:699–708, 1980.

Latham MJ, Brooker BE, Pettipher GL, Harris PJ. Adhesion of *Bacteroides succinogenes* in pure culture and in the presence of *Ruminococcus flavefaciens* to cell walls in leaves of perennial ryegrass (*Lolium perenne*). Appl Environ Microbiol 35:1166–1173, 1978.

Latter M, On having brucellosis. In: Freed DLJ, ed. Health Hazards of Milk. London: Bailliére, Tindall, 1984, pp xvi–xvii.

Littledike ET, Steudemann JA, Wilkinson SR, Horst RL. Grass tetany syndrome. In: Fontenot JP, Bunce GE, Webb KE Jr, Allen VG, eds. Role of Magnesium in Animal Nutrition. Blacksburg, VA: Virginia Polytechnic Institute and State University, 1983, pp 1–49.

Mackie RI, Bryant MP. Acetogenesis and the rumen: syntrophic relationships. In: Drake HL, ed. Acetogenesis. New York: Chapman and Hall, 1994, pp 331–364.

Mackie, RI, Heath S. Enumeration and isolation of lactate-utilizing bacteria from the rumen of sheep. Appl Environ Microbiol 38:416–421, 1979.

Martin SA, Nisbet DJ. Effect of direct-fed microbials on rumen microbial fermentation. J Dairy Sci 75:1736–1744, 1992.

McIntyre A, Gibson PR, Young GP. Butyrate production from dietary fibre and protection against large bowel cancer in a rat model. Gut 34:386–391, 1993.

Mead L, Jones GA. Isolation and presumptive identification of adherent epithelial bacteria ("epimural" bacteria) from the ovine rumen wall. Appl Environ Microbiol 41:1020–1028, 1981.

Meinert RA, Yang CMJ, Heinrichs AJ, Varga GA. Effect of monensin on growth, reproductive performance, and estimated body composition in Holstein heifers. J Dairy Sci 75:257–261, 1992.

Melville SB, Michel TA, Macy JM. Regulation of carbon flow in *Selenomonas ruminantium* grown in glucose-limited continuous culture. J Bacteriol 170:5305–5311, 1988.

Moerman A, Straver PJ, DeJong MCM, Quak J, Baanvinger T, Van Oirschot JT. Clinical consequences of a bovine diarrhoea virus infection in a dairy herd: a longitudinal study. Vet Quart 16:115–119, 1994.

Myers JA, Steele JH. Bovine Tuberculosis Control in Man and Animals. St. Louis, MO: Warren H. Green, 1969.

National Research Council. Ruminant Nitrogen Usage. Washington, DC: National Academy Press, 1985.

Neill SD, Pollock JM, Bryson DB, Hanna J. Pathogenesis of *Mycobacterium bovis* infection in cattle. Vet Microbiol 40:41–52, 1994.

Odenyo AA, Mackie RI, Stahl DA, White BA. The use of 16S rRNA-targeted oligonucleotide probes to study competition between ruminal fibrolytic bacteria: pure culture studies with cellulose and alkaline peroxide-treated wheat straw. Appl Environ Microbiol 60:3697–3703, 1994.

Odom JM, Singleton R Jr, eds. Sulfate-Reducing Bacteria: Current Perspectives. New York: Springer-Verlag, 1993.

Office of the Federal Register. Code of Federal Regulations, Title 21, Section 131.110. Washington, DC: National Archives and Records Administration, 1995.

O'Reilly LM, Daborn CJ. The epidemiology of *Mycobacterium bovis* infection in animals and man. Tubercle Lung Dis. 76(suppl 1):1–46, 1995.

Oremland RS. Biogeochemistry of methanogenic bacteria. In: Zehnder AJB, ed. Biology of Anaerobic Microorganisms. New York: John Wiley and Sons, 1988, pp 641–705.

Orpin CG. Studies on the rumen flagellate, *Neocallimastix frontalis*. J Gen Microbiol 91:249–262, 1975.

Ørskov ER. Protein Nutrition in Ruminants. New York: Academic Press, 1982.

Patterson DSP, Glancy EM, Roberts BA. The carryover of aflatoxin M_1 into the milk of cows fed rations containing a low concentration of aflatoxin B_1. Food Cosmet Toxicol 18:35–37, 1980.

Parodi PW. Milk fat components: possible chemopreventive agents for cancer and other diseases. Aust J Dairy Technol 51:24–32, 1996.

Pavlostathis SG, Miller TL, Wolin MJ. Kinetics of insoluble cellulose fermentation by continuous cultures of *Ruminococcus albus*. Appl Environ Microbiol 54:2660–2663, 1988.

Pavlostathis SG, Miller TL, Wolin MJ. Cellulose fermentation by continuous cultures of *Ruminococcus albus* and *Methanobrevibacter smithii*. Appl Microbiol Biotechnol 33:109–116, 1990.

Quirk MF, Bushell JJ, Jones RJ, Megarrity RG, Butler KL. Liveweight gains on *Leucaena* and native grass after dosing cattle with rumen bacteria capable of degrading DPH, a ruminal metabolite from *Leucaena*. J Agric Sci 111:165–170, 1988.

Rasmussen MA, Cray WC Jr, Casey TA, Whipp SC. Rumen contents as a reservoir of enterohemorrhagic *Escherichia coli*. FEMS Microbiol Lett 114:79–84, 1993.

Rulka J, Dacko J, Kozaczynski B, Reichert M, Kimentowski S. Milk production in cows infected with enzootic bovine leukemia virus evaluated on the basis of ELISA and AGAD tests. Med Weteynaryjna 49:408–411, 1993.

Russell JB. Fermentation of cellodextrins by cellulolytic and noncellulolytic ruminal bacteria. Appl Environ Microbiol 49:572–576, 1985.

Russell JB, Baldwin RL. Comparison of substrate affinities among several rumen bacteria: a possible determinant of rumen bacterial competition. Appl Environ Microbiol 37:531–536, 1979a.

Russell JB Baldwin RL. Comparison of maintenance energy expenditures and growth yields among several rumen bacteria grown in continuous culture. Appl Environ Microbiol 37:537–543, 1979b.

Russell JB, Cook GM. Energetics of bacterial growth: balance of anabolic and catabolic reactions. Microbiol Rev 59:48–62, 1995.

Russell JB, Dombrowski DB. Effect of pH on the efficiency of growth by pure cultures of rumen bacteria in continuous culture. Appl Environ Microbiol 39:604–610, 1980.

Russell JB, Forsberg N. Production of tricarballylic acid by rumen microorganisms and its potential toxicity in ruminant tissue metabolism. Br J Nutr 56:153–162, 1986.

Russell JB, Hino T. Regulation of lactate production in *Streptococcus bovis*: a spiraling effect that contributes to rumen acidosis. J Dairy Sci 68:1712–1721, 1985.

Russell JB, Wilson DB. Potential opportunities and problems for genetically engineered rumen organisms. J Nutr 118:271–279, 1988.

Shaver RD, Nytes AJ, Satter LD, Jorgensen NA. Influence of amount of feed intake and forage physical form on digestion and passage of prebloom alfalfa hay in dairy cows. J Dairy Sci 69:1545–1559, 1986.

Slater JH. Microbial populations and community dynamics. In: Lynch JM, Hobbie JE, eds. Microorganisms in Action: Concepts and Applications in Microbial Ecology. 2nd ed. Oxford, UK: Blackwell Scientific Publications, 1988, pp 51–74.

Slyter LL. Influence of acidosis on rumen function. J Anim Sci 43:910–929, 1976.

Smolenski WJ, Robinson JA. In situ rumen hydrogen concentrations in steers fed eight times daily, measured using a mercury reduction detector. FEMS Microbial Ecol 53:95–100, 1988.

Stahl DA, Flesher B, Mansfield HR, Montgomery L. Use of phylogenetically-based hybridization probes for studies of ruminal microbial ecology. Appl Environ Microbiol 54:1079–1084, 1988.

Stephen AM. Other plant polysaccharides. In: Aspinall GO, ed. The Polysaccharides. Vol 2. New York: Academic Press, 1983, pp 97-193.

Swain RA, Nolan JV, Klieve AV. Natural variability and diurnal fluctuations within the bacteriophage population of the rumen. Appl Environ Microbiol 62:994–997, 1996.

Swanepol, R. Crimean-Congo haemorrhagic fever. In: Coetzer JAW, Thomson GR, Tustin RC, eds. Infectious Diseases of Livestock, with Special Reference to Southern Africa. Cape Town, South Africa: Oxford University Press, 1994, pp 722–729.

Tamminga S. Influence of feeding management on ruminant fiber digestibility. In: Jung HG, Buxton DR, Hatfield RD, Ralph J, eds. Forage Cell Wall Structure and Digestibility. Madison, WI: American Society for Agronomy—Crop Science Society of America—Soil Science Society of America. 1993, pp 571–602.

Tanner GJ, Moate PJ, Davis LH, Laby RH, Yuguang L, Larkin P. Proanthocyanidins (condensed tannins) destabilise plant protein foams in a dose dependent manner. Aust J Agric Res 46:1101–1109, 1995.

Theodorou MK, Gill M, King-Spooner C, Beever DB. Enumeration of anaerobic chytridomycetes as thallus-forming units: quantification of fibrolytic fungal populations from the digestive tract ecosystem. Appl Environ Microbiol 56:1073–1078, 1990.

Thurston B, Dawson KA, Strobel HJ. Cellobiose versus glucose utilization by the ruminal bacterium *Ruminococcus albus*. Appl Environ Microbiol 59:2631–2637, 1993.

Trinci APJ, Davies DR, Gull K, Lawrence MI, Nielsen BB, Rickers A, Theodorou MK. Anaerobic fungi in herbivorous animals. Mycol Res 98:129–152, 1994.

United States Public Health Service. Grade A Pasteurized Milk Ordinance. Publication No. 229, Public Health Service, Food and Drug Administration. Washington, DC: U.S. Department of Health and Human Services, 1993.

Van der Poel WHM, Mourits MCM, Nielsen M, Frankena K, Van Oirschot JT, Schukken YH. Bovine respiratory syncytial virus reinfections and decreased milk yield in dairy cattle. Vet Quart 17:77–81, 1993.

van Egmond HP, ed. Mycotoxins in Dairy Products. London: Elsevier Applied Science Publishers, 1989.

van Gylswyk NO. *Succiniclasticum ruminis*, gen. nov., sp. nov. a ruminal bacterium converting succinate to propionate as the sole energy-yielding mechanism. Int J Syst Bacteriol 45:297–300, 1995.

Van Soest PJ. The uniformity and nutritive availability of cellulose. Fed Proc 32:1804–1810, 1973.

Van Soest PJ. Nutritional Ecology of the Ruminant. 2nd ed. Ithaca, NY: Cornell University Press, 1994.

Varel VH, Yen JT, Kreikemeier KK. Addition of cellulolytic clostridia to the bovine rumen and pig intestinal tract. Appl. Environ Microbiol 61:1116–1119, 1995.

Vogels GD, Hoppe WF, Stumm CK. Association of methanogenic bacteria with rumen ciliates. Appl Environ Microbiol 40:608–612, 1980.

Waldo DE, Smith LW, Cox EL. Models of cellulose disappearance from the rumen. J Dairy Sci 55:125–129, 1972.

Wang G, Zhao T, Doyle MP. Fate of enterohemorrhagic *Escherichia coli* O157:H7 in bovine feces. Appl Environ Microbiol 62:2567–2570, 1996.

Weimer PJ. Cellulose degradation by ruminal microorganisms. CRC Crit Rev Biotechnol 12:189–223, 1992.

Weimer PJ. Effect of dilution rate and pH on the ruminal cellulolytic bacterium *Fibrobacter succinogenes* S85 in cellulose-fed continuous culture. Arch Microbiol 160:288–294, 1993.

Weimer PJ, Why don't ruminal bacteria digest cellulose faster? J Dairy Sci 79:1496–1502, 1996.

Weimer PJ, Shi Y, Odt CL. A segmented gas/liquid delivery system for continuous culture of microorganisms on insoluble substrates and its use for growth of *Ruminococcus flavefaciens* on cellulose. Appl Microbiol Biotechnol 36:178–183, 1991.

Williams PEV, Tait CAG, Innes GM, Newbold CJ. Effects of the inclusion of culture (*Saccharomyces cerevesiae* plus growth medium) in the diet of dairy cows on milk yield and forage degradation and fermentation patterns in the rumen of steers. J Anim Sci 69:3016–3026, 1991.

Wilson JB, McEwen SA, Clarke RC, Leslie KE, Wilson RA, Waltner-Toews D, Gyles CL. Distribution and characteristics of verocytotoxigenic *Escherichia coli* isolated from Ontario dairy cattle. Epidemiol Infect 108:423–439, 1992.

Wilson JR, Mertens DR. Cell wall accessibility and cell structure limitations to microbial digestion of forage. Crop Sci 35:251–259, 1995.

Wiryawan KG, Brooker JD. Probiotic control of lactate accumulation in acutely grain-fed sheep. Aust J Agric Res 46:1555–1568, 1995.

Woese CR, Olsen GJ. Archaebacterial phylogeny: perspectives on the urkingdoms. Syst Appl Microbiol 7:161–177, 1986.

Wohlt JE, Finkelstein AD, Chung CH. Yeast culture to improve intake, nutrient digestibility, and performance by dairy cattle during early lactation. J Dairy Sci 74:1395–1400, 1991.

Wolin MJ. Rumen fermentation: biochemical interactions between populations of the microbial community. In: Akin DE, Ljungdahl LG, Wilson JR, Harris PJ, eds. Microbial and Plant Opportunities to Improve Lignocellulose Utilization in Ruminants. New York: Elsevier, 1990, pp 237–251.

Wood TM, Wilson CA, McCrae SI, Joblin KN. A highly active extracellular cellulase from the anaerobic rumen fungus *Neocallimastix frontalis*. FEMS Microbiol Lett 34:37–40, 1986.

Woodford ST, Murphy MR. Effect of physical form on chewing activity, dry matter intake, and rumen function of dairy cows in early lactation. J Dairy Sci 71:674–686, 1988.

Wu MC, Shanks RD, Lewin HA. Milk and fat production in dairy cattle influenced by advanced subclinical bovine leukemia virus infection. Proc Natl Acad Sci USA 86:993–996, 1989.

Zinn RA, Owens FN. Influence of level of feed intake on nitrogen metabolism in steers fed high concentrate rations (abstr). Am Soc Anim Sci p 448, 1981.

Ziolecki A. Isolation and characterization of large treponemes from the bovine rumen. Appl Environ Microbiol 37:131–135, 1979.

2

Raw Milk and Fluid Milk Products

RICHARD A. LEDFORD

Cornell University, Ithaca, New York

I. INTRODUCTION

This chapter deals with the microbiology of raw milk and fluid milk products with an emphasis on the microbiological flora that influences spoilage and safety of these products. In particular, the influence of processing techniques on the diverse flora is considered. A naturally occurring antibacterial system in raw milk is also discussed.

II. RAW MILK

Information in this chapter pertains to bovine milk, which, as it is secreted by the cow, is free of microorganisms. However, microorganisms associated with the teat move up the teat canal and into the interior of the udder (Olson and Mocquot, 1980). This causes even aseptically drawn milk to contain a small number of microorganisms, mostly bacteria. The surface of the teat may be contaminated with microorganisms from various sources including soil, bedding, manure, and feeds. Diverse microflora including spore formers are present in and on materials from these various sources and, depending on the sanitary practices used on a farm, microorganisms from these sources may contaminate the teat and udder and gain entrance into the milk. Bacteria in aseptically drawn milk are usually limited in number and include mostly micrococci, lactococci, and *Corynebacterium bovis* (Olson and Mocquot, 1980). Freshly drawn milk from healthy cows during a routine milking operation may contain from a few to several thousand microorganisms (Olson and Mocquot, 1980). The number is highly variable among animals and even from quarter to quarter in a given cow. Whereas a few microorganisms may enter milk while it is in the udder, most of the microorganisms in raw milk are contaminants from the outside of the udder, milking equipment, and human handlers (Ayres et al., 1980).

Outstanding progress has been made by the dairy industry in developed countries since the 1950s in improving the handling and processing of milk, and these technological changes have resulted in shifts in the microbial flora in milk, especially raw milk. Developments of closed milking systems, use of bulk tanks to store and transport raw milk, and improvements in refrigeration systems have caused the microflora naturally present in raw milk to change from predominantly gram-positive, acid-producing bacteria to gram-negative, psychrotrophic microorganisms, chiefly species of the genus *Pseudomonas* (Ayres et al., 1980). The definition of psychrotrophic bacteria as used widely in the dairy industry includes those bacteria that grow at 7°C or less, irrespective of their optimal growth temperature (Champagne et al., 1994). Whereas pseudomonads are psychrotrophs of primary importance in the dairy industry, other bacteria and yeasts and molds are psychrotrophic (Cousin, 1982). The emphasis of this chapter is on psychrotrophic bacteria because of their importance in milk and dairy products. Cousin (1982) published a comprehensive review on psychrotrophic microorganisms in milk and milk products.

Both gram-negative and gram-positive psychrotrophic bacteria are significant in the microbiology of milk. Gram-negative bacteria, the primary ones important in milk, include *Pseudomonas*, the most important and *Achromobacter, Aeromonas, Alcaligenes, Chromobacterium*, and *Flavobacterium*. Enzymes are produced especially by gram-negative psychrotrophs during refrigerated storage of raw milk and, because some of these enzymes are heat-stable, their activities can contribute to spoilage of pasteurized milk and milk products (Cousin, 1982). *Yersinia enterocolitica* is a gram-negative pathogenic psychrotroph that has caused illness from consumption of contaminated milk (see Chapter 11). *Bacillus cereus* and *Listeria monocytogenes* are gram-positive, psychrotrophic, pathogenic bacteria that are significant in the microflora of milk (see Chapter 11). Extensive occurrence of *B. cereus* in milk and milk products (except yogurt) has been reported (Ahmed et al., 1983). Seasonal occurrence of psychrotrophic species of *Bacillus* has been noted (Sutherland and Murdoch, 1994). The highest incidence occurred in the winter months and the lowest in the summer and fall (Sutherland and Murdoch, 1994).

Species of microorganisms, in addition to those discussed previously, that may be in raw milk include *Lactobacillus, Acinetobacter, Staphylococcus, Flavobacterium,* and *Micrococcus,* as well as the species that belong to the coliform group. Physiological groups of bacteria, that are classified in the above genera and commonly found in raw milk are those that produce lactic acid, propionic acid, butyric acid, and proteolytic and lipolytic enzymes. Furthermore, raw and pasteurized milk as well as milk products may contain microbial pathogens primarily as a result of contamination.

The lactoperoxidase system is a naturally occurring antibacterial property of raw milk (Davidson et al., 1983; Jay, 1992; Bjorck, 1992; Champagne et al., 1994). This system consists of three components: lactoperoxidase, thiocyanate, and hydrogen peroxide (Reiter and Härnulv, 1984). Even though lactoperoxidase is present in the milk of all species, bovine milk is considered rich in this enzyme (Reiter and Härnulv, 1984). Although it is generally felt that freshly secreted milk contains no hydrogen peroxide, the metabolic activity of the lactic acid bacteria produces sufficient hydrogen peroxide to activate the lactoperoxidase system. Thiocyanate is widely found in animal secretions including milk, and milk with elevated somatic cell counts (perhaps more than 750,000 per mL) contain higher levels.

Lactoperoxidase activity in the presence of thiocyanate and hydrogen peroxide produces short-lived intermediate oxidation products of thiocyanate, which are inhibitory to microbial growth (Reiter and Härnulv, 1984). The system can be brought to a higher level of activity by addition of minute amounts of thiocyanate and hydrogen peroxide. Hypothiocyanite is the primary intermediate oxidation product formed. To control growth of psychrotrophic bacteria in milk, considerable research has been directed toward activation of this system in milk (Champagne et al., 1994). Resistance to the lactoperoxidase system varies among different microorganisms. Gram-negative bacteria such as the pseudomonads, the coliforms, *Salmonella,* and *Shigella* are inhibited and may be killed depending on extrinsic conditions such as temperature. On the other hand, gram-positive bacteria such as the lactic acid bacteria are inhibited but not killed by exposure to the activated lactoperoxidase system (Reiter and Härnulv, 1984). The public health safety of the lactoperoxidase system was discussed by Reiter and Härnulv (1984) who argued against any undesirable side effects.

Catalase, reducing agents, heat, and storage reportedly are effective in inactivating the inhibitory properties of the lactoperoxidase system. Extended storage at 5°C (30°F) or 30°C (86°F) reduces the inhibitory activity, apparently because of the instability of an inhibitory intermediate compound of thiocyanate. The inhibitory activity is inactivated by heating to more than 70°C (158°F) for

20 minutes. Pasteurization and sterilization treatments partially or completely inactivate the lactoperoxidase system (Davidson et al., 1983).

A. Lactic Acid Bacteria

Lactic acid bacteria are widely distributed in nature (Alfa-Laval AB, undated) and include species of the genera *Lactococcus, Leuconostoc,* and *Lactobacillus,* which are commonly found in milk. Lactose is fermented by the lactococci almost entirely to lactic acid, whereas species of *Leuconostoc* and some species of *Lactobacillus* convert lactose to lactic acid and other products including acetic acid, carbon dioxide, and hydrogen. Some of the lactic acid bacteria are important as starter cultures in the production of various fermented dairy foods (see Chapter 6). In addition, numerous research studies have been directed toward understanding the inhibition of other bacteria, especially pathogens, by growth of lactic acid bacteria in milk (Champagne et al., 1994).

B. Coliform Bacteria

Coliform bacteria are widely used as indicators of unsanitary conditions in foods (see Chapter 13). The presence of these bacteria in milk indicates possible contamination with materials such as manure, soil, and contaminated water. Although bacteria in several genera are in the coliform group those often present in raw milk include species of *Enterobacter, Escherichia, Citrobacter,* and, occasionally *Klebsiella.*

C. Butyric Acid Bacteria

Butyric acid bacteria are often present on plants, in soil and in manure. Raw milk can easily be contaminated, especially from low-quality silage fed to lactating cows (Alfa-Laval AB, undated). *Clostridium tyrobutyricum* and *Clostridium butyricum* comprise this group of bacteria, which can spoil cheeses, especially the Swiss types (see Chapter 9). These bacteria are spore formers that are resistant to destruction by pasteurization. They thrive in cheeses in which the conditions are appreciably anaerobic. Carbon dioxide, hydrogen, and butyric acid are produced, reducing cheese quality and often leading to a condition known as "late gas blowing." Control of these bacteria in the manufacture of Swiss types of cheese represents a major challenge.

D. Proteolytic and Lipolytic Bacteria

Proteolytic and lipolytic bacteria such as *Pseudomonas* spp. are significant in the microflora of milk as important spoilage microorganisms, especially of processed milk products. These bacteria usually enter milk from surfaces of improperly cleaned equipment. They are heat sensitive so their presence in pasteurized products is an indication of postpasteurization contamination. Furthermore, sanitizers are effective in destroying them. Milk that has been in contact with inadequately cleaned and sanitized equipment may contain several hundred thousand microorganisms per milliliter, with the pseudomonads being one of the most common types.

III. INFLUENCE OF THE MILKING PROCESS ON THE MICROBIOLOGY OF MILK

The milking process, especially the equipment associated with it, introduces the greatest proportion of microorganisms in raw milk (Olson and Mocquot, 1980; Cousin, 1982). Although a number of pieces of equipment are used in handling milk on farms, the principal items are milking machines, pails, cans, strainers, pipelines, and coolers. Milking machines have several components; one of the main ones that may contaminate raw milk is the teatcup inflation. Tank trucks used to transport raw milk from farms to plants can also contribute to contamination of raw milk. Proper cleaning and sanitization of all milk handling equipment is a primary factor in the microbial quality and safety of raw milk (Olson and Mocquot, 1980; Cousin, 1982). Many states (e.g., New York) have established requirements for production, processing, and distribution of milk and milk products. These requirements include steps to minimize contamination by equipment, including tank trucks (NYS Department of Agriculture and Markets, 1996).

Bacteriological standards, established by the U.S. Public Health Service, stipulate that individual producer milk should not exceed 100,000 colony-forming units per milliliter before commingling with other producer milk and should not exceed 300,000 colony-forming units per milliliter as commingled milk before pasteurization (US Public Health Service, 1993).

Raw milk from cows with mastitis, a bacterial udder inflammation, is heavily contaminated with disease-causing bacteria (see Chapter 1). Several bacteria may cause mastitis, with the prevalent ones being *Streptococcus agalactiae*, *Staphylococcus aureus*, *Pseudomonas aeruginosa*, and members of the coliform group. Microorganisms in milk from cows with mastitis can be readily observed microscopically in stained preparations, which, depending on the severity of the case of mastitis, also contain somatic cells or leukocytes. Tests for somatic cells are widely used to detect milk from cows with mastitis (Richter et al., 1992). Frequently, raw milk payment contracts provide for premiums for milk with low somatic cell counts in the range of 200,000 mL. The "Grade A Pasteurized Milk Ordinance" indicates that individual producer milk should not exceed 750,000 somatic cells per milliliter (US Public Health Service, 1993).

A potable water supply for dairy farm operations is crucial. In the United States, according to the Northeast Dairy Practices Council (Dersan et al., 1990) a dairy farm water supply must meet the following conditions to be in compliance with the "Grade A Pasteurized Milk Ordinance" (US Public Health Service, 1993):

1. Bacteriologically, it must be safe and practically free of any type of bacterial contamination that may affect milk quality.
2. The chemical and physical quality of the water must be within the limits acceptable to health authorities and regulatory agencies.
3. There should be no impurities present in the water supply that might create problems in cleaning milk equipment, corrosion in the pipeline, or undesirable milk flavor.
4. The water supply shall be properly located, protected, operated, and accessible for inspection.

Milk from cows with disease may contain other infection-causing micro-organisms including species of the genera *Listeria, Mycobacterium, Brucella,* and *Coxiella* (see Chapter 1). During milk handling, processing, and packaging, other pathogens may be introduced especially if good manufacturing practices are neglected. Pathogens that have been involved in foodborne outbreaks associated with the consumption of milk produces include *L. monocytogenes, Salmonella Campylobacter, Staphylococcus aureus, Brucella, Y. enterocolitica, B. cereus,* and *Clostridium botulinum.* Since 1983, *L. monocytogenes* has caused several outbreaks that were associated with consumption of milk and milk products. This development has stimulated much research on this important pathogen (Kraft, 1992). Chapter 11 focuses on the safety of dairy foods and includes a complete discussion of the important pathogens that are significant in milk and milk products. In addition, a comprehensive review of microbial pathogens in milk and milk products is presented in *Standard Methods for the Examination of Dairy Products* (Flowers et al., 1992).

IV. INFLUENCE OF PROCESSING ON THE MICROBIOLOGY OF MILK

The principal milk processing operations include separation and clarification, homogenization, pasteurization, packaging, and cooling (Olson and Mocquot, 1980). Contamination with microorganisms is minimal to many thousands, depend-ing on whether the equipment is properly cleaned and sanitized. Vigorous agita-tion, such as with the use of pumps, may increase the number of microorganisms in milk by disrupting bacterial clumps into individual cells, each of which can give rise to a colony in plate-count procedures. Generally, municipalities do not add sufficient chlorine to water to destroy the gram-negative bacteria likely to be in city water. Because of this, equipment washed with city water supplies must be sanitized before use to control these bacteria (Olson and Mocquot, 1980).

During separation, many microorganisms are physically removed from milk and deposited along with somatic cells and other particulate matter in the slime that contains large numbers of microorganisms. Normally, cream also contains significant numbers of microorganisms.

In the United States, the "Grade A Pasteurized Milk Ordinance" (US Public Health Service, 1993) defines pasteurization temperatures and times. Among a number of combinations are 63°C (145°F) for 30 minutes for batch pasteurization or low temperature, long time or (LTLT), and 72°C (161°F) for 15 seconds for continuous flow systems, or high temperature, short time (HTST). If the fat content of the milk product is 10% or more (as in ice cream mix) or if it contains added sweeteners, these temperatures must be increased by 3°C (5°F). These are minimal temperatures and times, and, in practice, processors often use temperatures several degrees higher to obtain products with a longer shelf-life. With regard to dairy products, the term "ultra-pasteurized" means that the product has been heat processed at or above 138°C (280°F) for at least 2 seconds, either before or after packaging.

Pasteurization by either the batch or continuous procedure is sufficient to destroy non–spore-forming pathogens likely to be present. Unless temperatures and times higher than the minimums for pasteurization are used, however, a group of bacteria known as thermodurics survive. These are mesophilic microorganisms that grow in the range of 20°C (68°F) to 45°C (113°F) and include species of *Micrococcus, Bacillus*, and *Lactobacillus casei*. Occasionally, thermoduric strains of *Enterococcus fecalis* and *Enterococcus faecium* are present in milk. Thermoduric microorganisms generally grow slowly in milk at 5°C (41°F) and usually cause spoilage problems only if present in large numbers. However, they do contribute to numbers of bacteria determined by the plate-count procedure and so, if present in large numbers, could cause milk to exceed the established bacterial standard, which, for pasteurized milk, is 20,000 colony forming units per milliliter (US Public Health Service, 1993).

In processing milk, care is taken to minimize the risk of contamination of pasteurized product by unpasteurized product or the cooling medium. In pasteurization by the HTST procedure, the pressure on the pasteurized product during the pasteurization process is greater than that on the unpasteurized product. Any leakage then is in the direction from pasteurized product to unpasteurized product or cooling medium. Pasteurization equipment is carefully designed to include safeguards such as this important one.

The primary effect of homogenization on microorganisms in milk is to break up bacterial clumps that may be present (Olson and Mocquot, 1980). Microbial plate-counts may be affected by extensive breaking of clumps.

During cooling and packaging, microbial contaminants may be introduced from equipment such as pumps, pipes, valves, and the filling machine. Air and condensates may be significant sources of contamination. The effectiveness of cleaning and sanitizing practices greatly influences the level of contamination and the types of microorganisms introduced. In general, species of the genera *Pseudomonas, Alcaligenes*, and *Flavobacterium*, as well as coliforms and other bacteria in the family Enterobacteriaceae are likely contaminants (Olson and Mocquot, 1980). Over the years, postpasteurization contamination, especially at the filling machine, has been a major concern. Development of "clean" filling machines has reduced the potential for contamination and has led to longer shelf-lives of pasteurized milk and milk products.

V. FLUID MILK PRODUCTS

A. Pasteurized Milks and Creams

Pasteurized milks and creams include milk, skim milk, low-fat milks, cream, light cream, heavy cream, half-and-half, cultured milk, and cultured low-fat milks. Other specialized fluid milk products such as reduced-lactose milk are also available. This discussion applies collectively to these products.

Champagne et al. (1994) indicate that spoilage of milk and milk products arises from (a) growth of psychrotrophs before pasteurization, (b) activity of thermoresistant enzymes, (c) growth of thermoresistant psychrotrophs, and (d) postpasteurization contamination, by far the most common source. Thermoresistant enzymes are generally produced by psychrotrophic pseudomonads in milk before pasteurization (Kraft, 1992).

Pasteurized fluid milk under refrigeration is spoiled primarily by growth of psychrotrophic, gram-negative bacteria, which are the most likely postpasteurization contaminants. These microorganisms cause the development of off-flavors among which "unclean" is the term often mentioned (Flowers et al., 1992). Other terms frequently used to describe off-flavors of pasteurized fluid milk caused by microbial spoilage include "rancid," "fruity," "bitter," and "putrid." Bacteria in the genus *Pseudomonas* are usually the primary causes of spoilage; however, on occasion, species of *Flavobacterium, Alcaligenes*, and *Chromobacterium* are also present in spoiled pasteurized milk.

Postpasteurization contamination with gram-negative psychrotrophs that are heat sensitive is usually the cause of spoilage of pasteurized milk products, but, in some instances, psychrotrophic types that survive pasteurization are considered to be significant in spoilage. *Bacillus cereus* is the main microorganism of importance in this regard (Ahmed et al., 1983; Olson and Mocquot, 1980).

B. Concentrated Milk Products

Evaporated milk, condensed milk, and sweetened condensed milk are the primary concentrated milk products (see Chapter 3). These differ in the heat treatments used in concentration, the degree of concentration and the ingredients added to extend shelf stability or to influence final product characteristics (Richter et al., 1992).

Pasteurization, preheating, evaporation, and cooling are the unit operations used in manufacturing these products. Evaporated milk manufacturing differs from the processing used in condensed milk production in that a more intensive preheat treatment is given to evaporated milk to provide storage stability, a stabilizer may be added, and the finished product is sterilized in a can by retorting. Inadequate heat treatment or can leakage may result in spoilage of evaporated milk. An obligate thermophile, *Bacillus stearothermophilus* may grow and cause spoilage if cans of the product are subjected to abnormally high storage and marketing temperatures (Sanders, 1992).

A concentration of 3:1 is usually used for plain condensed milk products that contain no added ingredients. Plain condensed milk and plain condensed skim milk are susceptible to microbial spoilage and they must be stored under refrigerated conditions (Richter et al., 1992). Thermoduric bacteria, *Bacillus, Micrococcus, Lactobacillus, Microbacterium*, coryneform bacteria, *Streptococcus, Enterococcus*, and *Arthrobacter* spp., may survive pasteurization and the heat treatment used during concentration. Sources of these bacteria are unsanitary equipment and environmental contamination, and the level of contamination with these microorganisms is related to the quality of the milk used in manufacturing. Postprocessing contamination is indicated by the presence of psychrotrophic or coliform bacteria.

Sweetened condensed milk is different from plain condensed milk in that sugar is added and it is usually sold in cans. Lactose is crystallized in sweetened condensed milk to prevent the occurrence of large lactose crystals that cause an undesirable "grainy" or "sandy" texture (Richter et al., 1992). Pasteurization, low-water activity, and high sugar content are factors involved in preservation of sweetened condensed milks. The type of microbial spoilage between the two products is considerably different. Osmophilic, sucrose-fermenting yeasts and molds are the primary spoilage microorganisms of condensed milk sweetened with a sucrose concentration of 42% to 43% (Richter et al., 1992). Molds may grow on the surface when sufficient air is available. This is a problem that can be controlled by filling containers to eliminate air, by using good manufacturing practices, and by using sugar of good quality. Enumeration of yeasts and molds in the sugar used as an ingredient should be part of a quality control program.

REFERENCES

Ahmed, AA-H, Moustafa MK, Marth EH. Incidence of *Bacillus cereus* in milk and some milk products. J Food Prot 46:126–128, 1983.

Alfa-Laval AB. Dairy Handbook. S-722 13. Vasteras, Sweden: Teknisk Documentation AB, undated.

Ayres, JC, Mundt JO, Sandine WE. Microbiology of Foods. San Francisco: WH Freeman, 1980.

Bjorck L. Lactoperoxidase. In: Fox PF, ed. Advanced Dairy Chemistry. Vol. 1. Proteins. New York: Elsevier Applied Science, 1992, p 332.

Champagne CP, Laing RR, Roy D, Mafu AA, Griffiths MW. Psychrotrophs in dairy products: their effects and their control. Crit Rev Food Sci Nutr 34:1–30, 1994.

Cousin MA. Presence and activity of psychrotrophic microorganisms in milk and dairy products: a review. J Food Prot 45:172–207, 1982.

Davidson PM, Post LS, Branen AL, McCurdy AR. Naturally occurring and miscellaneous food antimicrobials. In: Branen AL, Davidson PM, eds. Antimicrobials in Foods. New York: Marcel Dekker, 1983, p 371.

Dersan P, Adams RS, Atherton HV, Johnson JR, Smith GB, Zimmermann AF. Guidelines for Potable Water on Dairy Farms. Syracuse, NY: Northeast Dairy Practices Council, 1990.

Flowers RS, Andrews W, Donnelly CW, Koenig E. Pathogens in milk and milk products. In: Marshall RT, ed. Standard Methods for the Examination of Dairy Products 16th ed. Washington, DC: American Public Health Association, 1992, p 103.

Jay JM. Modern Food Microbiology 4th ed. New York: Van Nostrand Reinhold, 1992.

Kraft AA. Psychrotrotrophic Bacteria in Foods: Disease and Spoilage. Boca Raton, FL: CRC Press, 1992.

New York State Department of Agriculture and Markets. Requirements for the production, processing and distribution of milk and milk products. Circular 958. Albany, NY: NYS Department of Agriculture and Markets, 1996.

Olson JC, Mocquot G. Milk and milk products. In: International Commission on Microbiological Specifications for Foods. Microbial Ecology of Foods. Food Commodities. Vol 2. New York: Academic Press, 1980, p 470.

Reiter B, Härnulv G. Lactoperoxidase antibacterial system: natural occurrence, biological functions and practical applications. J Food Prot 47:724–732, 1984.

Richter RL, Ledford RA, Murphy SC. Milk and milk products. In: Vanderzant C, Splittstoesser DF, eds. Compendium of Methods for the Microbiological Examination of Foods. 3rd ed. Washington, DC: American Public Health Association, 1992, p 837.

Sanders ME. Dairy products. In: Lederberg J, ed. Encyclopedia of Microbiology. Vol 2. New York: Academic Press, 1992, p 1.

Sutherland AD, Murdoch R. Seasonal occurrence of psychrotrophic *Bacillus* species in raw milk, and studies on the interactions with mesophilic *Bacillus* sp. Intern J Food Microbiol 21:279–292, 1994.

United States Public Health Service. Grade A Pasteurized Milk Ordinance. Publication No. 229, Public Health Service, Food and Drug Administration, Washington, DC: U.S. Department, 1993.

3

Concentrated and Dry Milks and Wheys

WARREN S. CLARK, JR.
American Dairy Products Institute, Chicago, Illinois

I. INTRODUCTION

Fluid milk and whey are perishable dairy products that require proper cooling and handling to maintain their freshness and quality. However, milk and whey solids may be preserved for future use by various methods, the most common of which is concentration by removing water, using either heat or membrane methodology, followed by drying. Dairy products commonly manufactured through the use of one or more of these processes are evaporated milks, condensed and sweetened condensed milks, dry milks, condensed whey products, and dry whey products. Emphasis in this chapter is given to the major products and similarities are made to other closely related products.

All evaporated milks and most condensed, sweetened condensed, and dry milk products are manufactured using grade A raw milk (U. S. Public Health Service, 1995b). In some areas of the United States, condensed and dry whey products are also made entirely from raw milk meeting grade A requirements, but, overall, a lesser quantity of condensed and dry whey products are manufactured using grade A milk. In those instances, milk that meets United States Department of Agriculture (USDA) requirements (U. S. Department of Agriculture, 1972) is used. Current estimates (D. R. Spomer, personal communication, 1996) are that 5% of the U. S. milk supply is non–grade A and that approximately 8.2% of domestic manufactured dairy products (condensed and dry milks, condensed and dry wheys, cheese, and butter) are made from milk meeting USDA requirements. All milk and whey used in the manufacture of concentrated and dry milk and whey products are pasteurized (see Chapter 2).

II. CONDENSED MILK

Bulk condensed milk may be manufactured using either whole or skim milk. Typically, the milk is pasteurized and then concentrated by heat in an evaporator until the product contains 40% to 45% total solids. Following concentration, the product may be dried or distributed for use as a concentrated milk. A detailed processing scheme for condensed milk is shown in Figure 1. Most condensed whole milk is used as an ingredient in the chocolate/confectionery, bakery, or dairy (frozen dessert) industries; condensed skim milk not subsequently dried is used primarily within the dairy industry (American Dairy Products Institute, 1996a). These products are not considered to be commercially sterile and, when intended for shipment as an ingredient, they are immediately cooled and continuously held at temperatures below 7°C [45°F]. Microorganisms surviving the heat treatments usually are thermoduric or thermophilic types. Under proper handling and storage conditions, these organisms grow slowly, if at all, and are not expected to create keeping quality problems. If spoilage occurs, it usually is attributed to postheating contamination. Psychrotropic bacteria, yeasts, or molds may cause spoilage if product is held for unusually long periods or under improper storage conditions.

III. SWEETENED BULK CONDENSED AND
   CANNED MILK

The primary difference between condensed and sweetened condensed milks is the addition of sugar. Sweetened condensed milk is preserved by addition of sugar, which increases the osmotic pressure to a point inhibitory to most microorganisms. The increased milk solids content also increases the osmotic pressure. The sugar-in-water concentration of sweetened condensed milk is called the "sugar ratio," which is calculated as follows:

$$\frac{\%\ \text{sugar in condensed milk}}{100 - \text{total milk solids in condensed milk}} \times 100 = \text{sugar ratio}$$

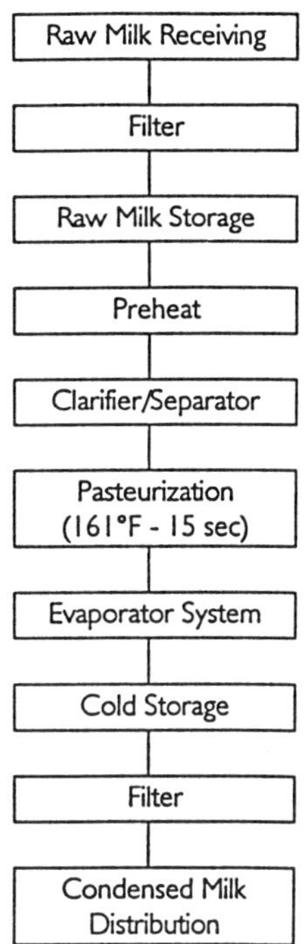

FIGURE 1 Processing scheme for condensed milk.

Like condensed milk, sweetened condensed milk may be as whole milk or skim milk and used either in bulk or consumer (canned) form. Most sweetened condensed milk is whole and used in bulk in the bakery and confectionery industries. With modern processing, storage, and handling practices, spoilage seldom is encountered. If the bulk product is improperly handled or held for extended periods before use, surface growth of yeasts or molds may occur. These microorganisms are the most common cause of spoilage of sweetened condensed milks. Their presence is indicative of unsanitary postpasteurization conditions. The consumer (canned) product has been thermally processed and is commercially sterile (see Sec. IV).

IV. EVAPORATED MILKS

A. History

Evaporated milk, like other processed canned foods, originated with the experiments of the French scientist Nicholas Appert (Flake and Clark, 1991). Appert, whose work on food preservation began in 1795, was the first person to evaporate milk by boiling it in an open container and then preserving it by heating the product in a sealed container. Fifty years later, another French scientist, Louis Pasteur, laid the scientific foundation for heat preservation through demonstrations that food spoilage could be caused by bacteria and other microorganisms.

Patents dealing with preservation of milk after evaporation in a vacuum were granted to Gail Borden by the United States and England in 1856. These patents applied to concentrating milk without addition of sugar. In 1884, U. S. patent number 308,421 was issued for "an apparatus for preserving milk" and, in 1885, the first commercial evaporated milk plant in the world was opened in a converted wool factory in Highland, IL where "evaporated cream" was manufactured and sold (Flake and Clark, 1991).

B. Products and Processing

Evaporated milk is a canned whole milk concentrate to which a specified quantity of vitamin D has been added and to which vitamin A may be added. It conforms to the Food and Drug Administration (FDA) Standard of Identity 21 CFR 131.130 (U. S. Department of Health and Human Services, 1995a), having a minimum of 6.5% milk fat, 16.5% milk solids-not-fat, 23% total milk solids, and 25 IU vitamin D per fluid ounce. Related evaporated milk products are evaporated skimmed milk, evaporated low-fat milk, evaporated filled milk, and evaporated goat's milk. Evaporated skimmed milk conforms to the FDA Standard of Identity 21 CFR 131.132 (U. S. Department of Health and Human Services, 1995a) and contains not less than 20% of total milk solids, not more than 0.5% milk fat, with added vitamins of 25 IU vitamin D and 125 IU vitamin A per fluid ounce. Standards of Identity have not been established for the other evaporated milk products. Their typical compositions are the following:

> Evaporated low-fat milk—2% milk fat, 18% nonfat milk solids, vitamins A and D added
>
> Evaporated filled milk—6% vegetable fat, 17.5% nonfat milk solids, vitamins A and D added
>
> Evaporated goat's milk—not less than 7% milk fat and 15% nonfat milk solids, vitamin D added

A typical processing scheme for evaporated milk (Fig. 2) begins with high-quality, fresh whole milk, to which vitamins, emulsifiers, and stabilizers are added. The product is then pasteurized, concentrated under reduced pressure in an evaporator, homogenized, cooled, and standardized to the composition desired in the final product. After cans are filled and sealed, they are sterilized in a three-phase continuous system consisting of preheater, retort, and cooler, then labeled and packed for shipment. In the United States, evaporated milk is packed in

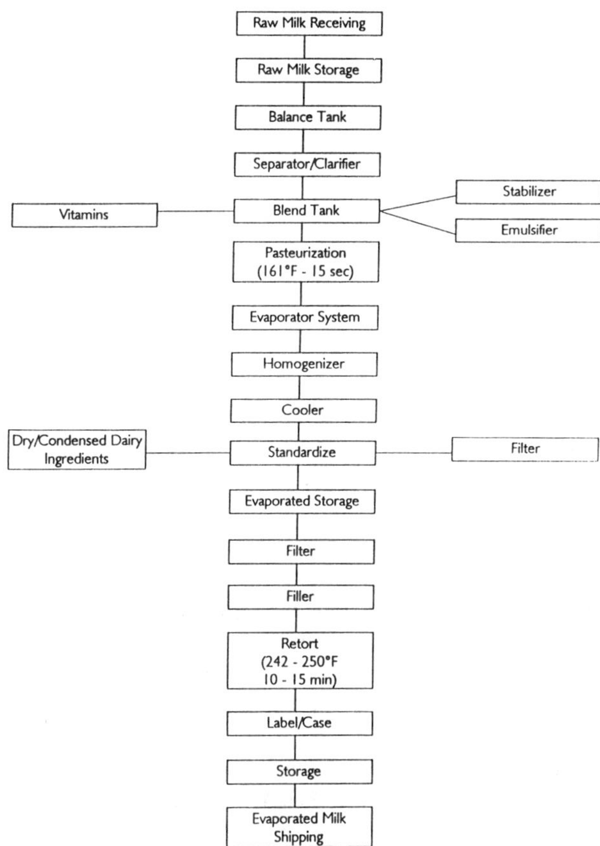

FIGURE 2　Processing scheme for evaporated milk.

5-, 12- and 97-fluid ounce lead-free cans. In 1995, production of evaporated milk and related products (evaporated skimmed milk, evaporated low-fat milk, and evaporated filled milk) was slightly more than 498 million pounds (American Dairy Products Institute, 1996b).

Evaporated milk processing is covered by FDA regulations dealing with thermally processed low-acid foods packaged in hermetically sealed containers (U. S. Department of Health and Human Services, 1995b). Therefore, manufacturers of evaporated milk and related products must comply with stringent processing regulations, including the establishment and filing of scheduled processes with FDA and the maintenance of strict processing records.

C. Microbiology

Because of the heat processes and packaging used in the manufacture of evaporated milks, the product is commercially sterile. This means that the product is free of all microorganisms of public health significance and does not show microbial defects during its intended shelf-life under normal conditions of handling, storage, and distribution. Whereas vegetative cells do not survive evaporated milk processing, and absolute sterility is obtained in most cans, small numbers of nonpathogenic spores occasionally may survive the heat treatment and, depending on the microorganism and its previous growth and heat exposure, subsequently may germinate (Curran and Evans, 1945). Kalogridou-Vassiliadou (1992) studied 40 strains of bacilli implicated in causing flat sour spoilage in evaporated milk. The microorganisms were identified as *Bacillus stearothermophilus* (five strains), *Bacillus licheniformis* (10 strains), *Bacillus coagulans* (15 strains), *Bacillus macerans* (five strains), and *Bacillus subtilis* (five strains). Species of the genus *Bacillus* (i.e., *cereus, coagulans, megatherium, stearothermophilus*, and *subtilis*) earlier were implicated in cases of evaporated milk spoilage (Foster et al., 1957; Hammer and Babel, 1957). Recently, Langeveld et al. (1996), in studies of *Bacillus cereus* naturally present in raw milk, reported no evidence that this organism would cause intoxication in healthy adult humans at levels less than 10^5 mL. Classic studies (Curran and Evans, 1945; Theophilus and Hammer, 1938) on the microbiology of evaporated milk have contributed significantly to the knowledge of the microbiology of this product.

Under current continuous processing conditions wherein heat treatments of 117°C to 121°C (242°F–250°F) for 10 to 15 minutes are common and batch retorting is uncommon, spoilage of evaporated milk is unlikely to be encountered. Specific methods for the microbiological examination of evaporated milk are contained in *Standard Methods for the Examination of Dairy Products* (Marshall, 1992).

V. DRY MILKS

A. History

The development of the dry milk industry stems from the days of Marco Polo in the 13th century. It is reported that Marco Polo encountered sun-dried milk on his journeys through Mongolia and that, from this beginning, dry milk products evolved (Clark, 1991a). Through early pioneering scientists, such as Appert and Borden, the basic methods were developed for emergence of processes for drying milk products. Ekenberg and Merrill have been acknowledged as the developers of the first commercial roller- and spray-process drying systems, respectively, in the United States (Beardslee, 1948). Since the initial development of commercial drying systems, significant technological advances have been made, resulting in the manufacture of a variety of dry milk products.

B. Products and Processing

The primary dry milk products manufactured domestically are nonfat dry milk, dry whole milk, and dry buttermilk. Nonfat dry milk is the product resulting from removal of fat and water from milk. It contains lactose, milk proteins, and milk minerals in the same relative proportions as the fresh milk from which it is made. Nonfat dry milk contains not more than 5% by weight of moisture. The fat content is not more than 1.5% by weight unless otherwise indicated. Dry whole milk is the product resulting from removal of water from milk and contains not less than 26% milk fat and not more than 4% moisture. Dry whole milk with milk fat contents of 26% and 28.5% are most commonly produced. Dry buttermilk is the product resulting from removal of water from liquid buttermilk derived from the manufacture of butter. It contains not less than 4.5% milk fat and not more than 5% moisture.

The steps in a typical dry milk processing operation include (a) receipt of fresh, high-quality milk delivered in refrigerated stainless-steel bulk tankers, (b) clarification, and, if nonfat dry milk is to be manufactured, (c) separation. The milk fat removed is usually churned into butter. If dry whole milk is to be manufactured, the separation step is omitted but may be replaced by a standardization procedure. Pasteurization by a continuous high-temperature short-time (HTST) process, whereby every particle of milk is subjected to a heat treatment of at least 72°C (161°F) for 15 seconds, is accomplished next. Holding the pasteurized milk at an elevated temperature for an extended period of time (85°C [185°F] for 20–30 minutes) is used in the manufacture of high-heat nonfat dry milk, which is commonly used as an ingredient in bakery or meat products. Following concentration of the milk by removing water in an evaporator until a milk-solids content of at least 40% is reached, the product enters the dryer for final moisture removal.

Commercial U. S. drying processes are of two types: spray and roller (drum). Currently, the latter is used to a limited extent and primarily for product intended for other than human consumption. Two basic configurations of spray dryers are in use: horizontal (box) and vertical (tower). In both, the pasteurized and concentrated milk is directed under pressure to a spray nozzle (horizontal dryer) or to either a spray nozzle or an atomizer (vertical dryer), where the dispersed liquid then comes into contact with a current of filtered, heated air. The droplets of milk are dried almost immediately and fall to the bottom of the fully enclosed stainless-steel drying chamber. The dry milk product continuously is removed from the drying chamber, transported through a cooling and collecting system, and

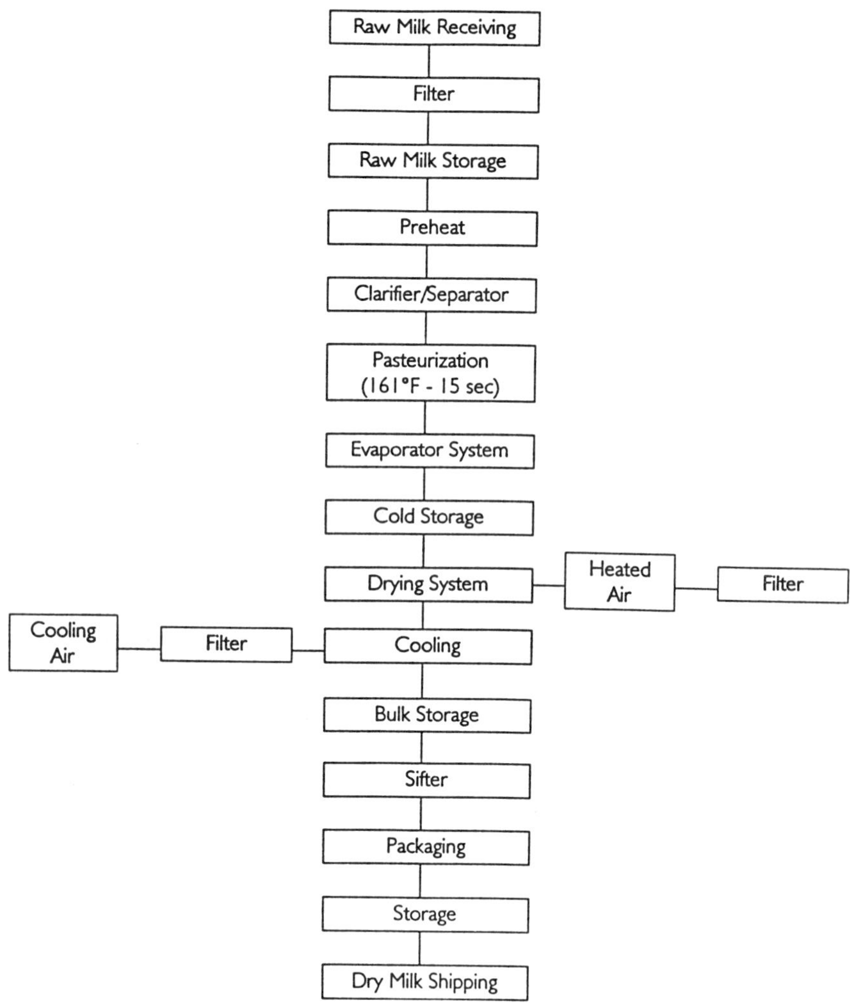

FIGURE 3 Processing scheme for dry milk.

finally conveyed into a hopper for packaging, usually in 50-pound bags or in tote bins. Figure 3 reflects a typical processing scheme for dry milk.

In the processing of nonfat dry milk, various heat treatments may be applied to give the finished dry milk product desirable functional characteristics. Three heat treatment classifications, based on use of the whey protein nitrogen test, are of practical importance in indicating the suitability of spray-process nonfat dry milk for specific purposes (American Dairy Products Institute, 1990). Instant-type dry milks are processed by special methods that result in products with improved solubility. Instant nonfat dry milk is defined by its solubility index value (American Dairy Products Institute, 1990).

The American Dairy Products Institute (1996a) publishes annual census figures that reflect markets of end use for dry milk products, which may be referenced for further information about quantities of dry milks processed and their use. In 1995, U. S. production of nonfat dry milk was 1.2 billion pounds, dry whole milk production was 165 million pounds, and dry buttermilk production was 47 million pounds (American Dairy Products Institute, 1996a).

C. Standards

Industry microbiological standards for dry milk products are established by the American Dairy Products Institute. In addition, government standards for these products also have been generated by the USDA and the FDA (U. S. Public Health Service, 1995a). Table 1 shows these standards by source, product, and, as applicable, grade.

D. Microbiology

Relatively few species of bacteria have been reported as being naturally occurring in dry milks. Hammer and Babel (1957) and Foster et al. (1957), in earlier texts covering the microbiology of dry milk products, summarized literature reports indicating microorganisms of the genera *Streptococcus, Micrococcus, Bacillus, Clostridium,* and *Sarcina* as comprising the primary microflora of dry milks. Rodriquez and Barrett (1986), based on a study of the microbial population and growth in reconstituted dry milk, confirmed the occurrence of viable cells of the genera *Bacillus* and *Micrococcus* in nonfat and dry whole milks.

Since initiation of the requirement that all milk be pasteurized before drying, current heat treatments used to process dry milks destroy all microorganisms of public health significance. Relatively low numbers of microorganisms survive the processing, and those heat-resistant organisms (both spore-forming and non–spore-forming types) rarely, if ever, are responsible for finished product deterioration. Because the drying process is accomplished in a completely closed system, postprocessing contamination also is rare. When such occurs, it usually is from an airborne source. Because of the low moisture levels in dry milks, those viable organisms that may be present are unable to grow and decrease in number during storage. Specific methods for the microbiological examination of dry milks are contained in *Standard Methods for the Examination of Dairy Products* (Marshall, 1992).

TABLE 1 Microbiological Standards for Condensed and Dry Milk Products[a]

Product	American Dairy Products Institute standards[b]	United States Department of Agriculture standards[b]	Food and Drug Administration (grade A) standards
Condensed milk	None	None	Bacterial estimate: 30,000/g Coliform: 10/g
Nonfat dry milk			
Extra grade	SPC: 40,000/g Coliform: 10/g	SPC: 40,000/g Coliform: 10/g	Bacterial estimate: 30,000/g Coliform: 10/g
Standard grade	SPC: 75,000/g Coliform: 10/g	SPC: 75,000/g Coliform: 10/g	
Dry whole milk			
Extra grade	SPC: 50,000/g Coliform: 10/g	SPC: 50,000/g Coliform: 10/g	None
Standard grade	SPC: 100,000/g Coliform: 10/g	SPC: 100,000/g Coliform: 10/g	
Dry buttermilk			
Extra grade	SPC: 50,000/g Coliform: 10/g	SPC: 50,000/g Coliform: 10/g	Bacterial estimate: 30,000/g Coliform: 10/g
Standard grade	SPC: 200,000/g Coliform: 10/g	SPC: 200,000/g Coliform: 10/g	

[a]All counts expressed as "not more than."
[b]DMC may not exceed 100 million/g for ADPI- and USDA-graded nonfat dry milk and dry whole milk.
Abbreviations: DMC, direct microscopic clump count; SPC, standard plate count.

Spray-dried milks have been implicated in outbreaks of staphylococcal food poisoning (Anderson and Stone, 1955; Armijo et al., 1957). In both instances, the illnesses were caused by a preformed enterotoxin that was not inactivated by the drying process. Miller et al. (1972), in a study of the effect of spray drying on survival of *Salmonella* and *Escherichia coli*, reported that heat treatments typically associated with spray drying could not be counted on to supplant adequate pasteurization and postdrying sanitary procedures. Bradshaw et al. (1987), in studies of the thermal resistance of disease-associated *Salmonella typhimurium* in milk, reported the organism did not survive pasteurization.

Doyle et al. (1985) studied survival of *Listeria monocytogenes* during the manufacture and storage of nonfat dry milk. Concentrated (30% solids) and unconcentrated skim milks were inoculated with 10^5 to 10^6 *L. monocytogenes*/mL. They reported reductions of 1 to 1.5 $\log_{10}$ *L. monocytogenes*/g occurred during the spray drying process and that the organism progressively died during storage. The inoculated milks were not pasteurized before drying. Bradshaw et al. (1985) and Donnelly et al. (1987) reported that *L. monocytogenes* did not survive in milk during pasteurization. Earlier studies (Nichols, 1939; Higginbottom, 1944) also reported on the destruction of microorganisms during drying and the fate of surviving organisms during storage.

VI. DRY WHEY PRODUCTS

A. History

Although spray and roller processes have been used to dry whey for many years, development of a whey processing industry in the United States did not fully materialize until the organization of the Whey Products Institute in 1971 (Clark, 1991b). At that time, development of product identity and quality standards was undertaken as a guide to production of uniformly high-quality whey products. In 1981, the FDA accepted industry-recommended common and usual names for a variety of whey products and affirmed the generally recognized as safe (GRAS) status of these products and their method of manufacture (U. S. Department of Health and Human Services, 1981). Technological changes associated with whey processing are dynamic. In no area of the modern dairy industry have changes of a technical nature been as innovative and rapid as in the whey products segment. Important applications to whey processing include use of selective membrane techniques that allow the various whey constituents to be separated into protein-, carbohydrate-, or mineral-rich streams, which then may be further processed and made available in concentrated functional forms. Significant developments, reflecting continuing changes, are anticipated in this area.

B. Products and Processing

The primary whey products currently manufactured in the United States are concentrated and dry whey, and the modified whey products, including reduced-lactose whey, reduced-minerals whey, and whey protein concentrate. Other modified whey products manufactured in smaller quantities include lactalbumin (minimum protein content, 80%) and whey protein isolate (minimum protein content, 90%). Lactose, the carbohydrate of milk, also is being produced in large quantities as a coproduct with the manufacture of modified wheys. Table 2 defines the commonly known whey products currently being manufactured.

A typical processing scheme for the manufacture of dry whey is shown in Figure 4. Some whey drying operations receive only condensed whey for processing; others receive condensed and fresh fluid whey. The solids concentration of the transported condensed whey and the time–temperature conditions of its shipment determine how the product is processed before entering the drying

TABLE 2 Composition of Whey Products[a]

Name of product	Major parameters (%)[b]				
	Protein	Fat	Ash	Lactose	Moisture
Whey	10–15	0.2–2.0	7–14	61–75	1–8
Concentrated whey	10–15	0.2–2.0	7–14	61–75	1–8
Dry or dried whey	10–15	0.2–2.0	7–14	61–75	1–8
Reduced-lactose whey	16–24	0.2–4.0	11–27	60 max	1–6
Reduced-minerals whey	10–24	0.2–4.0	7 max	85 max	1–6
Whey protein concentrate	25 min	0.2–10.0	2–15	60 max	1–6
Lactose	N/A	N/A	0.3	98 min	4–6

[a]FDA affirmation of direct food substances as generally recognized as safe.
[b]On a dry product basis.

system. Currently, the USDA requires all condensed whey containing less than 40% solids to be pasteurized or repasteurized in the processing plant where it is to be dried. The process of drying is similar to that used to manufacture dry milks and some processing plants may dry both products interchangeably.

Processing operations to manufacture modified whey products include reverse osmosis, ultrafiltration, and electrodialysis procedures, some of which may be proprietary in nature. For more information on these processes, various published texts (Sienkiewicz and Riedel, 1990; Gillies, 1974) may be consulted.

The American Dairy Products Institute (1996c) publishes data annually that reflect production and utilization trends for whey products. In 1995, 1.9 billion pounds of whey solids were processed in the United States as follows: 1.1 billion pounds of dry whey; 102 million pounds (solids) as condensed whey; 98 million pounds of reduced-lactose and reduced-minerals whey; 208 million pounds of whey protein concentrate; and 305 million pounds of lactose.

C. Standards

As for dry milk products, industry microbiological standards for whey products have been established by the American Dairy Products Institute, the USDA, and the FDA. Table 3 shows current microbiological standards for whey products.

D. Microbiology

As drying processes for whey are essentially the same as those for milk, the discussion of dry milk microbiological considerations also apply to dry whey. Microbiological methods to assay the quality of whey products are contained in *Standard Methods for the Examination of Dairy Products* (Marshall, 1992).

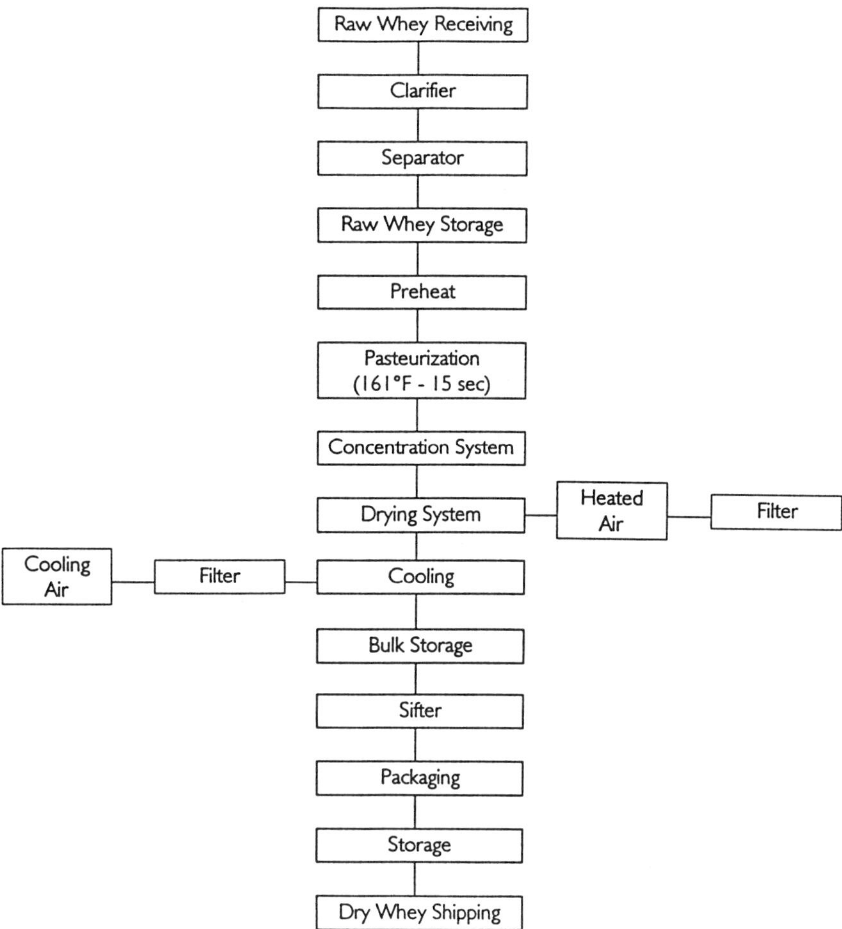

FIGURE 4 Processing scheme for manufacture of dry whey.

Cultural or direct microscopic (DMC) procedures may be used. If using the latter, it must be understood that most whey processed is derived from cheese manufactured using bacterial cultures; thus, large numbers of viable lactic organisms are present in fresh whey. Except for the more heat-resistant strains of lactic bacteria, these organisms are not expected to survive pasteurization and are not detected by cultural techniques. However, when freshly dried whey is examined by direct microscopic techniques, cells of nonviable bacteria often stain. Therefore, results of DMC techniques used to assess the quality of dry whey must be interpreted with care.

TABLE 3 Microbiological Standards for Whey Products[a]

Product	American Dairy Products Institute standards	United States Department of Agriculture standards	Food and Drug Administration (grade A) standards[b]
Condensed whey	None	None	Bacterial estimate: 30,000/g Coliform: 10/g
Dry whey			
Extra grade	SPC: 50,000/g Coliform: 10/g	SPC: 50,000/g Coliform: 10/g	Bacterial estimate: 30,000/g Coliform: 10/g

[a]All counts expressed as "not more than."
[b]Includes grade A dry whey and dry whey products.
Abbreviations: SPC, standard plate count.

Merin (1986), in a study of the microfiltration of whey using 1.2-μm pore size membranes, reported that the membranes reduced bacterial counts by one to three times and that increased fat content in the feed stream governed the decrease. Fat trapped on the membrane formed a barrier to microorganism penetration into the permeate.

REFERENCES

American Dairy Products Institute. Standards for grades of dry milks including methods of analysis. Chicago, IL, 1990, as revised.

American Dairy Products Institute. Dry milk products. Utilization and production trends, 1995. Chicago, IL, 1996a.

American Dairy Products Institute. Evaporated milk production. Chicago, IL, 1996b.

American Dairy Products Institute. Whey products. Utilization and production trends, 1995. Chicago, IL, 1996c.

Anderson PHR, Stone DM. *Staphylococcus* food poisoning associated with spray-dried milk. J Hyg 53:387, 1955.

Armijo R, Henderson DA, Timothee R, Robinson HB. Food poisoning outbreaks associated with spray-dried milk. An epidemiologic study. Am J Public Health 47:1093, 1957.

Beardslee CE. Dry milks. The story of an industry. Chicago, IL: American Dry Milk Institute, Inc., 1948.

Bradshaw JG, Peeler JT, Corwin JJ, Hunt JM, Tierney JT, Larkin EP, Twedt RM. Thermal resistance of *Listeria monocytogenes* in milk. J Food Prot 48:743, 1985.

Bradshaw JG, Peeler JT, Corwin JJ, Barnett JE, Twedt RM. Thermal resistance of disease-associated *Salmonella typhimurium* in milk. J Food Prot 50:95, 1987.

Clark WS Jr. Dry milk. In: Hui YH, ed. Encyclopedia of Food Science and Technology. New York: John Wiley & Sons, 1991a, pp 656–658.

Clark WS Jr. Whey processing: history and development. In: Hui YH, ed. Encyclopedia of Food Science and Technology. New York: John Wiley & Sons, 1991b, pp 2845–2847.

Curran HR, Evans FR. Heat inactivation inducing germination in the spores of the thermotolerant and thermophilic aerobic bacteria. J Bacteriol 49:335, 1945.

Donnelly CW, Briggs EH, Donnelly LS. Comparison of heat resistance of *Listeria monocytogenes* in milk as determined by two methods. J Food Prot 50:14, 1987.

Doyle MP, Meske LM, Marth EH. Survival of *Listeria monocytogenes* during the manufacture and storage of nonfat dry milk. J Food Prot 48:740, 1985.

Flake JC, Clark WS Jr. Evaporated milk. In: Hui YH, ed. Encyclopedia of Food Science and Technology. New York: John Wiley & Sons, 1991, pp 750–751.

Foster EM, Nelson FE, Speck ML, Doetsch RN, Olson JC Jr. Dairy Microbiology. Englewood Cliffs, NJ: Prentice-Hall, 1957.

Gillies MT. Whey Processing and Utilization. Park Ridge, NJ: Noyes Data Corp., 1974.

Hammer BW, Babel FJ. Dairy Bacteriology. 4th ed. New York: John Wiley & Sons, 1957.

Higginbottom C. Bacteriological studies of roller-dried milk powders, roller-dried buttermilk and of roller- and spray-dried whey. J Dairy Res 13:308, 1944.

Kalogridou-Vassiliadou D. Biochemical activities of *Bacillus* species isolated from flat sour evaporated milk. J Dairy Sci 75:2681, 1992.

Langeveld LPM, van Spronsen WA, van Beresteijn ECH, Notermans SHW. Consumption by healthy adults of pasteurized milk with a high concentration of *Bacillus cereus*: a double blind study. J Food Prot 59:723, 1996.

Marshall RT, ed. Standard Methods for the Examination of Dairy Products. 16th ed. Washington, DC: American Public Health Assoc. 1992.

Merin U. Bacteriological aspects of microfiltration of cheese whey. J Dairy Sci 69:326, 1986.

Miller DL, Goepfert JM, Amundson CH. Survival of salmonellae and *Escherichia coli* during the spray drying of various food products. J Food Sci 37:828, 1972.

Nichols AA. Bacteriological studies of spray-dried milk powder. J Dairy Res 10:202, 1939.

Rodriquez MH, Barrett EL. Changes in microbial population and growth of *Bacillus cereus* during storage of reconstituted dry milk. J Food Prot 49:680, 1986.

Sienkiewicz T, Riedel CL. Whey and Whey Utilization. 2nd ed. Gelsenkirchen-Buer, Germany: Verlag Th. Mann, 1990.

Theophilus DR, Hammer BW. Influence of growth temperature on the thermal resistance of some bacteria from evaporated milk. IA. Agr Exp Sta Res Bull No. 244, 1938.

U. S. Department of Agriculture, Agricultural Marketing Service. Milk for manufacturing purposes and its production and processing; requirements recommended for adoption by state regulatory agencies. Federal Register 37:7046, 1972, as amended by FR 50:34726, 1985, and FR 58:86, 1993, and FR 61:48120, 1996. Washington, DC: U. S. Government Printing Office, 1972.

U. S. Department of Health and Human Services, Food and Drug Administration. Code of Federal Regulations. Part 184. Direct food substances affirmed as generally recognized as safe. Federal Register 46:44439, 1981, as amended by FR 47:7410, 1982, and FR 54:24899, 1989. Washington, DC: U. S. Government Printing Office, 1981.

U. S. Department of Health and Human Services, Food and Drug Administration. Code of Federal Regulations. Part 131. Milk and cream. Washington, DC: U. S. Government Printing Office, 1995a, pp 274–276.

U. S. Department of Health and Human Services, Food and Drug Administration. Code of Federal Regulations. Part 113. Thermally processed low-acid foods packaged in hermetically sealed containers. Washington, DC: U. S. Government Printing Office, 1995b, pp 223–248.

U. S. Public Health Service. Grade A Condensed and Dry Milk Ordinance. Supplement 1 to the Grade A Pasteurized Milk Ordinance, 1995 Recommendations. Washington, DC: U. S. Department of Health and Human Services, Public Health Service, Food and Drug Administration, 1995a.

U. S. Public Health Service. Grade A Pasteurized Milk Ordinance. Publication No. 229. Washington, DC: U. S. Department of Health and Human Services, Public Health Service, Food and Drug Administration, 1995b.

4

Ice Cream and Frozen Yogurt

Robert T. Marshall

University of Missouri, Columbia, Missouri

I. INTRODUCTION

The temperatures at which ice cream is stored and served are normally below freezing, and microbial growth is of no concern. Because the viability of many microorganisms is preserved by freezing, this treatment is not expected to be lethal for microorganisms. Freezing and frozen storage are detrimental to some microorganisms, and these effects are discussed later in this chapter. Although ice cream itself does not suffer direct microbial spoilage, several ingredients of ice cream are susceptible to spoilage because they are held at temperatures suitable for microbial growth.

A major concern of the ice cream industry is the potential for frozen desserts to be carriers of pathogenic microorganisms and of toxins produced by such microorganisms. Sources of disease producers and methods of protecting consumers from them are important topics for discussion in this chapter.

Some frozen desserts depend on microbial growth to produce typical flavor and textural characteristics. The most common of these products is frozen yogurt. Some of the bacteria used in yogurt fermentation are thought to provide health benefits and are called probiotics. The beans used to produce vanilla and chocolate flavors are fermented by microorganisms under controlled conditions.

This chapter progresses through definition of frozen desserts, considers their major ingredients and the potential contribution of those ingredients to the microflora of the finished product, describes processes used to produce mixes, and explores the freezing, storage, distribution, and serving of frozen desserts. Finally, regulations and quality assurance are discussed.

II. ENVIRONMENTAL SOURCES OF CONTAMINANTS

Outbreaks of foodborne disease from pasteurized dairy foods in the 1970 to 1985 period prompted the U. S. Food and Drug Administration to launch the Dairy Product Safety Initiative in 1985. A part of this program was microbiological surveillance of finished products for pathogenic bacteria. Potential pathogens were isolated from samples collected in 70 (6.9%) of 1016 plants surveyed during the second year of the program. Among the isolates were *Yersinia enterocolitica* (3.2%), *Listeria* spp. (2.9%), and miscellaneous isolates of *Salmonella, Aeromonas hydrophila*, and other pathogenic species (0.8%). Positive test results were associated with postpasteurization contamination (U. S. Food and Drug Administration, 1987).

Klausner and Donnelly (1991) surveyed 34 dairy processing plants in Vermont, focusing on floors and other nonproduct surfaces. *Y. enterocolitica* and other strains of *Yersinia* were isolated from 10.5% and 2.5% of the sites, respectively. The incidence of *Listeria innocua* (16.1%) was high compared with that of *Listeria monocytogenes* (1.4%). Pathogens were significantly more likely to be found in wet than in dry areas ($P < .05$). This points to the importance of depriving microorganisms of water. Although sanitizing floor mats and foot baths are designed to reduce incidence of transmission of bacteria by personnel, data from the study by Klausner and Donnelly indicate these devices may actually be sources of bacteria if they are not properly cleaned and refreshed with sanitizer.

A survey for listeriae in frozen milk product plants in California by Walker et al. (1991) revealed an incidence of 12% among 922 samples. Among the 39 plants sampled, *L. monocytogenes* and *L. innocua* were the single species recovered in five and 13 plants, respectively, and both species were recovered from nine plants. No listeriae were isolated from 12 plants. Most sites (95%) from which *Listeria* were isolated yielded a single species, suggesting that a single species dominates at any particular size. Although floor drains have been major sources of *Listeria* in dairy plants, no isolates were made from drains in nine plants where they were present in other selected sites. The authors suggested that increased awareness of high risks of drain-associated *Listeria* may have directed much attention to them even though other areas in the plant were neglected.

Whereas confidential reports from industry laboratories indicate that it is not unusual to find listeriae in environmental samples, it is unusual to find them in finished product. Rationale for this is that hygienic practices common to the frozen desserts industry are effective in preventing transfer of pathogens from the environment to pasteurized product.

III. COMPOSITION AND CHARACTERISTICS

Ice cream is a frozen foam. The continuous phase is a viscous syrup, and the suspended phase consists of tiny air cells, fat globules, colloidal substances (principally casein and stabilizing gums), and ice crystals. Microorganisms are also suspended in the continuous phase. Their viability is mainly affected by the pH, osmotic pressure, and their abilities to withstand high concentrations of salts plus the physical forces of ice crystals.

Freezing results in concentration of dissolved substances in the syrup. Substances detrimental to microorganisms include acids, salts, and, for some bacteria, sugars. In general, the order of survival of microorganisms in frozen desserts, ranked from highest to lowest survivability, is (a) bacterial spores, (b) spores of molds and yeasts, (c) gram-positive bacteria, (d) vegetative cells of molds and yeasts, and (e) gram-negative bacteria.

Ice cream contains from approximately 34% to 44% total solids. The most abundant component is carbohydrates, especially sugars. A typical full-fat formula may include 12% sucrose and 6% lactose as well as approximately 2% glucose and maltose from corn sweeteners. (These mono and disaccharides are listed as sugars in current nutritional labeling practice.) Additionally, such a formula includes approximately 4% higher saccharides from hydrolyzed corn starch. These carbohydrates lower the freezing point of the mix to about –3°C (26.6°F). The characteristic mix also contains approximately 1% ash, which is made up of minerals, especially calcium, magnesium, and phosphorus.

As ice is frozen out of the continuous phase, dissolved substances become increasingly concentrated, and the freezing point of this phase decreases. If heat is steadily and continuously removed, the cryohydric point of the least soluble substance is reached ultimately. At this temperature, this substance starts to precipitate, and latent heat of fusion is released. Therefore, the rate of decline in temperature is slowed until that substance is precipitated. There is a large number of substances in ice cream that may precipitate; therefore, during freezing, rates of temperature decline are not expected to be constant once eutectic points begin to be reached.

An unstable rate of decrease in temperature of ice cream being frozen is not expected to be a factor in survival of microorganisms, but formation of crystals and increasing concentrations of salts are likely to be detrimental. Salts tend to destabilize proteins and lipoproteins, and renaturation of them on thawing does not always occur. This is especially important for permeases that are located on the exterior of the cell. Sugars, however, may protect microorganisms from injury by freezing. Luyet (1962) suggested that microorganisms that best survive freezing are those that are able to dehydrate themselves most rapidly. Such cells are able to reduce the number of intracellular ice crystals that form, crystals that may puncture the cytoplasmic membrane.

The quantity of milk fat in ice cream ranges from less than 0.5% to more than 16%. Milk fat is an insulator in that it slows the rate of heat transfer through the frozen foam. Air cells, which may constitute up to one-half of the volume of ice cream, are also insulators. Both fat globules and air cells restrict growth of ice crystals. In so doing they reduce the amount of damage done to microbial cells by extracellular ice.

Colloidal substances that associate with water through hydration reduce the amount of water to be frozen, thus reducing the size and number of extracellular ice crystals. It is expected, therefore, that chances of survival of microorganisms are enhanced as increasing concentrations of colloidal substances are included and free water content is decreased in ice cream mixes.

In frozen yogurt, the concentration of lactic acid is expected to vary from 0.1% to 0.2% of the total weight of the mix. As a percentage of the weight of the unfrozen aqueous phase at the temperature of storage of ice cream, –20°C, lactic acid may constitute 1% to 2%. Depending on the buffering capacity of the mix constituents, the pH in the microenvironment of the microbial cells of the ice cream may be detrimental to viability of the cells.

IV. INGREDIENTS

A. Milk and Milk Products

Raw milk and cream are likely to contain the following pathogens sporadically but on a consistent basis when milk is assembled from numerous farms to a single large facility: *Campylobacter jejuni* (and other campylobacteria), *Salmonella dublin* (and other salmonellae), *Escherichia coli* (at times including pathogenic strains), and *L. monocytogenes*. Animals used for food production are infrequent carriers of these bacterial pathogens and a few others.

Ryser and Marth (1991) summarized results of tests of raw milk in the United States, Canada, and Europe, finding 3.1%, 2.7%, and 4.1%, respectively, of the samples contaminated with *L. monocytogenes*. However, numbers commonly found in raw milk are seldom more than 10/mL. Sources of *Listeria* in raw milk include infected mammary glands, poorly fermented silage, and soil. This bacterium is generally considered to be transmitted by nonzoonotic means (Kozak et al., 1995).

Raw fluid milk and cream spoil relatively rapidly. In general, raw milk is delivered from producing farms to processors within 40 to 72 hours of production and is not permitted to be held for more than 72 hours in the receiving dairy before processing. Most manufacturers process raw milk much sooner than the maximal time the system permits. It is important to do so to minimize risks of spoilage by psychotropic bacteria, especially members of the genus *Pseudomonas*. These bacteria are prolific producers of hydrolytic enzymes, including proteinases (Mayerhofer et al., 1973), lipases (Christen and Marshall, 1983), phospholipases (Fox et al., 1976), and glycosidases (Marin and Marshall, 1983). Many of the proteinases, phospholipases, and lipases retain their activity after pasteurization. Some can be inactivated at the relatively low temperatures of 40°C to 60°C (Marshall and Marstiller, 1981; Christen and Marshall, 1985).

Concentrated milks, commonly known as condensed milk and condensed skim milk, are widely used as ice cream ingredients. Concentrated milk products are almost always pasteurized before or during the concentration operation. Therefore, the incidence of microbial pathogens in these products is practically nil, and they have the microbiological keeping quality of pasteurized milk. Concentrated whey has similar characteristics. Bulk sweetened condensed milk and skim milk are prepared with sufficient sugar (approximately 42%) to prevent outgrowth of most spoilage bacteria. Furthermore, the evaporative process by which they are concentrated uses heat sufficient to destroy most vegetative forms of microorganisms. Therefore, they can be shipped and stored for limited periods without refrigeration.

Dry dairy ingredients include nonfat dry milk, dry buttermilk, dry whey, and whey protein concentrate. Processing commonly involves three steps, that is, pasteurization, concentration, and drying. Heat of these processes kills most of the vegetative microorganisms; therefore, viable bacteria recoverable from them usually are mostly spore formers. Major advantages to the use of dried dairy ingredients are their storability and low weight per unit of solids. The latter factor reduces the cost of transportation and the former provides maximal flexibility in use and helps balance supply with demand.

B. Sweeteners

1. Crystalline and Granular Sweeteners and Bulking Agents

Sucrose, dextrose, and fructose are available in both crystalline and syrup forms. Few microorganisms are contained in crystalline sweeteners. Maltodextrins, polydextrose, and corn syrup solids are available in granular form. Some of these materials may carry viable microorganisms, usually yeasts. Bottler's standards for 10-g samples of granulated sugars are less than 200 mesophiles, 10 yeasts, and 10 molds (National Soft Drink Association, 1975).

2. Syrups

In addition to sucrose, dextrose, and fructose, corn sweeteners are available as syrups. Because syrups contain water and provide energy, they may support growth of osmophilic fungi. These microorganisms, usually being highly aerobic, grow on surfaces. They can be killed by exposure to ultraviolet light and their growth can be inhibited by sealing full containers in which they are packed. This is not practicable when the container is a tank into which air must be admitted to displace syrup as it is drawn out during use. For them to flow steadily, syrups must be kept warm in pipelines that are used to transfer the sweetener to the batching tank for making mixes. Therefore, it is critical that the concentration of solids in the syrup be so high as to inhibit growth of the most osmophilic yeasts that might be contained. The usual solids concentration of these syrups is 71% to 82%, making the water activity (a_w) approximately 0.80. Syrups with a high dextrose equivalent (DE) are significantly more microbiologically stable than those with a low DE, for example 62 DE versus 36 DE. The sugar concentration, measured in Brix, ranges from 67° to 86°, depending on the sweetener. Smaller sugar molecules exert greater osmotic pressure than larger ones, given the same weight concentration. Therefore, concentrations of glucose, fructose, sucrose, and maltose necessary to limit microbial growth are lower than for corn syrups, which contain polymers of glucose that are products of incomplete hydrolysis of starch. High-fructose corn syrups of 42% and 55% have a_w values of 0.75 and 0.68, respectively (L. True, personal communication, 1997).

Osmotolerant yeasts can grow at a_w of less than 0.85. Even syrups with an a_w as low as 0.65 have been found to support growth of osmophilic yeasts (Troller, 1979). Most of these are in the genus *Zygosaccharomyces* (Walker and Ayres, 1970). Other genera of yeasts reportedly found are *Candida, Pichia, Schizosaccharomyces*, and *Torula*.

Condensate formation in syrup storage tanks raises the a_w and gives fungi opportunities to grow. Condensate accumulation can be prevented by forcing filtered and ultraviolet-treated air over the surface of the syrup.

In the preparation of corn syrups, the steps of steeping, wet milling, washing, purifying, and drying have a potential effect on microbial growth and survival. During steeping, corn is soaked in water at 45°C to 50°C for 48 hours at a pH of approximately 4.0 (Whistler and Paschall, 1967). During this period, the mixture is susceptible to growth of microorganisms that produce alcohols and butyric acid. A common microbial inhibitor added during steeping is sulfur dioxide (0.1%–0.2%).

Typical manufacture's maximal standards for microorganisms in syrups follow: aerobic plate count—100/g, yeasts—20/g, molds—20/g, *Escherichia coli*—none in 30 g, and *Salmonella*—none in 100g.

C. Honey

Honey is sometimes used in frozen desserts in the dual role of sweetener and flavoring agent. A typical concentration of honey in honey-flavored ice cream is 9%. Yeasts are likely contaminants of honey, because flowers from which the nectar is derived are the habitat of yeasts. Several species of *Zygosaccharomyces* have been isolated from defect-free and fermented honeys (Walker and Ayres, 1970). Because of its high hygroscopicity and viscosity, unprotected honey tends to develop areas (gradients) in which the a_w is high enough to permit yeast growth.

D. Flavorings

Pure synthetic or natural flavorings vary widely in content of microorganisms. Flavorings that are heat sensitive cannot be given a lethal heat treatment. Those that are low in viscosity and contain no suspended matter can be filter sterilized. Some are naturally antagonistic to microbial growth, especially those that have an alcohol base. Most are used in such small quantities that their contribution to the bacterial load is insignificant. Most are added after pasteurization, making it critical that they contain no pathogens.

1. Extracts

Alcohol is used to extract flavorful substances, such as vanilla, that are used to add flavor to frozen desserts. Pure vanilla is required to contain at least 35% ethanol to be labeled vanilla extract. This concentration of alcohol is sufficient to dehydrate and destroy most vegetative microbial cells. Other extractants include ethylene and propylene glycols.

2. Chocolate

Cacao beans are fermented before being ground and pressed to separate some of the cocoa butter from the cocoa. Grinding alone produces chocolate liquor, whereas pressing and grinding yields cocoa and cocoa butter. The latter contains only minor flavor notes; whereas the chocolate flavor is carried in the cocoa. Cocoa powders contain from 10% to 24% cocoa butter (fat) unless they have been extracted with a solvent. The microflora of uncontaminated cocoa and chocolate liquor consists nearly exclusively of bacterial spores and numbers are usually less than 100/g.

E. Fruits

Fruit ice creams represent approximately 15% of the total market.

1. Fresh and Frozen

Frozen fruits, especially berries, have been widely used in the frozen desserts industry for many years. Freezing tends to disrupt the structure and destroy the turgidity of fruits. On thawing, fruits become soft, juices escape from the cells, and color fades.

Because of the relatively low pH of fruits, the microflora of fresh and frozen fruits is dominated by yeasts, including the genera *Saccharomyces* and *Cryptococcus*, and by molds, including species of *Alternaria, Aspergillus, Botrytis, Fusarium, Geotrichum, Mucor, Penicillium*, and *Rhizopus*. Small numbers of soilborne bacteria are present also, including species of *Bacillus, Pseudomonas*, and *Achromobacter*. These bacteria do not compete well with the fungi in the pH range common to fruits. However, some lactic acid bacteria as well as species of *Acetobacter, Gluconobacter*, and *Zymomonas* may develop in the acidic environment of the fruit processing plant.

A principal source of pathogens in fresh and frozen fruits is persons who pick and handle them. Insects also may contaminate fruits. Peeling, washing, and blanching are processes that lower numbers of microorganisms on raw fruits.

Freezing kills some microorganisms on fruits but is not a dependable lethal process. Furthermore, it is not feasible from a quality viewpoint to blanch most fruits (except peaches) to destroy microorganisms. However, bactericidal chemicals, such as hypochlorite, may be added to wash water to reduce numbers of microorganisms on surfaces. Antioxidant dips are frequently applied to minimize browning. These include ascorbic acid, sulfur dioxide, and sugar syrup. Sulfur dioxide has some antimicrobial effect, and syrups may kill organisms that are susceptible to high osmotic pressures.

2. Processed

With the advent of highly effective heating and aseptic packaging processes, mostly aseptically processed fruits are used. In general, steam under pressure is not needed to destroy the microflora of fruits because they are acidic, and heating at 100°C or less is adequate. The more acidic the fruit, the lower the heat treatment required to preserve it. Among the fruits often used in frozen desserts, peaches and apricots fall within the "acid foods" range of pH 3.7 to 4.5, whereas berries have a pH less than 3.7, placing them in the "high acid foods" group.

Fruits that are aseptically processed can be stored at room temperature for several months with no microbial spoilage. Processors frequently use swept-surface heat exchangers that heat the mixture of fruit, sugar, acid, and stabilizer to

88° to 121°C, depending on the fruit. After holding the mixture for approximately 3 minutes at the maximal temperature, it is cooled to approximately 27°C and pumped directly to an aseptic filling machine. It is filled into sterile containers that are usually made of laminated polyethylene and foil.

In an alternative processing system (Fig. 1) fruit is pumped through coils that heat, hold, and cool the product. The coils cause the fruit to mix well in the tubes so that scrapers are not needed; therefore, little damage is done to the integrity of the fruit. Yet, microbial cells are efficiently and effectively destroyed.

Processing in open kettles permits heating to a maximum of 100°C and extends time of holding to at least 20 minutes. Volatile substances are able to escape the fruit, changing the flavor, and color usually darkens. Shelf-life is often short and refrigeration is needed to preserve the product.

Typical microbial specifications for fruit flavorings follow:

Process	APC/g	Yeast and mold/g	Coliforms/g
Cool fill	5,000	100	10
Hot fill	1,000	100	<1
Aseptic pack	100	10	<1

APC, aerobic plate-count.

3. Candied

Sugar is added to fruits before they are added to ice cream. The usual fruit to sugar ratio ranges from 2:1 to as high as 9:1. Candied and glacéd fruits have sugar concentrations high enough to lower the a_w below the level that permits microbial growth. Candying is accomplished by treating fruits with syrups having progressively higher sugar concentrations to prevent the exterior from becoming tough or leathery while the interior remains soft. Following impregnation with sugar, the fruit is washed to prevent crystallization of sugar on the surface and is then dried. To make glacéd fruit, candied fruit may be dipped into syrup and dried again.

F. Nuts

Nuts carry with them from the fields a wide array of microorganisms, many of which have their origin in soil. Some are contaminated with excreta from animals, birds, and insects. Various treatments are given nuts and nut meats in separating the nut meats from the shells. Most of these treatments lower microbial numbers in the nut meats. For example, sorting of lightweight pieces from the heavier nuts removes much of the dust that carries microorganisms. Flotation in water is used with pistachios to remove immature fruits and with pecans to

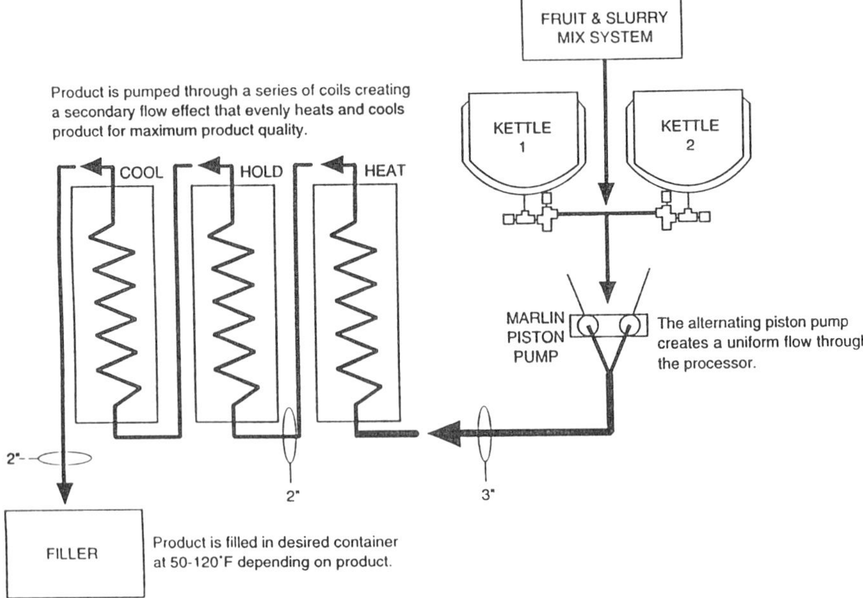

FIGURE 1 Aseptic processing system for fruits. (Courtesy of Lyons Magnus, Fresno, CA.)

remove fragments of shells. Blanching in hot water loosens pellicles from almonds and peanuts, and some nuts are salted in a brine solution. Treatments with water can remove microorganisms, but they can also elevate a_w, and reuse of water results in increasing populations of microorganisms that can be spread to other nuts. Therefore, frequent changes of water are needed.

Low a_w is the major limiting factor in preservation of nut meats; therefore, drying is required to prevent mold growth if harvested nuts are not sufficiently dry. Moisture content of tree nuts normally ranges from 3.8% to 6.7%. Thus, the a_w is usually less than 0.70 and microbial growth does not occur (Beauchat, 1978). Microorganisms usually die during storage. High temperatures and a_w just below the level sufficient for growth are factors that increase death rates (King and Shade, 1986).

Because of the wide variety of nuts, the different environments they come from, and the several treatments given them, types and numbers of microorganisms present on nuts vary widely. Counts range upward to several thousand per gram, and insect-damaged nuts carry more microorganisms than undamaged ones (King et al., 1970). Nuts harvested from orchards where farm animals have been kept have an increased likelihood of contamination with *E. coli*. However, neither tree nuts nor ground nuts are considered to be likely vehicles of pathogenic microorganisms. Treatments with propylene oxide, permitted on tree nuts but not on peanuts, destroys most of the residual microflora (Beauchat, 1973; USHEW, 1978). Roasting, a treatment given peanuts and some other nuts, destroys vegetative cells of microbes.

Mycotoxins, especially aflatoxins, are of concern because of the chance of mold growth on nuts that contain high amounts of moisture. Nut meats removed from refrigeration can have condensate form on them, especially when placed in areas of high humidity. If these nuts are not used soon and are stored at favorable temperatures, molds are likely to grow on them.

G. Confections and Bakery Products

Confections and baked goods are low in bacterial numbers and seldom carry pathogenic bacteria. Methods of preparation and very low a_w greatly limit survival and growth of microorganisms. However, they are usually added after freezing so that any contaminants they carry are given no positively lethal treatment.

H. Eggs and Egg Products

The ice cream industry uses egg yolks primarily for their flavor in the manufacture of French vanilla ice cream (also known as frozen custard; 1.4% egg yolk solids required) and in parfaits. Egg yolk is used also as a source of emulsifying and stabilizing agents, because egg yolk contains a high amount of lecithin. Sorbets usually contain 2.5% to 3% egg white. Pasteurized egg yolk is commercially available in three forms that are useful in manufacture of frozen desserts, that is, liquid, frozen, and dried. Egg white is available in dry and frozen forms. It is also possible to break and separate yolks from albumen of fresh shell eggs; however, this is usually feasible and economical only in production of small batches of ice cream.

Addition of 10% sucrose to egg yolks is effective in preventing gelation that occurs during storage of frozen yolks. Gelation of frozen plain yolk occurs most rapidly at approximately −18°C. Sugar is usually added to both the liquid and frozen forms. Salt also prevents gelation of egg yolk and is effective at approximately 2% concentration, but the salty flavor is undesirable in frozen desserts, making the sugar form the product of choice if frozen yolks are used.

The interiors of shell eggs (eggs in the shell) are usually sterile (with the possible exception of harboring certain salmonellae) at the time of laying (Brooks and Taylor, 1955; Morris, 1989). However, the shells of eggs become contaminated with several thousand to millions of bacteria during laying, collection, and processing.

Normally, 10 to 20 days pass between the time an egg is contaminated and the time when there is a significant increase in bacterial numbers. One reason is that little iron is available at the shell membranes and in the albumen, and most bacteria require iron for growth. Glycoproteins of the membrane fibers bind iron tightly. Ovotransferrin, a protein of the albumen (white), also chelates iron. Certain species of *Pseudomonas* produce an iron chelate, pyoverdine, that has been claimed to scavenge iron from ovotransferrin (Board and Tranter, 1995). Thus, they are able to overcome one of the major barriers to growth in egg albumen. Chemotaxis was observed to play a role in movement of *Pseudomonas putida* and *Salmonella enteritidis* toward yolk surfaces (Lock et al., 1992). The chemical attractant was not identified.

In addition to the hurdles that microbes face in the albumen of the shell egg, it has also been reported that avidin binds biotin (Chignell et al., 1975) and that ovoflavoprotein binds riboflavin (Clagett, 1971). Bacteria that require either or both of these vitamins would, therefore, be inhibited in the albumen of the egg. Furthermore, the highly alkaline (pH 9.5) albumen contains lysozyme, an enzyme that can lyse the cell membrane of certain gram-positive bacteria. Once a bacterium has reached the yolk of the egg, inhibitors are of no effect and nutrients abound, so growth can proceed rapidly.

Fresh eggs are seldom used in ice cream except in small operations. Because of the relatively high risk of the presence of salmonellae on and in fresh eggs, it is important that egg breaking be done in a room separate from the freezing and filling rooms. Furthermore, all eggs must be pasteurized if they are added to a frozen dessert after the mix is pasteurized.

Micrococci are nearly always present on freshly laid eggs, but spoilage of shell eggs is nearly always caused by gram-negative rods, especially species of *Pseudomonas* and *Proteus* (Board and Tranter, 1995).

Samples of unpasteurized liquid egg from commercial egg-breakers have been reported to range in aerobic plate count from 10^3 to 10^6 per gram (Froning et al., 1992). Although the number of salmonellae in unpasteurized liquid eggs is usually less than one per gram, the risk that these organisms may be present is high. Recently, the incidence of contamination of eggs with *S. enteritidis* through transovarian infection has caused considerable concern.

Most manufacturers use pasteurized egg products, including liquid whole egg, frozen sugar egg yolk, or dried egg yolk. Approved pasteurization standards for egg products produce 6 to 8 $\log_{10}$ reductions in numbers of *Salmonella* (Speck and Tarver, 1967; Shafi et al., 1970). All pasteurized egg products should meet the following microbiological limits: aerobic plate count, less than 10,000/g; coliform count, less than 10/g; yeast and mold count, less than 10/g; and salmonellae, negative in 25 g.

Freezing reduces numbers of viable microorganisms in egg products (Winter and Wilkin, 1947). Although most species of bacteria survive freezing in some numbers, the major survivors of both pasteurization and freezing are *Bacillus, Micrococcus,* and *Enterococcus* (Wrinkle et al., 1950; Froning et al., 1992). *Salmonella oranienburg* survived storage in frozen yolk (Cotterill and Glauret, 1972).

I. Coloring Materials

Coloring materials are often added to frozen dessert mixes after pasteurization; therefore, it is important that colorants be free of pathogens and low in total numbers of microorganisms. The following are typical microbiological specifications for food, drug, and cosmetic (FD&C) dry powders, blends, granulars, and FD&C lakes and lake blends: aerobic plate count less than 1000/g; coliforms, less than 10/g; yeasts and molds, less than 100/g; *E. coli* or *Salmonella*, negative in 25 g. Most firms do not test each batch for microbial counts but are willing to arrange for batch certification by an independent laboratory.

Colors and lakes provide very limited nutrients for growth of microorganisms, and, when sold in the liquid form, they contain low concentrations of benzoates as preservatives. When purchased in the powder or granular form, the water and containers used in hydrating them should be practically sterile and the water should be free of sources of nitrogen and energy that might enable microorganisms to grow. When rehydrated colorants are to be kept for several weeks, it is advisable to store them refrigerated.

J. Spices

Spices can carry widely varying numbers and types of microorganisms. Spore formers are especially prone to be present and to survive over long periods. Spices, like nuts, can be treated with ethylene oxide to reduce the microbial load. Furthermore, spices can be irradiated to kill microorganisms.

Cinnamon contains cinnaminic acid, a microbial inhibitor. However, dilution of cinnamon with ice cream mix greatly reduces this antimicrobial effect.

V. FROZEN YOGURT

A. Composition and Properties

Frozen yogurt has a composition similar to low fat ice cream. However, there is no Standard of Identity for frozen yogurt. The labeling regulations based on content of milk fat are the same as for ice cream. The unique characteristic of frozen yogurt is that it contains cultures of *Streptococcus thermophilus* and *Lactobacillus delbrueckii* ssp. *bulgaricus*. The shorter name, *L. bulgaricus* is usually used for the latter bacterium.

These two bacteria are typically grown together in skim milk fortified with 1% to 4% added nonfat milk solids. The skim milk is heated to approximately 85°C for 5 minutes and cooled before inoculation. Temperature of incubation is high, approximately 42°C, so generation time and, consequently, incubation time are short. From 10% to 20% of the finished and cooled yogurt is added to the processed and aged base mix at the time flavoring and coloring agents are added. Freezing follows.

It is also possible to add the yogurt culture to the base mix which is then incubated until the titratable acidity, expressed as lactic acid, reaches approximately 0.30%. However, this process involves cooling the mix after pasteurization to the incubation temperature, then completing the cooling of the full batch and holding it to permit aging. Therefore, time of production is longer and capacity of the fermentation tank must be larger than with the previously described method.

The product is frozen in the same way as ice cream and the overrun is typically in the range of 70% to 100%. Freezing kills many of the streptococci and lactobacilli of the yogurt culture. Sheu and Marshall (1993) observed that numbers of viable *L. bulgaricus* of two strains decreased approximately 45% and 90% during the continuous freezing of a simulated frozen yogurt milk. Viable cell numbers decreased approximately 5% more during storage at -20°C for 2 weeks after freezing. However, when the same two cultures were entrapped in beads (average diameter 18 μ) of calcium alginate gel, viable counts were approximately 45% higher than those of the nonentrapped cultures (Fig. 2). Cells of the strain of *L. bulgaricus* that were most susceptible to freeze damage were much larger than those of the smaller strain, suggesting that stresses of freezing are more damaging to large than to small cells.

Researchers have shown that exopolysaccharide (capsules) on bacteria renders cells comparatively resistant to thermal and physical shock (Robinson, 1981). Hong (1995) isolated three nonencapsulated mutants of *S. thermophilus* and compared them with the encapsulated parental strain for abilities to withstand freezing under a variety of conditions. The parent and mutant strains did not respond differently when frozen without agitation. However, freezing to -7°C with agitation in a batch freezer and hardening to -29°C resulted in survival of 28% of the encapsulated and only 17% of the nonencapsulated strains (Fig. 3). Early log phase cells were more sensitive to freezing than late log phase or stationary phase cells. Cell viability after batch freezing was unaffected by (a) culture growth temperatures between 40°C and 45°C, (b) fat content between 5% and 14%, or (c) neutralization of the acid produced by the cells during growth

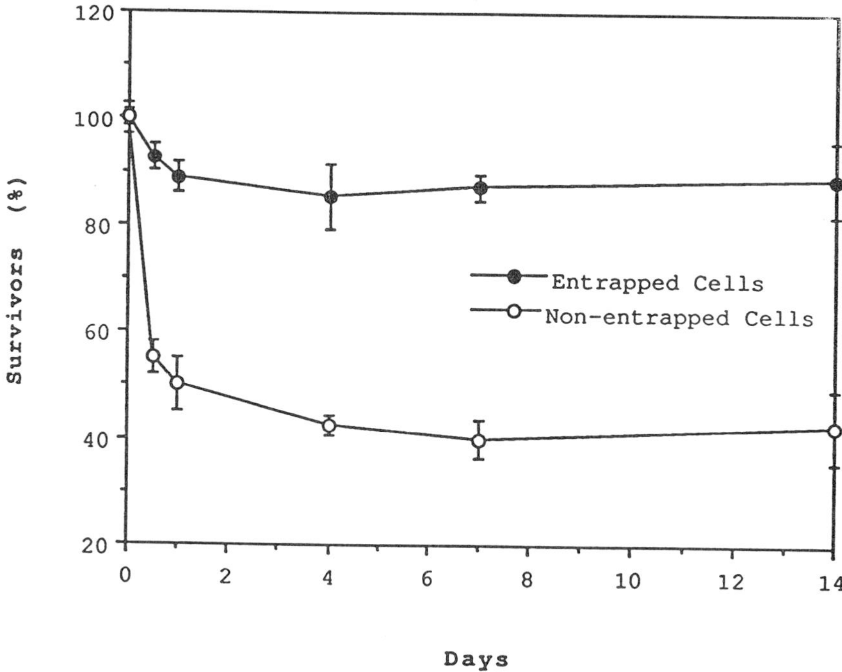

FIGURE 2 Numbers of survivors among *Lactobacillus delbrueckii* ssp. *bulgaricus* enrobed in calcium alginate. (From Sheu and Marshall, 1993.)

in skim milk. *S. thermophilus* survived significantly better in reduced fat ice cream frozen in a continuous freezer to 50% overrun than in the same mix frozen to 100% overrun. The added agitation and scraping of the freezer barrel walls needed to attain higher overrun may have been responsible for the lowered rate of survival. The additional oxygen whipped into the mix might have increased cellular exposure to free radicals and thus increased the death rate. However, no significant difference was found between numbers of survivors when the gas whipped into the ice cream was nitrogen or air. Storage of the frozen ice cream at –23°C or –29°C resulted in significantly more survivors than storage at –17°C.

B. Probiotic Nature

Although it was 1908 when Eli Metchnikoff suggested that certain bacteria in the human intestine could prolong the life of persons who consumed them in their foods, only recently have food microbiologists coined the term probiotic and have

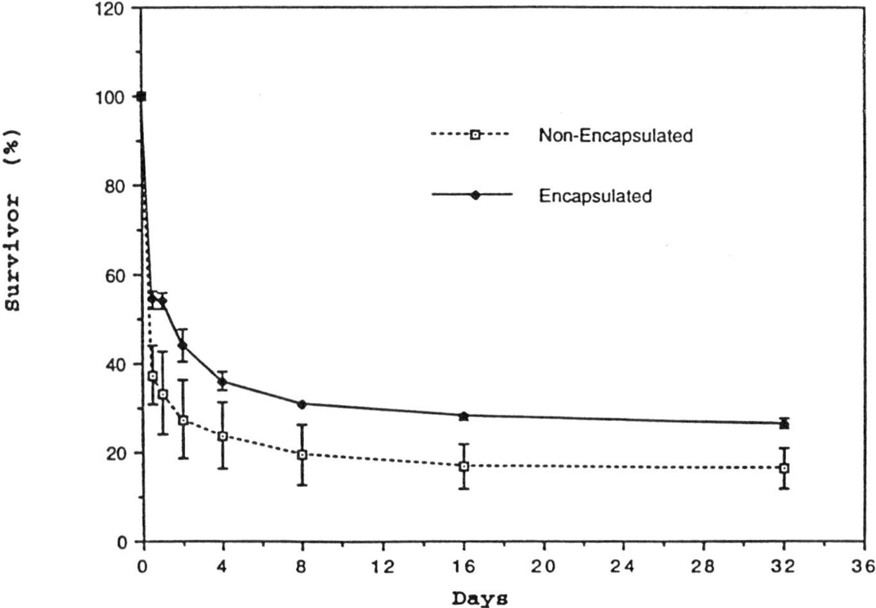

FIGURE 3 Numbers of survivors among encapsulated and nonencapsulated strains of *Streptococcus thermophilus* subjected to freezing in nonfat ice cream mix by a batch freezer. (From Hong, 1995.)

selected specific bacteria to add to foods as dietary adjuncts. The inference of the word probiotic is that a microorganism confers a positive effect on a biological entity, most importantly on human life. Most bacteria thought to have a probiotic effect are part of the natural microflora of the human intestine. Many of them are also useful as starter bacteria in food fermentations.

Benefits to consumers of fermented dairy foods and of those to which dietary adjunct bacteria are added include the following: (a) improved nutritional qualities (synthesized vitamins and enzymes as well as hydrolyzed proteins), (b) increased resistance to infection by competition, production of antibacterial substances (Shahani et al., 1977), and occupation of intestinal sites where pathogens may attach, (c) enhanced antibody production, (d) moderated response to endotoxin, and (e) anticarcinogenic activity.

Humans influence the nature of the intestinal microflora in several ways. Salivary, gastric, and intestinal secretions, bile, and mucus provide selective environmental factors. The stomach is strongly acidic, but pH increases as food moves to the distal end of the large intestine. Intestinal motility moves both food and microorganisms along the gastrointestinal tract, expelling billions of bacteria daily. Oxidation reduction potential is also a selective force. In general, the greater the distance intestinal contents travel from the stomach, the higher their microbial numbers.

Fermented dairy foods usually contain viable cells of the bacteria used as starter. Commonly used starter cultures contain lactococci, streptococci, lactobacilli, or leuconostocs. Some species of these genera have been shown to affect consumers favorably.

Frozen yogurt is the most popular dessert made from fermented milk. Most manufacturers produce frozen yogurt by adding 10% to 20% of plain yogurt to a pasteurized low fat ice cream mix. Flavoring is then added just before the mix is frozen. Assuming 5×10^8/g of viable *S. thermophilus* and *L. bulgaricus* in the plain yogurt and addition of 20% yogurt to the mix, the number of yogurt bacteria in the mix before freezing would be 10^8/g. If freezing were to kill 50% of the yogurt bacteria, the viable number remaining would be 5×10^7/g. This large number of viable cells may provide benefit to consumers.

Additionally, frozen desserts can be used as carriers of dietary adjuncts. Modler et al. (1990) used ice cream as a carrier for three species of bifidobacteria. At the end of 70 days of storage at $-17°C$, viable counts of these bacteria had decreased only 10%. Bifidobacteria have been receiving major attention as potential dietary adjuncts. These anaerobic, nonmotile, nonsporing, gram-positive, bifurcated (y-form) or curved rods produce acetic acid and L(+)-lactic acid as they ferment sugars. They comprise nearly 100% of the microflora in the stools of healthy breast-fed infants but only 30% to 40% of stool flora of formula-fed infants (Jao et al., 1978). As humans age, the percentage of bifidobacteria in stools decreases to low values. Their growth can be stimulated by oligosaccharides (Gyorgy et al., 1974), including β-linked *N*-acetylglucosaminides (Zilliken et al., 1955), glycoproteins (Bezkorovainy et al., 1979), and cysteine-containing peptides of kappa-casein (Poch and Bezkorovainy, 1991). Therefore, some foods are being supplemented with such substances with the intention of enhancing growth of bifidobacteria in the human intestine.

Another popular dietary adjunct that may be added to frozen desserts is *Lactobacillus acidophilus.* Certain strains of this bacterium were reported to assimilate cholesterol in a laboratory medium (Gilliland et al., 1984) as well as to lower serum cholesterol in rats (Grunewald, 1982). It is important that bacteria added to foods for probiotic effects be able to survive the effects of low pH and bile and to attach to and grow in a niche of the intestinal tract.

Each of the lactose-fermenting bacteria is a potential carrier of β-galactosidase. If these bacteria survive through the stomach and resist lysis by bile acids and enzymes, they may be permeated by lactose molecules. Intercellular β-galactosidase can then hydrolyze lactose to glucose and galactose so it can be absorbed through the human intestinal cell wall. Thus, symptoms of lactose malabsorption, a common malady among persons of Asian and African descent, can be reduced or eliminated.

VI. PROCESSING MIXES

The most important process in any dairy plant is pasteurization, because safety of the product depends on successful performance of this lethal heat treatment. Standards set for time and temperature of heating ice cream mixes (Table 1) are adequate to kill vegetative forms of pathogenic microorganisms that may be found in frozen dessert mixes. Residual spores of pathogenic bacteria are not considered dangerous, because they are unable to germinate and grow under conditions of storage of either the mix or frozen product.

Pathogens introduced into ice cream mixes by ingredients, equipment, personnel, or the environment are killed by pasteurization, but recontamination may occur in subsequent operations. The potential for amplification of the effects of pathogens increases as sizes of dairy processing facilities increase. This is true because large plants serve large numbers of consumers over a wide trade territory.

Controls are provided on continuous pasteurizers to ensure that minimal temperatures are maintained until mix reaches the end of the holding tube. Also, pasteurizers are required to be designed and operated to provide minimal times of holding mixes at the minimal temperature. However, research by Goff and Davidson (1992) revealed that mix viscosity is a major variable that can affect time of holding a mix in a pasteurizer. They found that laminar flow characteristics are likely to exist in holding tubes of high temperature, short time (HTST) pasteurizers when ingredients cause viscosities to become unusually high. Generalized Reynolds numbers, which are measures of turbulence in flowing liquids, ranged from 100 to 1700 in holding tubes of sizes common to the dairy industry. Laminar flow is likely to exist when Reynolds numbers are less

TABLE 1   Minimal Times and Temperatures
Required for Pasteurization of Frozen
Dessert Mixes

Method	Temperature	Time
LTLT	69°C (155°F)	30 min
HTST	80°C (175°F)	25 sec
	83°C (180°F)	15 sec

Abbrevations: LTLT, low temperature, long time, or batch (vat) method; HTST, high temperature, short time, or continuous method.

than 2100 (Denn, 1980). In true laminar flow, mix that is at the tube wall, flows one-half as fast as that in the center of the tube; whereas, in true turbulent flow, mixing is so thorough that particles travel at the same average rate in any cross-section of the pipe. Because of the high potential for laminar flow of ice cream mixes in pasteurizer holding tubes, special considerations should be given to their design.

The method approved in 3A Sanitary Standard No. 603-06 (3A Sanitary Standards Committee, 1992) provides that, for most pasteurizers, pumping rate is experimentally determined by timing the filling of a can of known volume and referencing this to a table of tube diameters and holding times of 15 and 25 seconds. Furthermore, holding time is confirmed by pumping water through the tube and detecting the time taken for an injected salt solution to pass conductivity sensors at each end of the holding tube. Whereas this method of testing provides reliable times for passage of products with the viscosity of milk, it is unlikely to be satisfactory for ice cream mixes that vary widely in viscosity. To overcome this problem, one approach is to design holding tubes to provide twice the holding time that would be applicable during turbulent flow. The 3A accepted practices provide that fully developed laminar flow is assumed when the desired holding tube length is calculated. This may result in more heated flavor than is desirable in the product. An alternative approach is to design pasteurizers with characteristics that ensure turbulent flow.

VII. FREEZING AND FROZEN STORAGE

In the freezing of ice cream, cold mix is admitted to a freezing chamber and subjected to whipping in the presence of air while the ice crystals that form on the wall of the freezing cylinder are scraped from the wall. Temperature drops rapidly and ice forms quickly in continuous freezers, but the process takes several minutes in batch freezers. These conditions place severe stresses on microorganisms in the mix. Ice crystals that form outside the cells reduce the amount of free water in which solute can be dissolved. Those that form inside cells have the potential to puncture cell membranes. Mazur (1966) concluded that viabilities of microorganisms subjected to subzero temperatures are affected primarily by solute concentration and intracellular freezing. Water that freezes in the cell is free water and this water forms ice crystals. Bound water remains unfrozen. As crystals form, the cytoplasm becomes more concentrated and viscous. Electrolytes and acids are concentrated. Colloidal constituents may be precipitated and proteins denatured. Intracellular ice is thought to be more harmful to microorganisms than extracellular ice. However, Ray and Speck (1973) concluded that, during freezing, formation of extracellular ice was the principal cause of bacterial death.

The result is that many microorganisms die. Generally, gram-negative rods and the vegetative cells of yeasts and molds are more easily killed than gram-positive bacteria, and bacterial and fungal spores are largely unaffected by sub-zero temperatures (Georgala and Hurst, 1963). Microorganisms in the logarithmic growth phase are more easily killed than are those in other phases of growth. Encapsulated bacteria survive freezing better than do the same strains that have lost the ability to express capsules because of mutations. The number of strains of encapsulated yogurt bacteria is limited, and it is important that yogurt bacteria survive freezing so they can deliver β-galactosidase to the human intestine of persons who are deficient in that enzyme and cannot, therefore, digest amounts of lactose they may ingest. During frozen storage at $-20°C$, the rate of death of yogurt bacteria in frozen yogurt was observed to be quite low. Ingram (1951) summarized the following effects of freezing on selected microorganisms: (a) many species experience an abrupt loss in viability on freezing, and (b) cells left viable after freezing die slowly during frozen storage, with death rate being highest when temperature approaches the melting point of the food and lowest at $-20°C$ and below.

VIII. SERVING FROZEN DESSERTS

All of the care in selecting and protecting ingredients, in cleaning and sanitizing equipment, in pasteurizing in a properly constructed and operated heat exchanger, and in packaging ice cream aseptically in containers that are practically sterile can be for naught if the product is contaminated with pathogens during serving.

Gould et al. (1948), in a survey of ice cream stores, found 11 of 20 hand-packed samples had coliform counts of more than 10/g, whereas only two of 14 factory-packed samples from the same stores had this high number of coliforms. Ice cream scoops and dippers as well as the hands of the store workers are likely sources of contaminants in dipped ice cream. Water should be kept flowing in dipper wells to ensure that bacterial growth is prevented in water used to warm and cleanse dippers and scoops.

Persons who are ill or infected should not dispense frozen desserts. All workers should wear clean clothing and hair restraints and should wash their hands before working in dispensing operations and every time there is a chance of their hands becoming contaminated.

IX. REGULATORY CONTROLS AND INDUSTRY STANDARDS

There is no federal standard for counts of bacteria in frozen desserts in the United States. However, most states enforce standards for coliform bacteria at less than or equal to 10/g and for standard plate count at 50,000/g. One state enforces a maximum standard plate-count of 20,000/g. Approximately 14 states permit coliform counts of up to 20/g for bulky flavored ice creams. These are products to which large amounts of flavorings, fruits, and nuts are added. Because many of these items are added after freezing, the chances of contamination with coliform bacteria is considerably greater than with plain ice creams. With the recent knowledge that microbial environmental contaminants include *Listeria*, it is prudent for manufacturers to consider the presence of coliform bacteria in ice cream as indicative of unsanitary practices and to increase the intensity of hygienic activities when coliform bacteria are found in finished products.

The U. S. Food and Drug Administration has tested finished ice cream products for pathogens, principally *L. monocytogenes*, and numerous recalls have ensued when samples have been positive (Anonymous, 1986a, 1986b, 1986c, 1986d, 1994). The U. S. Code of Federal Regulations, Title 21, Part 7.40 (21 CFR 7.40) provides recall policies, procedures, and industry responsibilities.

Recall is a voluntary act of manufacturers and distributors who seek to protect the health and welfare of consumers from products that may present a risk of injury or gross deception or are otherwise defective. Recall is an alternative to Food and Drug Administration–initiated court action to remove violative, distributed products. Recalls are assigned classes I, II, or III depending on the relative degree of health hazard with the greatest risk associated with class I recalls. The Food and Drug Administration may request a recall when a distributed product presents a risk of illness and the manufacturer or distributor has not initiated a recall. A recalling firm is expected to conduct checks of the effectiveness of the recall action.

A survey of 530 samples of ice cream mix (85), ice cream (394), and ice cream novelties (51) by Health and Welfare Canada revealed only two samples that contained *L. monocytogenes* (Farber et al., 1989). Furthermore, the WHO Working Group (1988) reported the incidence of *L. monocytogenes* in ice cream as varying from 0% to 5.5% with very low numbers (1 to 15 colony-forming units per gram) usually being observed.

The heat resistance of *L. monocytogenes* is higher than that of many vegetative bacteria (Doyle et al., 1987). Its heat resistance can be enhanced in milk and cream in which it is contained in white blood cells (leukocytes). As an agent of bovine mastitis (Gitter et al., 1980), *L. monocytogenes* is phagocytized by leukocytes of the mammary gland. If numbers of phagocytized *L. monocytogenes* are sufficiently high, the pathogen may survive minimal conditions of high temperature, short time pasteurization of milk (Garayzabel et al., 1985; Doyle et al., 1987). However, incidence of mastitis caused by *L. monocytogenes* is quite low. Furthermore, leukocytes and, consequently, phagocytized bacteria are mostly removed by clarification and separation during the preparation of cream for the manufacture of ice cream. No evidence has been forthcoming that these bacteria survive pasteurization of ice cream mix.

L. monocytogenes appears to survive well the freezing and frozen storage of ice cream (Golden et al., 1988; Palumbo and Williams, 1991; Dean and Zottola, 1996). Dean and Zottola (1996) inoculated ice cream mixes with an 18-hour-old culture of *L. monocytogenes* V7, froze the mix to –5°C to –6°C, and stored samples at –18°C for up to 3 months. One set of mixes contained 14 mg/L (535 IU/g) of the bacteriocin nisin (Nisaplin) and another set contained no nisin. Counts of *L. monocytogenes* decreased insignificantly in samples without nisin; however, counts decreased to near zero in the samples that contained nisin. Nisin was slightly less effective in ice cream containing 10% milk fat than in samples containing 3% milk fat. Jung et al. (1992) observed lowered nisin activities in high-fat–containing milk products.

L. monocytogenes is a gram-positive, non–spore-forming short rod that is motile with peritrichous flagella. This ubiquitous psychrotroph (Donnelly and Briggs, 1986; Rosenow and Marth, 1987) is pathogenic to humans and animals. Most persons who have contracted listeriosis have been pregnant women, neonates, or immunocompromised adults (Gray and Killinger, 1966; Seeliger, 1986).

An outbreak of gastrointestinal infections caused by *S. enteritidis* in ice cream was observed beginning in September, 1994. After a case-controlled study implicated a national brand of ice cream, much of the product was recalled by the manufacturer. Gastroenteritis developed in an estimated 224,000 persons (Hennessey et al., 1996), but fewer than 600 cases were reported to public health departments (Anonymous, 1996). The attack rate was estimated at 6.6% among consumers of the affected products. *Salmonella* was isolated from 8 of 266 ice cream products (3%) but not from environmental samples. The source of the pathogens was believed to be transport trucks used to haul both nonpasteurized liquid eggs and pasteurized ice cream mix. The mix was not repasteurized at the plant to which it was delivered in the tank trucks for freezing. The lesson learned was that repasteurization should be done when a mix is given any opportunity to be contaminated after pasteurization and especially when it is moved from one location to another in reusable containers. Such reusable containers should be dedicated to transport of mix only.

X. MICROBIOLOGICAL METHODS

Tests for microbiological quality and safety of frozen desserts and their ingredients are described in *Standard Methods for the Examination of Dairy Products* (Marshall, 1993), the *Compendium of Methods for the Microbiological*

Examination of Foods (Vanderzant and Splittstoesser, 1992), and the *Official Methods of Analysis* (Cuniff, 1996). Tests most relied on to reflect overall microbiological quality have been the standard plate count (aerobic plate count) and the coliform count. Petrifilm methods of performing each of these counts are given official status in standard methods and can be substituted for the plating methods. Furthermore, the spiral plating method for determination of the total aerobic plate count is an officially recognized method in standard methods.

Methods for enumeration of microorganisms are classified in *Standard Methods for the Examination of Dairy Products* (SMEDP) and in the official methods manual of AOAC International. Classification is based on three criteria: (a) research that thoroughly evaluates the method, (b) collaborative testing in qualified laboratories, and (c) demonstration of applicability based on extensive use. AOAC decides whether these qualifications have been met and awards a method first action status when it has been thoroughly evaluated and collaboratively tested; final action status is assigned when those methods have been proven in extensive use. The SMEDP classification terminology for these methods is A2 and A1, respectively.

As given in *Standard Methods for the Examination of Dairy Products* (Marshall, 1993), the agar method for enumerating coliform bacteria in ice cream products calls for making a 1:2 or 1:10 dilution and distributing 10 mL of this dilution equally into three petri dishes to which are added violet red bile agar. Matushek et al. (1992) showed that dilution of ice cream produced more accurate results than did direct plating. The major reason for inaccuracies with the direct plating method was atypical colonies produced with the directly plated samples. Non–lactose-fermenting bacteria can ferment sugar contained in plating media to which ice cream or frozen desserts are added, producing false-positive tests. Confirmation of suspect colonies as coliforms can be done by incubation in brilliant green bile lactose broth in which coliform bacteria produce gas when incubated at 32°C. False-negative results can occur when ingredients of frozen desserts inhibit growth. Inaccuracies may occur when excess product on a plate causes overcrowding (more than approximately 150 colonies), resulting in colonies that are less than 0.5 mm in diameter. Finally, pipeting undiluted sample cannot be done with precision because of the high and variable viscosities of frozen dessert mixes.

The official procedure (Marshall, 1993) for enumerating coliform bacteria with the Petrifilm method calls for making a 2:3 dilution of ice cream and plating 0.5 mL of this dilution onto one or each of three prehydrated coliform count plates. Experiments by Matushek et al. (1992) demonstrated that the Petrifilm coliform count method was highly satisfactory with higher confirmation rates (94%–100%) than any of the other methods tested.

XI. SUMMARY

Ice cream and other frozen desserts are protected from spoilage by very low temperatures of preparation and storage; however, major ingredients used to make these products are prone to spoilage and several ingredients are added after the last lethal process, pasteurization, has been completed. Therefore, microorganisms are of considerable importance to the frozen desserts industry. Pathogens of greatest importance are *L. monocytogenes* and *S. enteritidis*. The most threatening spoilage bacteria are psychrotrophs in the refrigerated dairy products and yeasts and molds in fruits and nuts. Dry ingredients and flavoring and colors are likely to contribute bacterial spores, but they seldom are of concern because of their low numbers and their inability to germinate and grow in the frozen products.

Ice cream is a relatively safe product, but failure to pasteurize it and to prevent environmental contamination can render it unsafe, especially to infants and immunocompromised adults.

REFERENCES

Anonymous. Food Chem News 28(19):31, 1986a.

Anonymous. Food Chem News 28(24):11, 1986b.

Anonymous. Food Chem News 28(27):17, 1986c.

Anonymous. Food Chem News 28(35):25, 1986d.

Anonymous. Food Chem News 36(33):34, 1994.

Anonymous. How safe is our food? Lessons from an outbreak of salmonellosis (editorial). N Engl J Med 334:1324, 1996.

Beuchat LR. *Escherichia coli* on pecans: survival under various storage conditions and disinfection with propylene oxide. J Food Sci 38:1036, 1973.

Beuchat LR. Relationship of water activity to moisture content in tree nuts. J Food Sci 43:754, 1978.

Bezkorovainy A, Grohlich D, Nichols JH. Isolation of a glycopolypeptide fraction with *Lactobacillus bifidus* subspecies *pennsylvanicus* growth-promoting activity from human milk casein. Am J Clin Nutr 32:1428, 1979.

Board RG, Tranter HS. The microbiology of eggs. In: Stadelman WJ, Cotterill OJ, eds. Egg Science and Technology. 4th ed. Binghamton, NY: The Haworth Press, 1995.

Brooks J, Taylor DI. Eggs and egg products. G. B. Dep Sci Ind Res Board, Spec Rep Food Invest. 60, 1955.

Chignell CF, Starkweather DK, Sinha BK. A spin label study of egg white avidin. J Biol Chem 250:5622, 1975.

Christen GL, Marshall RT. Effect of histidine on thermostability of lipase and protease of *Pseudomonas fluorescens* 27. J Dairy Sci 68:594, 1985.

Clagett CO. Genetic control of the riboflavin carrier protein. Fed Proc Fed Am Soc Exp Biol 30:127, 1971.

Cotterill OJ, Glauert J. Destruction of *Salmonella oranienburg* in egg yolk fractions; emulsion stabilizing power, viscosity and electrophoretic patterns. Poult Sci 48: 251, 1972.

Cunniff PA ed. Official Methods of Analysis. Gaithersburg, MD: AOAC International, 1996.

Dean JP, Zottola EA. Use of nisin in ice cream and effect on survival of *Listeria monocytogenes*. J Food Prot 59:476, 1996.

Denn MM. Process Fluid Mechanics. Englewood Cliffs, NJ: Prentice-Hall, 1980.

Donnelly CW, Briggs EH. Psychrotrophic growth and thermal inactivation of *Listeria monocytogenes* as a function of milk composition. J Food Prot 49:994, 1996.

Doyle MP, Glass KA, Beery GT, Garcia GA, Pollard DJ, Schultz RD. Survival of *Listeria monocytogenes* in milk during high temperature short-time pasteurization. Appl Environ Microbiol 53:1433, 1987.

Farber JM, Sanders GW, Johnston MA. A survey of various foods for the presence of *Listeria* species. J Food Prot 52:456, 1989.

Fox CW, Chrisope GL, Marshall RT. Incidence and identification of phospholipase C–producing bacteria in fresh and spoiled raw milk. J Dairy Sci 59:2024, 1976.

Froning G, Izat A, Riley G, Magwire H. Eggs and egg products. In: Vanderzant C, Splittstoesser DF, eds. Compendium of Methods for the Microbiological Examination of Foods. 3rd ed. Washington, DC: American Public Health Association, 1992.

Garayzabel JFF, Rodrigues LD, Boland JAV, Canulo JLB, Fernandez GS. *Listeria monocytogenes* dans let lait pasteurise. Can J Microbiol 32:149, 1985.

Georgala DL, Hurst A. The survival of food poisoning bacteria in frozen foods. J Appl Bacteriol 26:346, 1963.

Gilliland SE, Nelson CR, Maxwell C. Assimilation of cholesterol by *Lactobacillus acidophilus*. Appl Environ Microbiol 49:377, 1985.

Gitter M, Bradley R, Blampied PH. *Listeria monocytogenes* infection in bovine mastitis. Vet Rec 107:390, 1980.

Goff HD, Davidson VJ. Flow characteristics and holding time calculations of ice cream mixes in HTST holding tubes. J Food Prot 55:34, 1992.

Gould IA, Larsen PB, Doetsch RN, Potter FE. Maintaining quality in ice cream. Ice Cream Field 52:80, 1948.

Gray ML. Epidemiological aspects of listeriosis. Am J Public Health 35:554, 1963.

Gray ML, Killinger AH. *Listeria monocytogenes* and listeric infections. Bacteriol Rev 30:309, 1966.

Grunewald KK. Serum cholesterol levels in rats fed skim milk fermented by *L. acidophilus*. J Food Sci 47:2078, 1982.

Gyorgy P, Jeanloz RW, Von Nicolai H, Zilliken F. Undialyzable growth factors for *Lactobacillus bifidus* var *pennsylvanicus*. Eur J Biochem 43:29, 1974.

Hennessy TW, Hedberg CW, Slutsker L, White KE, Besser-Wiek JM, Moen ME, Feldman J, Coleman WW, Edmondson LM, MacDonald KL, Osterholm MT, Investigation Team. A national outbreak of *Salmonella enteritidis* infections from ice cream. N Engl J Med 334:1281, 1996.

Hong SH. Enhancing survival of lactic acid bacteria in ice cream by natural encapsulation and gene transfer. Doctoral Dissertation. University of Missouri, Columbia, MO, 1995.

Ingram M. The effect of cold on microorganisms in relation to food. Proc Soc Appl Bacteriol 14:243, 1951.

Jao YC, Mikolajcik EM, Hansen PMT. Growth of *Bifidobacterium bifidum* var *pennsylvanicus* in laboratory media supplemented with sugars and spent broth from *Escherichia coli*. J. Food Sci 43:1257, 1978.

Jung DS, Bodyfelt FW, Daeschel MA. Influence of fat and emulsifiers on the efficacy of nisin in inhibiting *Listeria monocytogenes* in fluid milk. J Dairy Sci 75:387, 1992.

King AD Jr, Miller MJ, Eldridge LC. Almond harvesting, processing and microbial flora. Appl Microbiol 20:208, 1970.

King AD Jr, Shade JE. Influence of almond harvesting, processing and storage on fungal population and flora. J Food Sci 51:202, 1986.

Klausner RB, Donnelly CW. Environmental sources of *Listeria* and *Yersinia* in Vermont dairy plants. J Food Prot 54:607, 1991.

Kozak J, Balmer T, Byrne R, Fisher K. Prevalence of *Listeria monocytogenes* in Foods: "Incidence in Dairy Products." International Dairy Foods Association, Washington, DC. Presented at the International Food Safety Conference, *Listeria*, The State of the Science, Rome, Italy, June 29–30, 1995.

Lock JL, Dolman J, Board RG. Observations on the mode of bacterial infection of hen's eggs. FEMS Microbiol Lett 100:701, 1992.

Luyet B. Recent developments in cryobiology and their significance in the study of freezing and freeze-drying of bacteria. In: Proceedings Low Temperature Microbiology. Camden, NJ: Campbell Soup Co., 1962, p 63.

Marin A, Marshall RT. Production of glycosidases by psychrotrophic bacteria. J Food Sci 48:570, 1983.

Marshall RT, ed. Standard Methods for the Examination of Dairy Products. 16th ed. Washington, DC: American Public Health Association, 1993.

Marshall RT, Marstiller J. Unique response to heat of extracellular protease of *Pseudomonas fluorescens* M5. J Dairy Sci 64:1545, 1983.

Mazur P. Physical and chemical basis of injury in single-celled microorganisms subjected to freezing and thawing. In: Meryman HT, ed. Cryobiology. New York: Academic Press, 1966.

Matushek MG, Curiale MS, McAllister JS, Fox TL. Comparison of various plating procedures for the detection and enumeration of coliforms in ice cream and ice milk. J Food Prot 55:113, 1992.

Mayerhofer HJ, Marshall RT, White CH, Lu M. Characterization of a heat-stable protease of *Pseudomonas fluorescens* P26. Appl Microbiol 25:44, 1973.

Metchnikoff E. The Prolongation of Life. New York: G P Putnam and Sons, 1908.

Modler HW, McKellar RC, Goff HD, Mackie DA. Using ice cream as a mechanism to incorporate bifidobacteria and fructoligosaccharides into the human diet. Cultured Dairy Prod J 25(3):4, 1990.

Morris GK. *Salmonella enteritidis* and eggs: Assessment of risk. Poultry Sci-Suppl. 1:1000, 1989.

National Soft Drink Association. Quality Specifications and Test Procedures for Bottlers' Granulated and Liquid Sugar. Washington, DC: National Soft Drink Association, 1975.

Poch M, Bezkorovainy A. Bovine milk κ-casein digest is a growth enhancer for the genus *Bifidobacterium*. J Dairy Sci 71:3214, 1991.

Ray B, Speck ML. Freeze injury in bacteria. In: King JK, Faulkner WR eds. Critical Reviews in Clinical Laboratory Science. Cleveland, OH: CRC Press, 1973, p 161.

Robinson DK. Yogurt manufacture—some considerations of quality. Dairy Indus Int 46:31, 1981.

Rosenow EM, Marth EH. Growth of *Listeria monocytogenes* in skim, whole and chocolate milk, and in whipping cream during incubation at 4, 8, 13, 21, and 35°C. J Food Prot 50:452, 1987.

Ryser ET, Marth EH. *Listeria*, Listeriosis and Food Safety. New York: Marcel Dekker, 1991.

3A Sanitary Standards Committee. 3A Accepted Practices for the Sanitary Construction, Installation, Testing and Operation of High-temperature Short-time and Higher-heat Short-time Pasteurizer Systems. Revised, Number 603-06. McLean, VA: 3A Sanitary Standards Committee, 1992.

Seeliger HPR. Diagnosis and treatment of listeriosis. In: Listeriosis. Joint WHO/ROI Consultation on Prevention and Control. West Berlin, Germany, December 10–12, 1986.

Shafi R, Cotterill OJ, Nichols ML. Microbial flora of commercially pasteurized egg products. Poultry Sci 49:578, 1970.

Shahani KM, Vakil JR, Kilara A. Natural antibiotic activity of *Lactobacillus acidophilus* and *bulgaricus*. II. Isolation of acidophilin from *L. acidophilus*. Cult Dairy Prod J 12(12):8, 1977.

Sheu J-S, Marshall RT. Improving survival of culture bacteria in frozen desserts by microentrapment. J Dairy Sci 76:1902, 1993.

Speck ML, Tarver FR. Microbiological populations in blended eggs before and after commercial pasteurization. Poultry Sci 46:1321, 1967.

Troller JA. Food spoilage by microorganisms tolerating low a_w environments. Food Technol 33:72, 1979.

US Food and Drug Administration. FDA dairy product safety initiatives. Second year status report. Washington, DC: Milk Safety Branch, Center for Food Safety and Applied Nutrition, 1987.

USHEW. Propylene oxide. 21 CFR 193.380. Public Health Service, Food and Drug Administration. Washington, DC: U. S. Department of Health, Education, and Welfare, 1978.

Vanderzant C, Splittstoesser DF, eds. Compendium of Methods for the Microbiological Examination of Foods. 3rd ed. Washington, DC: American Public Health Association, 1992.

Walker HW, Ayers JC. Yeasts as spoilage organisms. In: Rose AH, Harrison JS, eds. The Yeasts. Vol 3. Yeast Technology. New York: Academic Press, 1970.

Walker RL, Jensen LH, Kinde H, Alexander AV, Owens LS. Environmental survey for *Listeria* species in frozen milk product plants in California. J Food Prot 54:178, 1991.

Whistler RL, Paschall EF. Starch: Chemistry and Technology. Vol II. New York: Academic Press, 1967.

WHO Working Group. Foodborne listeriosis. Bulletin of the WHO 66(4):421, 1988.

Winter AR, Wilkin M. Holding, freezing and storage of liquid egg products to control bacteria. Food Freezing 2(4):338, 1947.

Wrinkle C, Weiser HH, Winter AR. Bacterial flora of frozen egg products. Food Res 15:91, 1950.

Zilliken F, Rose CS, Braun GA, Gyorgy P. Preparation of alkyl *N*-acetyl-α- and β-*D*-glucosaminides and their microbiological activity for *Lactobacillus bifidus* var. *penn*. Arch Biochem 54:392, 1955.

5

Microbiology of Butter and Related Products

JEFFREY L. KORNACKI

Silliker Laboratories of Wisconsin, Inc., Madison, Wisconsin

RUSSELL S. FLOWERS

Silliker Laboratories Group, Inc., Homewood, Illinois

I. INTRODUCTION AND DEFINITIONS

A. Volumes of Butter and Brief History

Worldwide consumption of butter and butterfat products is estimated at 2,420,000 tons in 1993 for countries where data are available (Table 1). Butter is one of the first dairy products manufactured by humans and has been traded internationally since the 14th century (Anderson, 1986; Varnam and Sutherland, 1994). All butter manufacture relies on cream. From ancient times through the latter part of the 1800s, cream was obtained by gravity separation from milk (Varnam and Sutherland, 1994). In the 1850s, factories began producing butter on a small scale. Large-scale manufacture became possible after development of mechanical cream separation in 1877 (Varnam and Sutherland, 1994).

TABLE 1 Total Consumption (1000 tons) of Butter and Butterfat Products (1993)

Country	1,000 tons
Austria	33.8
Australia	58.3
Belgium	70.0
Canada	84.776
Switzerland	37.5
Germany	555.9
Denmark	21.5
Estonia	8.91
Spain	9.0
Finland	27.0
France	389.8
United Kingdom	205
Hungary	9.7
India	58.51
Iceland	0.589
Italy	98
Japan	92
Netherlands	50.4
Norway	9.7
New Zealand	31.9
Sweden	19.9
United States	533
South Africa	14.748
TOTAL	2,420

Source: Adapted from Bulletin of IDF No. 301, 1995, p. 13.

B. Composition and Types of Butter

Butter is a water-in-oil emulsion, wherein butterfat forms the continuous phase. This is in contrast to cream, which is an emulsion of fat globules suspended in an aqueous phase. Thus, an emulsion phase reversal occurs during the manufacture of butter. This happens in churning of cream, and, as a result, fat is concentrated in the product. Butter contains more than 80% fat (typically approximately 81%), 15% to 16% moisture, less than 0.5% of carbohydrates and protein, and 2% sodium chloride (if added). The pH of sweet cream (unfermented) butter is between 6.1 and 6.5 (Milner, 1995). Many countries allow sodium chloride and lactic cultures as the only non-milk additives in butter (Milner, 1995). Some countries allow neutralization of cream and addition of natural coloring agents (e.g., annato, carotene, and turmeric).

There are essentially two kinds of butter: sweet cream, which may or may not be salted, and ripened-cream butter. Citrate in cream is fermented by lactic acid bacteria in ripened-cream butters to produce acetoin and diacetyl; the latter imparts characteristic flavor to the product (Adams and Moss, 1995). Ripened-cream butters are more popular in continental Europe, whereas unripened or sweet-cream butter is preferred in the United States, Ireland, England, Australia, and New Zealand (Adams and Moss, 1995).

Milk fat recovered from whey produced during cheese making can be passed through a separator to produce whey cream. This whey cream can be processed into butter, which is often indistinguishable from butter made from sweet cream (Halpin-Dohnalek and Marth, 1989b). Whey cream butter can be manufactured from neutralized or nonneutralized salted or unsalted whey cream.

II. MANUFACTURE OF BUTTER

The manufacture of butter (Fig. 1) is characterized by the following three processes:

1. **Concentration of the fat phase of milk**. This is done by milk *separation* and *standardization* of resultant cream to the desired fat content.
2. **Crystallization of the fat phase**. Large numbers of small solid fat crystals are required. *Pasteurization* temperature applied to cream yields fully liquefied butterfat. Thus, *cooling* (greater than 4 hours at approximately 5°C) is necessary to develop an extensive network of stable fat crystals. In ripened-cream butter, addition of lactic acid bacteria to pasteurized cream cooled to 16°C to 21°C is followed by incubation until a pH near 5 is obtained. Cooling to 3°C to 5°C is begun after the product reaches approximately pH 5.
3. **Phase separation and formation of a plasticized water-in-oil emulsion**. *Churning* and *working* break the oil-in-water (o/w) emulsion and result in a plasticized water-in-oil (w/o) emulsion. These processes occur in both batch and continuous churns. During churning, violent agitation is used to disrupt membranes on milk fat globules (Brunner, 1976). Effective clumping of fat globules to form a continuous matrix requires selection of an optimal temperature, usually 5°C to 7°C, for a batch churn. The optimal temperature is dependent on triglyceride composition and hence the physical state of the fat (Brunner, 1976). Churning is ineffective if all fat is either liquid or solid. The proper blend of liquid and solid fat is necessary. Continuous churn operations require temperatures that account for the fat content of cream and hardness of the butterfat. The optimum temperature for churning is often selected empirically by operators to minimize fat losses (Varnam and Sutherland, 1994).

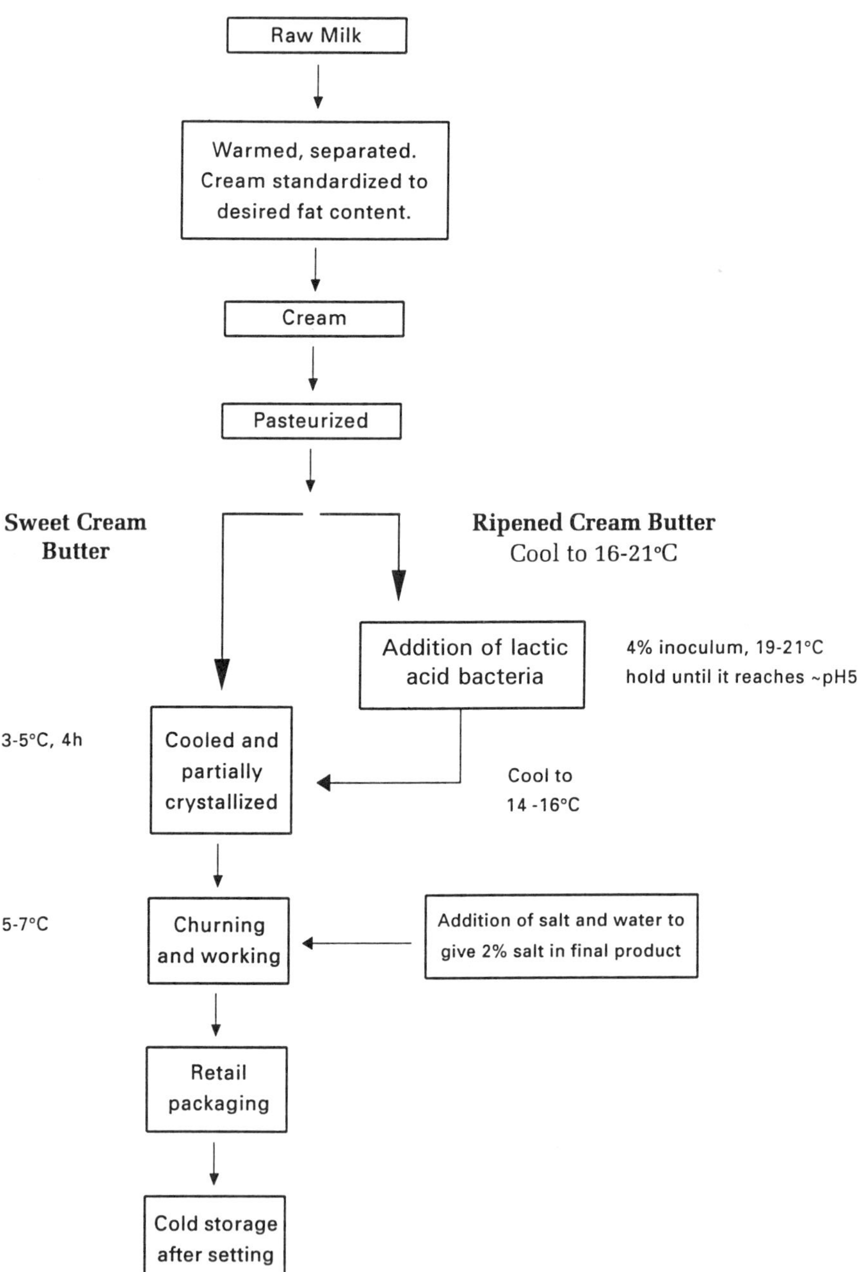

FIGURE 1 Production of butter. (Adapted from Milner, 1995.)

After churning and working, the butter is *washed* and *salted.* Salting is done after washing to prevent loss of salt. *Packaging* occurs after salting and may be done directly into retail portions or in bulk containers (25 kg is common) (Varnam and Sutherland, 1994). National intervention boards in the European Economic Community stipulate a *storage* temperature of –15°C; however, lower temperatures are frequently used. A temperature of –30°C was effective for storing butter in excess of 1 year (Varnam and Sutherland, 1994). Stored butter is then subjected to *repackaging* into retail containers. Butter from different manufacturers may be blended together during repackaging, and sometimes garlic, chopped herbs, and diacetyl concentrate are added for additional flavor (Varnam and Sutherland, 1994). Thus, butter manufacture involves separation of cream from raw milk, standardization to the desired fat content, pasteurization, possible addition of lactic acid bacteria (when ripened-cream butter is manufactured), churning, working, washing, salting, packaging, storage, and repackaging (see Fig. 1). All of these activities affect the microflora of the final product.

III. MICROBIOLOGICAL CONSIDERATIONS (BUTTER)

The microbiology of butter reflects the microflora present in cream from which it is made, fluid used to wash or salt the curd, sanitary conditions of process equipment, manufacturing environment, process hurdles that limit microbial growth and survival, and conditions under which the product is stored.

A. Cream

The main source of microorganisms in butter made under excellent sanitary conditions is cream. Fat globules rising in milk carry microbes present in raw milk (Milner, 1995). Raw milk can be contaminated with a wide variety of microbial pathogens and spoilage microorganisms. The microflora of raw milk is related to that found in the cow's udder and on the hide and milking utensils and lines (Jay, 1992). Proper handling and storage conditions should result in a predominantly gram-positive microflora in raw milk. Gram-negative microbes, along with yeasts and molds, are typically less heat resistant than gram-positive microorganisms and are more likely to be destroyed by pasteurization (Jay, 1992). Psychrotrophic *Bacillus* and *Clostridium* spp. have been found in 25% to 35% and 8% of raw milk samples, respectively (Jay, 1992; IDF/FIL, 1994). These organisms would be expected to survive pasteurization of cream. Raw milk is occasionally contaminated with *Mycobacterium* spp. (Jay, 1992; IDF/FIL, 1994), and controversy exists over survival of *Mycobacterium paratuberculosis* (the cause of Johne's disease in cattle and putative agent of Crohn's disease in humans) during milk pasteurization (Chiodini and Herman-Taylor, 1993; Grant et al., 1995, 1996; Keswani and Frank, 1996). A review of pathogenic micro-organisms in raw milk was prepared by the International Dairy Federation (IDF/FIL, 1994).

B. Importance of Pasteurization

Butter was contaminated with *Streptococcus agalactiae*, *Streptococcus pyogenes*, and two strains of *Staphylococcus aureus* when made with unpasteurized milk from a cow whose quarters were artificially infected with these organisms. These bacteria persisted 6 months both in salted (2%) and unsalted sweet-cream butter and in ripened (6 and 12 days at 21°C) cream butter. *Brucella abortus* inoculated into unpasteurized raw cream also survived 4 months at approximately 7°C in salted and unsalted sweet cream butter. This organism also survived 3 months in unsalted and salted, ripened-cream (6 and 12 days at 21°C) butter. All storage was at approximately 7°C. Butters made from contaminated cream that was subsequently pasteurized at 62.8°C for 30 minutes did not contain these pathogens (Bryan and Bryan, 1944). Pasteurization of cream from raw milk is designed to eliminate vegetative microbial pathogens and reduce numbers of potential spoilage organisms. In the United States, cream must contain not less than 18% fat. Dairy ingredients with more than 10% fat must be pasteurized at temperatures 3°C higher than milk (e.g., 66°C for 30 minutes for cream versus 63°C for 30 minutes for milk) (Code of Federal Regulations, 1995). However, heat-resistant microbes such as some strains of *Lactobacillus, Enterococcus*, and spores of *Bacillus* and *Clostridium* will survive. Temperatures between 95°C and 112°C are commonly used to inactivate them (Schweizer, 1986). Cream is also heated to deactivate lipases (which cause hydrolytic rancidity in butter), reduce the intensity of undesirable flavors (e.g., from feed ingredients), activate sulphydryl compounds (which can reduce the auto-oxidation of butter), and liquefy fat for subsequent control of crystallization (Schweizer, 1986).

C. Ripening

Many people in western and northern Europe prefer the flavor of butter manufactured from microbiologically ripened cream (Pesonen, 1986). Traditionally, pasteurized cream (after cooling to 6°C to 8°C for greater than 2 hours to initiate fat crystallization, followed by warming) at 19°C to 21°C is inoculated with lactic cultures composed of pure or mixed strains of *Lactococcus lactis* spp. *lactis*, *Lc. lactis* spp. *cremoris*, *Leuconostoc mesenteroides* ssp. *cremoris*, and *Lc. lactis* ssp. *lactis* biovar *diacetylactis*. Ripening occurs for 4 to 6 hours until a pH of 4.6 to 4.7 is achieved. The product is cooled to stop the fermentation. In this process, spoilage microorganisms are controlled primarily through the bacteriostatic effect of lactic acid produced by the starter culture.

D. NIZO Method

The NIZO method (Kimenai, 1986) for producing a cultured butter has been allowed in several countries and is used in many factories in western Europe. In the NIZO method, starter culture are not added to cream, but, instead, a mixture of diacetyl-rich permeate and starter cultures is worked into butter. Permeate is produced by fermentation of partly delactosed whey or other suitable media containing milk components by lactic acid bacteria (i.e., *Lactobacillus helveticus*). After incubation for 2 days at 37°C, the medium is subjected to ultrafiltration for removal of proteins and bacteria and further concentrated (Kimenai, 1986). During ultrafiltration, macromolecules are retained by a semipermeable membrane and concentrated in a retentate, whereas low molecular weight solutes pass through the membrane in a permeate stream. The pH of butter made with this process is more easily adjusted in the desired range of 4.8 to 5.3. It has been claimed that this permeate can be stored at 4°C for more than 4 months (Kimenai, 1986). Advantages cited for this process include the following (Kimenai, 1986):

1. Greater control of the manufacturing process by dividing it into more independent steps (e.g., greater flexibility for choosing the most appropriate temperatures for both butter consistency and propagation of cultures)
2. Lower copper content of butter made by this method, hence, reducing the risk of oxidative defects
3. Lower free fatty acid content, hence, lessening the chance of hydrolytic rancidity (soapy flavor) of the final product
4. Less starter culture needed to be produced
5. Greater freedom to choose different types of starter cultures to better control butter flavor
6. Elimination of problems associated with ripened cream that may become too viscous to pump to the churn
7. A yield of sweet, not cultured, buttermilk
8. Greater quality of butter after 3 years of cold storage than that of traditionally manufactured butter stored 3 years

Homofermentative lactic acid bacteria such as *Lc. lactis* ssp. *lactis* and *Lc. lactis* ssp. *cremoris* are used to produce lactic acid from lactose in dairy products. However, flavor production requires addition of a heterofermentative organism such as *L. mesenteroides* ssp. *cremoris* or *Lc. lactis* ssp. *lactis* biovar *diacetylactis* to produce diacetyl (Jay, 1992). Diacetyl, in addition to imparting flavor, inhibits gram-negative bacteria and fungi (Jay, 1992).

E. Churning and Working

The bacterial load of buttermilk is typically greater than that of cream or butter (Milner, 1995). When culture-ripened cream is used to manufacture butter, most of the starter culture organisms are retained in the buttermilk; however, some remain in the butter. In several studies, butter made from cultured cream retained 0.5% to 2.0% of the culture present in the cream (Hammer and Babel, 1957). Olsen et al. (1988) found that numbers of *Listeria monocytogenes* were 6.7 to 15 times higher in pasteurized but subsequently inoculated creams than in butter manufactured from the same cream. In an earlier study (Minor and Marth, 1972), *Staphylococcus aureus* behaved similarly. These organisms are gram-positive, and it is unclear how other microorganisms with different cell wall and membrane structures distribute themselves (e.g., *Mycobacterium* noted for its hydrophobicity) between cream and butter. Diacetyl content of cream increases during churning; agitation during churning favors the oxidative processes needed for diacetyl production (Foster et al., 1957). The pH of salted butter can prohibit formation of diacetyl (Foster et al., 1957).

F. Moisture Distribution During Churning and Working

It has been estimated that 10 to 18 billion droplets of water are dispersed in 1 g of the w/o emulsion that is butter (Hammer and Babel, 1957). Given the low microbial load expected in pasteurized sweet cream (less than 20,000/mL) (Jay, 1992), most of the droplets are sterile. This depends on the degree of dispersion of droplets and the microbial level in the cream (Hammer and Babel, 1957). The diameter of water droplets in conventionally made butter has been reported at <1–>30 μm. Water droplets in butter manufactured by the Fritz, Alfa, and Cherry-Burrell continuous processes have been reported to be less than 1 to more than 15 μm, less than 1 to less than 7 μm, and less than 1 to 30 μm, respectively (Brunner, 1976).

Fritz-type continuous churns are generally favored worldwide; however, many phase-inversion types of churns that use highly concentrated cream (e.g., 80% "plastic" cream) are used in the former Soviet Union and other eastern European countries (Varnam and Sutherland, 1994).

The Fritz butter making process is based on the same principles as traditional batch churning. For example, crystallization of fat is carried out in cream with phase inversion and fat concentration during churning and draining. However, phase inversion methods rely on concentration of cream to approximately 80% in a centrifugal separator, followed by phase inversion, cooling, and then crystallization of fat (Munro, 1986). Several modifications of this approach have been developed, including the Alfa and Cherry-Burrell processes mentioned earlier.

The Meleshin phase inversion process and modifications of it are in wide use in eastern Europe and the former Soviet Union (Munro, 1986). In the Meleshin process, pasteurized cream is concentrated to a fat content slightly above that of the final product (e.g., 61% to 85%). It is then standardized and cooled to 11°C to 13°C in a series of scraped surface heat exchangers, each called a "transmutator," wherein phase inversion and crystallization occur (Munro, 1986). Varnam and Sutherland (1994), Kimenai (1986), and Munro (1986) have provided more detailed descriptions of continuous butter manufacturing processes.

The number of water droplets greater than 30 μm in diameter is inversely proportional to the time of working during conventional (batch churn) butter manufacture (Hammer and Babel, 1957). A consequence of uneven distribution of droplets containing microorganisms is a high degree of nonhomogeneity regarding microbial distribution in butter. Thus, testing butter for quantitative microbial quality should be based on a three-class attribute plan (Smittle, 1992) to account for this variability. Inadequate working of the butter results in less dispersion of water droplets and promotes microbial spoilage (Hammer and Babel, 1957; Foster et al., 1957). Greater microbial growth was observed in serum when it was separated from unsalted butter than in butter before separation of serum. However, separation of serum from salted butter increased the rate of microbial destruction in the serum (Hammer and Babel, 1957, reporting on Hammer and Hussong, 1930). Thus, nutrients and salt in separated serum were more available for growth or inhibition of microorganisms, respectively, than when butter was in the normal physical state. This implies that the availability of nutrients or inhibitors is limited by the fine dispersion of water droplets (Foster et al., 1957). Droplet size is ideally less than 10 μm (Varnam and Sutherland, 1994).

G. Washing and Salting

Butter granules may be rinsed to remove excess buttermilk (Foster et al., 1957); however, this is not often done. If butter is washed, then salting is done after washing to prevent loss of salt. Salt added to butter can be expected to inhibit microbial growth. Salt must be distributed evenly in the moisture phase of the product to effectively inhibit microbial growth in contaminated water droplets. However, working does not result in a homogeneous distribution of salt in the water droplets (Hammer and Babel, 1957; Milner, 1995). Salting creates an osmotic gradient between salt granules and buttermilk. This tends to cause aggregation of water droplets and can lead to free moisture and a defect called "mottling." Adequate working and use of finely ground salt can minimize this defect (Varnam and Sutherland, 1994).

Use of brine is restricted to products with less than 1% salt because the brine cannot contain more than 26% salt (w/w). Sometimes, slurries of salt in saturated brine solutions containing up to 70% sodium chloride are used. Salt granules used for production of a slurry should be less than 50 μm in diameter. The salt slurry should also be of high purity, with less than 1 mg lead, 10 mg iron and 2 mg copper per liter (Varnam and Sutherland, 1994).

The microbiological quality of water used for washing or for brines is critical to production of a safe and stable product. Water with less than 100/mL total aerobic count when plates are incubated at 22°C and less than 10/mL total aerobic count when plates are incubated at 37°C has been deemed to be acceptable (Murphy, 1990).

It has been demonstrated that *Listeria* survive in a saturated brine solution held at 4°C for 132 days (Mitscherlich and Marth, 1984). Thus, brines used to salt butter must be free of *Listeria*. Water and brines used should also be free from pseudomonads. The most common form of spoilage in butter occurs with species of *Pseudomonas* (Jay, 1992; Milner, 1995). Addition of salt to butter lowers the freezing point so that psychrotrophic microorganisms present may be able to grow at less than 0°C. Some psychrotrophic organisms multiply in salted butter stored as low as 21.2°F (–19.4°C) (Hammer and Babel, 1957).

The distribution of salt in the moisture phase of butter has less impact on growth of yeasts and molds on the surface of butter as compared to bacteria (Hammer and Babel, 1957). Humid conditions appear to have a greater impact on mold growth than does the material on which they grow. Bacterial spoilage may occur in areas of low salt with large droplets of moisture. In the past, many instances of spoilage in butter were traced to contaminated water supplies (Hammer and Babel, 1957). These instances typically involved organisms of the genus *Pseudomonas*. *Micrococcus* sp., many of which are lipolytic, have also been implicated in butter spoilage, especially of temperature-abused sweet cream butter (Varnam and Sutherland, 1994).

H. Packaging

In smaller operations, butter is loaded directly from the churn onto trolleys and wheeled onto packaging machines by plant personnel (Varnam and Sutherland, 1994). This type of packaging of butter exposes the product to air, workers, the plant environment, and temperatures that may promote spoilage. Control of the microbiological quality of air during packaging is therefore important. Practices that result in standing water or entrapped wet residues facilitate growth of environmental contaminants. Practices that aerosolize these contaminants afford a level of microbiological contamination to the air. Thus, practices that promote dry conditions in the plant are preferred. Numerous approaches can be taken to monitor microbiological air quality, which include sedimentation, impaction on solid surfaces, impingement in liquids, centrifugation, and filtration (Hickey et al., 1992). Air quality is particularly important in the manufacture of whipped butter and in butter produced from continuous-type churns that may incorporate up to 5% air into butter (if a vacuum deaerator is not used) (Varnam and Sutherland, 1994).

Personnel hygiene is also critical at this point of butter manufacture because contaminants from hands, mouth, nasal passages, and clothing may be transmitted to butter during packaging. Some continuous churns are arranged to discharge product directly into the receiving hopper of packaging machinery (Varnam and Sutherland, 1994). However, to ensure uninterrupted operation, it is more common to transfer butter to a silo via a butter pump. Some silos are sealed, but others are open. Sealed silos minimize the risk of further contamination from the plant environment but are of limited capacity (approximately 900 kg), whereas open silos have a capacity of up to 10 tons (Varnam and Sutherland, 1994). Butter must be pumped from these silos to the packaging equipment. Direct packaging into consumer-size containers is preferable over bulk packaging because such butter must be repackaged before sale. Such reworking can result in poor distribution of water droplets and thus increase the risk of spoilage (Milner, 1995). A 1983 International Dairy Federation investigation of 79 butter factories indicated that 56% produced mainly 25-kg packages and 44% produced mainly consumer packages. However, two-thirds of these factories produced both types of packaged butter (Pointurier, 1986).

Cardboard boxes lined with vegetable parchment, aluminum foil, or a variety of plastic films are typically used for bulk packaging of butter (Varnam and Sutherland, 1994). Polyethylene is the preferred material based on its physical properties (low density, high impact, cost effectiveness, absence of copper, and nearly sterile condition) (Varnam and Sutherland, 1994). Parchment, which supports mold growth under humid conditions, is still frequently used (Varnam and Sutherland, 1994). Use of dry parchment or parchment treated 24 hours with 0.5% sorbic acid has been recommended (Varnam and Sutherland, 1994). Retail butter packs are typically wrapped in vegetable parchment or foil parchment laminate. Vegetable parchment is permeable to water. Thus, weight loss from water loss can occur during storage. Parchment allows penetration of ultraviolet light, which can accelerate onset of oxidative rancidity (Varnam and Sutherland, 1994). Individual butter packs, used in restaurants and food service establishments, are made by filling preformed polyvinyl chloride pouches.

I. Pathogen Survival and Growth in Butter

The following pathogenic microorganisms have been shown able to grow in butter products: *L. monocytogenes* in butter at 4°C and 13°C (from inoculated cream) (Olsen et al., 1988) and *S. aureus* in lightly salted (1% w/w) whey cream butter at 25°C and 30°C (Halpin-Dohnalek and Marth, 1989b) and inoculated whipped butter at 25°C (Halpin-Dohnalek and Marth, 1989a).

Inoculation of whipped butter promoted greater growth or survival of *S. aureus* than whipped butter made from inoculated cultured cream (Halpin-Dohnalek and Marth, 1989a). Data were not available regarding the size distribution of water droplets in the products just described (e.g., percentage of droplets <10 μm).

The following pathogenic bacteria were detected in butter before widespread use of pasteurized cream: *Salmonella, Streptococcus*, and *Mycobacterium* sp. (Berry, 1927). Abbar and Mohamed (1987) found several serotypes (not O157:H7) of enteropathogenic *E. coli* in butter manufactured in Iraq. *Listeria innocua* (not considered to be a pathogen but frequently associated with *L. monocytogenes* in environmental samples) was found in butter by Massa et al. (1990). Several authors found pathogen survival in butter that was artificially or directly inoculated. *Salmonella* survived 27 days in butter prepared from infected cream (Berry, 1927). Berry (1927) also reported on a study in which diphtheria bacilli survived for 1 month in butter. *Salmonella typhimurium, Salmonella enteritidis*, and *Salmonella schottmuelleri* survived 239, 228, and 212 days, respectively, in inoculated butter stored at 10.5°C (51°F). *Shigella* strains survived from 7 to 18 days, and *Streptococcus pyogenes* survived 17 days (Berry, 1927) under the same conditions. *Brucella abortus* survived 142 days in butter that was inoculated and stored at 46°F (Carpenter and Boak, 1928).

J. Food Poisoning Outbreaks

The incidence of documented food poisoning associated with butter consumption was low (Berry, 1927), even before widespread pasteurization of cream for butter manufacture. Berry (1927) summarized several outbreaks in which butter was thought to be the source. These included three outbreaks of typhoid fever and a rural outbreak of diphtheria. The widespread use of pasteurization resulted in a significant reduction in milk-borne illness (Bryan, 1983). Currently, postpasteurization environmental contamination of cream or butter represents the greatest risk to butter contamination and spoilage.

Several outbreaks of staphylococcal intoxication have been reported in the United States related to butter or cream (Centers for Disease Control, 1970, 1974, 1977). In one instance, gastrointestinal illness developed in 24 customers and employees of a department store restaurant and was traced to whipped butter manufactured from whey cream (Centers for Disease Control, 1970). The same butter used to manufacture the implicated whipped product also resulted in one case of gastroenteritis. This butter contained 10 ng of staphylococcal enterotoxin A/g.

In another instance, more than 150 people became ill from milkshakes sold at several fast food restaurants. Liquid milkshake mixes for two different eastern United States dairies were manufactured using the same dried whey cream (Centers for Disease Control, 1974). In 1977, more than 100 customers of pancake houses in the midwest became ill after consumption of whipped butter (Centers for Disease Control, 1977).

Halpin-Dohnalek and Marth (1989b) subsequently demonstrated the potential of temperature-abused neutralized cheddar cheese whey to support growth of *S. aureus*. Hence, the potential exists for toxin production in temperature-abused neutralized whey used for manufacture of whey cream butter.

K. Spoilage

The two principal types of microbial spoilage of butter are, in order, putridity, sometimes referred to as "surface taint," and hydrolytic rancidity (Jay, 1992). Both conditions can be caused by growth of *Pseudomonas* sp. Some *Pseudomonas* sp. are psychrotrophic (Kornacki and Gabis, 1990) and produce proteases and lipases (Cousin, 1982), which can spoil foods by hydrolysis of protein and fat, respectively. *Pseudomonas putrifaciens* can grow on butter surfaces at 4°C to 7°C and produce a putrid odor within 7 to 10 days (Jay, 1992). The putrid odor results from liberation of certain organic acids, especially isovaleric acid (Jay, 1992).

Rancidity, the second most common spoilage defect, is caused by both microbial and nonmicrobial lipases, which degrade butterfat to free fatty acids. *Pseudomonas fragi* and sometimes *Pseudomonas fluorescens* are associated with this defect (Jay, 1992). Mold growth on butter can also cause hydrolytic rancidity for the same reasons (Irbe, 1993). Molds that can cause this defect in butter include *Rhizopus, Geotrichum, Penicillium*, and *Cladosporium* (Irbe, 1993). Less common spoilage defects include malty flavor, skunklike odor, and black discoloration. These defects are caused by *Lc. lactis* var. *maltigenes, Pseudomonas mephitica*, and *Pseudomonas nigrifaciens*, respectively. Other microbially induced color changes result from surface growth of various fungi that produce colored spores (Jay, 1992). Heat-resistant proteases and lipases produced by pseudomonads that may grow during storage of raw milk can result in spoilage of butter after manufacture, even though spoilage organisms may have been destroyed by pasteurization.

L. Sources of Environmental Contamination

The necessity of milk and wash water to be of high microbial quality and the importance of cream pasteurization to public health have already been discussed. Yeasts and molds are particularly resistant to dry conditions when compared to bacteria. Unlike bacteria, many of these fungi can grow at water activities (a_w) below 0.84. A few can grow below an a_w of 0.65 (Troller and Christian, 1978). A study was reported in which molds would not grow on butter held at 70% humidity or below (Hammer and Babel, 1957). Therefore, to prevent growth of psychrotrophic yeasts and molds, a humidity of 60% or less should be maintained in the processing environment.

Ineffective sanitation of processing equipment could result in product contamination from equipment such as piping, pumps, coolers, or vats (Hammer and Babel, 1957). In our experience, the backplate of positive displacement pumps (e.g., from pasteurized cream storage tanks) may be neglected during sanitation and so can become a microbial growth niche, which, in turn, provides an inoculum to the product stream. Stress cracks in double-walled, insulated vats can also provide a source of product contamination when the insulating material between walls becomes wet. Early studies revealed numerous foci for yeast and mold contamination in wooden butter churns (Hammer and Babel, 1957). These churns were impossible to effectively clean and sanitize. Consequently, they were replaced by churns made of stainless steel or aluminum-magnesium alloys (Adams and Moss, 1995). However, published data validating effective cleaning and sanitation on these newer churns through use of microbiological swabs are lacking.

Personal hygiene of persons working with butter is also important. Cross-contamination from hands, mouths, nasal passages, and clothing must be precluded (Hammer and Babel, 1957). Product cross-examination with mold from workers handling raw cream cans has been documented (Hammer and Babel, 1957).

Handling butter after distribution to restaurants may also result in cross-contamination of a product, e.g., when 1-pound prints are divided with knives used for cutting meat. Whipping butter with improperly sanitized equipment can also be a source of contamination for the product (Halpin-Dohnalek and Marth, 1989a).

IV. MICROBIOLOGICAL CONTROL OF BUTTER

A. Factors Limiting Microbial Growth in Butter

We have seen that a variety of extrinsic (e.g., temperature) and intrinsic factors (e.g., salt in the moisture phase) combine to control the microflora of butter. Most important among these are (a) fine and uniform dispersion of the moisture phase, (b) addition and uniform dispersion of salt, (c) low temperature, and (d) use of lactic cultures (in ripened cream butter) (Hammer and Babel, 1957; Olsen et al., 1988). Microbial growth is proportional to the availability of nutrients and related to the radius of water droplets in a compartmentalized product, such as butter (Verrips, 1989). Thus, the smaller and more uniform the droplets, the less potential for microbial growth. Salt must also be distributed evenly in the moisture phase of the product to effectively inhibit microbial growth in contaminated water droplets. However, working does not result in a homogeneous distribution of salt in the water droplets (Milner, 1995; Hammer and Babel, 1957). Data suggest that dispersion of water droplets, salt, and bacteria in butter made by continuous churns may be more uniform than in butter made with batch churns (O'Toole, 1978). Aerobic plate-counts revealed a steady decrease in microbial contaminants compared with the sporadic counts obtained on butter made from batch churns (O'Toole, 1978). Salt-free droplets were found in freshly worked salted butter made with a batch churn (Hammer and Babel, 1957). Theoretically, butter with 16% moisture and 2% salt should have 12.5% salt in the moisture phase. If well dispersed, this salt content in the moisture phase should be adequate to inhibit most microbial growth.

The practice of adding milk to salted butter before whipping could decrease the salt content and increase moisture in the serum of whipped butter (Minor and Marth, 1972), which may have contributed to the food poisoning outbreak discussed previously (Centers for Disease Control, 1970). The importance of microbiological water quality used for washing or for brines to produce a safe and stable product has already been discussed. Technological developments that allow for uniform dispersion of moisture, salt, and bacteria enhance both the safety and shelf-life of butter.

Storage of butter at freezing temperatures is not adequate to guarantee complete cessation of microbial growth because of the depressed freezing point in the moisture phase of the product resulting from elevated salt content and presence of other dissolved solutes. O'Toole (1978) provided data that suggested the lowest temperature limit for microbial metabolic activity in butter was −8.7°C (16.3°F). In organoleptic trials, the flavor score of butter held at −6°C was marginally reduced after 12 weeks. Butter stored 8 weeks at 4°C or 10°C dropped about one point in flavor scores (O'Toole, 1978).

Some countries allow the use of preservatives, for example potassium sorbate is allowed in the United States, but not in the United Kingdom (Varnam and Sutherland, 1994). In some countries, benzoate is widely used; however, preservatives are not permitted in other countries, including France and Luxembourg (Varnam and Sutherland, 1994). Addition of 0.1% potassium sorbate inhibited growth of coliforms and molds in naturally contaminated butter (Kaul et al.,

1979). The inhibitory effect was enhanced when 2% salt was added along with 0.1% potassium sorbate. This inhibition occurred in all samples at the end of 4 weeks at −18°C and 5°C.

Caution should be exercised in the selection of any additives used to blend into butter products for flavor (e.g., spices) because these may contribute additional microflora to the product. Butter color ingredient that has not been mishandled has rarely contributed to the microflora of cream or butter (Foster et al., 1957).

B. Quality Assurance

Any quality assurance program should incorporate maintenance and documentation of good manufacturing practices (GMPs) and hazard analysis critical control points (HACCP).

C. Hazard Analysis Critical Control Points

Obvious critical control points for butter manufacturers include pasteurization or repasteurization of cream received at the plant and control of the microflora in the manufacturing environment. Each plant must evaluate its individual process and develop its own risk assessment and HAACP plan (Smittle, 1992). An environmental sampling protocol should be aimed at monitoring this critical control point for *L. monocytogenes* and *Salmonella*. Recalls of butter because of *L. monocytogenes* contamination were reported as recently as 1991 and 1992 (FDA Enforcement Report, 1991, 1992). Faust and Gabis (1988) have recommended areas of food plant environments that can be targeted for sampling. Presence of *Salmonella* and *Listeria* in the environment should result in corrective action with documentation of the success of that action. Irbe (1993) has recommended that manufacturers of whipped butter develop in-plant guidelines for aerobic plant-count and *S. aureus* at critical control points of manufacture. Finished products should be free of *Salmonella, E. coli* (Irbe, 1993), and *Listeria*.

Testing for these organisms can be done to validate success of the manufacturer's HACCP program. Manufacturers should also test for lipolytic and psychrotrophic spoilage organisms in the finished product and develop a three-class attribute sampling plan (Smittle, 1992) based on the principles of continuous quality improvement (Crosby, 1984). Sanitation of equipment used to manufacture product should be assessed on a daily basis by testing environmental swabs for their aerobic plate-count (incubated 3 days at 25°C). The lower temperature (25°C vs. 32°C or 35°C) allows for recovery of psychrotrophic bacteria as well as mesophiles.

Published microbiological criteria for butter include less than 50,000 aerobic plate-count per gram, less than 10 coliforms per gram, absence of *E. coli* type I in 1 gram, less than 10 yeasts and molds per gram, and absence of *S. aureus* in

1 gram (Anonymous, 1993). It has also been recommended that cream for butter making have an aerobic plate-count of less than 1000 per milliliter with less than 1 coliform, yeast, or mold per milliliter (Robinson, 1990).

V. MICROBIOLOGY OF RELATED PRODUCTS

A. Definitions

Margarines, like butter, contain approximately 81% fat, 15% moisture, 0.6% protein, 0.4% carbohydrate, and 2.5% ash (Irbe, 1993). In margarine, edible fats, oils, or mixtures of these whose origin is vegetable or rendered animal carcass fats are substituted for butterfat (Code of Federal Regulations, 1994). Both butter and margarine have fat contents in excess of 80%. This level is considered too high by many individuals concerned about their diets (Varnam and Sutherland, 1994). Consequently, a large number of spreads have been manufactured with lower fat contents. In many countries, there are no legal standards or definitions for these low-fat spreads. However, a working categorization has been made based on fat content (Varnam and Sutherland, 1994). **Full-fat spreads** are described as those with fat contents of 72% to 80%; **reduced-fat spreads** have 50% to 60% fat; **low-fat spreads** have 39% to 41% fat, and **very–low-fat spreads** have less than 30% fat. Vegetable fats, mixtures of vegetable fat and butterfat, and butterfat alone has been used to develop these spreads (Varnam and Sutherland, 1994). Another trend has been production of spreads in which fat has been replaced in part or completely by a variety of substances such as Neutrifat, Simplesse, and Stellar (Varnam and Sutherland, 1994). Olestra was recently (1996) approved by the Food and Drug Administration as a substitute for conventional fats and may appear in products in the future.

B. Dairy Spreads: Manufacture and Microbiological Considerations

Low-fat spreads are also w/o emulsions but contain a higher moisture content than butter. Consequently, these have increased likelihood of microbial growth unless preservatives are added. Preservative addition is allowed in some countries but not in others. Combining of ingredients occurs at 45°C in an emulsifying unit. This temperature may allow growth of thermoduric organisms (e.g., *Enterococcus faecium, Enterococcus faecalis*) and thermophiles. Higher fat dairy spreads are typically made using conventional continuous butter manufacturing methods. Margarine manufacturing techniques may be applied to high-fat vegetable oil spreads (Varnam and Sutherland, 1994).

Generally, these spreads are manufactured by adding moisture to the melted fat phase. The fat phase can be vegetable oil or a mixture of vegetable oil and butterfat alone (light butter) (Varnam and Sutherland, 1994). This is followed by addition of color, flavor, and vitamins. Addition of 1% to 1.5% salt results in the same perception of "saltiness" as for salted butter (Milner, 1995). Pasteurization (typically 80°C to 85°C, 2–3 seconds) of the blend occurs after this step, given the potential for microbial growth and difficulty of cleaning these emulsifying units (Varnam and Sutherland, 1994). However, pasteurization does not eliminate thermoduric or spore-forming microorganisms.

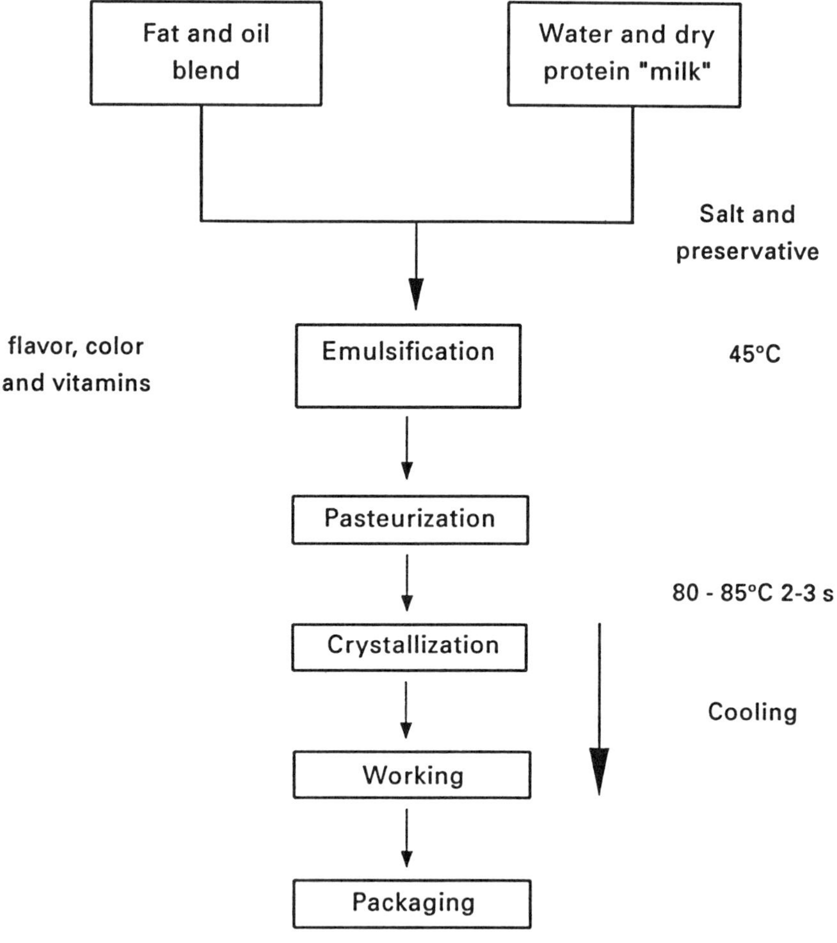

FIGURE 2 Production of dairy spreads. (Adapted from Milner, 1995.)

Crystallization of fat during working is critical to obtain the desired consistency of the finished product. Rapid supercooling (–10°C to –20°C) under high sheer conditions in a scraped surface heat exchanger initiates and maintains crystallization and disperses moisture within the fat matrix (Varnam and Sutherland, 1994). This cooling inhibits microbial growth. These products are often manufactured in an environment with filtered or sterile air. Control of cross-contamination during packaging is more critical than in butter manufacture because of the higher potential for microbial growth in spreads.

Microorganisms that cause spoilage in butter have also been implicated in margarine spoilage. However, vegetable fats are typically more resistant to lipolytic breakdown than butterfat (Varnam and Sutherland, 1994). *Yarrowia lipolytica, Bacillus polymyxa,* and *E. faecium* are spoilage organisms of concern in low-fat spreads (Varnam and Sutherland, 1994; Lanciotti et al., 1992). Lanciotti et al. (1992) showed that *L. monocytogenes* and *Yersinia enterocolitica* can grow in "light" butter at 4°C and 20°C. A class I recall of a 60% butter, 40% margarine product occurred in 1992 (FDA Enforcement Report, 1992). More detailed descriptions of margarines, spreads, and industrial milk fat products can be found in the report by Varnam and Sutherland (1994). An outline of margarine and spread manufacture is shown in Figure 2.

VI. CONCLUSION

The safety record of butter has improved considerably since the advent of widespread cream pasteurization and improvements in churn design, sanitation, and water quality. However, rigorous adherence to GMPs with appropriate environmental sampling and HACCP is necessary to ensure the safety and prolong the shelf-life of butter and spreads.

REFERENCES

Abbar FM, Mohamed MT. Occurrence of enteropathogenic *Escherichia coli* serotypes in butter. J Food Prot 50:829–831, 1987.

Adams MR, Moss MO. The microbiology of food preservation. In: Food Microbiology. Cambridge, UK: The Royal Society of Chemistry, 1995, pp 55–102.

Anonymous. Cleanliness takes the cream. Int. Food Hygiene 3(6):5–6, 1993.

Anderson RF. Introduction, Continuous Butter Manufacture. International Dairy Federation Bulletin No. 204. Brussels, Belgium: International Dairy Federation, 1986, p 2.

Berry AE. Viability of pathogenic organisms in butter. J Prev Med 1(6):429–442, 1927.

Brunner JR. Characteristics of edible fluids of animal origin: milk. In: Fennema OR, ed. Principles of Food Science. New York: Marcel Dekker, 1976, pp 619–658.

Bryan CS, Bryan PS. The viability of certain udder infection bacteria in butter made from raw cream. J Milk Technol 7(1):65–67, 1944.

Bryan F. Epidemiology of milk-borne diseases. J Food Prot 46:637–649, 1983.

Carpenter CM, Boak R. *Brucella abortus* in milk and dairy products. Am J Public Health 18:743, 1928.

Centers for Disease Control. Staphylococcal food poisoning traced to butter—Alabama. Morbid Mortal Weekly Rep 19:271, 1970.

Centers for Disease Control. Probable staphylococcal enterotoxin contamination of commonly produced milk shake mixes—Maryland, New Jersey, Pennsylvania, West Virginia. Morb Mortal Wkly Rep 23:155–156, 1974.

Centers for Disease Control. Presumed staphylococcal food poisoning associated with whipped butter. Morb Mortal Wkly Rep 19:271, 1977.

Chiodini R, Herman-Taylor J. The thermal resistance of *Mycobacterium paratuberculosis* in raw milk under conditions simulating pasteurization. J Vet Diagn Invest 5:629–631, 1993.

Code of Federal Regulations. 21 CFR 116.110. Washington, DC: Food and Drug Administration, 1994.

Code of Federal Regulations. 21 CFR 131.3. Washington, DC: Food and Drug Administration, 1995.

Cousin M. Presence and activity of psychrotrophic microorganisms in milk and dairy products: a review. J Food Prot 45:172–207, 1982.

Crosby PB. Quality without tears. New York: McGraw-Hill, 1984.

Faust RE, Gabis DA. Controlling microbial growth in food processing environments. Food Technol 42(12):81–82, 89, 1988.

FDA Enforcement Report. Recalls and field corrections: August 7. Foods—Class I, 1991, p 1.

FDA Enforcement Report. Recalls and field corrections: July 22. Foods—Class I, 1992, pp 1–2.

Foster EM, Nelson FE, Speck ML, Doetsch RN, Olson JC Jr. Microbiology of cream and butter. Dairy Microbiology. Englewood Cliffs, NJ: Prentice-Hall, 1957.

Grant IR, Ball HJ, Rowe M. Heat sensitivity of *Mycobacterium paratuberculosis* in cow's milk at pasteurization temperatures. In: Chiodini RJ, Collins MT, Bassey EOE, eds. Proceedings of the Fourth International Colloquium on Paratuberculosis. Rehoboth, MA: International Association for Paratuberculosis, Inc., 1995, pp 313–319.

Grant IR, Ball HJ, Neill SD, Rowe MT. Inactivation of *Mycobacterium paratuberculosis* in cow's milk at pasteurization temperatures. Appl Environ Microbiol 6(2):631–636, 1996.

Halpin-Dohnalek MI, Marth EH. Fate of *Staphylococcus aureus* in whipped butter. J Food Prot 52:863–866, 1989a.

Halpin-Dohnalek MI, Marth EH. Fate of *Staphylococcus aureus* in whey, whey cream, and whey cream butter. J Dairy Sci 72:3149–3155, 1989b.

Hammer BW, Babel FJ. Bacteriology of butter. In: Dairy Bacteriology. 4th ed. New York: Chapman & Hall, 1957, pp 442–509.

Hammer BW, Hussong RV. Bacteriology of butter, I: influence of the distribution of the non-fatty constituents on the changes in bacterial content during holding. Iowa Agr Expt Sta Res Bull 134, 1930.

Hickey PJ, Beckelheimer CE, and Parrow T. Microbiological tests for equipment, containers, water and air. In: Marshall RT, ed. Standard Methods for the Examination of Dairy Products. 16th ed. Washington, DC: American Public Health Association, 1992, pp 397–412.

IDF/FIL. The Significance of Pathogenic Microorganisms in Raw Milk. Brussels, Belgium: International Dairy Federation. 1994.

IDF/FIL. Consumption Statistics for Milk and Milk Products 1993. Bulletin No. 301. Brussels, Belgium: International Dairy Federation. 1995.

Irbe RJ. Microbiology of reduced-fat foods. Science for the food industry of the 21st century. In: Yalpani M, ed. Biotechnology, Supercritical Fluids and Other Advanced Technologies for Low Calorie, Healthy, Food Alternatives, Mount Prospect, IL: ATL Press, 1993, pp 359–382.

Jay JM. Modern Food Microbiology. 4th ed. New York: Chapman & Hall, 1992.

Kaul J, Singh J, Kuila RK. Effect of potassium sorbate on the microbiological quality of butter. J Food Prot 42:656–657, 1979.

Keswani J, Frank JF. Heat resistance of *Mycobacterium paratuberculosis* in milk, Session 205. Characterization of *Campylobacter, Vibrio, Bacillus cereus, Staphylococcus aureus*, and *Mycobacterium*, Abstracts: 96th General Meeting of the American Society for Microbiology. Washington, DC: American Society for Microbiology, 1996, p 384.

Kimenai MP. Cream crystallization, Section III—NIZO method. In: Continuous Butter Manufacture. Bulletin No. 204. Brussels, Belgium: International Dairy Federation, 1986, pp 11–15.

Kornacki JL, Gabis DA. Microorganisms and refrigeration temperatures. Dairy, Food and environmental Sanitation 10(4):192–195, 1990.

Lanciotti R, Massa S, Guerzoni M, Di Fabio G. Light butter: natural microbial population and potential growth of *Listeria monocytogenes* and *Yersinia enterocolitica*. Lett Appl Microbiol 15:256–258, 1992.

Massa S, Cesaroni D, Poda G, Trovatelli LD. The incidence of *Listeria* spp. in soft cheeses, butter and raw milk in the province of Bologna. J Appl Bacteriol 68:153–156, 1990.

Milner J. Butter and low-fat dairy spreads. In: LFRA Microbiology Handbook. Leatherhead, Surrey, UK: Leatherhead Food RA, 1995, pp B-1 to B-15.

Minor TE, Marth EH. *Staphylococcus aureus* and enterotoxin A in cream and butter. J Dairy Sci 55:1410–1414, 1972.

Mitscherlich E, Marth EH. Microbial Survival in the Environment: Bacteria and Rickettsiae Important in Human and Animal Health. New York: Springer-Verlag, 1984.

Munro DS. Alternative processes. In: Continuous Butter Manufacture. Bulletin No. 204. Brussels, Belgium: International Dairy Federation, 1986, pp 17–19.

Murphy MF. Microbiology of butter. In: Robinson RK, ed. Dairy Microbiology. Vol 2. The Microbiology of Milk Products. London, UK: Elsevier Applied Science, 1990, pp 109–130.

Olsen JA, Yousef AE, Marth EH. Growth and survival of *Listeria monocytogenes* during making and storage of butter. Milchwissenschaft 43:487–489, 1988.

O'Toole DK. Effect of storage temperature on microbial growth in continuously made butter. Aust J Dairy Technol 33:85–87, 1978.

Pesonen H. Cream crystallization, Section II—Ripened Cream. In: Continuous Butter Manufacture. Bulletin No. 204. Brussels, Belgium: International Dairy Federation, 1986, pp 12–13.

Pointurier H. Butter transfer systems. In: Continuous Butter Manufacture. Bulletin No. 204. Brussels, Belgium: International Dairy Federation, 1986.

Robinson RK, ed. Dairy Microbiology. Vol 2. The Microbiology of Milk Products. 2nd ed. London, UK: Elsevier Applied Science, 1990.

Schweizer M. Heat treatment. Continuous Butter Manufacture. Bulletin No. 204. Brussels, Belgium: International Dairy Federation, 1986, p 10.

Smittle RB. Foods, quality control. In: Encyclopedia of Microbiology. Vol 2. New York: Academic Press, 1992, pp 219–229.

Troller JA, Christian JHB. Water Activity and Food. New York: Academic Press, 1978.

Varnam AH, Sutherland JP. Butter, margarine and spreads. In: Milk and Milk Products. New York: Chapman & Hall, 1994, pp 223–274.

Verrips CT. Growth of microorganisms in compartmentalized products. In: Gould GW, ed. Mechanisms of Action of Food Preservation Procedures. London, UK: Elsevier Applied Science, 1989, pp 363–399.

6

Starter Cultures and Their Use

JOSEPH F. FRANK
University of Georgia, Athens, Georgia

ASHRAF N. HASSAN
Minia University, Minia, Egypt

I. INTRODUCTION

Modern dairy microbiology began with the study of the natural acidification process that occurs when milk, cheese whey, or buttermilk (from cultured butter manufacture) are held for a period of time. These acidified products had long been used as inocula for production of cheese, butter, and cultured milks, but the resulting fermentations were undependable and of uneven quality. Pasteur, in 1857, was the first to demonstrate that the lactic fermentation was of microbial origin, disputing the accepted theory of the time that chemical degradation of sugar to lactic acid resulted in spontaneous generation of microorganisms (Brock, 1961). It was not until 1878 that Lister isolated pure cultures of the lactic acid bacteria responsible for milk acidification (Brock, 1961). In the 1880s, Conn in the USA, Storch in Denmark, and Weigmann in Germany demonstrated the advantages of using selected lactic acid bacteria to culture cream for butter manufacture (Knudson, 1931; Cogan, 1996). Commercial production and use of starter cultures grew rapidly and was widespread at the beginning of the 20th century. The advantages of using starter cultures to initiate fermentation were convincing. Before the use of commercial starter cultures, cheddar cheese took 6 to 7 hours to produce, and much of the product was of too poor a quality to be sold (Conn, 1895). Slow fermentation was also a public health threat, because milk for cheese manufacture was not pasteurized. Currently, most cultured dairy products are produced using commercial starter cultures that have been selected for a variety of desirable properties in addition to rapid acid production. These may include flavor production, lack of associated off-flavors, bacteriophage tolerance, ability to produce flavor during cheese ripening, salt tolerance, polysaccharide production, bacteriocin production, and heat sensitivity.

A starter culture is any active microbial preparation intentionally added during product manufacture to initiate desirable changes. These microbial preparations can consist of lactic acid bacteria, propionibacteria, surface ripening bacteria, and yeasts, and molds. Starter cultures have a multifunctional role in dairy fermentations. Their ability to rapidly produce acid aids in separation of curd from whey during cheese manufacture, modifies texture of cheeses and cultured milks, and enhances preservation. Production of low molecular weight compounds such as diacetyl contributes to flavor and aroma. Gas production can cause eye formation in cheese. Development of flavor and changes in texture during ripening of cheeses is associated with enzymes originating from bacterial and fungal cultures, depending on the cheese variety.

Lactic starter cultures may consist of single strains used alone or in combinations, or undefined mixtures of strains (mixed strain cultures). Cultures can also be either mesophilic (optimal growth at approximately 26°C) or thermophilic (optimal growth at approximately 42°C) (Cogan, 1996). Mesophilic mixed strain starter cultures can be grouped by composition, O (or N) cultures consist of lactococci that do not ferment citrate, B (or L) cultures contain *Leuconostoc* spp. and lactococci that do not ferment citrate, D cultures contain both citrate fermenting and nonfermenting lactococci but no *Leuconostoc* spp, BD cultures contain *Leuconostoc* spp. as well as lactococci found in D cultures (Lodics and Steenson, 1993). The type of mixed strain culture used for a specific cheese variety depends primarily on the amount of gas production (if any) that is desired. Thermophilic starter cultures consist of a mixture of *Streptococcus thermophilus* and a *Lactobacillus* sp., usually either *Lactobacillus helveticus, Lactobacillus delbrueckii* subsp. *bulgaricus*, or *Lactobacillus delbrueckii* subsp. *lactis*. These cultures are used to produce Italian and Swiss cheese varieties and yogurt.

This chapter discusses characteristics of lactic acid bacteria and other microorganisms found in dairy starter cultures, preparation of these cultures, inhibitors of their activity, and microbial inhibitors that they produce.

II. STARTER CULTURE MICROORGANISMS

A. General Characteristics of Lactic Acid Bacteria

All dairy fermentation use lactic acid bacteria for acidification and flavor production. Although lactic acid bacteria are genetically diverse, common characteristics of this group include being gram-positive, non–spore forming, nonpigmented, and unable to produce iron-containing porphyrin compounds (catalase and cytochrome), having anaerobic growth but being aerotolerant, and having obligate fermentation of sugar with lactic acid as a major end product. Lactic acid bacteria tend to be nutritionally fastidious, often requiring specific amino acids, B vitamins, and other growth factors, while being unable to use complex carbohydrates.

1. Taxonomy

There are currently 11 genera of lactic acid bacteria, of which four—*Lactobacillus, Streptococcus, Lactococcus*, and *Leuconostoc*—are commonly found in dairy starter cultures. A fifth genus, *Enterococcus*, is occasionally found in mixed strain (undefined) starter cultures. Important phenotypic taxonomic criteria include morphologic appearance (rod or coccus), fermentation end products (homofermentative or heterofermentative), carbohydrates fermented, growth temperature range, optical configuration of lactic acid produced, and salt tolerance (Axelsson, 1993). rRNA sequences are used to accurately determine phylogenetic relationships among bacteria. This and other genetic methods have led to the reorganization of some genera of lactic acid bacteria (e.g., the reclassification of lactic streptococci to *Lactococcus* spp.).

2. Natural Habitat

Lactic acid bacteria are generally associated with nutrient-rich habitats containing simple sugars. These include raw milk, meat, fruits, and vegetables. They grow with yeast in wine, beer, and bread fermentations. In nature, they are found in the dairy farm environment and in decomposing vegetation, including silage. Some species colonize animal organs including the mouth, intestine, and vagina. They are also part of the normal microflora of the streak canal of the mammary gland. Lactic acid bacteria isolated from natural habitats are often physiologically distinct from their starter culture variants. For example, lactococci isolated from plants ferment lactose slowly, if at all (Chassy and Murphy, 1993).

B. Characteristics of Starter Culture Genera and Species

1. *Lactococcus*

Lactococci (formerly group N streptococci) are the major mesophilic micro-organisms used for acid production in dairy fermentations. Although five species are recognized, only one, *Lactococcus lactis*, is of significance in dairy fermentations. *Lc. lactis* cells are cocci that usually occur in chains, although single and paired cells are also found. They are homofermentative; when grown in milk, more than 95% of their end product is lactic acid (of the L isomer). Lactococci grow at 10°C but not at 45°C. They are weakly proteolytic and can use milk proteins. There are two subspecies, *Lc. lactis* subsp. *lactis* and *Lc. lactis* subsp. *cremoris*. Differential characteristics for these subspecies are presented in Table 1. *L. lactis* subsp. *lactis* is more heat and salt tolerant than *Lc. lactis* subsp. *cremoris*. A variant of *Lc. lactis* (*Lc. lactis* subsp. *lactis* var. *diacetylactis*) converts citrate to diacetyl, carbon dioxide, and other compounds. Some lactococci produce exopolysaccharide (Cerning, 1990). These variants are used to produce Scandinavian cultured milks having a ropy texture (viilli, taettamilk, and langmjolk). Another variant of *Lc. lactis* produces malty off-flavor caused by aldehyde production from amino acids (Morgan, 1976).

2. *Streptococcus*

The only *Streptococcus* sp. useful in dairy fermentation is *Streptococcus thermophilus*. This microorganism is genetically similar to oral streptococci

TABLE 1 Differentiation of Lactococci Used in Starter Cultures

	Lactococcus lactis subsp.	
Characteristic	*Lactis*	*Cremoris*
Acid from		
Lactose	+	+
Galactose	+	+
Maltose	+	–
Ribose	+	–
Growth in 4% salt	+	–
Arginine hydrolysis	+	–

Source: Schleifer et al., 1985.

(*Streptococcus salivarius*) but can still be considered a separate species (Axelsson, 1993). *S. thermophilus* is differentiated from other streptococci (and lactococci) by its heat resistance, ability to grow at 52°C, and ability to ferment only a limited number of carbohydrates (Axelsson, 1993). Most dairy products subjected to high temperatures during fermentation (>40°C) are acidified by the combined growth of *S. thermophilus* and *Lactobacillus* spp. *S. thermophilus* has limited proteolytic ability.

3. *Leuconostoc*

Leuconostoc spp. are distinguished from other lactic acid bacteria by being mesophilic heterofermentative cocci. They do not hydrolyze arginine and require various B vitamins for growth. *Leuconostoc* spp. used in the dairy industry produce diacetyl, carbon dioxide, and acetoin from citrate. Some also produce exopolysaccharide (dextran) from sucrose. Only two species of *Leuconostoc* are associated with dairy starter cultures, *Leuconostoc mesenteroides* subsp. *cremoris* (previously, *Leuconostoc citrovorum*) and *Leuconostoc lactis*. These are differentiated by their ability to ferment various carbohydrates. *Leuconostoc* spp. grow poorly in milk, probably because they are adapted to growth on vegetables and roots (Vedamuthu, 1994) and therefore lack sufficient proteolytic ability to grow in milk. *Leuc. mesenteroides* subsp. *cremoris* does not produce sufficient acidity in milk to coagulate it, but *Leuc. lactis* may (Thunell, 1995). In starter cultures, *Leuconostoc* spp. are combined with lactococci when production of diacetyl and carbon dioxide is desired in addition to acidification. When used in cultured milk starters, they convert excess acetaldehyde to diacetyl, thus reducing undesirable "green" flavor (Lindsay et al., 1965). *Leuconostoc* spp. do not grow well in high-phosphate phage-inhibitory media (Vedamuthu, 1994).

4. *Lactobacillus*

The *Lactobacillus* genus consists of a genetically and physiologically diverse group of rod-shaped lactic acid bacteria. The genus can be divided into three groups based on fermentation end products. Species in each of these groups can be found in dairy starter cultures, as listed in Table 2. Homofermentative lacto-bacilli exclusively ferment hexose sugars to lactic acid by the Embden-Meyerhof pathway. They do not ferment pentose sugars or gluconate. These are the lacto-bacilli (*Lb. delbrueckii* subsp. *bulgaricus, Lb. delbrueckii* subsp. *lactis*, and *Lb. helveticus*) commonly found in starter cultures. They grow at higher tempera-tures (>45°C) than lactobacilli in the other groups and are thermoduric. Another member of this group, *Lactobacillus acidophilus* is not a starter culture organism, but is added to dairy foods for its nutritional benefits.

Facultatively heterofermentative lactobacilli ferment hexose sugars either only to lactic acid or to lactic acid, acetic acid, ethanol, and formic acid when

TABLE 2 Characteristics of *Lactobacillus* spp. Associated with Dairy Products

Species	Products	Growth at 15°C	Growth at 45°C	Lactic acid isomer	Mole %G+C	Fermentation of					
						Glu	Gal	Lac	Mal	Suc	Rib
Homofermentative											
L. delbrueckii subsp. *bulgaricus*	Yogurt, kumiss, kefir, Italian, and Swiss cheeses	–	+	D	49–51	+	–	+	–	–	–
subsp. *lactis*	Hard cheese	–	+	D	49–51	+	d[a]	+	+	+	–
L. acidophilus	Acidophilus milk, laban	–	+	DL	34–37	+	+	+	+	+	–
L. helveticus	Yogurt, Swiss cheese	–	+	DL	38–40	+	+	+	d	–	–
Facultatively heterofermentative											
L. casei											
subsp. *casei*	Hard cheese	+	–	L	45–47	+	+	d	+	+	+
Obligately heterofermentative											
L. kefir	Kefir	+	–	DL	41–42	+	–	+	+	–	+

[a]Some strains are positive.
Abbreviations: Glu, glucose; Gal, galactose; Lac, lactose; Mal, maltose; Suc, sucrose; Rib, ribose.
Source: Cogan, 1996.

under glucose limitation. Pentose sugars are fermented to lactic and acetic acid via the phosphoketolase pathway. This group includes *Lactobacillus casei*, which is not usually found in starter cultures but is associated with beneficial secondary fermentation during cheese ripening.

Obligately heterofermentative lactobacilli ferment hexose sugars to lactic acid, acetic acid (or ethanol), and carbon dioxide using the phosphoketolase pathway. Pentose sugars are also fermented using this pathway. These lactobacilli can cause undesirable flavor and gas formation during ripening of cheese. One species, *Lactobacillus kefir*, is associated with kefir cultures.

Lactobacilli are the most acid tolerant of the lactic acid bacteria, preferring to initiate growth at acidic pH (5.5–6.2) and lowering the pH of milk to below 4.0. Lactobacilli are slow to grow in milk in pure culture. For this reason, they are generally used in combination with *S. thermophilus*.

5. Propionibacteria

Propionibacterium spp. are non–spore-forming, pleomorphic, gram-positive rods that produce large amounts of propionic and acetic acid and carbon dioxide from sugars and lactic acid. They are anaerobic to aerotolerant mesophiles. They are not considered belonging to the lactic acid bacteria, but are closely related to coryneform bacteria in the Actinomycetaceae group. Four species of *Propionibacterium* are found in cheese (Table 3), but *Propionibacterium freudenreichii* subsp. *freudenreichii* and *P. freudenreichii* subsp. *shermanii* are most often used in cheese manufacture (Lyon and Glatz, 1995). Although *Propionibacterium* spp. are found in raw milk, they may be present in insufficient numbers to produce an adequate fermentation, so they are often added along with the lactic culture.

TABLE 3 Differentiation of *Propionibacterium* spp. Associated with Dairy Products

Characteristic	*P. freudenreichii*	*P. jensenii*	*P. thoenii*	*P. acidipropionici*
Acid from				
Sucrose	–	+	+	+
Maltose	–	+	+	+
Mannitol	–	+	–	+
Rhamnose	–	–	–	+
Nitrate reduction	–	–	–	+
β-hemolysis	–	–	+	–
Colony color	Cream	Cream	Red-brown	Cream to orange-yellow

Source: Commins and Johnson, 1984.

Propionibacteria can use both inorganic and organic nitrogen sources, and their requirements for amino acids vary. Most strains require biotin. Cultures for cheese manufacture are grown on complex media, including hydrolyzed protein and yeast extract with lactic acid as a carbon source (Glatz, 1992).

Propionibacteria grow on the lactic acid produced during cheese fermentation. The lactate is oxidized to pyruvate, which then is either converted to acetate and carbon dioxide or propionate. The carbon dioxide forms the large eyes found in Swiss and similar types of cheese, and the other metabolic products including amino acids and fatty acids contribute to the flavor of these cheeses.

6. *Brevibacterium*

Brevibacterium cells are aerobic, gram-positive, pleomorphic rods that grow on the surface of surface-ripened cheese varieties. The species most often isolated from these cheeses is *Brevibacterium linens*. *B. linens* produces a yellow-orange carotenoid pigment that colors the surface of the cheese. Color formation is enhanced by exposure to light. Older cultures are primarily coccoid, but slender rods are produced in exponential growth. *B. linens* does not use lactose or citrate but can grow on the lactate produced during cheese manufacture. It also grows best at neutral pH, so it does not grow well on the cheese surface until lactic acid is neutralized or metabolized by yeasts or micrococci. Surface-ripened cheeses are surface-salted and *B. linens*, like yeasts and micrococci, grows well at high salt concentrations. *B. linens* is highly proteolytic with ability to degrade whey proteins and casein (Fringa et al., 1993; Holtz and Kunz, 1994). The ability of *B. linens* to degrade amino acids to ammonia and methionine to methanethiol is partially responsible for production of strong flavors and odors during surface ripening of cheese. Other volatile compounds produced by *B. linens* that contribute to the typical flavor of surface-ripened cheese include butyric acid, caproic acid, phenylmethanol, dimethyldisulphide, and dimethyltrisulphide (Jollivet et al., 1992). *B. linens* grows well in media containing hydrolyzed protein, glucose, yeast extract, potassium phosphate, and magnesium sulphate (Haysahi et al., 1990).

7. *Penicillium*

Penicillium spp. are molds in the class Hyphomycetes in the division Deuteromycota. Molds in this class produce conidia directly on mycelium or on conidiophores. The conidiophores of *Penicillium* spp. arise erect from the hyphae and branch near the tip to produce a brushlike ending (Beneke and Stevenson, 1987). Two groups of *Penicillium* spp. are used in cheese manufacture, the white mold (*Penicillium camemberti* Thom, formerly two species, *Penicillium caseicolum* and *Penicillium camemberti*), which grows on the surface of Camembert, Brie, and similar varieties, and the blue mold (*Penicillium roqueforti*, formerly

P. roqueforti var. *roqueforti*), which grows in the interior of blue-veined cheeses such as Roquefort, Gorgonzola, and Stilton. *P. camemberti* is closely related to *Penicillium commune*, a common cheese contaminant that produces various toxins (Frisvad and Filtenborg, 1989), whereas *P. camemberti* produces only one mycotoxin, cyclopiazonic acid. *P. roqueforti* is closely related to *Penicillium carneum* (formerly *P. roqueforti* var. *carneum*), a producer of the mycotoxin patulin, and *Penicillium paneum* (formerly *P. roqueforti* var. *carneum*), a producer of patulin and the mycotoxin botryodiploidin (Boysen et al., 1996).

P. camemberti and *P. roqueforti* are lipolytic and proteolytic. Both produce methyl ketones and free fatty acids, but the much higher levels produced by *P. roqueforti* give blue cheeses their distinctive flavor and aroma (Kinsella and Hwang, 1976; Jollivet et al., 1993). *P. camemberti* contributes to the flavor of Camembert and Brie cheeses by producing a complex mixture of compounds, the major ones being 2-heptanone, 2-heptanol, 8-nonen-2-one, 1-octen-3-ol, 2-nonanol, phenol, butanoic acid, and methyl cinnamate (Moines et al., 1975).

C. Desirable Properties of Lactic Cultures

Properties desired of lactic cultures for industrial use may differ from those found in typical wild-type microorganisms. For example, most dairy fermentations require rapid acid production and the lack of off-flavor production, whereas wild-type organisms are often slow acid producers and produce such off-flavors as fruity, bitter, and malty. Buchenhüskes (1993) summarized selection criteria for lactic acid bacteria to be used for food fermentations. These include (a) lack of pathogenic or toxic activity (e.g., production of biogenic amines), (b) ability to produce desired changes, (c) ability to dominate competitive microflora, (d) ease of propagation, (e) ease of preservation, and (f) stability of desirable properties during culturing and storage. Specific properties desired in a dairy starter culture depend on the product being produced.

1. Cheddar Cheese

The four main selective criteria for cheddar cheese cultures are rapid acid production, bacteriophage resistance (see Sec. V), salt sensitivity, and ripening activity (Strauss, 1997). Rapid acid production should occur at a steady rate throughout curd making. This ensures suppression of undesirable microflora, timely cheese manufacture, and the presence of sufficient ripening enzymes from the starter microorganisms. Rapid lactose fermentation in lactococci is associated with the presence of a phosphoenol pyruvate–dependent phosphotransferase system (see Chapter 7 for discussion of acid production).

In cheddar manufacture, salt is added after most of the desired acidity has developed. However, some acid-producing activity is still needed after salting to ensure that all lactose is metabolized. Residual lactose can serve as a substrate for salt-tolerant organisms such as heterofermentative lactobacilli that produce gas and undesirable flavors (Olson, 1990). Growth of starter microflora after salt addition also produces a low oxidation–reduction potential that has a beneficial impact on flavor development and inhibits some spoilage microorganims.

Ripening activity is related to production of proteases and other enzymes. These enzymes must be produced in sufficient quantity to develop typical cheddar flavor without off-flavors. Peptidase activity is more important than proteinase activity. In fact, starter culture proteinases are associated with development of bitter flavor (Visser et al., 1983). Cheese made using 45% to 75% proteinase-negative cells developed less bitter flavor than cheese made using all proteinase-positive culture (Mills and Thomas, 1980). However, proteinase-negative strains cannot use proteins, so their growth in milk is limited. Starter culture peptidases hydrolyze peptides (including those with bitter flavor) produced by action of rennet, and, in combination with other microbial enzymes, produce a chemical environment conducive to development of typical cheddar flavor. Starter cultures for cheddar cheese can include strains that specifically enhance ripening but take no or little part in initial acid production (Trepanier et al., 1991). See Chapter 7 for additional information on starter culture protease systems.

2. Mozzarella Cheese

Cultures for mozzarella cheese manufacture are combinations of *S. thermophilus* and either *Lb. delbrueckii* subsp. *bulgaricus* or *Lb. helveticus*. American-style mozzarella is manufactured for use as a food ingredient, especially on pizza. The starter culture contributes to functional properties related to this use, such as stretchability and heat-induced browning. The typical starter culture for American mozzarella manufacture has a 1:5 rod to coccus ratio (McCoy, 1997). This results in rapid initial acid production (by the streptococci) and shortens make time. Lactobacilli produce acid late in manufacture. Lactobacilli are much more proteolytic than streptococci, and proteolysis during storage increases meltability and decreases stretchability (Oberg et al., 1991a, 1991b). Rod to coccus ratio only slightly influences textural changes during storage (Yun et al., 1995); level of initial inoculum has a greater influence on texture.

Starter culture also affects color development during cooking. Many thermophilic cultures use only the glucose portion of the lactose molecule, excreting galactose (see Chapter 7). High-browning cheeses contain nearly five times more galactose than low-browning cheeses (Matzdorf et al., 1994). If low-browning cheese is desired, galactose-utilizing cultures such as *Lb. helveticus* can be used. Using *Lb. helveticus* instead of *Lb. delbrueckii* subsp. *bulgaricus* results in mozzarella cheese with lower galactose levels, improved melting, and decreased make time (Oberg et al., 1991a). Excessive heat during stretching (curd temperature >66°C) can inactivate starter culture enzymes and reduce galactose metabolism and proteolysis during storage (Chen et al., 1994).

3. Swiss Cheese

Starter cultures for Swiss cheese manufacture must survive the high temperatures used in manufacture (50°C–52°C). The starter culture is also responsible for development of typical Swiss cheese flavor and eye development. The typical Swiss cheese starter culture consists of *S. thermophilus, Lb. helveticus*, and *P. freudenreichii* subsp. *shermanii*. Mesophilic lactococci are sometimes added to increase acid production early in manufacture. A consistent rate of acid production by the starter is important, because more rapid acid production results in lower moisture content (Turner et al., 1983). Lactose fermentation occurs primarily during the first 24 hours of manufacture. Streptococci initially predominate, using lactose and excreting galactose. Subsequent growth of lactobacilli is required for complete utilization of galactose (Hutkins et al., 1986). If all residual sugars are not used, defects from growth of gas-forming microorganisms or brown pigment formation can occur (Harrits and McCoy, 1996).

Propionibacteria grow on the lactate produced by the lactic culture, converting it to carbon dioxide, propionic acid, acetic acid, and small amounts of other compounds. Propionibacteria can reach 10^9 colony-forming units per gram and use more than 50% of the lactate at the center of the cheese (Fryer and Peberdy, 1977). Swiss cheese is ripened at 21°C for eye formation and then aged at 10°C for flavor development. Therefore, the *Propionibacterium* culture should grow well at 21°C but not at 10°C (so the eyes do not split) (Harrits and McCoy, 1997). A predictable rate of gas formation at 21°C is required because too rapid gas formation results in split eyes (Hettinga et al., 1974).

High-moisture baby Swiss is manufactured using lower cook temperatures (approximately 40°C) and therefore is produced, not with thermophilic cultures, but with heat-tolerant lactococci. Propionibacteria are still used for eye formation.

4. Cultured Buttermilk and Sour Cream

Cultures for buttermilk, sour cream, and similar products must both acidify the substrate and produce flavor and aroma compounds. Using citrate-fermenting bacteria such as *Leuc. mesenteroides* subsp. *cremoris* or *Lc. lactis* subsp. *lactis* var. *diacetylactis* are combined with *Lc. lactis* subsp. *lactis* or *Lc. lactis* subsp. *cremoris*. Citrate fermentation is discussed in Chapter 7. Diacetyl, the major aroma compound in these products, can be reduced to acetoin by diacetyl reductase. Cultures should be selected that are low in diacetyl reductase activity. Acetaldehyde is often produced during fermentation, giving the product an undesirable "green apple" or yogurt flavor. Leuconostocs (but not lactococci) can metabolize acetaldehyde to ethanol, with a resulting flavor improvement (Peterson, 1997).

5. Yogurt

Acidification

Yogurt is made using a combination of *S. thermophilus* and *Lb. delbrueckii* subsp. *bulgaricus*. These organisms grow in a cooperative relationship resulting in rapid acidification. The presence of lactobacilli stimulates growth of the more weakly proteolytic *S. thermophilus* because lactobacilli liberate free amino acids and peptides from casein (Rajagopal and Sandine, 1990). *S. thermophilus*, in turn, stimulates growth of *Lb. delbrueckii* subsp. *bulgaricus*, possibly by removing oxygen, lowering pH, and producing formic acid and pyruvate (Radke-Mitchell and Sandine, 1984). Strains can be selected for the degree to which their growth depends on the presence of the other microorganisms (Vedamuthu, 1991). Yogurt may also contain *Lb. acidophilus* or other nutritionally beneficial cultures. The most important characteristics for yogurt cultures are (a) rapid acidification, (b) production of characteristic balanced flavor, and (c) ability to produce the desired texture. As with other thermophilic rod–coccus dairy fermentations, initial acidification is from growth of *S. thermophilus* with the lactobacilli growing later in the fermentation. Excessively rapid acidification can result in overacidification and a harsh flavor. Acidification of yogurt is controlled by refrigeration, but the culture may continue to slowly acidify at cold temperatures.

Flavor

The ideal yogurt flavor is a balanced blend of acidity and acetaldehyde. This is achieved through culture selection, balance of rod to coccus ratio, and fermentation control. The main source of acetaldehyde is the conversion of threonine to acetaldehyde catalyzed by threonine aldolase of *Lb. delbrueckii* subsp. *bulgaricus* (Hickey, 1983). Lactobacilli, such as *Lb. acidophilus*, which produces alcohol dehydrogenase, convert acetaldehyde to ethanol (Marshall and Cole, 1983). Therefore, yogurt produced with *Lb. acidophilus* does not have a typical yogurt flavor.

Texture

The texture of yogurt results from a complex interaction between milk proteins, acid, and exocellular polysaccharide produced by the starter culture. Important physical properties include firmness, smoothness, viscosity, and gel stability (susceptibility to syneresis). The starter culture, by production of exopolysaccharides, can influence each of these properties.

Yogurt cultures produce exopolysaccharide in a ropy or capsular form (Ariga et al., 1992). Capsular polysaccharides are formed as a discrete structure surrounding the cell (Fig. 1A) with no apparent interaction with casein (Hassan et al., 1995a, 1995b). Ropy polysaccharides are produced as filaments that are not visualized as discrete structures by light microscopy. Hassan et al. (1996a) classified yogurt cultures into three types, those that do not produce exopolysaccharide, those that produce capsular polysaccharide, and those that produce both capsular and ropy polysaccharide. Cultures that produce only ropy polysaccharide may exist, but an extensive survey has not been reported.

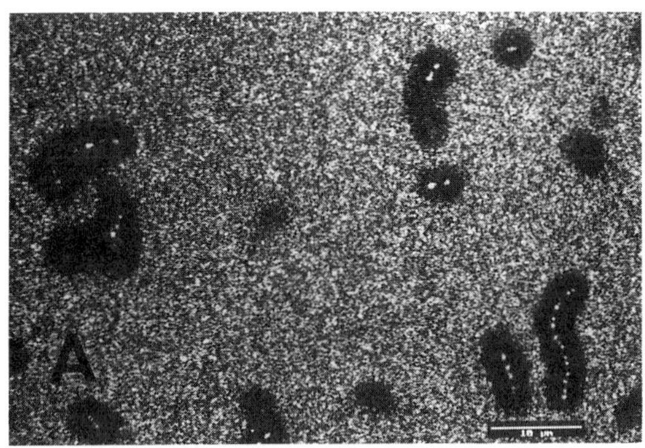

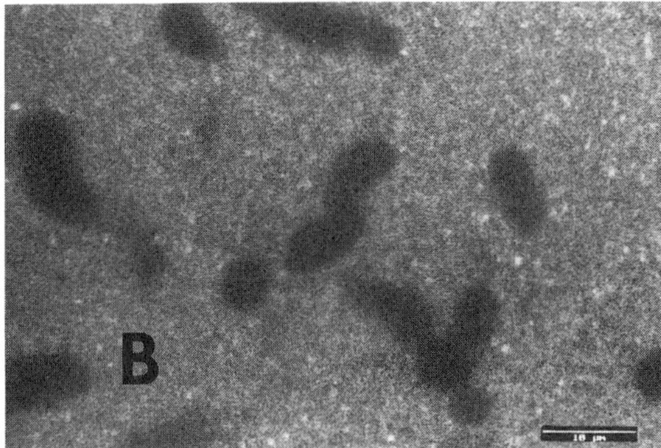

FIGURE 1 Confocal scanning laser micrographs of encapsulated *Streptococcus thermophilus*. A, Encapsulated cells in milk visualized using reflected light. B, pH gradient surrounding encapsulated cells in milk visualized using a pH-sensitive fluorochrome. Dark areas indicate low pH.

Yogurt cultures produce heteropolysaccharides that consist of different sugar residues in a repeating pattern. Their production does not depend on the presence of a specific substrate. In contrast, homopolysaccharides, such as dextran produced by *Leuc. mesenteroides*, consist of one sugar residue type (in this instance, glucose), and a specific substrate is required for their production (in this instance, sucrose). Cerning et al. (1986) found that the heteropolysaccharide produced by *Lb. delbrueckii* subsp. *bulgaricus* was comprised primarily of galactose, glucose, and rhamnose in a molar ratio of 4:1:1. Garcia-Garibay and Marshall (1991) found evidence that this exopolysaccharide is closely associated with protein and may be better classified as a glycoprotein. The exopolysaccharide of *S. thermophilus* is composed mainly of galactose and glucose with a small amount of other sugars (Cerning et al., 1988). Cerning (1990) stated that there is little agreement as to the precise composition of these polysaccharides.

The influence of ropy polysaccharide on yogurt texture is well documented, but reported effects must be interpreted with caution, because, until recently, nonropy cultures used as controls were not examined for capsule production. There is general agreement that ropy cultures can benefit yogurt texture by increasing viscosity and gel stability (Cerning, 1990). However, overproduction of ropy polysaccharide yields a product with an undesirable slippery mouth feel and pronounced ropiness. Capsular polysaccharide cannot be overproduced, because capsule size is limited (Hassan et al., 1995a). Bacterial capsules disrupt the yogurt gel microstructure, producing a softer texture (Hassan et al., 1995b). Encapsulated cultures with no ropy characteristic produce yogurt that is more viscous, structurally more stable, and less susceptible to syneresis than do cultures that do not produce capsules (Hassan et al., 1996a, 1996b). The capsule also slows diffusion of lactic acid away from the cell, causing the cells to stop acid production sooner (Hassan et al., 1995a). This helps prevent overacidification of the yogurt. The pH gradient resulting from encapsulation can be visualized using confocal scanning laser microscopy as shown in Figure 1B.

III. STARTER CULTURE PROPAGATION

A. Growth Media

The objective of starter culture propagation is to attain a preparation of active cells at high density so that fermentation is initiated as rapidly as possible. Providing adequate nutrients and controlling pH and incubation temperature are necessary to achieve this objective. Even though milk and whey are traditional growth media for lactic cultures, they provide neither optimal nutrition nor needed pH control. Consequently, various media formulations and culture growth systems have been devised to improve on traditional culture propagation.

1. Nutritional Requirements of Lactic Acid Bacteria

Lactic acid bacteria cannot synthesize various vitamins and amino acids. Lactococci require niacin, pantothenic acid, pyridoxine, and biotin for growth. *S. thermophilus* requires these vitamins plus nitroflavin, whereas lactobacilli require pantothenic acid, niacin, and nitroflavin, with some species also requiring cobalamin (Mäyrä-Mäkinen and Bigret, 1993). In regard to amino acids, lactococci and *S. thermophilus* cannot synthesize branched chain amino acids (isoleucine, leucine, valine) or histidine, and some strains also require arginine and methionine (Monnet et al., 1996). Lactobacilli require these amino acids in addition to several others. *Leuconostoc* spp. require valine and glutamate, and some species may have additional requirements. The presence of amino acids other than those required often stimulates growth. Although milk contains many of the essential amino acids for starter culture microorganisms, these are not present in sufficient quantity to sustain maximal growth rates (Monnet et al., 1996). Lactic acid bacteria with greater proteolytic ability have less need for amino acid supplementation of milk-based growth media.

2. Growth Media Formulations

Ingredients commonly used to formulate starter culture media have been described by Whitehead et al. (1993) and are presented in Table 4. Lactose is always used as the major carbohydrate, although low concentrations of maltose, sucrose, or glucose are sometimes added to stimulate growth (Sandine, 1996). Yeast extract is a source of nitrogen as well as a supplier of vitamins, minerals, and other growth stimulants. Casein hydrolysates are added to provide readily available amino acids. Corn steep liquor, although a good source of vitamins, is not often used because its supply is limited (Sandine, 1996). Sandine (1996)

TABLE 4 Ingredients Used in Formulating Bulk Starter Media for Lactic Acid Bacteria

Carbohydrate	Nitrogen source	Vitamins and minerals	Phage inhibitory agents	Antioxidants	Neutralizers
Lactose	Milk protein	Yeast extract	Phosphates	Ascorbic acid	Carbonates
Maltose	Whey protein	Corn steep liquor	Citrates	$FeSO_4$	Phosphates
Sucrose	Hydrolized casein				Hydroxides
Glucose					Oxides

Source: Whitehead et al. (1993).

questioned the need for added antioxidants in media formulations, because acceptable growth can often be achieved in their absence. Neutralizers, such a ammonium or potassium hydroxide, help prevent excessive acidity. Phosphates are commonly used in culture media because they act both as acid-neutralizing and phage-inhibitory agents.

3. Phage-Inhibitory Media

One of the first improvements in whey- and milk-based culture media was development of phage inhibitory media (see Sec. V).

B. pH Control During Culture Propagation

Although lactobacilli grow best under slightly acidic conditions, other starter culture microorganisms prefer conditions near neutrality. For example, the optimal pH for growth of *S. thermophilus* is 6.5, whereas for *Lb. delbrueckii* subsp. *bulgaricus*, it is 5.8 (Beal et al., 1989). The optimal pH for growth of lactococci ranges from 6.0 to 6.5 (Mäyrä-Mäkinen and Bigret, 1993). As the pH decreases below the optimal range, growth slows, and, as the pH continues to decrease, cells become susceptible to sublethal acid injury and gradually lose their activity. The greater the loss of activity, the longer the ripening time required before rennet addition when making cheddar cheese. Therefore, maintaining the pH of culture media high enough to avoid acid injury is critical for producing cultures that consistently have sufficient activity for timely cheese manufacture. Acid injury in lactococci occurs when the pH declines below 5.0 (Harvey, 1965). Limited pH control can be achieved by addition of buffers to culture media. Buffering agents, such as phosphates and carbonate, allow development of higher cell concentrations because the pH of the medium stays above 5.0 for a longer time. However, the neutralizing ability of the buffering agent is eventually overcome, exposing cells to excessive acidity. Also, high concentrations of buffers inhibit growth of some starter strains. Two approaches, internal and external pH control, are currently used to maintain growth media above pH 5.0 during culture preparation.

1. External pH Control

External pH control refers to a culture preparation system in which neutralizing agent is added to the medium during fermentation either manually or mechanically. There may be one or multiple additions of neutralizer. For one-step control, the pH of the medium is allowed to decrease to approximately 5.0, after which, sodium or potassium hydroxide is added to obtain a pH of 6.5 to 7.0 (Limsowtin et al., 1980). The culture is then allowed to incubate an additional 2 hours before cooling. Multiple step neutralization uses a mechanical system consisting of a pH electrode mounted in the bottom of the culture tank, a pump for adding ammonia to the tank, and a controller. When the pH declines below 5.8 to 6.2, the controller activates the pump to add ammonia until the pH is raised a certain amount (usually to 6.0–6.2). When acid production ceases because of lactose limitation, the culture is cooled (Thunell, 1988). External pH control has an additional advantage of requiring less phosphate for phage inhibition, because calcium is less soluble at higher pH. A disadvantage of external pH control is that the higher pH allows growth of nonstarter microflora, even after lactose is depleted (Thunell, 1988). Therefore, a high degree of sanitation is required to implement this system. External pH control systems produce starter culture with 10 times greater cell concentration than phosphate-buffered media (Thunell, 1988). These cells are also healthier (i.e., they have no acid injury). The result is that a lower volume of starter can be used and milk ripening times are reduced. In addition, the culture produces acid more rapidly after salting (for cheddar manufacture). More culture strains produce acceptable activity during cheese manufacture when external pH control is used for culture propagation as compared with conventional buffered media (Thunell, 1988).

2. Internal pH Control

Internal pH control describes a culture production system in which an insoluble neutralizing agent is added to the culture medium. The neutralizing agent is released in response to acid production. One means of achieving internal pH control is to use sodium carbonate encapsulated in magnesium stearate (Whitehead et al., 1993). The magnesium stearate dissolves at pH 5.2 to 5.3, releasing the sodium carbonate. A similar effect is obtained by using buffer salts that are insoluble above a pH of 5.2 (Mermelstein, 1982). Sandine (1996) considered trimagnesium phosphate to be the most effective agent for this purpose. Internal pH control media have similar advantages to external pH control systems. In addition, a mechanism for adding neutralizing compound to the medium does not need to be installed. However, the fermentation tank must be stirred to keep the insoluble neutralizing agent suspended during fermentation. Agitation may lead to the incorporation of sufficient oxygen into the medium to stimulate hydrogen peroxide production, resulting in autoinhibition (Mäyrä-Mäkinen and Bigret, 1993).

C. Incubation Conditions

Incubation temperature can affect the activity and strain balance of the starter culture. Mesophilic cultures are grown at 21°C if growth of leuconostocs is desired; otherwise, higher temperatures (up to 27°C) are used (McCoy and Leach, 1997). Incubation at 26°C helps maintain strain balance (Collins, 1976).

Incubation is usually for 14 to 16 hours or until a pH of 5.0 is reached. If pH control is not used, the final pH should be 4.8. Thermophilic cultures are incubated from 30°C to 46°C for 8 to 10 hours. A final pH value as low as 4.7 is acceptable, but this favors growth of lactobacilli (McCoy and Leach, 1997). Lower incubation temperatures favor growth of *S. thermophilus* and higher temperatures favor lactobacilli. Once the target pH is reached, the culture is cooled. Most cultures continue to produce acid during cooling. Mesophilic starters should be cooled to 5°C to 7°C and thermophilic cultures to below 12°C (McCoy and Leach, 1997).

IV. COMMERCIAL STARTER CULTURE PREPARATIONS

Manufacturers of cultured dairy foods have several options for meeting their culture needs. The simplest (and usually most expensive) is to purchase frozen concentrated cultures that can be used to directly inoculate milk from which product will be manufactured. Using these "direct-to-vat" or "direct-vat-set" cultures avoids the possibility that starter culture will become contaminated with phage during preparation within the plant. Also, appropriate strain balance is assured. Alternatively, culture can be prepared at the plant. This culture, called bulk culture, can be prepared from commercially available frozen concentrated or freeze-dried cultures, or the inoculum can be prepared at the plant. Preparing inoculum at the plant involves starting with a "mother" culture maintained in small amounts (approximately 100 mL) of medium. The mother culture is used to inoculate successively larger amounts of medium (using a 1% inoculum) until sufficient inoculum volume is obtained to prepare the bulk culture. Preparing bulk culture inoculum at the plant carries an increased risk of phage contamination, so most plants purchase an inoculum either as frozen concentrated or freeze-dried preparations.

A. Frozen Concentrated Cultures

Frozen concentrated cultures contain 10^{10} to 10^{11} colony-forming units per gram, a sufficient concentration to allow 70 mL to inoculate 1000 L of media for bulk culture preparation (Sandine, 1996). Preparation of frozen concentrated cultures involves (a) growing cultures under optimal conditions using pH control, (b) harvesting the cells via centrifugation or ultrafiltration, (c) standardizing the cell suspension to a specific activity, (d) adding a cryoprotectant, (e) packaging, and (f) rapid freezing using liquid nitrogen. The pH of the cell concentrate should be 6.6 for lactococci and 5.4 to 5.8 for lactobacilli (Stadhouders et al., 1971). There are many cryoprotective agents that can be used, including glycerol, monosodium glutamate, sucrose, and lactose (Mäyrä-Mäkinen and Bigret, 1993).

Rapid freezing can also be accomplished using a dry ice–alcohol mixture (Sandine, 1996). The frozen concentrate should be stored at –196°C (liquid nitrogen) for best retention of activity, although storage at –40° (dry ice) is also acceptable. Rapid thawing minimizes cell injury. This is accomplished by immersing the unopened can of cell concentrate in cool chlorinated water immediately before use.

B. Freeze-Dried Cultures

When transportation and storage of cultures at –40°C is not possible, freeze-dried cultures are a good alternative to frozen concentrates. Current technology can provide highly active freeze-dried cultures that, like some frozen concentrated cultures, can be added directly to milk in the cheese vat. The major disadvantage of using freeze-dried preparations in this manner is the longer lag phase they exhibit, adding an additional 30 to 60 minutes to the time required to make cheddar cheese (Sandine, 1996). Preparation of freeze-dried cultures is initially similar to preparation of frozen concentrates. After freezing, the culture concentrate is placed under high vacuum to dehydrate by sublimation. The dry cells are then packaged under aseptic conditions, preferably in the absence of oxygen. Exposure to oxygen rapidly damages the cells (Yang and Sandine, 1979).

V. BACTERIOPHAGE

A. Introduction

Bacteriophages (phages) are viruses that infect bacteria. Bacteriophage infection of starter cultures can result in failure of the fermentation and loss of product. Whitehead and Cox (1935) first recognized bacteriophage infection as a cause of failure of single-strain starter cultures used for cheddar cheese production. Excellent conditions for development of bacteriophages were created in the 1950s when cheese production increased, resulting in more intensive use of facilities and the preparation of larger amounts of lactic cultures (Huggins, 1984). Despite the implementation of control measures, bacteriophage infection still causes production problems in the modern dairy fermentation industry. Adoption of control strategies based on the use of lactic acid bacteria genetically engineered for bacteriophage resistance should provide substantial improvements in dependability of starter cultures (Dinsmore and Klaenhammer, 1995).

B. Characteristics of Bacteriophage

1. Morphology/Taxonomy

Bacteriophages that infect lactic acid bacteria usually consist of a head and tail section. The head can be either isometric or prolate (Fig. 2). An isometric head

                                                    Frank and Hassan

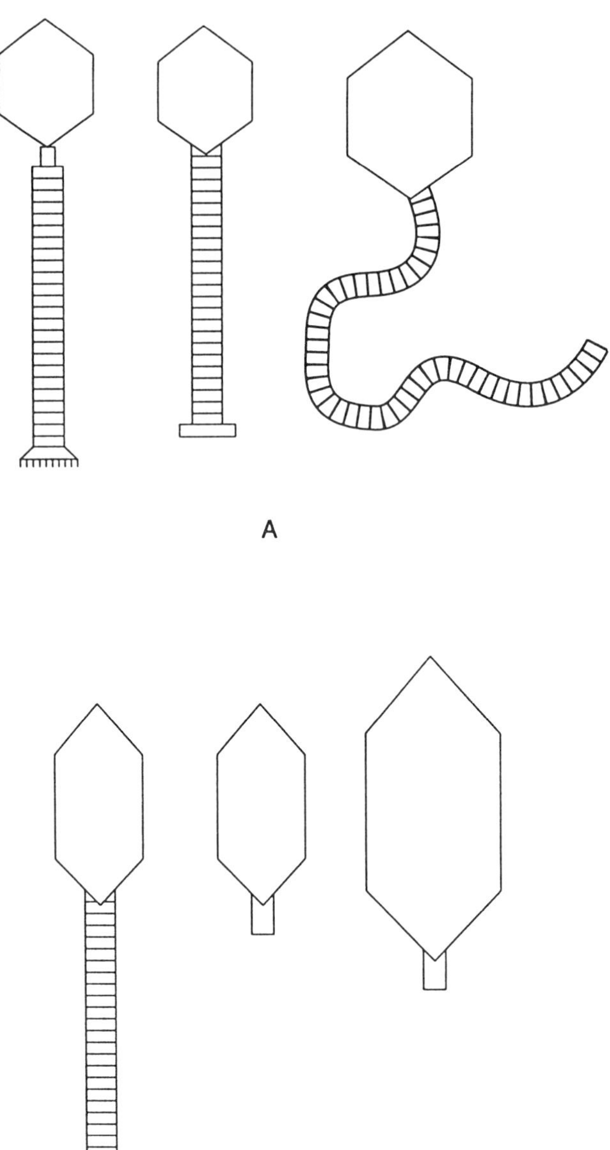

FIGURE 2 Morphology of common bacteriophages of lactic acid bacteria. A, Isometric phages with long tails. B, Prolate phages with short and long tails.

consists of 20 equal-size proteins that form an icosohedron. A prolate head has elongated side units. Phage DNA is enclosed by the head proteins. Phages attach to the host by their tail sections, through which the DNA passes into the bacteria. Tail sections are of variable length and may have collars, sheaths, and base plates. Base plates can be seen at the end of the tail of the phage illustrated in Figure 3.

Bacteriophages of lactic acid bacteria can be classified by morphology, serology, and DNA–DNA homology. These classification criteria generally produce consistent groupings (Lodics and Steenson, 1993). Six morphological types of lactic phages are commonly encountered. These include small isometric, collared small isometric, short-tailed small isometric, long-tailed small isometric, large isometric, and prolate (Lodics and Steenson, 1993). Each morphological type may include several distinct genotypes of which there are 12 (Neve, 1996). Bacteriophages of *S. thermophilus* form one homologous grouping as opposed to bacteriophages of mesophilic lactococci and *Lb. delbrueckii*, which are genetically diverse (Jarvis, 1989; Brussow et al., 1994).

2. Phage-Host Interactions

Host Range

Host range reflects the ability of a specific bacteriophage to infect different strains of bacteria. Host range varies widely between bacteriophages. In addition,

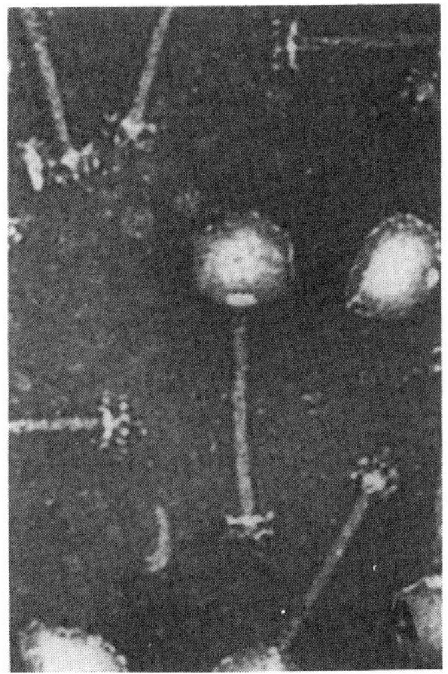

FIGURE 3 Electron micrograph of isometric phage of *Lactococcus lactis*. (From Moineau et al., 1994.

the susceptibility of specific strains of lactococci to phage attack is to some degree based on plasmid-associated resistance factors and is therefore highly variable. Bacteriophage of *Lc. lactis* subsp. *cremoris* tend to have a more limited host range than bacteriophage of *Lc. lactis* subsp. *lactis* (Jarvis, 1989). Isometric phages of lactococci tend to have limited host ranges, whereas prolate phages have broader host ranges. Some phages can attack both subspecies of *Lc. lactis* (*lactis* and *cremoris*). Several phages can attack both *Lb. delbrueckii* subsp. *bulgaricus* and *Lb. delbrueckii* subsp. *lactis* (Jarvis, 1989).

Lytic Cycle

Bacteriophage infections are caused by either lytic or temperate phages. Infection with lytic (virulent) phages results in the release of infectious virus particles (virions) into the environment, whereas temperate phages incorporate their DNA into the host chromosome and do not immediately produce new virions. The sequence of events in the lytic cycle are described by Neve (1996) and are illustrated in Figure 4. Phage infection is initiated by adsorption of the virion onto the surface of the host cell. Only bacteria with specific adsorption sites serve as host for the bacteriophage; the presence of these sites determines to a great extent the host range of a particular phage. Recognition of an appropriate site and adsorption to it are mediated by the base plate, spikes, or fibers at the end of the phage tail. Many phages require Ca^{2+} for adsorption.

After adsorption, the phage injects its DNA into the host. The DNA passes from the head through the tail into the bacterial cell, while the "empty" virion remains outside. Normal metabolism of the infected cell then ceases as the host first replicates phage DNA and then phage proteins. This process, called "maturation," ends with self-assembly of virions within the host cell. Initially, heads form around viral DNA followed by the attachment of tails. Finally, the lytic cycle is completed when a lytic enzyme (lysin), encoded on viral DNA, is produced, resulting in cell lysis and release of infective phage particles into the surrounding environment. Lysin released from infected cells can also lyse noninfected cells. The time from initial adsorption to the release of phages is called the latent period. For lactococcal phages, this period ranges from 10 to 140 minutes. The number of virulent particles released per infected cell is called the burst size. This ranges from less than 10 to more than 300 for lactococcal phages (Klaenhammer and Fitzgerald, 1994).

Temperate Cycle

Infection with a temperate phage does not necessarily lead to the immediate production of new virions. The DNA of a temperature phage may instead be incorporated into the chromosome of the host cell or maintained as a plasmid within the cell (Cogan and Accolas, 1990). This DNA, referred to as a prophage,

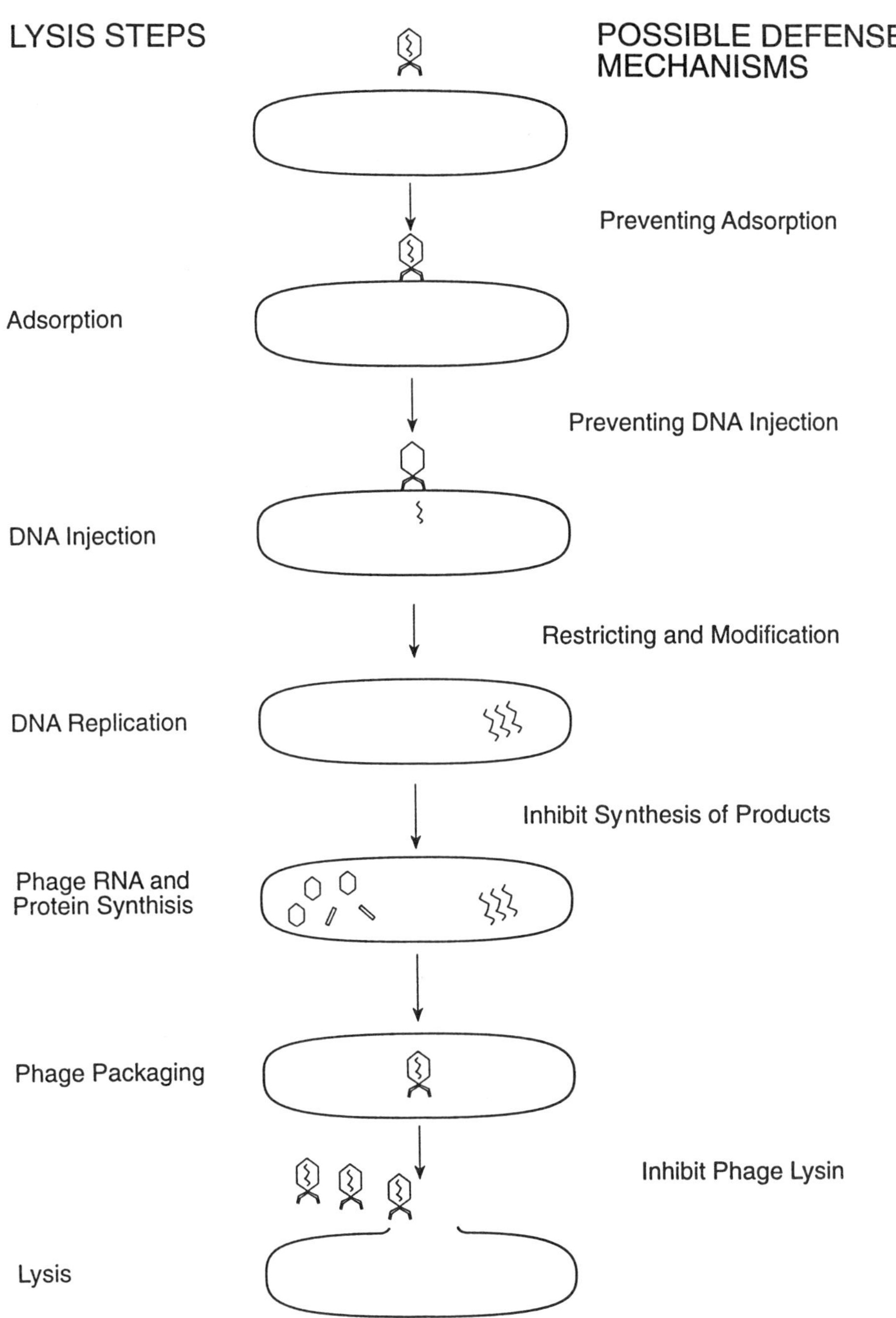

FIGURE 4 Stages in the lytic cycle wherein bacteriophage defense mechanisms are active.

replicates with the bacterium without affecting its metabolism. The resulting condition, lysogeny, is common in lactococci (Davidson et al., 1990) and lactobacilli (Sechaud et al., 1988), rare in *S. thermophilus* (Brussow et al., 1994) and unreported in *Pediococcus, Leuconostoc*, and *Propionibacterium* spp. (Davidson et al., 1990). Lysogeny can be maintained indefinitely. Lysogenous bacteria are immune to the infecting and other closely related phages. They maintain potential to produce virulent phage and can spontaneously realize this potential. Phage production can also be induced by exposing cells to ultraviolet (UV) light or mitomycin C to inactivate the repressor protein that blocks expression of the growth genes (Lodics and Steenson, 1993).

The extent to which lysogenic bacteria in starter cultures pose a threat to industrial fermentations is still uncertain (Jarvis, 1989; Davidson et al., 1990). Temperate phages can mutate to become virulent, resulting in fermentation failure (Shimizu-Kodata et al., 1983), although spontaneous induction of virulent phages from lysogenic strains appears to be rare (Teuber and Lembke, 1983). Surveys of lactococcal phage DNA homology indicate that, although some lytic phages appear to be variants of temperate phages, this is generally not the case (Davidson et al., 1990).

Pseudolysogeny

Pseudolysogeny (phage carrier state) occurs when a bacterial culture carries lytic phages while maintaining an active cell population. The culture remains active because only a portion of the total population is sensitive to the phage, the remaining population retaining the ability to grow rapidly and produce acid. Establishment of pseudolysogeny depends on the ability of a culture to produce variants having different degrees and types of phage sensitivity (Lodics and Steenson, 1993). Unlike true lysogeny, phages can be eliminated from a pseudolysogenous culture by growing it in the presence of phage-specific antibodies or by repeated culture purification (selection of isolated colonies on agar plates).

3. Phage Resistance Mechanisms

Phage resistance in lactic acid bacteria is based on at least four different mechanisms (Dinsmore and Klaenhammer, 1995; Hill, 1993): adsorption inhibition, DNA injection inhibition, DNA restriction and modification systems, and abortive infection. Stages in the lytic cycle where these mechanisms are active are illustrated in Figure 4. Many lactococci used in starter cultures exhibit one or more of these resistance mechanisms. Adsorption inhibition is the failure of phage to attach to the bacterial surface. This can result from spontaneous mutation modifying the attachment site or from a plasmid-linked factor (Dinsmore and Klaenhammer, 1995). Plasmids can encode for production of polymers that coat attachment sites, preventing phage adsorption.

DNA injection inhibition occurs when phage adsorbs to the cell surface but phage DNA stays inside the head section, failing to enter the host cell cytoplasm. This resistance mechanism has not been well characterized and appears to be rare (Dinsmore and Klaenhammer, 1995).

Phage resistance based on DNA restriction and modification enzymes is common in the lactococci. The restriction enzyme hydrolyzes phage DNA at a specific site. Host DNA is modified by methylation at this site and is therefore unaffected by the restriction enzyme. Restriction and modification enzymes are linked to the same plasmid. It is possible, but rare, for phage DNA to be methylated by the host modification system before it is hydrolyzed by the restriction enzyme. When this happens, the phage is able to cause a normal infection. Phages whose DNA does not contain the targeted restriction site are also unaffected by this resistance mechanism.

Abortive infection is a type of phage resistance resulting in decreased production of virulent phages by infected cells, but not involving restriction or modification. Abortive infection results in cell death, but because phage replication is much reduced, the phage population does not increase sufficiently to affect culture activity. Abortive infection does not induce genetic changes in the infecting phage. Numerous (at least seven) nonhomologous plasmids encode for abortive infection resistance, indicating that many different types exist (Dinsmore and Klaenhammer, 1995; Neve, 1996).

When a host cell with phage resistance is exposed to sufficiently high numbers of phages, it is possible for the phage to mutate to overcome the resistance mechanism. Also, if phage inhibition is not complete, resistant phages are selected (Hill, 1993). If phage DNA is modified by the host enzyme to become resistant to the restriction enzyme, the resulting resistance is lost when the phage infects a cell that lacks the methylase enzyme. More lasting insensitivity occurs when phages mutate at the hydrolysis site of the restriction enzyme. Some lactococcal bacteriophages have evolved to have very few sites available for restriction endonuclease hydrolysis (Dinsmore and Klaenhammer, 1995). Phages also develop insensitivity to abortive injection mechanisms, apparently through point mutations.

4. Phage Survival

Many bacteriophages have good survival characteristics. Some can survive high-temperature, short-time pasteurization, so media for starter preparation are usually heated to at least 85°C for 30 minutes to ensure inactivation of the phage (Neve, 1996). Phages can also survive spray drying and storage of milk powder (Chopin, 1980). Phage particles on surfaces are readily inactivated by chlorine, but not by iodine or acid sanitizers (Anonymous, 1990). Sanitizer inactivation depends on the elimination of organic matter through effective cleaning.

C. Characteristics of Phage Infection

Bacteriophages are primarily a problem in cheese manufacture. This is probably because cheese milk (as compared to cultured milks) is given only a mild heat treatment and because cheese milk and whey are often exposed to a phage-contaminated environment. Bacteriophages do not proliferate in cheese curd because virions cannot move through the protein matrix. However, cells infected with phage before coagulation become inactivated during cheese manufacturing. Because latency periods are normally approximately 30 minutes (but may be much longer), a culture may initially show normal growth in cheese milk, but then reduce or stop acid production during manufacture. If one culture preparation is used to inoculate a series of vats of milk, increasing numbers of phages active against this culture may develop within the manufacturing plant. The result is that acid production proceeds normally in the vats of milk inoculated initially but is delayed later in the production day.

D. Preventing Phage Inhibition

Preventing inhibition of acid production resulting from phage infection requires implementation of control measures throughout the manufacturing process. These should include selection, preparation, and maintenance of cultures free of virulent phage; controlling entry of phages into the processing facility, and controlling the spread of phages within the facility.

1. Phage Inhibitory Media

Growth of phages during the production of bulk starter can be controlled by using phage-inhibitory media. These media rely on the ability of phosphate and citrate salts to bind ionic calcium, thus inhibiting the phage absorption (Reiter, 1956). The chelating agents can slow the growth of the starter culture. Phage-control media often contain deionized whey, protein hydrolysates, ammonium and sodium phosphate, and citrate salts, and other growth stimulants such as yeast extract (Whitehead, 1993). Commercial phage-inhibitory media vary widely in their ability to prevent phage proliferation, the most effective being those that contain sufficient nutrients to overcome the inhibitory nature of the media and contain citrate buffers (Gulstrum et al., 1979). Not all bacteriophages are inhibited by the absence of calcium (Sozzi, 1972), so, to be effective, phage-inhibitory media should be used as only one part of an overall phage-control strategy. Proliferation of phages during starter preparation can also be avoided by using cell concentrates designed to be added directly to the cheese milk in the vat or by preparing cultures under strict aseptic conditions.

2. Use of Phage-Resistant Cultures

Lactic acid bacteria vary widely in their susceptibility to bacteriophage infection, so the use of resistant strains is an important aspect of phage control. Phage-resistant strains have been isolated from mixed culture systems that maintain activity while carrying low levels of phages (Lodics and Steenson, 1993). Strains can also be genetically altered to contain plasmids coding for phage resistance (Klaenhammer, 1991).

Phage-resistant variants can be selected by exposure to factory whey containing phages that have developed during cheese manufacture (Sandine, 1989). Resistant variants are tested for rapid acid production and added back to the starter in use in that factory. The use of such a system requires daily monitoring of whey for phages, but allows use of a single mixture of five or six defined strains over a long time. This approach to phage control is often used in North America and elsewhere.

Protease-negative strains of lactococci are resistant to phage infection because of their slow growth rates (Richardson, 1984). Although more cells must be used to compensate for lack of growth during cheese manufacture, these variants offer other advantages, including lowered sensitivity to antibiotics, lowered heat sensitivity (allowing use of higher cook temperatures), greater yield because of lowered casein solublization, and decreased risk of bitter flavor development in cheese.

3. Culture Rotation

Culture rotations control bacteriophage infection by limiting the length of time that a specific strain or mixture of strains is used. Cultures following each other in the series are susceptible to different phage types and are therefore unaffected by phages that may have infected the previous culture. Cultures can be rotated on a daily basis or after each vat of milk is inoculated. Short rotations over 2 to 3 days using 6 to 12 strains and long (5–10 days) rotations of up to 30 strains are used (Huggins, 1984). Culture rotation does not eliminate phage growth in cheese milk in vats, but if phage numbers are kept to less than 10,000 plaque-forming units per milliliter of cheese whey, the acid production is not affected (Huggins, 1984). Success of a culture rotation is limited by the availability of phage-unrelated strains with acceptable fermentation properties. In addition, using many different cultures can result in lack of product uniformity.

A new type of culture rotation system has been developed by Sing and Klaenhammer (1993) and Durmaz and Klaenhammer (1995). This system uses genetic derivatives of a single strain, each with a different phage resistance mechanisms. When used in rotation or as mixtures, resistant phages fail to develop, because they cannot overcome the multiple resistance mechanisms. This type of rotation avoids the lack of product uniformity associated with conventional culture rotations and allows continuous use of strains with special properties.

4. Sources of Bacteriophages in the Dairy Plant

Bacteriophages in the dairy plant probably are of farm origin, although, as discussed previously, lysogenic bacteria may also be a source. Although the major means by which a phage enters the plant is in the raw milk, trucks and personnel having had contact with the farm environment could also be carriers. It is not practical to eliminate entry of phages into the dairy plant because raw milk continually enters the facility. However, growth of phages within the plant and dissemination of phages to milk in the cheese vat can be controlled. The main growth niches for bacteriophages in a cheese plant are raw milk, whey, spilled product, pools of water, stagnant floor drains, equipment, and soiled walls (Anonymous, 1990). Phage development in these growth niches is controlled by effective sanitation. Phages are disseminated throughout the dairy plant by aerosol and human carriers. Air entering cheese manufacturing rooms should be under positive pressure of high-efficiency particulate air (HEPA) filtered air. When preparing bulk starter, air drawn into the tank when the culture medium cools should be filter sterilized. Milk in cheese vats is most susceptible to phage contamination during ripening and setting, so these processes should be accomplished in closed systems. Whey should be removed to a physically separate facility, because whey processing produces aerosols that can carry phage particles. Plant personnel with exposure to whey should not be allowed access to the milk ripening or bulk starter facilities.

VI. OTHER CULTURE INHIBITORS

A. Raw Milk–Associated Inhibitors

Lactic starter cultures grow more slowly in raw than in heated milk, a phenomenon caused by the presence of natural inhibitors. The lactoperoxidase system is the most significant microbial inhibitor in raw milk, but the presence of agglutinins is an important problem in acid-coagulated cheeses. Other naturally occurring microbial inhibitors in milk include lysozyme and lactoferrin. Mastitic milk has increased levels of microbial inhibitors and increased phagocytic activity that are part of the cow's response to infection. However, mastitic milk is also higher in protease activity, and the resulting casein fragments can counteract inhibitor effects and even stimulate the growth of weakly proteolytic lactics such as *S. thermophilus* (Marshall and Bramley, 1984; Okello-Uma and Marshall, 1986).

1. Lactoperoxidase System

Microbial inhibition by the lactoperoxidase system derives from the interaction of three components: lactoperoxidase, an enzyme native to milk; thiocyanate, derived from the hydrolysis of cyanogenic glucosides found in certain feeds; and hydrogen peroxide, generated by leukocytes and through the oxygen metabolism of lactic acid bacteria (Limsowtin, 1992). The inhibitor, hypothiocyanite, is produced when lactoperoxidase catalyzes the oxidation of thiocyanate and the simultaneous reduction of hydrogen peroxide. Hydrogen peroxide is usually the limiting component in raw milk, but thiocyanate is also often present in sub-optimal concentrations (Limsowtin, 1992). Lactoperoxidase is only partially inactivated by pasteurization (Wolfson and Sumner, 1993). The lactic starter cultures most sensitive to lactoperoxidase inhibition are those that generate hydrogen peroxide. This includes some strains of *Lb. delbrueckii* subsp. *bulgaricus* and *Lb. acidophilus* (Guirguis and Hickey, 1987b). Other lactic acid bacteria, including *S. thermophilus* and some strains of lactococci are sensitive to lactoperoxidase inhibition when combined with cultures that produce hydrogen peroxide. The inhibitory effects of the lactoperoxidase system can be controlled by limiting aeration of milk, avoiding use of hydrogen peroxide–generating cultures, using cultures that degrade hydrogen peroxide, and using heat treatments more severe than pasteurization.

2. Immunoglobulins (Agglutinins)

Lactic starter cultures can interact with immunoglobulins in milk to form aggregations or clumps. As the cells produce acid, casein coagulates around these clumps and they settle out of the milk forming a sludge (Grandison et al., 1986). Acid production is inhibited because diffusion of acid out of the sludge is limited, causing acid inhibition of the culture before the milk is properly acidified (Hicks and Ibrahim, 1992). This type of inhibition is of significance when acid coagulation is desired, as for cottage cheese, which exhibits a loss of curd. Culture agglutination can be reduced by selecting agglutination-resistant cultures, using whey-based culture media with the agglutinins removed by protease treatment (Ustunol and Hicks, 1994) and by homogenization of the milk before culturing (Hicks and Hamzah, 1992).

3. Lysozyme

Lysozyme inactivates bacteria by cleaving the glycosidic bond between *N*-acetylmuramic acid and *N*-acetylglucoseamine in the peptidoglycan of the cell wall. Gram-positive bacteria are highly susceptible to lysozyme activity because of the high peptidoglycan content of their cell wall and a lack of protective lipopolysaccharide. Bovine milk contains only approximately 0.1 µg/L of lysozyme, a very low level that is unlikely to inhibit starter culture micro-organisms (Vakil et al., 1969).

4. Lactoferrin

Lactoferrin is an iron-binding protein that inhibits bacteria by denying them access to iron. Cow's milk contains only 20 to 200 µg/mL of lactoferrin (Masson and Heremans, 1971) and its activity is limited because it competes with citrate for binding the iron (Batish et al., 1988). Inhibition of starter cultures by lacto-ferrin is unlikely to be significant.

B. Antibiotics

Treatment of mastitis in cows involves application of antibiotics. Milk from treated cows cannot be legally sold, but, occasionally, it becomes mixed with salable product. The resulting low-level antibiotic contamination may be suf-ficient to inhibit starter culture microorganisms. As antibiotic levels in milk increase, acid production decreases. Lactic acid bacteria are very sensitive to antibiotics commonly used for mastitis treatment. These include penicillin, cloxacillin, streptomycin, and tetracycline. Milk that tests negative for antibiotics, using *Bacillus stearothermophilus* as an indicator, can still have sufficient anti-biotic to cause starter culture inhibition (Valladao and Sandine, 1994a). When antibiotics other than penicillin are present, available methods may not be suffi-ciently sensitive to detect residues that could cause a 20% reduction in lactic acid production (Schiffmann et al., 1992).

Sensitivity of starter cultures to antibiotics is highly strain and species dependent. *S. thermophilus* is more susceptible to penicillin and cloxacillin (β-lactam antibiotics) than are the lactococci, but lactococci are more sensitive to streptomycin and tetracycline (Dezmazeaud, 1996). Swiss (Emmenthal) cheese made with antiobiotic-contaminated milk (0.005 IU/mL) exhibited abnormal eye formation, presumably from inhibition of propionibacteria (Mäyrä-Mäkinen and Migret, 1993).

C. Chemical Sanitizers

Occasionally, chemical sanitizers may contaminate milk, usually as a result of human error. Chlorine- and iodine-based sanitizers lose their activity in milk and are, therefore, unlikely to cause starter culture inhibition. Quaternary ammonium compounds present more potential problems, because they maintain activity in milk, and lactic acid bacteria are sensitive to low concentrations. Valladao and Sandine (1994b) observed that all tested *Lactococcus* strains were inhibited by 20 µg/mL and some were inhibited by only 10 µg/mL quaternary ammonium compound. Thermophilic starter cultures are inhibited at 0.5 to 2.0 µg/mL quarternary ammonium compound (Guirguis and Hickey, 1987a).

Peracetic acid and acid anionic sanitizers can also maintain some activity in milk (Dunsmore, 1985). Relatively high concentrations of hydrogen peroxide or quaternary ammonium compound are required to give positive results in antibiotic screening tests (Richard and Kerhavé, 1973). The amount of chemical sanitizer that might enter milk through lack of rinsing should not be sufficient to cause culture inhibition (Desmazeaud, 1996). However, problems can be encountered when sanitizer solution is not drained from tanks or trucks.

VII. INHIBITORY COMPOUNDS PRODUCED BY STARTER CULTURES

One of the valuable properties of starter cultures is their ability to inhibit growth of undesirable microorganisms. The main preservative action of lactic starter cultures is a result of acid production. Acids produced by lactic acid bacteria include not only lactic acid but also lesser amounts of acetic and formic acids. The production of acids other than lactic acid increases the preservative effect of the culture because, at equivalent pH, acetic and formic acids have greater inhibitory power than lactic acid.

Lactic starter cultures also produce nonacidic microbial inhibitors. These include hydrogen peroxide (which can act by itself or in concert with the lactoperoxidase system as previously discussed), carbon dioxide, low molecular weight carbonyl compounds, and bacteriocins. Production of nonacidic inhibitors by lactic starter cultures is not necessarily advantageous. Undesirable effects include autoinhibition resulting from hydrogen peroxide (produced when oxygen is present in the milk) and an inability to be used in multiple strain cultures as a result of bacteriocin production.

A. Low Molecular Weight Nonacidic Metabolites

Kulshrestha and Marth (1974) observed that many nonacidic low molecular weight metabolites of lactic acid bacteria have antimicrobial activity, but at concentrations higher than produced in cultured milk. The metabolite with greatest inhibitory activity is the flavor compound, diacetyl (2,3-butanedione). Jay (1982) found that yeasts and gram-negative bacteria are inhibited by 200 ppm diacetyl and that gram-positive bacteria are inhibited by 300 ppm. Although such levels are not found in cultured dairy products, diacetyl may act in combination with other compounds to enhance the preservative effect of starter cultures.

B. linens, when growing in a cheese-containing medium, produces an antimicrobial agent with a broad spectrum of activity, being active against yeasts and molds, *Clostridium botulinum, Staphylococcus aureus, Salmonella* spp., *Bacillus cereus*, and many yeasts and molds (Grecz, 1964). Volatile sulphur compounds are at least partially responsible for this activity (Beattie and Torrey, 1984).

B. Bacteriocins

Bacteriocins are proteins or polypeptides with potent bactericidal activity. They typically have a narrow spectrum of activity against species closely related to the producing organism. Their production and immunity to their action is plasmid encoded (with some exceptions). Some bacteriocins are especially interesting because their broad spectrum of activity may make them useful for inhibiting specific pathogenic or spoilage microorganisms. Activity of some bacteriocins against *Listeria* spp. is presented in Table 5. Production of bacteriocins by lactic acid bacteria is common, as shown by data presented in Table 6. However, strains of lactic acid bacteria selected for use in multiple-strain cultures generally do not

TABLE 5 Activity of Some Bacteriocins Against *Listeria* Species

Producer organism	Bacteriocin	Inhibition
Lactobacillus		
acidophilus	Lacticin F	–
acidophilus	Lacticin M	–
acidophilus	Lacticin B	–
helveticus	Helveticin J	–
plantarum	Plantaricin A	–
Leuconostoc		
mesenteroides	Mesentericin Y105	+
Lactococcus lactis		
subsp. *lactis*	Nisin	+
subsp. *cremoris*	Diplococcin	–
Streptococcus		
thermophilus	Thermophilin 347	+
thermophilus	Thermophilin A	–
Propionibacterium		
thoenii	Propionicin PLG-1	+
Pediococcus		
acidilactici	Pediocin PA-1	+
pentosaceus	Pediocin A	+

Sources: Harris et al., 1989; Lyon et al., 1993; Stiles, 1994; Villani et al., 1995; Ward, 1995.

TABLE 6 Frequency of Bacteriocin Production by Lactic Acid Bacteria

Organism	No. positive/ no. tested	% positive
Lactobacillus spp.	11/189	6
Lactobacillus		
fermenti	11/121	15
acidophilus	33/152	63
Lactococcus spp.	65/280	23
Streptococcus		
thermophilus	13/41	32
mutans	97/130	75

Sources: Tagg et al., 1976; Klaenhammer, 1988; Ward, 1995.

produce bacteriocins so they do not dominate the mixture. Bacteriocins produced by lactic starter cultures can be divided into three biochemical groups (Barefoot and Nettles, 1993): lanthionine-containing peptides such as nisin and lacticin 481; small non–lanthionine-containing proteins or peptides such as lacticin F, lactacin B, and lactococcin A; and large heat-labile proteins such as helveticin and caseicin 80.

1. Lactococci

Lc. lactis subsp. *lactis* produces the bacteriocin nisin. Nisin was isolated by Mattick and Hirsch (1947) and is the only bacteriocin widely approved for use as a food additive. (Other bacteriocins can be present in foods as a natural part of the culturing process.) Nisin is a polypeptide containing 34 amino acids and usually occurs as a dimer with molecular weight of 7000 daltons (Jarvis et al., 1968). It is the best known of the group of bacteriocins called lantibiotics, which contain the unusual amino acids, lanthionine, β-methyl lanthionine and dehydroalanine (Vanenbergh, 1993). Nisin has a relatively broad spectrum of activity for a bacteriocin, with activity against many lactic acid bacteria, spore-forming bacteria, and *Listeria monocytogenes* (Davidson and Hoover, 1993). Its ability to prevent outgrowth of bacterial endospores has led to its use in preventing the late gas defect in hard cheeses and as an inhibitor of *Clostridium botulinum* and spoilage microorganisms in canned foods and processed cheese (Daeschel, 1989). Nisin is heat stable and has greatest activity under mildly acidic conditions. Like other bacteriocins, the site of action of nisin is the cytoplasmic membrane (Sahl, 1991).

Lacticin 481, produced by *Lc. lactis* subsp. *lactis*, has activity against lactococci and some lactobacilli, leuconostocs, and clostridia. If produced by starter cultures used for cheese manufacture, lactocin 481 eliminates the sensitive microflora from the resulting cheese (Paird et al., 1991).

Lc. lactis subsp. *cremoris* produces diplococcin, which, unlike nisin, has a narrow spectrum of activity (primarily against other lactococci) and lacks stability. Producers of diplococcin rapidly predominate in multiple-strain starter cultures.

2. Lactobacilli

Lactobacilli used in starter cultures can produce many different bacteriocins, most with limited range of activity. *Lb. helvecticus* produces helveticin J and lactocin LP27. Helveticin J is an unusual bacteriocin because it is coded for on chromosomal DNA and is active at neutral pH (Joerger and Klaenhammer, 1986). *Lb. acidophilus* produces numerous bacteriocins including lactins B and F (Muriana and Klaenhammer, 1991), *Lb. casei* produces caseicin 80 (Rammelsberg and Radler, 1990), and *Lb. delbrueckii* subsp. *lactis* produces lacticins A and B (Toba et al., 1991). Properties of these compounds have been summarized by Davidson and Hoover (1993).

3. Leuconostocs

Most information on bacteriocins produced by *Leuconostoc* spp. has been obtained from meat rather than dairy isolates. Stiles (1994) concluded that bacteriocins of leuconostocs are active against *L. monocytogenes* but not necessarily against other lactic acid bacteria.

4. Propionibacteria

Propionicin PLG-1, a bacteriocin produced by *Propionibacterium thoenii*, is unusual because of its broad range of activity, which includes some gram-negative bacteria, including *Escherichia coli*, *Pseudomonas fluorescens*, and *Vibrio parahaemolyticus* (Lyon and Glatz, 1991). It is also active against other propionibacteria, lactic acid bacteria, and some yeasts and molds. It is inactivated at temperatures above 80°C, unlike jensenin G, which is produced by *Propionibacterium jensenii* and is stable at 100°C. Jensenin has a narrow range of activity, but is active against microorganisms commonly found in Swiss cheese (Grinstead and Barefoot, 1992).

5. Streptococci

Ward (1995) found that 13 of 41 strains of *S. thermophilus* produced bacteriocin-like substances that were active mainly against other *S. thermophilus* strains. He purified the bacteriocin, thermophilin A, which is heat stable and acid tolerant.

Villani et al. (1995) isolated thermophilin 347 and determined it to be heat stable and inhibitory toward *L. monocytogenes.*

6. Applications and Commercial Preparations

The dairy industry can take advantage of the preservative properties of bacteriocins either by using bacteriocin-producing cultures in the manufacturing process or by adding the bacteriocin-containing preparations directly to a product. Nisin can be purchased for use as a food additive under the brand name Nisaplin (Aplin and Barret, Ltd, England). It is mainly used in dairy foods for its ability to inhibit bacterial spore germination.

Skim milk fermented with a bacteriocin-producing strain of *P. freudenreichii* subsp. *shermanii* and then pasteurized can be purchased under the brand name Microgard (Wesman Food, Inc., Beaverton, OR). The presence of propionic acid, diacetyl, and acetic acid in Microgard enhances the preservative effect of the bacteriocin (Al-Zoreky, 1988). Microgard is used extensively in the United States as a preservative in cottage cheese (Daeschel, 1989). Other fermented milk and whey products containing bacteriocins are also commercially available.

REFERENCES

Al-Zoreky N. Microbiological control of food spoilage and pathogenic microorganisms in refrigerated foods. MS thesis, Oregon State University, Corvallis, OR, 1988.

Anonymous. Bacteriophage. Milwaukee, WI: Chr. Hanson's Laboratory, Inc., 1990.

Ariga H, Urashima T, Michihata E, Manabu I, Morizono N, Kimura T, Takahashi S. Extracellular polysaccharide from encapsulated *Streptococcus salivarius* subsp. *thermophilus* OR 901 isolated from commercial yogurt. J Food Sci 57:625, 1992.

Axelsson LT. Lactic acid bacteria: classification and physiology. In: Salminen S, von Wright A, eds. Lactic Acid Bacteria. New York: Marcel Dekker, 1993, p 1.

Barefoot SF, Nettles CG. Antibiosis revisited: bacteriocins produced by dairy starter cultures. J Dairy Sci 76:2366, 1993.

Batish VK, Chandler H, Zumdigni KC, Bhatia KL, Singh RS. Antibacterial activity of lactoferrin against some common food-borne pathogenic organisms. Aust J Dairy Technol 43:16, 1988.

Beal C, Llouvet P, Corrieu G. Influence of pH, lactose and lactic acid on growth and acidification of pure cultures of *Streptococcus thermophilus* 404 and *Lactobacillus bulgaricus* 398. Appl Microbiol Biotechnol 32:148, 1989.

Beattie SE, Torrey GS. Volatile compounds produced by *Brevibacterium linens* inhibit mold spore germination (abstr). J Dairy Sci 67(suppl 1):84, 1984.

Beneke ES, Stevenson KE. Classification of food and beverage fungi. In: Beuchat LR, ed. Food and Beverage Mycology. New York: Van Nostrand Reinhold, 1987, p 1.

Boysen M, Skouboe P, Frisvad J, Rossen L. Reclassification of the *Penicillium roqueforti* group into three species on the basis of the molecular genetic and biochemical profiles. Microbiol 142:541, 1996.

Brock TD. Milestones in Microbiology. Englewood Cliffs, NJ: Prentice-Hall, 1961.

Brussow H, Fremont M, Bruttin A, Sidoti J, Constable A, Fryder V. Detection and classification of *Streptococcus thermophilus* bacteriophages isolated from industrial milk fermentation. Appl Environ Microbiol 60:4537, 1994.

Buchenhüskes H. Selection criteria for lactic acid bacteria to be used as starter cultures for various food commodities. FEMS Microbiological Rev 12:253, 1993.

Cerning J. Exocellular polysaccharides produced by lactic acid bacteria. Fed European Microbiol Soc Microbiol Rev 87:113, 1990.

Cerning J, Bouillanne C, Desmazeaud MJ, Landon M. Isolation and characterization of exocellular polysaccharide produced by *Lactobacillus bulgaricus*. Biotechnol Lett 8:625, 1986.

Cerning J, Bouillanne C, Landon M, Desmazeaud MJ. Exocellular polysaccharide production by *Streptococcus thermophilus*. Biotechnol Lett 10:255, 1988.

Chassy B, Murphy C. *Lactococcus* and *Lactobacillus*. In: Sonenshein AL, Hoch JA, Losick R. eds. *Bacillus subtilis* and Other Gram-Positive Bacteria. Washington, DC: American Society for Microbiology, 1993, p 65.

Chen CM, Bogenrief DB, Jaeggi JJ, Tricomi WA, Johnson ME. Evaluation of mixer molder temperature and speed on composition, functional and sensory characteristics of 50% reduced-fat mozzarella cheese. J Dairy Sci 77(suppl 1):15, 1994.

Chopin MC. Resistance of 17 mesophilic lactic *Streptococcus* bacteriophages to pasteurization and spray-drying. J Dairy Res 47:131, 1980.

Cogan TM. History and taxonomy of starter cultures. In: Cogan TM, Accolas JP, eds. Dairy Starter Cultures. New York: VCH Publishers, 1996, p 1.

Cogan TM, Accolas JP. Starter cultures: types, metabolism and bacteriophage. In: Robinson RK, ed. Dairy Microbiology: The Microbiology of Milk. London: Elsevier Applied Science, 1990, p 77.

Collins EB. Influence of medium and temperature on end products and growth. J Dairy Sci 60:799, 1976.

Conn HW. Dairy Bacteriology. Bulletin No. 25. Washington, DC: U.S. Department of Agriculture Office of Experiment Stations, 1895.

Cummins CS, Johnson JL. *Propionibacterium*. In: Butler JP, ed. Bergey's Manual of Systematic Bacteriology. Vol II. Baltimore: Williams & Wilkins, 1984, p 1346.

Daeschel MA. Antimicrobial substances from lactic acid bacteria for use as food preservatives. Food Technol 43:164, 1989.

Davidson BE, Powell IB, Hillier AJ. Temperate bacteriophages and lysogeny in lactic acid bacteria. FEMS Microbiol Rev 87:79, 1990.

Davidson MP, Hoover DG. Antimicrobial components from lactic acid bacteria. In: Salminen S, von Wright A, eds. Lactic Acid Bacteria. New York: Marcel Dekker, 1993, p 127.

Desmazeaud M. Growth inhibitors of lactic acid bacteria. In: Cogan TM, Accolas JP, eds. Dairy Starter Cultures. New York: VCH Publishers, 1996, p 131.

Dinsmore PK, Klaenhammer TR. Bacteriophage resistance in *Lactococcus*. Molecular Biotechnol 4:297, 1995.

Dunsmore D. Effect of residues of five disinfections in milk on acid production by strains of lactic starters used for cheddar cheesemaking and on organoleptic properties of the cheese. J Dairy Res 52:287, 1985.

Durmaz E, Klaenhammer TR. A starter culture rotation strategy incorporating paired restriction/modification and abortive infection bacteriophage defences in a single *Lactococcus lactis* strain. Appl Environ Microbiol 61:1266, 1995.

Fringa E, Holtz C, Kunz B. Studies about casein degradation by *Brevibacterium linens*. Milchwissenschaft 48:130, 1993.

Frisvad JC, Filtenborg O. Terverticillate penicillia: chemotaxonomy and mycotoxin production. Mycologia 81:837, 1989.

Fryer TF, Peberdy MF. Growth of propionibacteria in Swiss and Egmont cheese. N Z J Dairy Sci Technol 12:133, 1977.

Garcia-Garibay M, Marshall, VME. Polymer production by *Lactobacillus delbrueckii* ssp. *bulgarious*. J Appl Bacteriol 70:325, 1991.

Glatz BA. The classical propionibacteria: their past, present and future as industrial organisms. ASM News 58:197, 1992.

Grandison AS, Brooker BE, Young P, Ford GD, Underwood HM. Sludge formation during the manufacture of cottage cheese. J Soc Dairy Technol 39:119, 1986.

Grecz N. Natural antibiosis in some varieties of soft surface ripened cheese. In: Melin N, ed. Microbial Inhibitors in Food. Stockholm: Almqvist and Wiksell, 1964, p 307.

Grinstead DA, Barefoot SF. Jensenin G, A heat-stable bacteriocin produced by *Propionibacterium jensenii* P126. Appl Environ Microbiol 58:215, 1992.

Guirguis N, Hickey MW. Factors affecting the performance of thermophilic starters. 1. Sensitivity to dairy starters. Aust J Dairy Technol 41:11, 1987a.

Guirguis N, Hickey MW. Factors affecting the performance of thermophilic starters. 2. Sensitivity to the lactoperoxidase system. Aust J Dairy Technol 41:14, 1987b.

Gulstrum TJ, Pearce LE, Sandine WE, Elliker PR. Evaluation of commercial phage inhibitory media. J Dairy Sci 62:208, 1979.

Harris LJ, Daeschel MA, Stiles ME, Klaenhammer TR. Antimicrobial activity of lactic acid bacteria against *Listeria monocytogenes*. J Food Prot 52:384, 1989.

Harrits J, McCoy D. Swiss cheese. In: Cultures for the Manufacture of Dairy Products. Milwaukee, WI: Chr. Hansen, Inc. 1997, p 79.

Harvey RJ. Damage to *Streptococcus lactis* resulting from growth at low pH. J Bacteriol 11:1330, 1965.

Hassan AN, Frank JF, Farmer MA, Schmidt KA, Shalabi SI. Observation of encapsulated lactic acid bacteria using confocal scanning laser microscopy. J Dairy Sci 78:2624, 1995a.

Hassan AN, Frank JF, Farmer MA, Schmidt KA, Shalabi SI. Formation of yogurt microstructure and three-dimensional visualization as determined by confocal scanning laser microscopy. J Dairy Sci 78:2629, 1995b.

Hassan AN, Frank JF, Schmidt KA, Shalabi SI. Rheological properties of yogurt made using encapsulated nonropy lactic cultures. J Dairy Sci 78:2091, 1996a.

Hassan AN, Frank JF, Schmidt KA, Shalabi SI. Textural properties of yogurt made using encapsulated nonropy lactic cultures. J Dairy Sci 78:2098, 1996b.

Haysahi K, Cliffe AJ, Law BA. Culture conditions of *Brevibacterium linens* for production of proteolytic enzymes. Nippon Shokuhin Kogyo Gakkaishi 37:737, 1990.

Hettinga DG, Reinbold GW, Vedamuthu ER. Split defect of Swiss cheese. I. Effect of strain of *Propionibacterium* and wrapping material. J Milk Food Technol 37:322, 1974.

Hicks CL, Hamzah B. Effect of culture agglutination on cottage cheese yield. Cult Dairy Prod J 27:4, 1992.

Hicks CL, Ibrahim SA. Lactic bulk starter homogenization affects culture agglutination. J Food Sci 57:1086, 1992.

Hickey MW, Hillier AJ, Jago GR. Enzymatic activities associated with lactobacilli in dairy products. Aust J Dairy Technol 38:154, 1983.

Hill C. Bacteriophage and bacteriophage resistance in lactic acid bacteria. FEMS Microbiol Rev 12:87, 1993.

Holtz C, Kunz B. Studies on degradation of whey proteins by *Brevibacterium linens.* Milchwissenschaft 49:130, 1994.

Huggins AR. Progress in dairy starter technology. Food Technol 38:41, 1984.

Hutkins R, Halambeck SM, Morris HA. Use of galactose-fermenting *Streptococcus thermophilus* in the manufacture of Swiss, mozzarella, and short-method cheddar cheese. J Dairy Sci 69:1, 1986.

Jarvis AW. Bacteriophages of lactic acid bacteria. J Dairy Sci 72:3406, 1989.

Jarvis B, Jeffcoat J, Cheeseman GC. Molecular weight distribution of nisin. Biochim Biophys Acta 168:153, 1968.

Jay JM. Antimicrobial properties of diacetyl. Appl Environ Microbiol 44:525, 1982.

Joerger MC, Klaenhammer TR. Characterization and purification of helveticin J and evidence for a chromosomally determined bacteriocin produced by *Lactobacillus helveticus* 481. J Bacteriol 167:439, 1986.

Jollivet N, Belin JM, Vayssier Y. Comparison of volatile flavor compounds produced by ten strains of *Penicillium camemberti* Thom. J Dairy Sci 76:1837, 1993.

Jollivet N, Bezenger MC Vayssier Y, Belin JM. Production of volatile compounds in liquid cultures by six strains of coryneform bacteria. Appl Microbiol Biotechnol 36:790, 1992.

Kinsella JE, Hwang D. Biosynthesis of flavors by *Penicillium roqueforti.* Biotechnol Bioeng 18:927, 1976.

Klaenhammer TR. Bacteriocins of lactic acid bacteria. Biochemie 70:337, 1988.

Klaenhammer TR. Development of bacteriophage-resistant strains of lactic acid bacteria. Food Biotechnol 19:675, 1991.

Klaenhammer TR, Fitzgerald GF. Bacteriophage and bacteriophage resistance. In: Gasson MJ, De Vos WM, eds. Genetics and Biotechnology of Lactic Acid Bacteria. London: Chapman and Hall, 1994, p 106.

Knudsen S. Starters. J Dairy Res 2:137, 1931.

Kulshrestha DC, Marth EH. Inhibition of bacteria by some volatile and nonvolatile compounds associated with milk. II. *Salmonella typhimurium.* J Milk Food Technol 37:539, 1974.

Limsowtin G. Inhibition of starter cultures. Aust J Dairy Technol 47:100, 1992.

Limsowtin GKY, Heap HA, Lawrence RC. A new approach to the preparation of bulk starter in commercial cheese plants. N Z J Dairy Sci Technol 15:219, 1980.

Lindsay RC, Day EA, Sandine WE. Green flavor defect in starter cultures. J Dairy Sci 46:863, 1965.

Lodics TA, Steenson LR. Phage-host interactions in commercial mixed-strain dairy starter cultures: practical significance—a review. J Dairy Sci 76:2380, 1993.

Lyon WJ, Glatz BA. Partial purification and characterization of a bacteriocin produced by *Propionibacterium thoenii*. Appl Environ Microbiol 57:701, 1991.

Lyon WJ, Glatz BA. Propionibacteria. In: Hui YH, Khachatourians GG, eds. Food Biotechnology Microorganisms. New York: VCH Publishers, 1995, p 703.

Lyon WJ, Sethi JK, Glatz BA. Inhibition of psychrotrophic organisms by propionicin PLG-1, a bacteriocin produced by *Propionibacterium thoenii*. J Dairy Sci 76:1506, 1993.

Marshall VM, Bramley J. Stimulation of *Streptococcus thermophilus* growth in mastitic milk. J Dairy Res 51:17, 1984.

Marshall VM, Cole WM. Threonine aldolase and alcohol dehydrogenase activities in *Lactobacillus bulgaricus* and *Lactobacillus acidophilus* and their contribution to flavour production in fermented foods. J Dairy Res 50:375, 1983.

Masson S, Heremans JF. Lactoferrin in milk from different species. Comp Biochem Physiol 39B:119, 1971.

Matzdorf B, Cuppett S, Keeler L, Hutkins R. Browning of mozzarella cheese during high temperature pizza baking. J Dairy Sci 77:2850, 1994.

Mattick ATR, Hirsch A. Further observation on an inhibitor (nisin) from lactic streptococci. Lancet 2:5, 1947.

Mäyrä-Mäkinen A, Bigret M. Industrial use and production of lactic acid bacteria. In: Salminen S, von Wright A, eds. Lactic Acid Bacteria. New York: Marcel Dekker, 1993, p 65.

McCoy DR. Italian type cheeses. In: Cultures for the Manufacture of Dairy Products. Milwaukee, WI: Chr. Hansen, Inc., 1997, p 70.

McCoy DR, Leach RD. Culture propagation and handling. In: Cultures for the Manufacture of Dairy Products. Milwaukee, WI: Chr. Hansen, Inc. 1997, p 32.

Mermelstein NH. Advanced bulk starter medium improves fermentation process. Food Technol. 34(8):69, 1982.

Mills OE, Thomas TD. Bitterness development in cheddar cheese: effect of the level of starter proteinase. N Z J Dairy Sci Technol 15:131, 1980.

Moineau S, Pandian S, Klaenhammer TR. Evolution of lytic bacteriophage via DNA acquisition from the *Lactococcus lactis* chromosome. Appl Environ Microbiol 60:1832, 1994.

Moines M, Groux M, Horman I. La flaveur des fromages. 3. Mise en évidence de quelques constituents mineurs de l'arôme du Camembert. Lait 547:414, 1975.

Monnet V, Condon S, Cogan TM, Gripon JC. Metabolism of starter cultures. In: Cogan TM, Accolas JP, eds. Dairy Starter Cultures. New York: VCH Publishers, 1996, p 47.

Morgan ME. The chemistry of some microbially induced flavor defects in milk and dairy foods. Biotechnol Bioeng 18:953, 1976.

Muriana PM, Klaenhammer TR. Purification and characterization of lactacin F, a bacteriocin produced by *Lactobacillus acidophilus* 11088. Appl Environ Microbiol 57:114, 1991.

Neve H. Bacteriophage. In: Cogan TM, Accolas JP, eds. Dairy Starter Cultures. New York: VCH Publishers, 1996, p 157.

Oberg CJ, Merrill RK, Moyes LV, Brown R, Richardson GH. Effects of *Lactobacillus helveticus* culture on physical properties of mozzarella cheese. J Dairy Sci 74:4101, 1991a.

Oberg CJ, Wang A, Moyes LV, Brown R, Richardson GH. Effects of proteolytic activity of thermolactic cultures on physical properties of mozzarella cheese. J Dairy Sci 74:389, 1991b.

Okello-Uma I, Marshall V. Influence of mastitis on growth of starter organisms used for the manufacture of fermented milks. J Dairy Res 53:631, 1986.

Olson NF. The impact of lactic acid bacteria on cheese flavor. FEMS Microbiol Rev 87:131, 1990.

Paird JC, Delorme F, Giraffa G, Commissaire J, Desmazeaud M. Evidence for a bacteriocin produced by *Lactococcus lactis* CNRZ 481. Neth Milk Dairy J 44:143, 1991.

Peterson LW. Buttermilk, sour cream, and related products. In: Cultures for the Manufacture of Dairy Products. Milwaukee, WI: Chr. Hansen, Inc., 1997, p 101.

Radke-Mitchell L, Sandine W. Associative growth and differential enumeration of *Streptococcus thermophilus* and *Lactobacillus bulgaricus*: a review. J Food Prot 47:245, 1984.

Rammelsberg M, Radler F. Antibacterial polypeptides of *Lactobacillus* species. J Appl Bacteriol 69:177, 1990.

Rajagopal SN, Sandine WE. Associative growth and proteolysis of *Streptococcus thermophilus* and *Lactobacillus bulgaricus* in skim milk. J Dairy Sci 73:894, 1990.

Reiter B. Inhibition of lactic *Streptococcus* bacteriophage. Dairy Ind 21:877, 1956.

Richard J, Kerhervé L. Influence of disinfectant residues on two methods of antibiotic detection in milk. Revue Laitère Française 306:127, 1975.

Richardson GH. Proteinase negative cultures for cheese making. Cult Dairy Prod J 19:6, 1984.

Sahl HG. Pore formation in bacterial membranes by cationic lantibiotics. In: Jung G, Sahl HG, eds. Nisin and Novel Lantibiotics. Netherlands: Escom, Leiden, 1991, p 347.

Sandine WE. Use of bacteriophage-resistant mutants of lactococcal starters in cheesemaking. Neth Milk Dairy J 43:211, 1989.

Sandine WE. Commercial production of dairy starter cultures. In: Cogan TM, Accolas JP, eds. Dairy Starter Cultures. New York: VCH Publishers, 1996, p 191.

Schiffmann AP, Schutz M, Wiesner HU. False negative and positive results in testing for inhibitory substances in milk. 1. The influence of antibiotics residues in bulk milk on lactic acid production of starter cultures. Milchwissenschaft 47:712, 1992.

Schleifer KH, Kraus J, Dvorak C, Kilpper-Balz R, Collins MD, Fischer W. Transfer of *Streptococcus lactis* and related streptococci to the genus *Lactococcus* gen. nov. System Appl Microbiol 6:183, 1985.

Sechaud L, Cluzel PJ, Rousseau M, Baumgertner A, Accolas JP. Bacteriophages of lactobacilli. Biochimie 70:401, 1988.

Shimizu-Kadota M, Sakurai T, Tsuchida N. Prophage origin of a virulent phage appearing on fermentations of *Lactobacillus casei* S-1. Appl Environ Microbiol 45:669, 1983.

Sing WD, Klaenhammer TR. A strategy for rotation of different bacteriophage defences in a lactococcal single strain starter culture system. Appl Environ Microbiol 59:365, 1993.

Sozzi PD. Calcium requirements of lactic starter phages. Milchwissenschaft 27:503, 1972.

Stadhouders J, Hup G, Jansen LA. A study of optimum conditions of freezing and storing concentrated mesophilic starters. Neth Milk Dairy J 25:229, 1971.

Stiles ME. Bacteriocins produced by *Leuconostoc* species. J Dairy Sci 77:2718, 1994.

Strauss K. American-type cheeses. In: Cultures for the Manufacture of Dairy Products. Milwaukee, WI: Chr. Hansen, Inc., 1997, p 65.

Tagg JR, Dajanii AS, Wannamaker LW. Bacteriocins of gram-positive bacteria. Bacteriol Rev 40:722, 1976.

Teuber M, Lembke J. The bacteriophages of lactic acid bacteria with emphasis on genetic aspects of group N lactic streptococci. Antonie van Leewenhoek 49:283, 1983.

Thunell RK. pH-controlled starter: a decade reviewed. Cult Dairy Prod J 23(3):10, 1988.

Thunell RK. Taxonomy of the leuconostocs. J Dairy Sci 78:2514, 1995.

Toba T, Yoshioka E, Itoh T. Lacticin, a bacteriocin produced by *Lactobacillus delbrueckii* subsp. *lactis*. Lett Appl Microbiol 12:43, 1990.

Trepanier G, Simard RE, Lee BH. Lactic acid bacteria relation to accelerated maturation of cheddar cheese. J Food Sci 56:1238, 1991.

Turner KW, Martley FG, Gilles J, Morris HA. Swiss-type cheese III. The effect on cheese moisture of varying the rate of acid production by *S. thermophilus*. N Z J Dairy Sci Technol 18:125, 1983.

Ustunol Z, Hicks CL. Use of an enzyme-treated, whey-based medium to reduce culture agglutination. J Dairy Sci 77:1479, 1994.

Vakil JR, Chandan RC, Parry RM, Shahani KM. Susceptibility of several microorganisms to milk lysozymes. J Dairy Sci 52:1192, 1969.

Valladao M, Sandine WE. Standardized method for determining the effect of various antibiotics on lactococcal cultures. J Food Prot 57:235, 1994a.

Valladao M, Sandine WE. Quaternary ammonium compounds in milk: detection by reverse phase high performance liquid chromatography and their effect on starter growth. J Dairy Sci 77:1509, 1994b.

Vanenbergh PA. Lactic acid bacteria: their metabolic products and interference with microbial growth. FEMS Microbiol Rev 12:221, 1993.

Vedamuthu ER. The dairy leuconostocs: use in dairy products. J Dairy Sci 77:2725, 1994.

Villani F, Pepe O, Mauriello G, Salzano G, Moschetti G, Coppola S. Antilisterial activity of thermophilin 347, a bacteriocin produced by *Streptococcus thermophilus*. Int J Food Microbiol 25:179, 1995.

Visser S, Hop G, Exterkate FA, Stadhouders J. Bitter flavour in cheese. 2. Model studies on the formation and degradation of bitter peptides by proteolytic enzymes from calf rennet, starter cells, and starter cell fractions. Neth Milk Dairy J 37:169, 1983.

Ward DJ. Characterization of a bacteriocin produced by *Streptococcus thermophilus* ST134. Appl Microbiol Biotechnol 43:330, 1995.

Whitehead HR, Cox GA. The occurrence of bacteriophage in cultures of lactic streptococci. N Z J Dairy Sci Technol 16:319–320, 1935.

Whitehead WE, Ayres JW, Sandine WE. A review of starter media for cheese making. J Dairy Sci 76:2344, 1993.

Wolfson LM, Sumner SS. Antibacterial activity of the lactoperoxidase system: A review. J Food Prot 56:887, 1993.

Yang NL, Sandine WE. Acid producing activity of lyophilized streptococcal cheese starter concentrates. J Dairy Sci 62:908, 1979.

Yun JJ, Barbano DM, Keily L, Kindstedt PS. Mozzarella cheese: impact of rod:coccus ratio on composition, proteolysis, and functional properties. J Dairy Sci 78:751, 1995.

7
Genetics and Metabolism of Starter Cultures

JAMES L. STEELE
University of Wisconsin—Madison, Madison, Wisconsin

I. INTRODUCTION

Starter cultures are the microorganisms intentionally added to initiate a fermentation. In dairy fermentations, these organisms include lactic acid bacteria, propionibacteria, surface ripening bacteria, yeasts, and molds (see Chap. 6). Lactic acid bacteria are responsible for the acidification that occurs during the manufacture of essentially all fermented dairy products. Because of the central role that these organisms play, this chapter primarily focuses on the genetics and physiological properties of lactic acid bacteria. A section on propionic acid fermentation by propionibacteria is also included because of the importance of this fermentation in the production of Swiss-type cheeses.

Although there is no one definition of lactic acid bacteria, they are generally considered a diverse group of gram-positive, nonsporulating rods or cocci, incapable of respiration, having complex nutritional requirements, and whose major metabolic end product is lactic acid. They are used in a wide variety of food fermentations including vegetable, meat, and milk fermentation. In general, they produce lactic acid from carbohydrates present in the raw material, resulting in a reduction in the pH, thereby yielding protection against spoilage. Their metabolic end products and enzymes also play a significant role in determining the texture and flavor of the final product. In dairy fermentations, the genera of lactic acid bacteria used as starter cultures include *Lactobacillus, Lactococcus, Leuconostoc*, and *Streptococcus*. Additionally, species from the genus *Pediococcus* can be isolated from some varieties of ripening cheese.

Over the past 15 years, the genetics and physiological properties of lactic acid bacteria have received considerable research attention. This has been particularly true for the genera used in dairy fermentations. The impetus for this attention has been the belief that, as a greater understanding of the basic biological properties of these organisms is developed, so will be the ability to isolate or construct strains of lactic acid bacteria with enhanced industrial utility. For a comprehensive review on the physiological properties and genetics of lactic acid bacteria, see Gasson and de Vos (1994) and Salminen and von Wright (1993).

II. GENETICS OF LACTIC ACID BACTERIA

Research on the genetics of lactic acid bacteria began in the early 1970s and initially focused on plasmids and natural gene transfer systems in lactococci. Interest in genetics of lactic acid bacteria has expanded rapidly and there are currently hundreds of researchers active in this area. This interest has resulted in development of a relatively detailed understanding of the basic genetics of these bacteria and their gene transfer systems, and development of tools required for application of recombinant DNA techniques. For a more complete description of the genetics of lactic acid bacteria, see Chassy and Murphy (1993), de Vos and Simons (1994), Gasson and Fitzgerald (1994), Mercenier et al. (1994), and von Wright and Sibakov (1993).

A. Genetic Elements

Genetic elements of lactic acid bacteria that have been characterized include chromosomes, transposable elements, and plasmids. Chromosomes of lactic acid bacteria are relatively small compared to those of other eubacteria, ranging from 1.1 to 2.6 megabase pairs (Bourgeois et al., 1993). Several chromosomal genetic and physical maps have been constructed for this group of bacteria. These maps have revealed the genetic organization of lactic acid bacteria and will be useful in future genetic characterization. Transposable elements, genetic elements capable of moving as discrete units from one site to another in the genome, have been described for lactic acid bacteria (Gasson and Fitzgerald, 1994). Insertion sequences, the simplest of transposable elements, are widely distributed in bacteria and have also been identified in numerous lactic acid bacteria. Their ability to mediate molecular rearrangements and affect gene regulation has had both positive and negative implications for dairy product fermentations. In one instance, incorporation of an insertion sequence into a prophage of *Lactobacillus casei* resulted in conversion of a temperate bacteriophage into a virulent bacteriophage. Alternatively, insertion sequence-mediated cointegration has played a pivotal role in dissemination of many beneficial characteristics via conjugation. More complex transposable elements, such as self-transmissible conjugal transposons, which encode for production of the bacteriocin nisin, the ability to metabolize sucrose, and a bacteriophage defense mechanism, have also been described in detail (Gasson and Fitzgerald, 1994). Plasmids, that is, autonomous replicating extrachromosomal circular DNA molecules, have been identified in several lactic acid bacteria. These are of particular importance in lactococci, in which they encode numerous characteristics essential for dairy product fermentations, including lactose metabolism, proteinase activity, oligopeptide transport, bacteriophage-resistance mechanisms, bacteriocin production and immunity, bacteriocin resistance, exopolysaccharide production, and citrate utilization (von Wright and Sibakov, 1993).

An indication of the extent of research activity in the molecular biology of lactic acid bacteria is that there have been 893 nucleotide sequences from this group submitted to the GenBank Genetic Sequence Data Bank, most of which are from lactobacilli and lactococci. Research has resulted in identification of sequences responsible for initiation and termination of transcription and initiation of translation in lactobacilli and lactococci. The cumulative knowledge gained by characterization of these genes, their regulatory signals, and their excretion mechanisms has set the stage for use of molecular biological techniques to construct strains of lactic acid bacteria that express proteins of interest in a controlled fashion.

B. Gene Transfer Mechanisms

1. Transduction

Transduction, the transfer of bacterial genetic material by a bacteriophage, has been demonstrated with lactococci, lactobacilli, and *Streptococcus thermophilus*. Transduction was used to demonstrate that lactose metabolism and proteinase activity are plasmid encoded in lactococci. However, usefulness of transduction in construction of strains for industry is limited because of the relatively narrow host range of transducing bacteriophages.

2. Conjugation

Conjugation, the transfer of genetic material from one bacterial cell to another, which requires cell-to-cell contact, has been characterized in detail in lactococci. Most plasmid-encoded characteristics important in the manufacture of fermented dairy products can be transferred by conjugation. Utilizing an approach that does not require antibiotic-resistance markers, conjugation has been used to transfer plasmids, which encode bacteriophage-resistance genes, into commercial lactococcal strains (Sanders et al., 1986). These strains have enhanced resistance to infection by bacteriophages and have been used successfully in the dairy industry for several years. Conjugation is likely to be used in the future to construct other lactococcal strains with enhanced industrial utility.

3. Transformation

Transformation, the introduction of free DNA into a bacterial cell, was the greatest obstacle to application of recombinant DNA techniques to lactic acid bacteria. Protoplast transformation was the first method successfully used in lactic acid bacteria. However, this technique is highly strain specific and yields inconsistent results. Development of electrotransformation procedures has provided an approach that has largely overcome difficulties associated with protoplast transformation. Electrotransformation is accomplished by exposing the bacterial cell to a high-voltage electric field pulse, which causes transient pores in the cytoplasmic membrane. Free DNA enters the cell through these transient pores (Chassy et al., 1988). Electrotransformation has been successfully applied to all genera of lactic acid bacteria.

C. Genetics of Bacteriophages that Infect Lactic Acid Bacteria

Significant progress has been made on characterization of bacteriophages from lactic acid bacteria. All of the bacteriophages examined contain double-stranded linear DNA genomes with either cohesive or circularly permuted terminally redundant ends. Both lytic and temperate bacteriophages have been characterized. A major outcome of this characterization has been a clear classification, based on DNA homology and morphological studies, of lactococcal bacteriophages (Jarvis et al., 1991). Twelve species have been described and type phages have been proposed for each species. Less comprehensive DNA homological data are also available for bacteriophages of *Lactobacillus, S. thermophilus*, and *Leuconostoc* species. Additionally, complete nucleotide sequence information is available for several lactococcal bacteriophages.

   This has made it possible to determine how bacteriophage and host interactions evolve over time and to construct novel bacteriophage defense mechanisms.

Two mechanisms have been identified by which bacteriophages have evolved resistance to restriction/modification systems. Characterization of lactococcal bacteriophages has revealed that they contain far fewer restriction endonuclease sites than expected for genomes of their size, suggesting that evolutionary pressure has selected for bacteriophages with few restriction endonuclease sites. This view is supported by the observation that lactococcal bacteriophages that have recently emerged in the dairy industry are more sensitive to restriction/modification systems. Characterization of bacteriophages that have evolved to overcome a specific restriction/modification system has revealed that they have acquired a functional copy of the modification enzyme from that system. Some bacteriophages can overcome abortive infection mechanisms. These phages have either a relatively small change in their genome or have acquired a piece of chromosomal DNA from their host. In those instances in which bacteriophages have acquired a piece of chromosomal DNA, it is believed that they have acquired a new origin of replication. These examples illustrate that bacteriophage–host interactions are continually changing, with bacterial strains acquiring new defense mechanisms and bacteriophages evolving mechanisms to overcome them.

Durmaz and Klaenhammer (1995) proposed using a culture system wherein derivatives of a single strain containing different abortive bacteriophage infection and restriction/modification systems are rotated. This strategy would reduce the selection for bacteriophages that have evolved resistance to a specific defense mechanism.

D. Genetic Modification of Lactic Acid Bacteria Using Recombinant DNA Techniques

Approaches using recombinant DNA techniques have led to most of the recent advances in understanding the physiological characteristics of lactic acid bacteria; these are being used to construct strains with unique industrially significant characteristics. The advantage of using recombinant DNA approaches for studying physiology is that strains can be constructed that differ in a single defined genetic alteration, such as inactivation of a specific gene. By comparing the wild-type culture to its isogenic derivative, the role of that gene in the phenotype being examined can be unequivocally determined. This general approach has resulted in a detailed understanding of how these bacteria use lactose, obtain essential amino acids, produce diacetyl, and resist bacteriophage infection.

Use of recombinant DNA approaches holds great promise for the construction of dairy starter cultures with enhanced industrial utility. The inactivation or overexpression of genes in a metabolic pathway can be used to alter end products that accumulate from a given pathway. One example of such an approach is described for citrate metabolism (see Sec. IV). Regulated promoters can be used to express genes in a controlled manner (Kok, 1996). One application of such promoters is the controlled expression of lytic genes resulting in autolysis of the starter culture. This would result in rapid release of enzymes (i.e., peptidases) into the cheese matrix and potentially accelerate cheese flavor development. Several recombinant DNA approaches have been applied to constructing starter cultures with novel bacteriophage-resistance mechanisms (Dinsmore and Klaenhammer, 1995). These include high copy presentation of native phage defense systems (restriction/modification systems and abortive infection systems) (see Chapter 6), antisense RNA, cloning of bacteriophage origins of replication on plasmids, and construction of a phage-inducible suicide system.

Of these approaches, the phage-inducible suicide system appears to have the greatest potential for the construction of starter cultures with enhanced resistance to infection by bacteriophage. The system consists of a phage-inducible promoter that activates a lethal gene after bacteriophage infection (Djordjevic et al., 1996). The system presents a genetic trap, wherein the objective is to invite phage infection and then trigger a lethal device that destroys the host bacterium and phage, before progeny phages are replicated within the culture. In the single construct described to date, the phage-inducible promoter was isolated from a bacteriophage, and a restriction endonuclease cassette was used as the lethal gene of the suicide system.

III. CARBOHYDRATE METABOLISM

Lactic acid bacteria, as nonrespiring organisms, principally generate adenosine triphosphate (ATP) by fermentation of carbohydrates coupled to substrate-level phosphorylation. The two major pathways for metabolism of hexoses (i.e., glucose) in these organisms are the homofermentative (Embden-Meyerhof) and heterofermentative (phosphoketolase) pathways. The homofermentative pathway (Fig. 1) theoretically yields 2 moles of lactate and 2 ATPs per hexose. This pathway used by all lactic acid bacteria except *Leuconostoc* spp. and hetero-fermentative lactobacilli. These organisms use the heterofermentative pathway (Fig. 2), which theoretically yields 1 ATP and 1 mole each of lactate, ethanol, and CO_2 per hexose. Differences in metabolic end products have a significant impact on the final product. For example, production of CO_2 can result in formation of holes (eyes) within the cheese and, depending on which cheese variety is being produced, this can be either a quality attribute or a defect. Additionally, production of ethanol, via esterification with short-chain fatty acids, can result in fruity flavor defects in cheddar cheeses. For more detailed descriptions of carbohydrate fermentation and energy transduction in lactic acid bacteria, see Axelsson (1993), Monnet et al. (1996), and Poolman (1993).

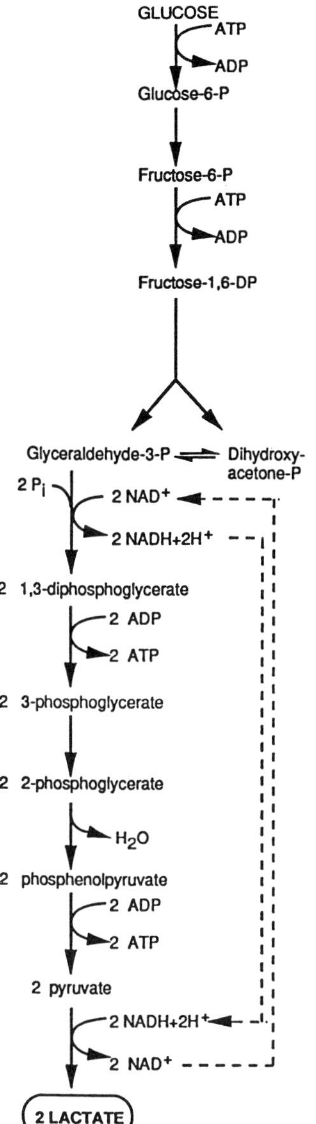

FIGURE 1 The homofermentative (Embden-Meyerhof) pathway.

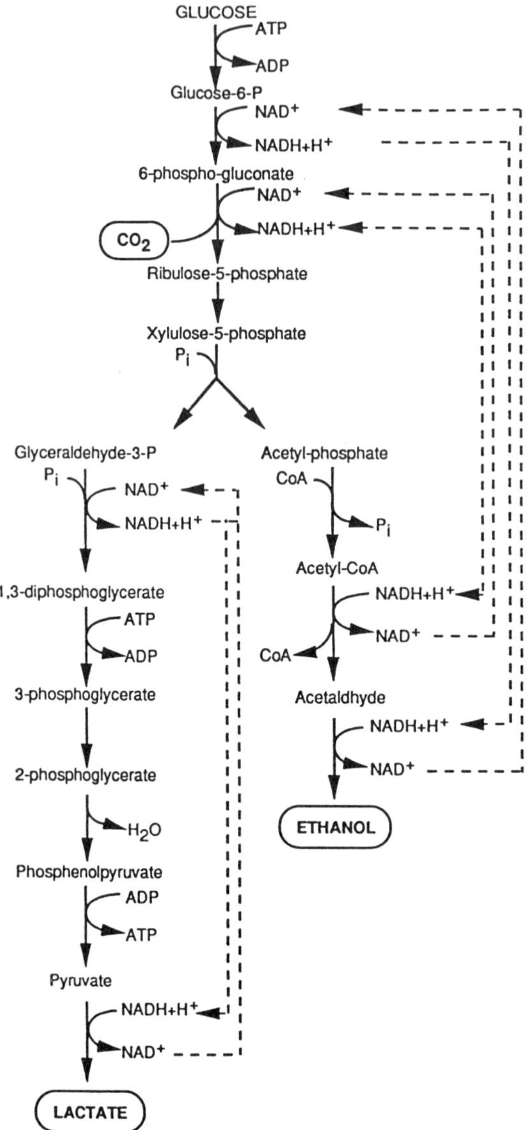

FIGURE 2 The heterofermentative (phosphoketolase) pathway.

A. Lactose Metabolism

Lactose, a disaccharide composed of glucose and galactose, is the principal carbohydrate present in milk (45–50 g/L). Therefore, lactose metabolism is primarily responsible for production of energy by lactic acid bacteria growing in milk. The pathway used to metabolize lactose differs among lactic acid bacteria. These pathways can be divided into two groups based on whether the lactose is transported by a phosphoenolpyruvate (PEP)-dependent phosphotransferase system (PTS) or a lactose permease.

1. Organisms that Transport Lactose via a PEP-Dependent PTS

Lactococcus lactis and *Lb. casei* translocate lactose via a PEP-dependent PTS. In these organisms, lactose enters the cytoplasm as lactose-phosphate and is hydrolyzed by phospho-β-galactosidase to yield galactose-6-phosphate and glucose. Galactose-6-phosphate is metabolized by the tagatose-6-phosphate pathway (Fig. 3) and the glucose moiety is metabolized by the Embden-Meyerhof

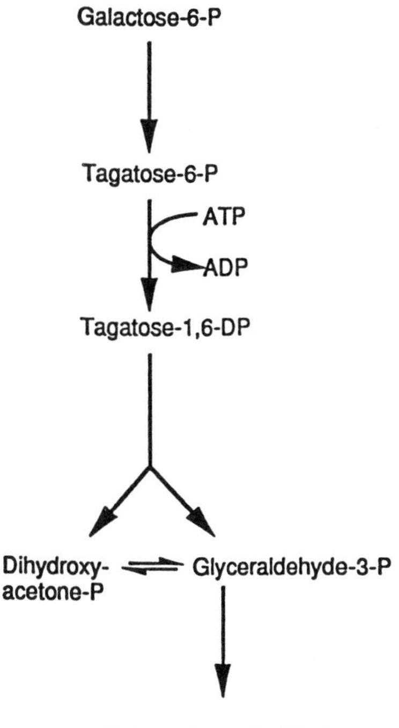

FIGURE 3   The tagatose-6-phosphate pathway.

pathway (see Fig. 1). In both pathways, an aldolase cleaves a diphosphate intermediate compound to produce the triose phosphate—dihydroxyacetone phosphate and glyceraldehyde phosphate. These triose pathways are converted to pyruvate at the expense of NAD^+. Regeneration of NAD^+ is usually accomplished by reducing pyruvate to lactate. Two enzymes, L-lactate dehydrogenase (L-LDH) and D-lactate dehydrogenase (D-LDH), are responsible for conversion of pyruvate to L-lactate and D-lactate, respectively. Lactococci produce only L-lactate whereas *Lb. casei* forms either only L-lactate or both L- and D-lactate, depending on the subspecies.

The efflux of lactate ions and protons via a symport mechanism provides a secondary mechanism by which energy can be generated. This efflux results in formation of a proton gradient across the membrane which can be used for transport of nutrients or to synthesize ATP via a reversible ATPase present in the cytoplasmic membrane.

To ensure the availability of PEP for use in translocating lactose and other carbohydrates into the cytoplasm, the intracellular PEP pool (PEP, 2-phosphoglycerate, and 3-phosphoglycerate) is regulated. This is accomplished by regulation of pyruvate kinase activity, the enzyme that catalyzes the conversion of PEP to pyruvate, by inorganic phosphate (P_i) and fructose, 1.6 diphosphate (FDP), an intermediate compound in the Embden-Meyerhof pathway. These compounds allosterically regulate pyruvate kinase, with FDP being an activator and P_i being an inhibitor. When cells are rapidly fermenting lactose, the intracellular concentration of level of FDP is high and P_i level is low, resulting in an active pyruvate kinase and formation of pyruvate. However, when lactose becomes limiting, the level of FDP is greatly reduced and P_i increases resulting in inhibition of pyruvate kinase and accumulation of a PEP pool.

In the 1930s, the observation was made that the ability of lactococcal strains to metabolize lactose was an unstable trait. In the late 1970s, McKay and coworkers were able to demonstrate that the high rate of spontaneous loss of the ability to ferment lactose by lactococci resulted from the loss of plasmid DNA. Similar results have demonstrated that the ability to metabolize lactose is also plasmid encoded in strains of *Lb. casei*. Subsequently, a DNA fragment coding for lactose metabolism was cloned from lactococci and its nucleotide sequence determined.

Analysis of this nucleotide sequence revealed that the fragment encoded for lactose-specific enzymes of the PTS, enzymes of the tagatose-6-phosphate pathway, phospho-β-galactosidase, and a repressor. The genes encoding the enzymes involved in lactose metabolism are organized in an operon, and the repressor (LacR) is divergently transcribed. Binding of LacR to operators in the intercistronic region between the *lac* operon and *lacR* activates transcription of *lacR* and represses transcription of the *lac* operon. Binding of tagatose-6-phosphate by LacR results in release of LacR from the operator and consequently the lac operon is transcribed. Transcription of the *lac* operon is also decreased by presence of glucose. This is most likely caused by lower intracellular levels of cyclic adenosine monophosphate (AMP) resulting in catabolite repression, as has been described in numerous other microorganisms. Regulation at the level of transcription allows the organism to avoid synthesizing enzymes required for lactose metabolism when lactose is not present, or when a more readily used carbohydrate such as glucose is present. Additionally, this mechanism of regulation allows the organism to rapidly produce these enzymes in case lactose becomes available.

2. Organisms that Transport Lactose via a Permease

Lactobacillus delbrueckii, Lactobacillus helveticus, Leuconostoc spp., and *S. thermophilus* transport lactose without modification via a lactose permease. There are two distinct types of lactose permeases. *Lb. helveticus* nad *Leuconostoc* spp. are believed to transport lactose in symport with a proton. *S. thermophilus* and *Lb. delbrueckii* transport lactose via a lactose–galactose antiport system. Regardless of the type of permease, the intracellular lactose is hydrolyzed by β-galactosidase to yield glucose and galactose. The glucose moiety is metabolized via the Embden-Meyerhof pathway (see Fig. 1) by *S. thermophilous, Lb. delbrueckii*, and *Lb. helveticus. Lb. helveticus* metabolizes the galactose moiety via the Leloir pathway (Fig. 4). Regeneration of NAD$^+$ typically is accomplished via an LDH. *S. thermophilous, Lb. delbrueckii*, and *Lb. helveticus* produce L-lactate, D-lactate, and both L- and D-lactate, respectively.

Leuconostoc spp. convert galactose to glucose-6-phosphate via the Leloir pathway (see Fig. 4), which then enters the heterofermentative pathway (see Fig. 2). The glucose moiety also is metabolized by the heterofermentative pathway. *Leuconostoc* spp. express only D-LDH and hence only D-lactate is formed.

IV. CITRATE METABOLISM

Relatively small quantities (0.15%–0.2%) of citrate are present in milk and the ability to metabolize citrate among dairy starter cultures is limited to *Leuconostoc* spp. and biovariants of *Lc. lactis*, designated Cit$^+$ *Lc. lactis* strains or *Lc. lactis* subsp. *lactis* biovar *diacetylactis* (Monnet et al., 1996). The metabolic end products of citrate metabolism are diacetyl, acetoin, 2,3-butanediol, acetic acid, and carbon dioxide. Diacetyl is an important flavor compound in a variety of fermented milks and sour cream, and is believed to be important in cheddar cheese flavor. Acetoin and 2,3-butanediol, both flavorless compounds, are produced from diacetyl by 2,3-butanediol dehydrogenases (Crow, 1990). The carbon dioxide produced is responsible for the holes (eyes) in Gouda and Edam cheeses and the effervescent quality of buttermilk.

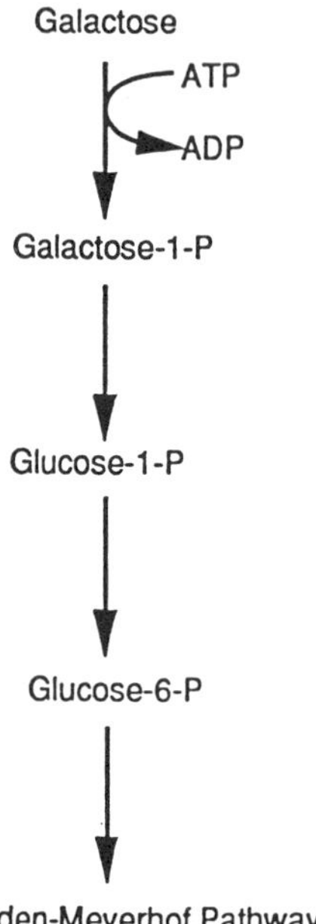

FIGURE 4 The Leloir pathway.

Citrate is transported without modification into the cell by a citrate permease, which is plasmid encoded in lactococci and *Leuconostoc* spp. (Kempler and McKay, 1981; Vaughan et al., 1995). The intracellular citrate is then hydrolyzed by citrate lyase to yield oxaloacetate and acetate; the oxaloacetate is then decarboxylated to pyruvate. Five pathways exist for the metabolism of pyruvate: lactate production via an LDH, formate and acetate or ethanol production via pyruvate-formate lyase, acetate and CO_2 production via pyruvate dehydrogenase, acetate and CO_2 production via pyruvate oxidase, and the diacetyl/acetoin pathway (Fig. 5). In the diacetyl/acetoin pathway, a pyruvate molecule is decarboxylated to acetaldehyde-thiaminpyrophosphate (TPP). α-Acetolactate synthase catalyses the condensation of acetaldehyde-TPP with another molecule of pyruvate to form α-acetolactate. It is generally recognized that diacetyl is formed by the nonenzymatic decarboxylative oxidation of α-acetolactate. However, some researchers postulate that an enzyme (diacetyl synthase) exists that is capable of forming diacetyl directly from acetaldehyde-TPP. For more detailed descriptions of citrate metabolism, see Hugenholtz, 1993, and Monnet et al., 1996.

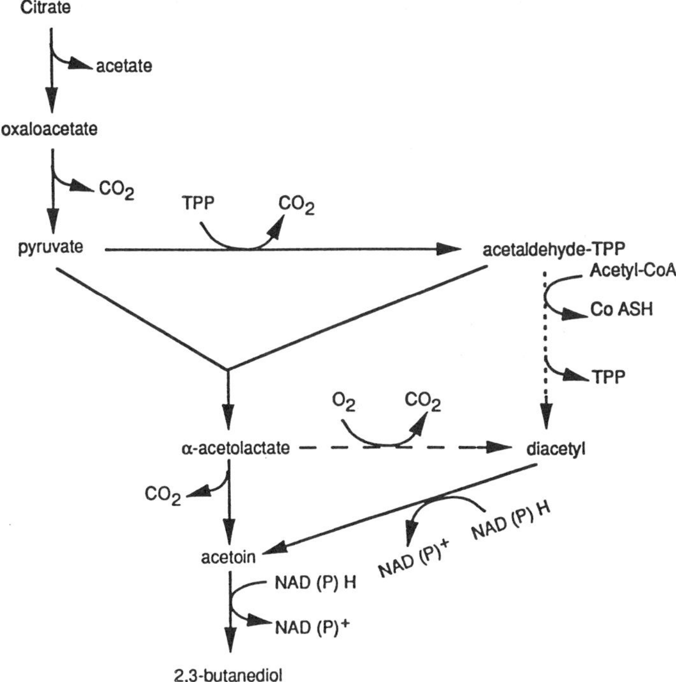

FIGURE 5   The diacetyl/acetoin pathway. The line with longer dashes indicates a nonenzymatic reaction (α-acetolactate to diacetyl). The line with shorter dashes indicates that the reaction is hypothetical (acetaldehyde-TPP to diacetyl).

Hugenholtz (1993) described strategies for use of genetic engineering to construct strains of lactococci that produce elevated levels of diacetyl. These include inactivation of genes coding for enzymes involved in the metabolism of pyruvate that do not lead to diacetyl (i.e., LDH), increasing the activity of α-acetolactate synthase and inactivating the gene coding for the enzyme (α-acetolactate decarboxylase), which converts α-acetolactate to acetoin. The intent of these manipulations is to increase the quantity of pyruvate that enters the diacetyl/acetoin pathway and to eliminate branches in the pathway that do not lead to diacetyl.

V. PROPIONIC ACID FERMENTATION

Propionibacterium freudenreichii subsp. *shermanii* is a component of the starter culture used to manufacture Swiss-type cheeses (see Chap. 6). The maturation process of Swiss-type cheeses includes holding the cheese at 7°C for a few days and then elevating the ripening temperature to 20°C to 26°C (see Chap. 9). This increase in temperature is required to permit growth of *P. freudenreichii* subsp. *shermanii*. This organism metabolizes lactate produced by fermentation of lactose by other components of the starter culture, *S. thermophilus* and *Lb. helveticus*. The metabolism of lactate by propionibacteria results in formation of propionate, acetate, and CO_2; the fermentation is typically described by the following equation (Frank and Marth, 1988):

$$3 \text{ lactate} \rightarrow 2 \text{ propionate} + 1 \text{ acetate} + 1 \text{ } CO_2$$

The metabolic pathway used by propionibacteria in the fermentation of lactate is complex and involves two connected cycles (Fig. 6), with the organism generating 1 mole of ATP per mole of lactate.

Fermentation of lactate by propionibacteria in the cheese matrix is more complicated than that described in Figure 6 and summarized in the equation given previously. Crow (1986) demonstrated that metabolism of aspartate to succinate

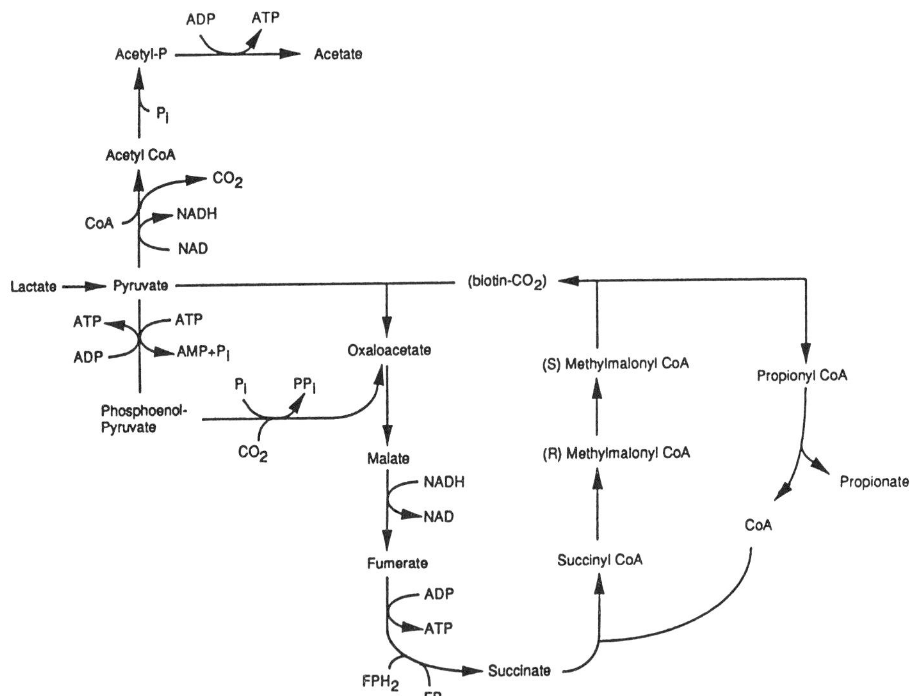

FIGURE 6 Metabolic pathway for the fermentation of lactate by propionibacteria.

by *P. freudenreichii* subsp. *shermanii* influenced the fermentation of lactate to propionate, acetate, and CO_2. The generation of reducing equivalents from aspartate metabolism results in an increase in the amount of lactate being fermented to acetate and CO_2 rather than propionate. That aspartate is catabolized in Swiss-type cheeses is supported by the presence of succinate in those cheeses and by the observation that the molar ratio of propionate to acetate is typically significantly lower than 2:1.

VI. PROTEOLYTIC SYSTEMS OF LACTIC ACID BACTERIA

Lactic acid bacteria used in dairy fermentations are amino acid auxotrophs that typically require several amino acids for growth. The quantities of free amino acids present in milk are not sufficient to support growth of these bacteria to high cell density; therefore, they require a proteolytic system capable of using the peptides present in milk and hydrolyzing milk proteins (α_{S1}-, α_{S2}-, and κ-, and β-caseins) to obtain essential amino acids. Additionally, the activity of proteolytic enzymes derived from lactic acid bacteria is necessary for development of flavor in ripened cheeses. Peptides and amino acids formed by proteolysis may impart flavor directly or serve as flavor precursors in fermented dairy products. The resulting flavors may have either a positive or negative impact on the quality of the finished product.

A. Proteolytic System of Lactococci

The proteolytic system of lactococci has been characterized in significant detail. Many of the enzymes have bene purified and characterized, and numerous genes encoding components of proteolytic systems have been sequenced. The proteolytic systems of other lactic acid bacteria are not as well characterized; however, many individual components have been examined. These components, in most instances, have homologs in the lactococcal system. Therefore, the lactococcal system is believed to serve as a general model for the proteolytic systems of other lactic acid bacteria. For extensive reviews of the proteolytic enzyme systems of lactic acid bacteria, see Kunji et al. (1996) or Kok and de Vos (1994).

A model of the lactococcal proteolytic system is presented in Figure 7. The system can be separated into three components: extracellular enzymes, transport systems, and intracellular enzymes. While growing in milk, lactic acid bacteria obtain essential amino acids in a variety of ways. They first use non-protein nitrogen sources such as free amino acids and small peptides. Casein,

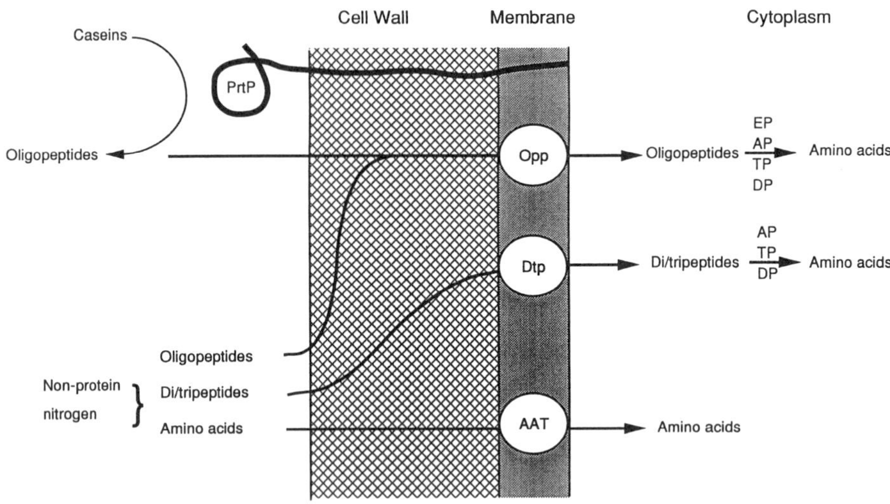

FIGURE 7 Schematic representation of the lactococcal proteolytic system. Abbreviations: PrtP, cell-envelope–associated proteinase; Opp, oligopeptide transport system: Dtp, di/tripeptide transport systems; AAT, amino acid transport systems; EP, endopeptidases; AP, aminopeptidases; TP, tripeptidases; DP, dipeptidases.

which composes 80% of all proteins present in milk, becomes the primary nitrogen source after nonprotein nitrogen is depleted.

Extensive investigations have revealed that a cell-envelope–associated proteinase, designated PrtP, is the only extracellular proteolytic enzyme present in lactococci. The enzyme is a serine-protease, which is expressed as a pre-pro-proteinase. A signal peptidase removes the signal peptide upon transport across the cytoplasmic membrane. Subsequently, a lipoprotein maturase (PrtM) is thought to cause a conformational change in the pro-proteinase, resulting in release of the pro-region via autoproteolysis, and an active PrtP. The activated enzyme remains associated with the cell because of the presence of a C-terminal membrane anchor sequence. Genes encoding lactococcal proteinases with different substrate specificities have been sequenced and the amino acid residues involved in substrate binding and catalysis determined. A critical feature of the enzyme is its broad cleavage specificity, which results in release of more than 100 oligopeptides from soluble-β-casein, 20% of which are small enough to be transported by the oligopeptide transport system (Juillard et al., 1995). Loss of PrtP, which is typically plasmid encoded in lactococcal strains, results in derivatives capable of reaching only approximately 10% of the final cell density of the parental strain, indicating that PrtP is essential for growth of lactococci to high cell density in milk.

Transport of nitrogenous compounds across the lactococcal cytoplasmic membrane takes place via group-specific amino acid transport systems, dipeptide/tripeptide transport systems, and an oligopeptide transport system (Opp). Of these systems, Opp is of greatest importance during growth of lactococci in milk. Growth studies with Opp derivatives have shown that Opp is essential for the uptake of PrtP-generated peptides from β-casein and oligopeptides present in the nonprotein nitrogen component of milk (Juillard et al., 1995). This system, which is organized in an operon, is comprised of two ATP-binding proteins, two integral membrane proteins, and a substrate-binding protein. The system has broad specificity, being able to transport at least 18 distinct peptides generated by PrtP from β-casein (Kunji et al., 1995). The system can transport peptides containing between four and eight amino acid residues, although it is possible that longer peptides may also be transported. Two distinct dipeptide/tripeptide transport systems, designated DtpT and DtpP, have been characterized. Characterization of DtpT derivatives has revealed that DtpT is not essential for lactococcal growth on the nonprotein nitrogen component of milk or casein-derived peptides. To date, the physiological role or DtpP has not been defined.

Once inside the cell, peptides are hydrolyzed by peptidases. Peptidase classes that have been identified in lactococci include expopeptidases and endo-peptidases. The greatest variety of enzymes are from the exopeptidase class, which includes aminopeptidases, tripeptidases, and dipeptidases. No carboxy-peptidases have been detected in lactococci. Peptidases identified in lactococci and their cleavage specificity is summarized in Table 1. This combination of endopeptidases, aminopeptidases, tripeptidases, and dipeptidases converts the transported peptides into free amino acids, which lactococci require for growth. Examination of the growth of strains lacking one of the peptidases (PepX, PepN, PepC, PepO, or PepT) in milk has revealed that only the PepN single mutant grows significantly slower. Characterization of lactococcal derivatives lacking more than one of these peptidases indicated that, in general, the more peptidases that were inactivated, the slower the strain grew in milk. The only possible exception to this general rule is with PepX, which does not have a significant effect on growth rate in milk. Inactivation of all five of these peptidases results in a derivative that grows more than 10-fold slower that the wild-type strain (Mierau et al., 1996). These results confirm the hypothesis that a variety of peptidases, with overlapping specificities, is required to hydrolyze the casein-derived oligo-peptides transported by Opp to liberate essential amino acids.

Proteolytic systems of other lactic acid bacteria have not been as extensively studied. However, numerous components of proteolytic enzyme systems of *Lb. helveticus* and *Lb. delbrueckii* have been characterized. In general, these

TABLE 1 Peptidases Purified and Characterized from Lactococci

Peptidase	Abbreviation	Specificity[a]
X-prolyl dipeptidyl aminopeptidase	PepX	X-Pro↓Y...
Aminopeptidase N	PepN	X↓Y-Z...
Aminopeptidase C	PepC	X↓Y-Z...
Aminopeptidase A	GAP	Asp(Glu)↓Y-Z...
Pyrrolidone carboxylyl peptidase	PCP	pGlu↓Y-Z...
Prolyl iminopeptidase	PIP	pGlu↓Y-Z...
Depeptidase	DIP	Pro↓Y-Z...
Prolidase	PRD	X↓Y
Tripeptidase	PepT	X↓Pro
Endopeptidases	PepO	X↓Y-Z
	PepF	...W-X↓Y-Z...
	LEPI	
	MEP	

[a] ↓ indicates which peptide bond is hydrolyzed.

components have homologs in the lactococcal system. However, unlike the lactococcal system, inactivation of PepX in *Lb. helveticus* does significantly reduce the growth of this bacterium in milk (Yüksel and Steele, 1996).

B. Proteolysis and Cheese Flavor Development

Even though flavor development in various types of cheese remains a poorly defined process, it is generally agreed that proteolysis is essential for flavor development in bacterial ripened cheeses (Fox et al., 1993; Visser, 1993). Proteolytic enzymes present in this group of products include chymosin, plasmin, and proteolytic enzymes from starter cultures, adjunct cultures, and nonstarter lactic acid bacteria. The specificities and relative activities of the proteolytic enzymes present in the cheese matrix determine which peptide and amino acids accumulate and, hence, how flavor develops. The use of isogenic strains, differing only in the specificity of activity of a single proteolytic enzyme, will have great utility in elucidating the role of individual enzymes in cheese flavor development.

Free amino acids and peptides in the cheese matrix can contribute to cheese flavor either directly or indirectly and with positive or negative effects. Cheese flavor development has been the subject of numerous comprehensive reviews (Fox et al., 1993; Urbach, 1995). A major negative effect of proteolytic products is bitterness, which is believed to be caused by hydrophobic peptides ranging in length from 3 to 27 amino residues (Lemieux and Simard, 1992). They can be generated either from casein directly by chymosin or via the joint action of chymosin and cell-envelope–associated proteinases from lactic acid bacteria; they can be hydrolyzed to nonbitter peptides and amino acids by peptidases of lactic acid bacteria. Therefore, accumulation of bitter peptides is dependent on the relative rates of their formation and hydrolysis. A variety of volatile compounds can be derived from catabolism of amino acids (Hemme et al., 1982). Numerous sulfur-containing compounds, particularly methanethiol, are thought to be important in cheese flavor. Formation of methanethiol by lactococci is believed to result from the action of either cystathionine β-lyase (Alting et al., 1995) or cystathionine γ-lyase (Bruinenberg et al., 1997) on methionine. Alternatively, amino acid catabolism can give rise to compounds that have a negative impact on cheese flavor. For example, catabolism of aromatic amino acids can give rise to compounds such as indole and skatole, which contribute to unclean flavors in cheese. Overall, proteolysis is believed to be essential for development of characteristic flavor compounds in bacterial ripened cheeses; however, the mechanisms by which products of proteolysis give rise to beneficial flavor compounds remains unknown. Additionally, other than bitterness, the mechanisms by which proteolysis impacts on the development of undesirable flavor compounds are unknown.

REFERENCES

Alting AC, Engels WJM, van Schalkwijk S, Exterkate FA. Purification and characterization of cystathionine β-lyase from *Lactococcus lactis* subsp. *cremoris* B78 and its possible role in flavor development in cheese. Appl Environ Microbiol 61:4037, 1995.

Axelsson LT. Lactic acid bacteria: classification and physiology. In: Salminen S, von Wright A , eds. Lactic Acid Bacteria. New York: Marcel Dekker, 1993, p. 1.

Bourgeois PL, Lautier M, Ritzenthaler P. Chromosome mapping in lactic acid bacteria. FEMS Microbiol Rev 12:109, 1993.

Bruinenberg PG, de Roo G, Limsowtin GKY. Purification and characterization of cystathionine γ-lyase from *Lactococcus lactis* subsp. *cremoris* SK11: possible role in flavor compound formation during cheese maturation. Appl Environ Microbiol 63:561, 1997.

Chassy BM, Mercenier A, Flickinger JL. Transformation of bacteria by electroporation. TIBTECH 6:303, 1988.

Crow VL. Metabolism of aspartate by *Propionibacterium freudenreichii* subsp. *shermanii*: effect on lactate fermentation. Appl Environ Microbiol 52:359, 1986.

Crow VL. Properties of 2,3-butanediol dehydrogenases from *Lactococcus lactis* subsp. *lactis* in relation to citrate fermentation. Appl Environ Microbiol 56:1656, 1990.

de Vos WM, Simons GFM. Gene cloning and expression systems in lactococci. In: Gasson MJ, de Vos WM, eds. Genetics and Biotechnology of Lactic Acid Bacteria. London: Chapman and Hall, 1994, p 52.

Dinsmore PK, Klaenhammer TR. Bacteriophage resistance in *Lactococcus*. Mol Biotechnol 4:297, 1995.

Djordjevic GM, O'Sullivan JJ, Walker SA, Conkling MA, Klaenhammer TR. A phage-triggered suicide system designed for bacteriophage defense of *Lactococcus lactis*. 5th Symposium on Lactic Acid Bacteria: Genetics, Metabolism and Applications, Abstract F26, 1996.

Durmaz E, Klaenhammer TR. A starter culture rotation strategy incorporating paired restriction/modification and abortive infection bacteriophage defenses in a single *Lactococcus lactis* strain. Appl Environ Microbiol 61:1266, 1995.

Fox PF, Law J, McSweeney PLH, Wallace J. Biochemistry of cheese ripening. In: Fox PF, ed. Cheese: Chemistry, Physics and Microbiology. Vol 1. London: Chapman and Hall, 1993, p 389.

Frank JF, Marth EH. Fermentations. In: Wong NP, Jenness R, Keeney M, Marth EH, eds. Fundamentals of Dairy Chemistry. 3rd ed. New York: Van Nostrand Reinhold, 1988, p 655.

Gasson MJ, de Vos WM. Genetics and Biotechnology of Lactic Acid Bacteria. London: Chapman and Hall, 1994.

Gasson MJ, Fitzgerald GF. Gene transfer systems and transposition. In: Gasson MJ, de Vos WM, eds. Genetics and Biotechnology of Lactic Acid Bacteria. London: Chapman and Hall, 1994, p 1.

Hemme D, Boulianne C, Métro F, Desmazeaud M-J. Microbial catabolism of amino acids during cheese ripening. Sci Aliments 2:113, 1982.

Hugenholtz J. Citrate metabolism in lactic acid bacteria. FEMS Microbiol Rev 12:165, 1993.

Jarvis AW, Fitzgerald GF, Mata M, Mercenier A, Neve H, Powell IB, Ronda C, Saxelin M, Teuber M. Species and type phages of lactococcal bacteriophages. Intervirology 32:2, 1991.

Juillard V, Le Bars D, Kunji ERS, Konings WM, Gripon J-C, Richard J. Oligopeptides are the main source of nitrogen for *Lactococcus lactis* during growth in milk. Appl Environ Microbiol 61:3024, 1995.

Kempler GM, McKay LL. Biochemistry and genetics of citrate utilization in *Streptococcus lactis* spp. *diacetylactis*. J Dairy Sci 64:1527, 1981.

Kok J. Inducible gene expression and environmentally regulated genes in lactic acid bacteria. Antonie van Leeuwenhoek 70:129, 1996.

Kunji ERS, Hagting A, De Vries CJ, Juillard V, Haandrikman AJ, Poolman B, Konings WN. Transport of β-casein–derived peptides by the oligopeptide transport system is a crucial step in the proteolytic pathway of *Lactococcus lactis*. J Biol Chem 270:1569, 1995.

Kunji ERS, Mierau I, Hagting A, Poolman B, Konings WN. The proteolytic systems of lactic acid bacteria. Antonie van Leeuwenhoek 70:187, 1996.

Lemieux L, Simard RE. Bitter flavour in dairy products. II. A review of bitter peptides from caseins: their formation, isolation and identification, structure masking and inhibition. Lait 72:335, 1992.

Mercenier A, Pouwels PH, Chassy BM. Genetic engineering of lactobacilli, leuconostocs and *Streptococcus thermophilus*. In: Gasson MJ, de Vos MD, eds. Genetics and Biotechnology of Lactic Acid Bacteria. London: Chapman and Hall, 1994, p 252.

Mierau I, Kunji ERS, Leenhouts KJ, Hellendoorn A, Haandrikman AJ, Poolman B, Konings WN, Venema G, Kok J. Multiple-peptidase mutants of *Lactococcus lactis* are severely impaired in their ability to grow in milk. J Bacteriol 178:2794, 1996.

Monnet V, Condon S, Cogan TM, Gripon JC. Metabolism of starter cultures. In: Cogan TM, Accolas J-P. Dairy Starter Cultures. New York: VCH Publishers, 1996, p 47.

Poolman B. Energy transduction in lactic acid bacteria. FEMS Microbiol Rev 12:125, 1993.

Ramos A, Jordan KN, Cogan TM, Santos H. ^{13}C Nuclear magnetic resonance studies of citrate and glucose cometabolism by *Lactococcus lactis*. Appl Environ Microbiol 60:1739, 1994.

Salminen S, von Wright A. Lactic Acid Bacteria. New York: Marcel Dekker, 1993.

Sanders ME, Leonhard PJ, Sing WD, Klaenhammer TR. Conjugal strategy for construction of fast acid-producing, bacteriophage-resistant lactic streptococci for use in dairy fermentation. Appl Environ Microbiol 52:1001, 1986.

Urbach G. Contribution of lactic acid bacteria to flavor compound formation in dairy products. Int Dairy J 5:877, 1995.

Vaughan EE, David S, Harrington A, Daly C, Fitzgerald GF, de Vos WM. Characterization of plasmid-encoded citrate permease (*citP*) genes from *Leuconostoc* species reveals high sequence conversation with the *Lactococcus lactis citP* gene. Appl Environ Microbiol 61:3172, 1995.

Visser S. Proteolytic enzymes and their relation to cheese ripening and flavor: an overview. J Dairy Sci 76:329, 1993.

von Wright A, Sibakov M. Genetic modification of lactic acid bacteria. In: Salminen S, von Wright, A, eds. Lactic Acid Bacteria. New York: Marcel Dekker, 1993, p 161.

Yüksel GÜ, Steele JL. DNA sequence analysis, expression, distribution, and physiological role of the X-prolyl dipeptidyl aminopeptidase (PepX) gene from *Lactobacillus helveticus* CNRZ32. Appl Microbiol Biotechnol 44:766, 1996.

8
Fermented Milks and Probiotics

STANLEY E. GILLILAND
Oklahoma State University, Stillwater, Oklahoma

I. INTRODUCTION

A. Reasons for a Variety of Fermented Milks

From the historical perspective, fermented milks, as they are currently known, probably originated from the manner in which raw milk was stored in times before development of mechanical refrigeration. Storage of milk under certain conditions would have favored growth and action of lactic acid bacteria, which converted the milk into a new form having different organoleptic characteristics than the original milk. In addition to the milk's having a pleasing taste and flavor, it was likely observed that the fermented product did not spoil (i.e., develop undesirable flavor) and thus the product was preserved. The selection of storage conditions that favored the desired fermentation was most likely discovered accidentally or through trial and error, all before bacteria were even identified as being involved in such conversions of milk into the new products.

The fermentation of milk in the new products not only added variety to the diets but also resulted in foods with improved keeping quality. Thus, one of the main benefits of the early milk fermentations was to preserve the milk.

B. Health Benefits

Since the time of Eli Metchnikoff, much attention has been focused on the possible health benefits that can be derived from certain fermented milk products. Metchnikoff (1908) theorized that people should consume milk fermented with lactobacilli to exert control over the microbial activity in the intestinal tract. He theorized that some microorganisms that grew there produce materials toxic to the human. However, when consumers drank milk that had been fermented with the lactobacilli, these undesirable microorganisms were inhibited, thus improving the health of the consumer. Since then, there has been much interest and research activity focused on the potential health and nutritional benefits of many of the lactic acid bacteria.

C. Nomenclature of Starter Culture Bacteria

The nomenclature of the bacteria used as starter cultures for fermented milk products has changed drastically in recent years. Table 1 summarizes some of these changes.

II. FERMENTED MILK AND CREAM PRODUCTS

A. Buttermilk

1. Characteristics of the Product

Buttermilk is an acid-tasting fluid product having a smooth body and texture with diacetyl as a major aroma component (Kosikowski, 1977). The overall flavor of this product results from lactic acid, diacetyl, acetic acid, and some carbon dioxide. The flavor of the product is enhanced by addition of a small amount (usually ≤ 1%) of sodium chloride. The lactic acid results primarily from fermentation of lactose in the milk by *Lactococcus lactis* subsp. *lactis* and the diacetyl primarily from the action of leuconostocs, often referred to as citric acid fermenters, on citrate.

Cultured buttermilk is manufactured from pasteurized skim or low-fat milk. Often, it is supplemented with dried skim milk solids before pasteurization. This is to improve the body and texture of the product.

2. Starter Culture and Its Functions

The starter culture traditionally used to manufacture cultured buttermilk is composed of two species. *Lc. lactis* subsp. *lactis* and *Leuconostoc mesenteroides* subsp. *cremoris*. Thus, it is considered a mixed species starter culture. Once inoculated, the product is incubated at 22°C until a titratable acidity of 0.9% is reached (final acidity may vary depending on the tartness desired). Incubation at

TABLE 1 Changing Names of Dairy Starter Culture Bacteria

Old name	1986 Bergey's Manual[a]	More recent name
Streptococcus lactis	*Streptococcus lactis* subsp. *lactis*	*Lactococcus lactis* subsp. *lactis*
Streptococcus diacetilactis	*Streptococcus lactis* subsp. *diacetilactis*	*Lactococcus lactis* subsp. *lactis* biovar *diacetylactis*
Streptococcus thermophilus	*Streptococcus thermophilus*	*Streptococcus salivarius* subsp. *thermophilus*
Leuconostoc cremoris[b]	*Leuconostoc mesenteroides* subsp. *cremoris*	
Lactobacillus bulgaricus	*Lactobacillus delbrueckii* subsp. *bulgaricus*	

[a]Sharpe, 1986
[b]Older literature may use *Leuconostoc citrovorum* or *Streptococcus citrovorus*

this temperature is necessary so that both species in the starter culture can carry out their function in producing the desirable characteristics of cultured butter-milk. Incubation at a higher temperature would favor growth of *Lc. lactis* subsp. *lactis*. This would result in excess acid production and minimize the flavor production by the *Leu. mesenteroides* subsp. *cremoris*.

Neither of these two species of bacteria alone can produce cultured butter-milk with its typical flavor. The *Lc. lactis* produces primarily lactic acid as an end product of the fermentation of lactose. This bacterium also produces some acetal-dehyde, which can contribute to the flavor; overproduction of acetaldehyde can result in an undesirable flavor (sometimes described as a "green" flavor). The *Leu. mesenteroides* subsp. *cremoris* produces the major aroma component of cultured buttermilk, diacetyl, and produces small amounts of acetic acid and carbon dioxide, which contribute to the overall flavor of the product. The leuconostocs produce very little total acid. Both species in the starter culture grow well in 22°C, but, as sufficient acid is produced by *Lc. lactis* subsp. *lactis* to lower the pH to about 5, the leuconostocs stop growing. However, it is at a pH below 5 that the leuconostocs produce diacetyl. Diacetyl production results from fermen-tation of citrate, thus the name "citric acid fermenter" for the leuconostocs. The final pH of cultured buttermilk in most instances is approximately 4.4, thus the lactococci continue growing beyond the point at which the leuconostocs stop growing. The body and texture of the product primarily results from coagulation of casein in the milk by the lactic acid produced by the lactococci. The flavor is produced by the leuconostocs, which cannot happen unless the acid is present to lower the pH sufficiently for diacetyl synthesis to occur.

Lc. lactis subsp. *lactis* biovar *diacetylactis* is an organism that produces both lactic acid and diacetyl. It could be used as a starter culture for cultured butter-milk; however, it has not been well accepted because the quality of the resulting product has not been as good as the one prepared using the mixture of *Lc. lactis* subsp. *lactis* and *Leu. mesenteroides* subsp. *cremoris*.

B. Sour Cream

1. Characteristics of the Product

Sour cream normally contains not less than 18% butterfat (Kosikowski, 1977). The cream is pasteurized and homogenized before inoculation with a starter culture. The product is then incubated at 22°C until the titratable acidity reaches at least 0.5%. This results in a product having a gel-like structure, which is considerably more viscous than cultured buttermilk. In the modern market place, several products are marketed as low-fat sour cream. These products con-tain much lower levels of butterfat and yet, through the addition of suitable stabilizers, it is possible to produce a cultured product having body and texture characteristics closely resembling those of traditional sour cream. Whereas the body and texture of sour cream is much more viscous than that of cultured buttermilk, the desired flavor in sour cream is much the same as that in cultured buttermilk. It is difficult to compare the flavors directly because the two product differ tremendously in fat content and viscosity.

2. Starter Culture and Its Functions

The starter culture traditionally used for sour cream manufacture is the same as the one used to make cultured buttermilk. The functions of the two species in the starter culture are much the same as in cultured buttermilk. It is possible to produce sour cream using only *Lc. lactis* subsp. *lactis*; however, the flavor of the product would not be as pleasing because the major flavor component, diacetyl, would be missing.

C. Yogurt

1. Characteristics of the Product

Yogurt is manufactured from pasteurized milk having the desired percentage of butterfat, which may range from 0% to 3.2% (Rasic and Kurman, 1983). If the product contains butterfat, it normally is homogenized before being cultured to ensure that cream does not rise to the surface during the fermentation process. In most instances, the yogurt mix is fortified with nonfat dried milk solids before pasteurization to improve the body and texture of the product. In some instances, stabilizers are added to enhance the water-holding capacity of the product so that a firmer body and texture may be achieved. However, if the product is prepared properly, addition of stabilizer should not be necessary to produce yogurt having the traditional body and texture. The pasteurization time and temperatures used for the yogurt mix normally are higher than those used for fluid milk. Yogurt mix normally is pasteurized at a temperature of 85°C to 91°C (185°F–195°F) for 30 minutes. This helps ensure interaction of the whey protein with casein to increase water-holding capacity, thereby improving the custardlike consistency of the gel in the finished yogurt. It also results in less wheying off of the product, thus improving the appearance.

In the United States, most yogurt is made with added fruit flavoring. The acid or tart taste of the yogurt enhances the fruit flavor. In addition to the traditional custardlike yogurt product, there are yogurt drinks, particularly in Europe, that have flavor that resembles that of custard-type yogurt. Yogurt might be characterized as having an acid or tart taste. The typical flavor of yogurt is somewhat dependent on acetaldehyde produced by the starter culture.

2. Starter Culture and Its Functions

The typical or traditional starter culture used to manufacture yogurt is a mixture *Streptococcus thermophilus* (*S. salivarius* subsp. *thermophilus*) and *Lactobacillus delbrueckii* subsp. *bulgaricus*. Some yogurt makers think that the best starter culture for the process is one containing a 1:1 ratio of rods and cocci; however, the optimal ratio is likely to depend on the characteristics of the individual strain of each species involved in the starter. It has been said, however, that excess numbers of *Lb. delbrueckii* subsp. *bulgaricus* can result in a sharper flavor of the yogurt, a characteristic that is not desirable in the United States, where most consumers prefer a product that is not quite as tart as that encountered in Europe.

The two species of bacteria in the starter culture work together to produce the necessary acidity and flavor for yogurt. Oftentimes, one culture produces substances that stimulate the growth of the other. For example, *S. thermophilus* can produce small amounts of formic acid, which, in some instances, stimulates the growth of *Lb. delbrueckii* subsp. *bulgaricus* (Bottazzi et al., 1971). *Lb. delbrueckii* subsp. *bulgaricus*, on the other hand, may stimulate growth of *S. thermophilus* through proteolytic action on milk proteins. The major end product of metabolism necessary for manufacture of yogurt is lactic acid resulting from the fermentation of lactose by both of the starter culture organisms.

New cultured yogurt products also may include cultures of *Lactobacillus acidophilus* or *Bifidobacterium* species. This provides consumers with a source of microorganisms thought to have health and nutritional benefits through their growth and action in the intestinal tract. The traditional yogurt starter cultures (i.e., *S. thermophilus* and *Lb. delbrueckii* subsp. *bulgaricus*) do not grow in the intestinal tract. They may provide some benefits, however, particularly in improving the utilization of lactose in persons who are lactose maldigestors. The topic of potential health and nutritional benefits derived from bacteria involved in fermented dairy products is discussed later in this chapter.

D. Acidophilus Milk

1. Characteristics of the Product

Fermented acidophilus milk is a coagulated milk product having a strong acid taste combined with a severe cooked flavor (Kosikowski, 1977). Milk of the desired percent fat is heated at extreme temperatures to enhance the ability of *Lb. acidophilus* to grow. Because of the severe cooked flavor and the very tart taste, this product is sometimes referred to as "aufuldophilus" rather than acidophilus. Fermented acidophilus milk is no longer available in most parts of the United States. In the past, it was sold for its "therapeutic properties."

2. Starter Culture and Its Functions

Fermented acidophilus milk is cultured with *Lb. acidophilus*. The primary function of this organism is to produce lactic acid from lactose. This lactic acid combined with the intense cooked flavor from the pasteurization process results in the rather undesirable flavor of the product. In addition to having a role in the manufacture of fermented milk, *Lb. acidophilus* also has the potential of providing several health or nutritional benefits for consumers (Gilliland, 1989). One characteristic that it possesses, which most of the other starter cultures lack, is the ability to grow in the presence of bile (Albus, 1928), which enables it to establish or grow in the intestinal tract.

E. Koumiss

1. Characteristics of the Product

Koumiss is a product resulting from an acidic and alcoholic fermentation of milk (Kosikowski, 1977). It is traditionally made from mares' milk. Because mares' milk contains much less protein than cows' milk, the protein precipitates without coagulating the product. Alternatively, the product can be made from cows' milk supplemented with liquid whey to reduce the amount of protein that precipitates, thus leaving a fluid product. The product has been characterized as being a frothy drink with a distinct flavor. The frothy characteristic is indicative of gas production during its manufacture.

2. Starter Culture and Its Function

Traditionally, koumiss is prepared using a starter culture containing *Lb. delbrueckii* subsp. *bulgaricus* and a lactose-fermenting yeast. The lactobacilli are responsible for acid production and the yeast is responsible for production of ethanol and CO_2 required for the frothy appearance of the product. Koumiss has been primarily used in Russia for therapeutic purposes. Some starter cultures may include *Lb. acidophilus, Lb. delbrueckii* subsp. *bulgaricus*, and *Saccharomyces lactis*.

F. Kefir

1. Characteristics of the Product

Kefir has been described as having a sour cream–like consistency with a distinct flavor (Kosikowski, 1977). In most instances, however, it does not have as much viscosity as sour cream. It is traditionally made from pasteurized whole milk that is inoculated with kefir grains. Following incubation, the kefir grains are filtered out of the product and used to inoculate a new batch of milk. The distinct flavor from the product results from the acid and alcohol fermentation.

2. Starter Culture and Its Function

The starter culture for kefir manufacture is kefir grains, which resemble cauliflower florets (1–2 mm to 3–6 cm in diameter). The kefir grains are complex mixtures of polysaccharides of microbial origin containing streptococci, leuconostocs, lactobacilli, yeast, and often acetic acid bacteria. The composition of the microbial flora of the kefir grains may vary depending on protocols used for production of the product. Some "kefir-like" products may be manufactured using a starter culture containing a mixture of the group of microorganisms contained in kefir grains without actually using the grains.

III. PROBIOTICS

A. Background

Eli Metchnikoff (1908) theorized that *Lb. delbrueckii* subsp. *bulgaricus* could grow in the intestinal tract of humans and displace any putrefying bacteria that could grow there. Displacement of this group would reduce production of toxic compounds that adversely affects the human body, thus enabling humans to live longer. Research done since Metchnikoff's period has shown that *Lb. delbrueckii* subsp. *bulgaricus* neither survives nor establishes itself in the gastrointestinal tract. Other species of lactobacilli have been reported to provide some beneficial effects through growth and action in the gastrointestinal tract. This group of bacteria and others comprise the group referred to as probiotics. Cultures most often mentioned as probiotics for humans include *Lb. acidophilus, Lactobacillus casei*, and *Bifidobacterium* species. These all can survive and grow in the intestinal tract. Thus, through their growth or action in the intestinal tract, they have the potential to provide benefits.

Cultures normally used as starter cultures for some fermented milk products, such as *Lb. delbrueckii* subsp. *bulgaricus* and *S. thermophilus* used to manufacture yogurt, also may provide benefits, but not through the ability to survive and grow in the intestinal tract. Their function comes primarily from providing a source of enzymes necessary for improving the digestion of nutrients in the gut. For example, β-galactosidase is needed for the initial breakdown of lactose in the small intestine (Gilliland and Kim, 1984).

Whereas there are reports that indicate that the nutritional value of milk can be improved by some fermentations (Hargrove and Alford, 1980), this chapter focuses on the potential health and nutritional benefits that rely on the growth or action of probiotic bacteria following their ingestion. Several benefits are possible from probiotic cultures, including control of intestinal infections, control of serum cholesterol levels, beneficial influences on the immune system, improvement of lactose utilization in persons who are classified as lactose maldigestors, and anticarcinogenic action. Research is continuing in each of these areas to provide definite scientific evidence that could permit specific health claims for dairy products containing one or more of this group of probiotic organisms. Several publications have focused on these in more detail. Foods containing such microorganisms may be promoted as functional foods in the future.

B. Potential Benefits

1. Control of Growth of Undesirable Organisms in the Intestinal Tract

Lb. acidophilus, Lb. casei, and species of *Bifidobacterium* have been reported to inhibit growth of undesirable microorganisms that might be encountered in the gastrointestinal tract. Most of the older reports dealing with this type of control focused on a therapeutic approach, in that, cultured products made with these organisms were used to treat infections of various types (Gordon et al., 1957; Winkelstein, 1955). Some of the studies involving these organisms were poorly done, and improper controls were used, so it is difficult to draw definite conclusions concerning the benefit of the probiotic organisms. The newer approach is to provide consumers with products containing the probiotic organisms as a preventive treatment in controlling intestinal infections. Studies using chickens as an animal model in which the birds were dosed with specific intestinal pathogens following consumption of cells of *Lb. acidophilus* have shown that the lactobacilli do exert control over *Salmonella* infections (Watkins et al., 1982). Feeding the birds the lactobacilli before challenge with the pathogens, followed by continued consumption of *Lb. acidophilus* after the challenge dose, resulted in best control of the pathogens, suggesting that continuous consumption of the probiotic organism is desirable. The researchers conducting this study also showed that *L. acidophilus* was effective in controlling *Escherichia coli* in the intestinal tract of chickens (Watkins et al., 1982).

Just how the probiotic bacteria function in inhibiting the growth of intestinal pathogens in the intestinal tract is not clear. Many of the probiotic organisms produce substances that are inhibitory in vitro; however, it is difficult to confirm the activity of these compounds in vivo. The probiotic bacteria in question all produce large amounts of acids during their growth because they rely on fermentation to obtain energy for growth. However, the antagonistic action that they produce toward undesirable microorganisms apparently is not caused just by acids produced during their growth. Several of these organisms produce "antibiotic like" substances, some of which have been classified as bacteriocins, which may be involved in the antagonistic action toward the pathogens. Bacteriocins, according to the classical definition, are proteins produced by bacteria that are active against organisms closely related to the producer organism (Tagg et al., 1976). This may limit the breadth of action of these inhibitory substances produced by probiotic bacteria toward a wide range of intestinal pathogens. They would not be expected to have any effect on gram-negative intestinal pathogens. Furthermore, because of their sensitivity to proteolytic enzymes, bacteriocins may not survive to function in the intestines.

Competitive exclusion by the probiotic bacteria is another mechanism that has been suggested as being important in controlling intestinal infections (Watkins and Miller, 1983). Competitive exclusion involves the ability of the lactobacilli or bifidobacteria to occupy binding sites on the intestinal wall, thereby preventing attachment and growth of enteric pathogens.

Definitive scientific data showing the mechanism of action whereby these probiotic bacteria may exert inhibitory actions toward pathogens in the intestinal tract would make it easier to select the most effective strains of probiotic bacteria for use in dairy products to help control intestinal infections. Most likely, the antagonistic actions produced by the probiotic bacteria toward intestinal pathogens result from a combination of factors.

2. Improved Immune Response

Enhancement of the body's immune response by consuming cells of certain lactobacilli increases the resistance of the host to intestinal infections (Lessard and Brisson, 1987; Perdigon et al., 1990a; Satos et al., 1988). Of the lactobacilli, *Lb. casei* seems to be the primary one involved (Perdigon et al., 1990b). As with other characteristics of the lactic acid bacteria, the relative ability of probiotic bacteria to cause such an effect probably varies tremendously among strains of individual species. Researchers in this area have suggested that this action involves activation of macrophages, which, in turn, destroy pathogenic organisms in the body. This enhancement of the immune system increases the host defense mechanisms and could be very important in the control of foodborne illnesses. This may be a key explanation as to how certain probiotic microorganisms used as dietary adjuncts can exert control over intestinal infections.

3. Improved Lactose Digestion

People who lack the ability to adequately digest lactose are classified as lactose maldigestors. (In the past, terms such as "lactose intolerance" or "lactose mal-absorption" have been used to describe these individuals.) The problem results from inadequate levels of β-galactosidase in the small intestine to adequately hydrolyze ingested lactose. Once a lactose maldigestor ingests sufficient lactose, it passes into the large intestine where it undergoes an uncontrolled fermentation that results in symptoms of cramps, flatulence, and diarrhea. These symptoms often follow consumption of milk by such individuals. Because lactose maldigestion is a result of inadequate levels of an enzyme to hydrolyze lactose in the small intestine, the possibility exists for providing such an enzyme in the diet. Inclusion of a purified enzyme such as β-galactosidase in a diet would be rather expensive, and survival of the enzyme during passage through the stomach likely would be minimal. Research has shown that the presence of viable starter cultures in yogurt can be beneficial to lactose maldigestors (Gilliland and Kim, 1984; Kolars et al., 1984). This beneficial action has been shown to result from the presence of β-galactosidase in bacterial cells. Apparently, being inside the bacterial cells protects the enzyme during passage through the stomach so that it is present and active when the yogurt reaches the small intestine. Once the yogurt culture reaches the small intestines, it interacts with bile, which increases the permeability of cells of these bacteria and enables the substrate to enter and be hydrolyzed (Noh and Gilliland, 1992). The enzyme remains inside the cell upon exposure to bile rather than leaking out into the surrounding medium. As discussed, the starter cultures used for yogurt manufacture (*Lb. delbrueckii* subsp. *bulgaricus*, and *S. thermophilus*) do not have bile resistance and thus are not expected to survive and grow in the intestinal tract. Despite this limitation, consumption of these bacteria does provide a means of transferring β-galactosidase into the small intestines where it can improve lactose utilization in lactose maldigestors.

Nonfermented milk containing cells of *Lb. acidophilus* also can be beneficial for lactose maldigestors (Kim and Gilliland, 1983). This organism, unlike the yogurt starter cultures, can survive and grow in the intestinal tract. However, a similar mechanism in improving lactose utilization in lactose maldigestors to that observed of yogurt bacteria is probably involved. β-galactosidase activity of *Lb. acidophilus* is greatly increased in the presence of bile (Noh and Gilliland, 1993). As with the yogurt culture, cells of *Lb. acidophilus* do not lyse in the presence of bile but their permeability is increased, permitting lactose to enter the cells and be hydrolyzed. Because *Lb. acidophilus* can survive and grow in the intestinal tract, it is reasonable to expect, however, that additional β-galactosidase may be formed after ingestion of milk containing this organism.

There has been some controversy over the benefit of acidophilus milk in being effective in improving lactose utilization in lactose maldigestors; however, if the cells contain sufficient levels of β-galactosidase before ingestion, it is reasonable to assume they would provide such a benefit. Some studies that have suggested that milk containing *Lb. acidophilus* is ineffective (Payne et al., 1981; Saviano et al., 1984) in improving lactose digestion might be questioned because no evidence was provided concerning the cultures used or the procedure by which they were produced. It is possible, in those studies, that insufficient β-galactosidase was present in the milk containing cells of *Lb. acidophilus* at the time of consumption. One of the studies (Saviano et al., 1984) indicated that no β-galactosidase activity was detected in milk containing *Lb. acidophilus*.

Based on the proposed mechanism for improving lactose digestion by yogurt cultures, it seems reasonable that any product containing bacterial cells having adequate intracellular β-galactosidase activity could provide a benefit such as improving lactose utilization. Because this enzyme usually is inducible in most microorganizations, it is important that the organisms be grown in the medium containing lactose before ingestion. This becomes particularly important when cells of probiotic bacteria grown in some medium other than milk are added to nonfermented milk. The level of β-galactosidase activity also varies among strains of *Lb. acidophilus* as well as among commercial yogurt cultures. Therefore, it is important to consider the level of β-galactosidase activity in probiotic or starter cultures to be used for improving lactose digestion in lactose maldigestors. It also is important for the activity to remain high during transportation and storage of such products so that the consumer receives the product containing enough of the enzyme to provide a benefit.

4. Anticarcinogenic Actions

Anticarcinogenic or antimutagenic activities have been reported for several cultures used to manufacture various fermented milk products (Goldin and Gorbach, 1984; Oda et al., 1983; Reddy et al., 1983; Shahani et al., 1983). Some of these have involved products containing probiotic bacteria expected to survive and grow in the intestinal tract, whereas others have involved the use of fermented products manufactured when organisms not normally expected to survive and grow in the intestinal tract. For instance, consumption of yogurt by mice inhibited development of certain tumors (Reddy et al., 1983). In other studies involving human subjects, a culture of lactobacilli exhibited potential in controlling cancer of the colon (Goldin and Gorbach, 1984). The lactobacillus used in this study was later identified as *Lb. casei*.

Lb. acidophilus, Lb. casei, and *Lb. delbrueckii* subsp. *bulgaricus* are the species most often mentioned as having potential to produce or provide anticarcinogenic actions. For *Lb. delbrueckii* subsp. *bulgaricus*, the anticarcinogenic action apparently is associated with substances formed by the organism during fermentation of the yogurt as opposed to being produced in the body following consumption of the yogurt. However, for *Lb. acidophilus* and *Lb. casei*, growth or action in the gastrointestinal tract seems to be important. Part of the benefit may involve direct effects on inhibiting tumor formation; however, the main effect may result indirectly through inhibition of the growth of undesirable bacteria that may form carcinogens in the large intestine (Goldin and Gorbach, 1984). Thus, this may represent another benefit from being able to control growth of undesirable organisms in the gastrointestinal tract.

5. Control of Serum Cholesterol

In the 1970s, two studies revealing that organisms such as *Lb. acidophilus* can have potential in reducing serum cholesterol levels in humans were published. One of these involved milk fermented with what was described as a "wild" strain of *Lactobacillus* and then fed to a group of men on a high-cholesterol diet (Mann and Spoerry, 1974). The study was designed to study the influence of a surfactant (Tween 20) on serum cholesterol levels. The researchers theorized that the surfactant would increase absorption of cholesterol from the intestine and thus increase serum cholesterol levels. Conversely, the serum cholesterol level in both groups of men, that is, those receiving the surfactant and those who did not, decreased. This was one of the first studies that suggested consumption of a fermented dairy product could reduce serum cholesterol levels in humans. However, neither the organism involved in the fermentation nor the mechanism was identified. In another study, cells of *Lb. acidophilus* added to infant formula reduced serum cholesterol in infants receiving the formula (Harrison and Peat, 1975), whereas infants receiving the formula without cells of *Lb. acidophilus* exhibited increased serum cholesterol levels. The researchers concluded that *Lb. acidophilus*, through its growth in the intestine, in some way influenced the serum cholesterol level, although no mechanism was suggested.

Several animal feeding studies have shown that consumption of milk containing cells of *Lb. acidophilus* by animals resulted in lower serum cholesterol levels than in animals which did not receive milk containing the lactobacilli (Danielson et al., 1989; Gilliland et al., 1985; Grunewald, 1982). There is variation among strains of this organism in their ability to exert control over serum cholesterol levels (Gilliland, 1989). Some strains of *Lb. acidophilus* can actively assimilate or take up cholesterol during growth in laboratory media. This occurs when the organisms are grown anaerobically in the presence of bile, conditions that occur in the intestinal tract. Pigs on a high-cholesterol diet fed a strain of *Lb. acidophilus* that actively assimilated cholesterol during growth in laboratory media had significantly lower serum cholesterol levels than did pigs receiving a strain of *Lb. acidophilus* that did not actively assimilate cholesterol in laboratory media. This suggests that the ability to assimilate cholesterol in laboratory media provides an indication of the potential of this organism to exert some control over serum cholesterol levels.

Another activity of *Lb. acidophilus* that may be important is its ability to deconjugate bile acids. This provides yet another mechanism whereby *Lb. acidophilus* might exert control on serum cholesterol levels. The deconjugation of bile acids by lactobacilli can occur in the small intestines. Free bile acids are less well absorbed from the small intestines than are conjugated ones and thus more are excreted through feces (Chickai et al., 1987). Excretion of bile acids through feces represents one of the major mechanisms whereby the body eliminates cholesterol. This is because cholesterol is a precursor for synthesis for bile acids and any bile acids that are excreted from the body are replaced by synthesis of new ones. Thus, there is a potential for reducing cholesterol pool in the body. Furthermore, free bile acids do not support the absorption of cholesterol from the intestinal tract as well as the conjugated ones (Eyssen, 1973), thus deconjugation of bile acids in the intestinal tract may reduce the efficiency of absorption of cholesterol from the intestinal tract.

Research into the potential of *Lb. acidophilus* to exert hypocholesterolemic effects in humans has indicated tremendous variation among strains of *Lb. acidophilus* isolated from the human intestinal tract in their ability to assimilate cholesterol (Buck and Gilliland, 1994). Evaluation of strains of *Lb. acidophilus* currently used commercially in cultured or culture-containing dairy products in the United States has revealed that none of them are particularly active with regard to actively assimilating cholesterol in laboratory media (Gilliland and Walker, 1990). On the other hand, new strains that are very active in this regard have been isolated from the human intestinal tract and thus they may provide greater potential for use as dietary adjuncts to assist in controlling serum cholesterol levels (Buck and Gilliland, 1994). From 122 isolates of *Lb. acidophilus* obtained from human intestinal sources, several were identified as having great potential for exerting control over serum cholesterol levels because they were very active in assimilating cholesterol during growth in a laboratory medium. They were far more active in this regard than were the currently commercially available strains of *Lb. acidophilus*. One of these strains has been used by a company in Holland to produce a fermented yogurt named *Fysiq*, which has been promoted as being useful in helping maintain a healthy cholesterol level. This strain has been used in a human feeding trial of hypercholesterolemic individuals and effectively caused a significant reduction in serum cholesterol levels (Anderson et al., 1997).

There may be possibly other probiotic organisms that can exert benefits in controlling serum cholesterol levels. Some of these include *Lb. casei* as well as *Bifidobacterium* species. Cultures in both of these species can remove cholesterol from laboratory media during anaerobic growth in the presence of bile. They also are effective in deconjugating bile acids. Currently, there is great interest throughout the world in the potential of these bacteria exerting some control over serum cholesterol levels in hypercholesterolemic individuals.

C. Characteristics Needed for Probiotic Cultures

There is potential for probiotic cultures to provide health and nutritional benefits for consumers. However, data are insufficient in most instances to permit specific health claims in the United States for dairy products containing them. The improvement of lactose utilization in lactose maldigestors is the possible exception. Before specific health claims can be made for most of these products, it is necessary for clinical trials to establish that the benefit indeed occurs. Such trials should be conducted using only probiotic bacteria that have been selected for a specific activity. In other words, they should be selected in some manner to produce the desired health or nutritional benefit (Gilliland and Walker, 1990). It is unreasonable to expect one strain of any of the species involved to provide all of the potential health or nutritional benefits. In the past, most of the knowledge gained concerning variations among strains of species of starter culture bacteria has focused on the ability of these organisms to produce desired organoleptic properties in cultured products and to do so as rapidly as possible. Very little, if any, attention has focused on the potential health or nutritional benefits possible from these cultures. Most strains of the probiotic bacteria commercially available have not been selected for any specific activity except to, perhaps, have their identity confirmed as the indicated organism. To be successful as probiotic cultures, they must be selected for their ability to provide the targeted benefit for the consumer.

If cultured or culture-containing dairy products are to be useful as functional foods in providing health or nutritional benefits for consumers, it is necessary to alter the basis used for selecting starter culture bacteria. The cultures not only must be selected for their ability to produce the desired organoleptic properties in the cultured product but factors related to their potential health or nutritional benefits must also be considered (Gilliland and Walker, 1990). The primary factor to be considered in the selection is that the culture must be able to produce the desired benefit. Furthermore, the culture should retain that ability during production, manufacturing, distribution, and storage of the product before reaching the consumer. If the desirable action requires that the organism grow in the intestinal tract, then the characteristics that enable the organism to grow well under these conditions must be considered. To help ensure the ability of the organism to establish or grow in the intestines, it is important to consider the bile tolerance of the strain selected. The probiotic bacteria under consideration tend to be host specific; therefore, it is necessary to consider the source from which the organism came. In other words, it is desirable to select the strain that is compatible with the host (i.e., humans) for which the product is intended.

In some instances, a product such as yogurt, which is made with the traditional yogurt culture, *Lb. delbrueckii* subsp. *bulgaricus* and *S. thermophilus*, also is supplemented with cells of *Lb. acidophilus* and or *Bifidobacterium* species. If such a procedure is implemented to provide the consumer with the beneficial organisms, then care must be exercised to ensure that adequate numbers of probiotic organisms are present. Some of the probiotic cultures with the potential for providing health and nutritional benefits may not grow as well in milk during the manufacture of fermented milk products as some of those that have been traditionally used for producing such products. Thus, research may be necessary to determine ways to improve the growth of the probiotic organisms in the milk being fermented so that the consumer is provided with adequate numbers of these potentially beneficial bacteria.

For any of the potential health or nutritional benefits that might be derived from probiotic cultures, it is necessary to develop proper testing of the cultures and the products containing them to ensure that the consumer receives the product that is most likely to provide the intended benefit. If such products as functional foods are to be effective, it is necessary to use the properly selected probiotic cultures. This may result in the requirement of several types of yogurt or other fermented milk products, each containing a different selected strain or strains of the probiotic culture to provide the specific desired health or nutritional benefits.

REFERENCES

Albus WR. The effect of surface tension upon the growth of the lactobacilli. J Bacteriol 16:197, 1928.

Anderson JW, Gilliland SE, Tijssens R, Elbers FJ. Effect of fermented milk (yogurt) containing *Lactobacillus acidophilus* L1 on serum cholesterol in hypercholesterolemic humans: report of a placebo-controlled study. Am J Clin Nutr (submitted).

Bottazzi V, Battistotti B, Vescovo M. Continuous production of yoghurt cultures by stimulation of *Lactobacillus bulgaricus* by formic acid. Milchwissenschaft 26:214, 1971.

Buck LM, Gilliland SE. Comparisons of freshly isolated strains of *Lactobacillus acidophilus* of human intestinal origin for ability to assimilate cholesterol during growth. J Dairy Sci 77:2925, 1994.

Chickai T, Nakao H, Uchida K. Deconjugation of bile acids by human intestinal bacteria implanted in germ-free rats. Lipids 22:669, 1987.

Danielson AD, Peo ER Jr, Shahani KM, Lewis AJ, Whalen PJ, Amer AM. Anticholesteremic property of *Lactobacillus acidophilus* yogurt fed to mature boars. J Am Sci 67:966, 1989.

Eyssen H. Role of the gut microflora in metabolism of lipids and sterols. Proc Nutr Soc 32:59, 1973.

Gilliland SE. Acidophilus milk products: a review of potential benefits to consumers. J Dairy Sci 72:2483, 1989.

Gilliland SE, Kim HS. Effects of viable starter culture bacteria in yogurt on lactose utilization in humans. J Dairy Sci 67:1, 1984.

Gilliland SE, Nelson CR, Maxwell C. Assimilation of cholesterol by *Lactobacillus acidophilus*. Appl Environ Microbiol 59:377, 1985.

Gilliland SE, Walker DK. Factors to consider when selecting a culture of *Lactobacillus acidophilus* as a dietary adjunct to produce a hypocholesterolemic effect in humans. J Dairy Sci 73:905, 1990.

Goldin BR, Gorbach SL. Alterations of the intestinal microflora by diet, oral antibiotics, and *Lactobacillus*: Decreased production of free amines from aromatic nitro compounds, azo dyes, and glucuronides. J Nat Cancer Inst 73:689, 1984.

Gordon D, Macrae J, Wheater DM. A lactobacillus preparation for use with antibiotics. The Lancet 272:889, 1957.

Grunewald KK. Serum cholesterol levels in rats fed skim milk fermented with *Lactobacillus acidophilus*. J Food Sci 47:2078, 1982.

Hargrove RE, Alford JA. Growth response of weaning rats to heated, aged, fractionated, and chemically treated yogurts. J Dairy Sci 63:1065, 1980.

Harrison VC, Peat G. Serum cholesterol and bowel flora in the newborn. Am J Clin Nutr 28:1351, 1975.

Kim HS, Gilliland SE. *Lactobacillus acidophilus* as a dietary adjunct for milk to aid lactose digestion in humans. J Dairy Sci 66:959, 1983.

Kolars JC, Levitt MD, Aouji M, Saviano DA. Yogurt—an autodigesting source of lactose. N Engl J Med 310:1, 1984.

Kosikowski F. Cheese and Fermented Milk Foods. 2nd ed. Ann Arbor, MI: Edwards Brothers, 1977.

Lessard M, Brisson GJ. Effects of *Lactobacillus* fermentation product on growth, immune response and fecal enzyme activity in weaned pigs. Can J Anim Sci 67:509, 1987.

Mann GV, Spoerry A. Studies of a surfactant and cholesteremia in the Maasai. Am J Clin Nutr 27:464, 1974.

Metchnikoff E. The Prolongation of Life Optimistic Studies. New York: GP Putnam's Sons, 1908.

Noh DO, Gilliland SE. Influence of bile on β-galactosidase activity of *Lactobacillus bulgaricus*. Okla Agri Experiment Station MP-136:48, 1992.

Noh DO, Gilliland SE. Influence of bile on cellular integrity and β-galactosidase of *Lactobacillus acidophilus*. J Dairy Sci 76:1253, 1993.

Oda M, Hasegawa H, Komatsu S, Kambe M, Tsuchiya F. Antitumor polysaccharide from *Lactobacillus* species. Agri Biol Chem 47:1623, 1983.

Payne DL, Welch JD, Manion CV, Tsegaye A, Herd LD. Effectiveness of milk products in dietary management of lactose malabsorption. Am J Clin Nutr 34:2711, 1981.

Perdigon G, Aalvarez S, Demacias MEN, Roux ME, Holgado APR. The oral administration of lactic acid bacteria increase the mucosal intestinal immunity in response to enteropathogens. J Food Prot 53:404, 1990a.

Perdigon G, Demacias MEN, Alvarez S, Oliver G, Holgado AAPR. Prevention of gastrointestinal infection using immunobiological methods with milk fermented with *Lactobacillus casei* and *Lactobacillus acidophilus*. J Dairy Res 57:255, 1990b.

Rasic JL, Kurman JA. Bifidobacteria and Their Role. Basel, Switzerland: Birkbauser Verlag Basel, 1983.

Reddy GV, Friend BA, Shahani KM, Farmer RE. Antitumor activity of yogurt components. J Food Prot 46:8, 1983.

Sato K, Saito H, Tomioka H, Yokokura T. Enhancement of host resistance against *Listeria* infection by *Lactobacillus casei*: efficacy of cell wall preparation of *Lactobacillus casei*. Microbiol Immunol 32:1189, 1988.

Saviano DO, El Ansuar AA, Smith DE, Levitt MD. Lactose malabsorption from yogurt, pasteurized yogurt, sweet acidophilus milk and cultured milk in lactose-deficient individuals. Am J Clin Nutr 40:1219, 1984.

Shahani KM, Friend BA, Bailey PJ. Antitumor activity of fermented colostrum and milk. J Food Prot 46:385, 1983.

Sharpe ME, ed. Bergey's Manual of Systemic Bacteriology. Vol 2, Baltimore: Williams & Wilkins, 1986.

Tagg JR, Dajani AS, Wannamaker LW. Bacteriocins of gram-positive bacteria. Bacteriol Rev 40:722, 1976.

Watkins BA, Miller BF. Competitive gut exclusion of avian pathogens by *Lactobacillus acidophilus* in gnotobiotic chicks. Poultry Sci 62:1772, 1983.

Watkins BD, Miller BF, Neil DH. In vivo inhibitory effects of *Lactobacillus acidophilus* against pathogenic *Escherichia coli* in gnotobiotic chicks. Poultry Sci 61:1298, 1982.

Winkelstein A. *Lactobacillus acidophilus* tablets in the therapy of various intestinal disorders: a preliminary report. Am Pract Digest Treatment 1:1637, 1955.

9
Cheese Products

MARK E. JOHNSON
University of Wisconsin—Madison, Madison, Wisconsin

The origin of cheese is lost in antiquity. But, most assuredly, milk was contaminated with lactic acid bacteria, which through acidification of the milk, created conditions unfavorable for growth of other bacteria. As the story goes, milk held in storage vessels (animal stomachs) clotted, making cream cheese, the "mother of all cheeses." Centuries later, humans attributed this action to rennet found in animal stomachs. It has since become known that the milk would have clotted anyway because acid develops in the presence of lactic acid bacteria. It was a great leap forward when humans discovered the use of coagulating enzymes that led to production of less sour cheeses. Imagine, a long time ago, when humans first tasted that odorous morsel covered with colorful molds, yeasts, and bacteria. But now consider a world without Roquefort, Stilton, Limburger, or Gruyere. Boring! Unthinkable!

Modern cheese making is controlled and has been refined through strict adherence to manufacturing guidelines and careful selection of specific lactic acid bacteria and ripening microorganisms. Even so, sometimes there are problems. No cheese is produced in a sterile environment, so contamination is inevitable. One of the chief causes of poor flavor quality in cheese is the undesirable metabolism of contaminating microorganisms. A *preventable* cause of poor quality flavor is that many retailers sell products long after they have reached their period of vulnerability. The ability of a cheese to age well with regard to undesirable microbial growth depends on cheese composition, manufacturing protocol, level of contamination, and the ability of the contaminants to grow in the cheese. Therefore, the cheese maker, retailer, and consumer must be aware of the limitations of the product with regard to growth of contaminants and defects that they cause.

Although there are more than 1000 named varieties of cheese worldwide, this chapter discusses only the major types.

I. DAIRY CHEMISTRY AND THE CHEESE MAKING PROCESS

For many cheese makers, there is an art to making cheese. To cheese manufacturers, it is commonly a routine, strictly controlled process. No matter how it is made, cheese is a complex entity in a constant state of change, which has been likened to an ecological community of living organisms in which microbiological activities affect and are influenced by chemical changes.

Production of cheese involves two phases: the first is to develop the desired composition and pH, and the second is to develop the desired physical and flavor characteristics. The first is controlled through milk composition and rate and extent of acid development by the starter during the manufacturing process. The second is influenced by the first but is dictated by metabolism of a variety of microorganisms and by enzymatic and chemical reactions. This process is called ripening, curing, or maturation, and depending on the cheese variety, may take many months to complete.

In concept, the manufacture of unripened cheese is simple. In reality, it is governed by a series of interrelated chemical and physical phenomena. During cheese making, a coagulum is formed in which milk proteins are clotted, entrapping the milk fat, water, and water-soluble components. Further manipulations of the coagulum (cutting, heating, stirring) and development of acid result in controlled moisture expulsion. The resulting curd and whey mixture is separated, with the curd formed into blocks, wheels, or other shapes.

Development of the desired flavor, body, and texture is completed through a combination of the activity of specific introduced microflora and enzymes as well as naturally occurring or contaminating bacteria and enzymes. Part of the initial maturation process involves physical changes to the protein brought about through interrelated changes in pH, loss of minerals, and changes in the state of water (bound or absorbed). Without the ripening process it would be impossible to distinguish one variety of cheese from another except to note that different unripened cheeses may have different physical characteristics.

Milk solids are composed of protein (casein and whey protein), milk fat, lactose, citric acid, and mineral salts (usually associated with the casein), collectively called ash. The composition of milk varies considerably between species and individual animals. It is affected by breed and genetics of the animal, feed, environment conditions, lactation number, stage of lactation, and animal health. All of these factors can also influence cheese making and cheese characteristics. An average composition of cow milk is as follows: 87.6% water, 3.9% milk fat, 3.2% protein (78% caseins, 22% whey proteins), 4.6% lactose, 0.7% ash (also see Chapter 1).

There are three basic ways to make cheese, but a given variety is made with only one method. All methods involve development of acid by a select group of lactic acid bacteria called the starter. All involve some means of concentrating the milk solids (mostly milk fat and protein) by expelling a portion of the aqueous phase of milk (serum or whey).

Rennet curd cheeses (most varieties) are made by clotting the milk with a coagulating enzyme such as chymosin (the most active ingredient in rennet). Acid curd cheeses (cottage, cream) are made with acidification of the milk sufficient to cause the casein to form a clot. Heat-precipitated curd cheeses (ricotta, queso blanco) are made with a combination of low pH and high heat to precipitate the proteins (both casein and whey proteins).

Fresh cheeses such as cottage and mozzarella can be made by direct addition of acid (acetic, lactic, or citric). Cheeses made by this method are called direct acid cheeses (e.g., direct acid mozzarella).

A. Rennet Curd Cheese Manufacture

Rennet curd cheeses are those in which the coagulum is formed by the activity of a coagulant, an enzyme mixture with particular proteolytic activity. Coagulants are commonly called rennets. Calf rennet is derived from an extract of calf stomachs but there are microbial rennets, other animal rennets, and plant rennets. All contain proteolytic enzymes, which, through their activity, help to destabilize the casein micelles in milk, an event that subsequently transforms milk from a liquid to a semisolid (coagulum). Chymosin is the desired coagulating enzyme in calf rennet but, because of cost, demand and the lack of calf stomachs, most chymosin used in the United States is produced by genetically engineered bacteria, yeasts, or molds. Fermentation-derived chymosin is highly purified (100% purity) and is used in liquid or tablet form. Chymosin is the preferred coagulant because it has specificity toward one peptide bond in κ-casein. Although chymosin hydrolyzes bonds in casein molecules at other sites when they are accessible, the specific site of hydrolysis that occurs during coagulation is the Phe_{105}-Met_{106} of κ-casein. The nonspecific proteolytic activity of some coagulants (not chymosin) causes concern over excessive proteolysis, leading to a soft body, bitter flavor defects, and loss of cheese yield.

Caseins exist in complexes of discretely arranged molecules called micelles, which are thought to be held together by various types of bonding including hydrophobic interactions and complexes of calcium-phosphate. There are three types of casein molecules, α- β- and κ-caseins. The exact molecular arrangement of molecules is not known but it is hypothesized that the micelles are composed of submicelles linked together through calcium phosphate bridges and hydrophobic interactions. A hydrophilic portion of the κ-casein molecules is thought to protrude from the micelle surface, giving rise to the "hairy micelle" model.

At the normal pH of milk (pH 6.6–6.7), micelles are negatively charged. Through electrostatic repulsion and stearic hindrance via the "hairs" of κ-casein, micelles are stable (show no tendency to flocculate or gel) and remain as individual entities. The activity of the coagulant removes the hydrophilic region on the κ-casein molecule. This eliminates stearic hindrance and reduces the charge at the micelle surface. With loss of these barriers, micelles begin to come together (clot formation). Ionic calcium (added as $CaCl_2$ or released from the micelles through acidification of milk) allows adjacent micelles to join through hydrophobic and electrostatic interactions. Eventually (20–30 minutes), the casein micelles form a continuous network called the clot or coagulum. Milk fat, water, and water-soluble components (serum) are entrapped within the casein network. Whey proteins are water soluble and do not participate in forming the network but are trapped in the pores that form between aggregates of micelles.

Once the desired firmness of the coagulum has been reached, it is cut into small cubes or pieces (curd). The firmer and larger the curd particles, the higher the moisture content of the cheese. After the coagulum is cut, casein molecules continue to interact and squeeze out serum trapped between them, and curds shrink and become firmer. This process is called syneresis and is enhanced by lowering the pH, increasing the temperature, and stirring the curd. Therefore, the rate of acid development by the starter has a great influence over moisture content of cheese and its control is a key process in cheese manufacture.

Each variety of cheese has a desired rate and extent of acid development and, if not met or compensated for, may result in too much or too little moisture creating undesirable physical characteristics. At the proper time, curd is separated from whey and treated appropriately as dictated by the variety of cheese. The curd may be continuously stirred as whey is being removed or it may be allowed to mat without agitation. The curd may be salted first and then formed into the desired shape or formed first and then salted by placing the cheese into a brine. Pressing of the blocks, cylinders, or wheels of cheese removes trapped whey from the cheese, but not all cheeses require pressing. The unripened cheese is then ready for maturation.

B. Acid Curd Cheeses

Acid curd cheeses do not rely on the activity of a coagulating enzyme to clot the milk. Instead, milk is acidified by direct addition of acid or through lactic acid developed by starter bacteria. At a pH of approximately 5.2, caseins in milk begin to gel. Gelation is the consequence of acidification-induced physicochemical changes to caseins. At neutral pH, casein micelles remain as individual entities and are unable to interact or form aggregates. This is, in part, caused by charge repulsion (micelles are negatively charged). In addition, hydrophilic regions of κ-casein molecules protrude from the micelle core and prevent hydrophobic cores of adjacent micelles from interacting (stearic repulsion).

As the pH is lowered, the calcium-phosphate complex disintegrates and some of the casein molecules dissociate from the micelles. There is also a reduction of charge on the casein molecules, an increase in hydrophobicity, and it is thought that the protruding portion of the casein molecules falls back onto the casein micelle core. The net result is that micelles and solubilized casein molecules begin to form aggregates, eventually leading to formation of a continuous network and visible gel (pH ~4.95). In cottage cheese, the gel is cut into small cubes at a pH of 4.65 to 4.75. In cream cheese manufacture, the gel is stirred at pH 4.4 to 4.5, rather than cut as in cottage cheese, and whey is removed by centrifugation. Traditionally, fermented milk was put into bags of cheese cloth and hung to filter out the serum. The separated cheese is packaged (cold-pack cream cheese) or processed. Hot-pack cream cheese is made by blending cold-pack cream cheese with cream, whole milk, salt, stabilizers, and skim milk solids and heating the mixture to (72°C–74°C). The homogenized blend is packaged hot. Microbiologically induced defects are similar to those in cottage cheese.

C. Acid-Heat Coagulated Cheese

The premise for manufacture of acid-heat coagulated cheeses is to heat the milk to 78°C to 80°C and then acidify the milk by direct addition of citric, acetic, or lactic acid to the desired pH (5.8–5.9 for ricotta, 5.2–5.3 for queso blanco). Milk for queso blanco can also be first acidified by lactic acid bacteria (*Lactococcus* spp.) and then heated. Heating of the milk (ricotta milk is usually a mixture of sweet whey and milk) causes coagulation and flocculation of caseins and whey proteins. In ricotta cheese manufacture, proteins and entrapped fat are removed or filtered from the remaining serum and drained until packaged. In queso blanco cheese manufacture, curds are allowed to settle and whey is drained. The curds are then salted and pressed. Both cheeses are consumed fresh, and, because denatured whey protein is present, the cheeses resist melting during frying or baking. Because of the high heat treatment under acid conditions, survival of bacteria other than spore formers is minimal, but contamination after the curd is separated is of concern. The microbiologically induced defects are comparable to those of cottage cheese.

II. INFLUENCES OF MICROBIOLOGICAL QUALITY AND MILK COMPOSITION ON CHEESE QUALITY

The microbiological quality and composition of milk play an integral part in the quality of the cheese made from it. Cheese can be made from grade A or grade B milk, but cottage cheese must be made from grade A milk only. The bacterial count of grade A milk as determined by a standard bacterial count or loop count, cannot exceed 100,000/mL at the time of receipt or collection. The bacterial count of grade B milk cannot exceed 300,000/mL (Wisconsin Administrative Code). Processors often pay premiums for low bacterial count milk as an enticement to farmers to produce high-quality milk. In practice, processors have recorded that milk from greater than 90% of producers has a bacterial count of less than 20,000/mL. The bacteria found in the milk arise from contamination or from the animal itself (see Chapter 2).

The level of contamination is reflective of the cleanliness of the entire milking operation, including that of the animal before milking. Improper cooling rates or final holding temperatures of the milk result in high numbers of bacteria reflective of an environment conducive to microbial growth. Most bacteria in milk are, not surprisingly, psychrotrophic bacteria and they are the contaminants likely to grow at the low temperature at which milk must be stored (not to exceed 7°C for grade A and 10°C for grade B within 2 hours after milking). *Pseudomonas* sp. are usually the dominant psychrotrophic organisms found in milk. Although these bacteria are easily killed by pasteurization, they produce lipases and proteases, which are not totally inactivated by this heat treatment (Griffiths et al., 1981). The enzymes are active in milk and can cause bitterness (protein hydrolysis) and rancidity (milk fat hydrolysis) in products made from milk if the level of activity is high enough (Cousin, 1982). Milk may be held for 2 days (legally) after receipt at the factory and microbial counts will undoubtedly increase. It is the growth of *Pseudomonas* sp. during milk storage that concerns the cheese maker.

A more important cause of rancidity in milk and cheese is the activity of endemic animal lipases (milk lipase). The level of activity of this enzyme is increased in milk obtained from animals with mastitis (udder infection). In this instance, lipase activators and somatic cells are secreted from blood into the milk. Somatic cells are used as an indicator of cow health and limits have been set by individual states (not to exceed 750,000/mL, Wisconsin Administrative Code). Milk from mastitic animals has decreased casein, the major protein found in milk, although the total amount of all proteins (whey proteins increase) may decrease only slightly, if at all (see Chapter 1).

The composition, quality, and amount of cheese produced is greatly affected by the casein content of milk. The other proteins, collectively called whey proteins, are water soluble and contribute much less to cheese yield. The lower the casein contents of milk, the lower the yield of cheese. Cheese makers do not routinely directly measure casein in milk because the test is expensive and takes too long to complete. Instead, they use fast, inexpensive, automated tests to measure total protein. Casein content is calculated by multiplying the percentage of total protein by 0.82. In mastitic milk, however, the amount of casein as a percentage of total protein decreases. The cheese maker cannot predict this value. Rather, a high somatic cell count indicates that the casein content of the milk is reduced. The cheese maker commonly pays premiums for low somatic count milk.

III. MILK PRETREATMENT: CLARIFICATION, STANDARDIZATION, AND HEAT TREATMENT

All milk received by the cheese plant is first tested for presence of antibiotics. Milk containing antibiotics must be dumped (liquid manure or landspread), even though, if diluted with other milk, a negative test could be obtained. Raw milk, as it is received by the cheese maker, is almost universally filtered to remove extraneous matter (straw, hay, large clumps of bacteria). The Code of Federal Regulations establishes fat (milk fat content by weight of the cheese solids or fat in the dry matter [FDM]) and moisture limits for some cheeses. These values are called the standard of identity. The casein-to-milk fat ratio in milk determines the FDM of the cheese, whereas moisture is controlled by the manufacturing process. Use of whole milk almost always results in cheese with an FDM of at least 50%. To manufacture cheeses with a lower FDM, such as part-skim mozzarella or Swiss cheese, milk fat is removed or skim milk is added to whole milk. The process of manipulating the composition of the milk is called standardization and is becoming more popular for all cheese types because of economic considerations and the desire for uniformity of cheese composition.

A. Heat Treatment

Heat treatment given milk before cheese making varies from country to country and cheese to cheese. Pasteurization of milk is a legal requirement in the United States for fresh cheeses such as cottage, mozzarella, and reduced-fat varieties. Cheeses made from unpasteurized milk must be held for 60 days at a temperature not less than 1.7°C (Code of Federal Regulations, 1995). It is thought that pathogens will die out during this time period because of the acid conditions in the cheese and the growth of nonstarter lactic acid bacteria. However, this may not be true. Manufacturers who do not pasteurize milk use another heat treatment (65°C–70°C for 16–20 seconds), but the trend is toward full pasteurization. A main argument against pasteurization is that cheeses made from pasteurized milk tend to have a milder flavor (the flavor takes longer to develop or the flavor is atypical of raw milk cheese). Research into development of flavor in cheese may provide means to overcome this obstacle but the question of safety of raw milk cheeses remains. Pasteurization is not a guarantee of safety because milk or cheese can be contaminated after the milk has been pasteurized. When cases of illness can be attributed to consumption of cheese containing pathogens (a rare event), often the cheese is manufactured under poor hygienic conditions, is a fresh cheese, is made from unpasteurized milk, or the rate and extent of acid development were curtailed (Johnson et al., 1990a).

Although rarely used in the United States, a specially designed centrifuge called a Bactofuge (bactofugation) is used to remove most of the bacterial cells and spores (empirically 98%) from milk. Two streams of milk result from bactofugation, the "cleaned" milk and the bactofugate containing the bacterial spores and cells. If used, the bactofugate is heated to 130°C for a few seconds but the milk is pasteurized. The two fractions are then recombined. Bactofugation is used in Europe in lieu of sodium nitrate in controlling outgrowth of *Clostridium tyrobutyricum* spores, whose metabolism results in gassy, split cheese. Use of sodium nitrate is not permissible in the United States for the manufacture of cheese.

After heat treatment, the milk is cooled and pumped into retaining vessels called vats. Cheese vats vary in size, with the larger vats holding as much as 22,700 kg and the smaller commercial vats holding approximately 4500 to 6800 kg. The vats are generally double walled to permit controlled indirect heating of the milk. If starter is used, it can be added while milk is entering the vat or after the vat is filled. The temperature of milk at the time starter is added is determined by the type of cheese to be made, type of starter, and the desired temperature at the time of coagulant addition, but it is generally between 31°C and 34°C.

B. Starters

The strains and balance of strains of bacteria used in starters is often dictated by tradition as much as it is by manufacturing protocol and desired cheese characteristics. The choice of starter depends on the desired rate and extent of acid development (pH) during manufacture, proteolytic activity of the strains, flavor (and gas formation if desired), and conditions encountered during manufacture and storage such as pH, acidity, salt, and temperature profiles. Mesophiles are sometimes used in the manufacture of mozzarella and Swiss varieties instead of the traditional thermophilic starters. A lower cook temperature is used and the resultant cheese is higher in moisture. The amount of starter used is based on the rate of acid development desired by the manufacturer but is influenced by strain and how the culture was propagated (conditions of growth such as media, pH control, and age). This is an important concept because the amounts of starter listed in literature for cheese manufacture can be misleading (e.g., use of 1% w/w starter grown with no pH control may be equivalent to using 0.2% w/w starter grown with pH control). Additional information about starter cultures is given in Chapters 6 and 7.

Use of artisinal cultures is not common in the United States. These cultures are mixtures (unknown composition) of several genera, species, and strains of lactic acid bacteria. They may contain lactococci, lactobacilli, leuconostocs, streptococci, and enterococci and probably give the cheese special flavor characteristics.

IV. CHEESE MICROBIOLOGY

A. Introduction

The diversity of cheese manufacturing protocols, ripening regimens, and composition makes cheese a complex subject microbiologically. It is a misconception to think of cheese microflora in terms of the type of cheese, for example, all cheddars, blue cheeses, and so on. Each individual cheese (not type) has its own unique microflora regardless of the starter or any deliberately added secondary ripening microorganisms (e.g., molds or yeasts). There is an extensive list of adventitious microorganisms that can grow in or on cheese. These nonstarter, nondeliberately added microbes are contaminants. Cheese microflora should be looked at as a situation of chance. Thus, the contaminants that are found in any cheese result because the specific microbes happen to have had chance contract with the cheese or milk from which it was made.

Microorganisms that grow in cheese or at least maintain viability follow the same set of criteria (pH, moisture, salt, acidity/type of acid, redox potential, nutrient availability, competition, temperature, anaerobic/aerobic conditions) as in any food product. Two factors determine the microflora of cheese: the presence of the microorganism and the ability of the microorganism to grow.

During cheese maturation, the environmental conditions can change sufficiently to allow growth of initially inhibited contaminants, or conditions may become even more inhospitable. The cheese environment is dynamic, thus the microflora in cheese can be considered a dynamic ecological system. Few studies on bacterial viability in cheese have been completed in which the changes in cheese chemistry during ripening are correlated with its affect on the microflora.

A complicating factor in the study of cheese microflora is the methodology used to isolate microorganisms. Selective media may provide too harsh an environment for recovery and growth of injured or stressed cells. Microorganisms may be viable and metabolically active but not culturable with current methods. Nonselective media may not be appropriate to detect low numbers in a competitive environment.

Why is it important to study the microorganisms in cheese? Pathogens in cheese are of utmost importance. However, flavor quality (both desirable and undesirable) of cheese is also a consequence of the metabolism of microorganisms. Additionally, some textural defects can be directly attributed to growth and metabolism of microorganisms.

Molecular techniques are being applied to selectively determine the presence of individual species and strains of bacteria in cheese. The polymerase chain reaction (PCR) is a rapid procedure for in vitro enzymatic amplification of a defined segment of DNA (Atlas and Bej, 1994). It is particularly useful in identifying the proverbial needle-in-the-haystack and individual strains of bacteria. A unique oligonucleotide sequence (probe) can be used to specifically identify (through amplification) the presence of DNA from particular bacteria in cheese. Enumeration of the bacteria is not necessary, but the bacteria may no longer be alive. DNA extracted from individual bacteria isolated using traditional techniques can also be tested to determine the exact species or strains of bacteria. Of particular interest is rapid detection of low levels of pathogens in milk and cheese (Herman et al., 1995). The technique has also been applied to identify species of *Clostridium* in cheese (Klijn, 1995), new strains of *Lactococcus lactis* subsp. *cremoris* (Salama et al., 1993), and individual strains of *Lactobacillus helveticus* (Drake et al., 1996).

Many of the adjuncts used to enhance the flavor of cheese are *Lactobacillus* spp. and are often not easily differentiated from other strains by biochemical tests. Complicating the situation is that lactobacilli are the dominant nonstarter lactic acid bacteria found in cheese. Selective media for lactobacilli cannot differentiate between adjunct and contaminant lactobacilli. This makes it difficult to determine numbers of individual strains of lactobacilli in mixed populations of lactobacilli. It is important to follow numbers of individual strains of lactobacilli (or other bacteria) to study the cause and effect of *Lactobacillus* spp. (or other bacteria) on flavor development in cheese. In addition, the ability to unequivocally determine the presence of patented or licensed strains of adjuncts can be useful for legal purposes.

B. Cottage Cheese

Cottage cheese curd is made from grade A pasteurized skim milk. Fortification of milk low in casein (<2.4% casein of <9% total milk solids) with very low heat-treated nonfat dry milk can improve cheese yield and quality (Emmons and Tuckey, 1967).

Milk is inoculated with *L. lactis* subsp. *lactis* and *cremoris*, with the latter being generally preferred. Commercially, cottage cheese is usually made with a "short set," that is, 4 to 5 hours elapse between time of starter addition (milk pH 6.60, 31°C–32°C) and time of cutting (coagulated milk pH 4.70 to 4.8). However, a "long set" method is also used. To promote efficiency, when the long set method is used, vats are filled with milk (20°C to 22°C) and starter is added so that overnight (9–12 hours) the pH of the milk declines to 4.90. Thus, when the cheese maker returns at the start of the work day, the milk coagulum is almost ready to cut. In the short set method, the inoculation rate of the starter is 3% to 5% w/w of the milk; whereas, in the long set method, much less (0.5%–2% w/w) starter is used.

The pH at which the curd is cut and the final pH of the curd after processing is critical for yield and cheese quality but varies among processing plants. This variability results from a variety of factors, including casein content of milk, heat treatment of milk, and rate of acid development by the starter. Overacidification or underacidification leads to brittle curd that shatters when stirred. The tiny pieces of curd may be lost in subsequent manufacturing steps, causing a loss in yield, or, if retained, may cause graininess (many hard bits of curd) and lack of uniform curd size (a visual defect that downgrades the product). Most manufacturers use a very small amount of coagulant. This enables the curd to be cut at slightly higher pH. The curd is less fragile and the yield is higher.

Once the coagulum is cut, the curd and whey mixture is heated to 54°C to 57°C (approximately 2 hours) and held (15–20 minutes) until proper firmness is reached. The rate of heating and the final temperature can prevent over acidification and firms the curd (removes whey). Although strain dependent, most lactococci do not produce significant amounts of acid at temperatures above 40°C and are reduced in number by the cooking procedure (Collins, 1961). *Pseudomonas* sp. and *Enterobacteriaceae*, common spoilage bacteria of cottage cheese, are also sensitive to the cooking procedure, which greatly reduces their number. The lethality of the cooking procedure is time and temperature dependent and is determined by the initial bacterial load. Therefore, the lower the bacterial population at the time the curd reaches the final cooking temperature, the more effective a given heat treatment is.

After the correct curd firmness is reached, most of the whey is removed and the curd is washed two or three times with progressively colder water. The wash step removes lactic acid and lactose, and helps to control the level of acidity (acid taste) in the finished cheese. The water is acidified (pH 4.5–6.0) and chlorinated (5–10 ppm) or pasteurized to kill bacteria in the water. Washing cools the curd rapidly to less than 5°C, which is essential to keep the growth of contaminating bacteria to a minimum. After the last wash water is removed, pasteurized cold cream dressing is added and the product is packaged. The amount of fat in the cream dressing determines the fat content of the final cheese, so reduced fat cottage cheese is made by adjusting the solids and fat content of the cream dressing.

Contaminated equipment and air are the most likely sources of spoilage bacteria in creamed cottage cheese. Although cottage cheese curd is acid (approximately 4.5–4.7), the pH of the final commercial product, creamed cottage cheese, is higher (5.0–5.3). The pH of the creamed cheese can be manipulated by the acidity of the cream dressing. Low product pH (<5.0) may lead to free whey accumulation (clotting and syneresis of cream dressing) during storage, whereas a higher pH allows for increased growth of contaminating bacteria. Wash treatments of the curd lower the acid content and rapidly cool the curd to less than 5°C. The final product should be stored at less than 5°C. Although salt is added in the dressing, the salt-in-moisture ratio (S/M) of creamed cottage cheese (1%–2%) is not high enough to hinder growth of contaminating bacteria. The dressing also contains lactose, which can be fermented by undesirable microorganisms.

In properly manufactured creamed cottage cheese, the environmental conditions within the cheese (low acid, relatively high pH, low S/M), are not harsh enough to strongly inhibit growth of most psychrotrophic contaminants. Thus, similar to conditions in raw or pasteurized milk, the microorganisms able to grow the fastest at the low storage temperatures are the dominant ones found in cottage cheese (Cousin, 1982). Gram-negative psychrotrophic bacteria such as pseudomonas (particularly *Pseudomonas fluorescens, Pseudomonas fragi,* and *Pseudomonas putida*), *Enterobacteriaceae* (coliforms, especially *Enterobacter aerogenes, Enterobacter agglomerans,* and *Escherichia coli*), *Alcaligenes, Achromobacter,* and *Flavobacterium* are the contaminants most likely to be found in cottage cheese (Brocklehurst and Lund, 1985; Marth, 1970). All are destroyed by pasteurization. *Pseudomonas* spp. are obligately aerobic and predominate at the surface, whereas the coliforms are aerobic and facultative anaerobes and can be found throughout the cheese. Their growth and metabolism, as well as that of yeasts and molds, result in undesirable flavors (called unclean, putrid, rancid, fruity, yeasty), surface film (Brockelhurst and Lund, 1985; Davis and Babel, 1954) and discoloration.

As with other cheeses, consumer acceptance of cottage cheese flavor varies considerably. Cottage cheese is consumed as a fresh product (a few days to 4 weeks old) and the ingredients (milk, nonfat dry milk, cream) can all influence the flavor (Bodyfelt et al., 1988). However, three main concerns can be controlled microbiologically: level of acidity, diacetyl (aroma), and level of undesirable flavors. The wash treatment and cream dressing can be used to adjust the pH of the cheese. Diacetyl can be added directly as a starter distillate or can be formed in the cream dressing through the metabolism of citric acid by *L. lactis* subsp. *lactis* biovar. *diacetylactis* and *Leuconostoc* spp. However, in addition to development of undesirable flavors, *Pseudomonas* spp., *Alcaligenes*, and *Enterobacter aerogenes* can oxidize diacetyl to acetoin, a flavorless compound (Seitz, 1963). This results in cheese that is bland or flat in flavor.

Growth of microorganisms in cottage cheese is inhibited most effectively by low storage temperature (<5°C), but it is also affected by pH and antimicrobials. Potassium sorbate is added to control yeasts, molds, and certain bacteria (Liewen and Marth, 1985; Sofos and Busta, 1982), although it may impart bitterness in creamed cottage cheese at levels greater than 0.075% (Bodyfelt, 1981). As with other acids, the effectiveness of sorbate depends on the sensitivity of the spoilage organisms and is a function of antimicrobial concentration of the undissociated form of the acid in the aqueous phase (pKa). It is enhanced by lower pH, the symbiotic effect of other antimicrobials, lower initial microbial load, and lower storage temperature. Thus, the degree of shelf-life extension resulting from use of sorbate is directly related to the quality of the initial product and subsequent handling (Bodyfelt, 1981).

Microgard (Wesman Foods, Inc., Beaverton, OR) is grade A skim milk that has been fermented by *Propionibacterium freudenreichii* and then pasteurized. It is widely used by the cottage cheese industry to inhibit growth of gram-negative bacteria, some yeasts, and some molds. The actual inhibitory compound is a bacteriocin (700 Da, heat stable, and proteinaceous in nature) (Daeschel, 1989).

Recently, direct injection of CO_2 into cream dressing has been shown to inhibit growth of *Pseudomonas* (Chen and Hotchkiss, 1991), *Listeria monocytogenes*, and *Clostridium sporogenes* (Chen and Hotchkiss, 1993). The technique has been used commercially (Mans, 1995) without the side effect of "carbonation" flavor in the cheese. It is claimed to substantially improve the shelf-life. It is believed that the CO_2 enters the cells and inhibits growth or kills the cells by lowering the pH within the cell. The technique is more effective at 4°C than 7°C.

The relatively short storage time (2–4 weeks) and rapid attainment and maintenance of low temperature during storage (<5°C) probably preclude growth of contaminating lactobacilli. However, if the temperature of storage is high enough (>7°C), as may occur in retail outlets, metabolism of lactobacilli may be a potential problem. Of particular concern is acid development through metabolism of lactose by either the nonstarter lactobacilli or surviving starter bacteria, or lactococci used in fermentation of the cream dressing. Poor acidification results in free whey or watery cheese, and an acid-tasting product. Growth and metabolism of psychrotrophic microorganisms is also increased.

In the past, mixed strain cultures, which included high levels of *Lc. lactis* subsp. *lactis* biovar *diacetylactis* were inadvertently used. These bacteria produce gas (CO_2) from the cometabolism of lactose and citric acid. The gas causes the curd to float, but the curd structure is also disrupted and weakened, leading to curd that is more easily shattered as the curd is stirred (Sandine et al., 1957).

C. Internally Ripened Blue Mold Cheeses

Roquefort, Stilton, blue and Gorgonzola are examples of cheeses in which development of flavor is dominated by metabolism of *Penicillium roqueforti* or *Penicillium glaucum*. These molds grow throughout the cheese (internally ripened) and are able to grow in the low-oxygen, high-salt conditions that are typical of these cheeses (Godinho and Fox, 1981; Golding, 1937). To facilitate the exchange of air with CO_2 produced in the cheese (via mold metabolism), the cheese is manufactured to produce an open texture and is pierced or punched with large bore needles. If the texture is too tight, the mold only grows near the puncture. In addition, internally mold-ripened cheeses may also be surface ripened with yeasts and bacteria (e.g., Stilton), a process that provides for distinctive taste sensations in a number of cheeses, including Limburger.

By international agreement, Roquefort cheese must be made from sheep's milk, in the Roquefort Valley, and ripened in naturally air-conditioned, high-humidity caves near the town of Roquefort, France (Bertozzi and Panari, 1993). Similarly manufactured cheese is produced from cow's milk in the United States and other countries is called blue (bleu) cheese. Morris (1981) provids an excellent technical description of the manufacture of blue-veined cheeses.

Blue cheese is usually made from a blend of heat-treated (raw) or pasteurized skim milk and homogenized cream, whereas Roquefort is made from non-homogenized raw, whole sheep's milk. The purpose of homogenization is to break up large fat globules. Sheep's milk naturally contains more smaller globules. Homogenization results in a whiter curd, (and increased contrast with the blue mold), increased flavor development through enhanced lipase activity (Morris et al., 1963), and a more porous, crumbly texture. Pasteurization destroys the milk lipase, which is believed to aid in ripening of the cheese and kills most of the nonstarter bacteria, especially lactobacilli, which might play an important role in the overall development of characteristic flavor. Milk is inoculated with spores (10^{3-4}/mL milk) of *P. roqueforti*. Some manufacturers prefer to inoculate the curd instead. Either method ensures that spores and thus flavor development will occur evenly throughout the cheese. During manufacture, steps are taken to produce a porous or open texture. The starter is *Lc. lactis*, and citrate-metabolizing strains (*Lc. lactis* subsp. *lactis* biovar *diacetylactis* and *Leuconostoc* sp.) are sometimes added. The carbon dioxide produced through metabolism of citric acid expands mechanical openings in the cheese, which, in turn, allows for more intrusive growth of the mold. The coagulum is cut when very firm into large cubes (0.95 cm diameter). The whey and curd mixture is heated to 35°C to 37°C, held for 30 minutes with agitation, and then whey is removed. The curds may be salted. The dry curds are put into hoops (drained vessels) and allowed to sit for several days at 21°C to 27°C. This encourages complete fermentation of the lactose, results in a cheese with a pH of 4.8 to 4.9 and permits full drainage of the whey. The body is desirably brittle and crumbly. Improper whey drainage may result in soft, mushy surface areas during storage. These areas are ideal for growth of yeasts and putrefactive bacteria such as *Pseudonomas* sp. and *Acinetobacter* (Smith et al., 1987).

Cheeses are brine salted or rubbed with salt for several days. Roquefort cheese must be dry salted by regulation. After salting, cheeses are pierced with 0.24-cm diameter needles and placed in a curing room (10°C at 90% to 95% humidity) for 2 to 4 weeks or until mold growth begins to appear at the openings of the holes. The piercing allows for transfer of oxygen and CO_2 to stimulate growth and metabolism of *P. roqueforti*. *P. roqueforti* is more tolerant of low oxygen, high CO_2, and high salt than most other species of molds. After sufficient mold growth, cheeses are wrapped and stored (matured) for 2 to 4 months at (10°C). In France, the later curing of the cheese occurs in the famous caves of Roquefort. After proper curing, cheeses are cleaned of surface growth (molds, yeasts) and repackaged for sale.

Metabolism of the molds (lipolysis and proteolysis) during maturation is essential for development of the distinctive blue cheese flavor (Goghill, 1979). A distinctive yeast flora also develops on Roquefort, including *Debaromyces hansenii, Candida* sp. and *Kluyveromyces lactis* (Besancon et al., 1992). The intensity of mold-derived flavors is so strong that, while other microorganisms are present in such high numbers (yeasts, micrococci, and lactobacilli), their contribution to the flavor of blue cheese cannot be ignored, neither can it be ascertained.

The S/M in the interior of blue cheeses can be as high as 6% to 8%. This inhibits growth of lactococci and *Leuconostoc* sp. Free fatty acids released through lipolysis and via oxidative decarboxylation are converted to methyl ketones. Of particular importance in blue-veined cheeses are 2-heptanone and 2-nonanone without which there is no distinctive blue cheese flavor. Secondary alcohols, methyl and ethyl esters derived from fatty acid metabolism and proteolysis, are essential for well-balanced and distinctive flavor of blue-veined cheeses (Kinsella and Hwang, 1976). Molds require oxygen to grow albeit at low levels for *P. roqueforti*. With *P. roqueforti*, too little oxygen can result in a change in color from blue-green to greenish-yellow. This situation can occur if the cheese is vacuum packaged. The proper color returns when the cheese is exposed to air.

During the initial salting and ripening of blue cheese, there is a conspicuous lack of visible growth of mold and yeasts on the surface. The low pH and high salt content keep the level of these microorganisms in check (Godinho and Fox, 1981). Some manufacturers also use a hot brine treatment (72°C for 20 seconds) to kill microorganisms at the cheese surface. However, as the cheese matures, salt diffuses in and the pH rapidly increases (>5.8 up to 6.5). Yeasts and molds metabolize lactic acid and hydrolyze protein, releasing ammonia and amino acids. Both metabolic activities result in pH increase. During maturation, microorganisms once held in check by adverse conditions (low pH) can begin to grow. Salt-tolerant bacteria such as *L. monocytogenes* and *Staphylococcus aureus* (de Boer and Kuik, 1987) are of particular concern (see Chapter 11).

D. Externally Mold-Ripened Cheeses: Camembert and Brie

Camembert and Brie are essentially the same cheeses, but in France are made in different regions. Brie cheese wheels are also larger in diameter (Masui and Yamada, 1996). Whole milk, sometimes with cream added (double-cream Brie), is inoculated with *Lactococcus* sp. After considerable acid development by the starter, coagulant is added. The coagulum is very firm when cut. This results in a higher moisture cheese. The coagulum is cut into large pieces, 1.6 cm diameter, stirred, and dipped into forms. Alternatively, the uncut coagulum may be dipped directly into the forms. The height of the finished cheese is important because the degree of ripening of the cheese depends on its thickness. The curd settles in the forms, which are turned approximately 6 to 8 hours after being filled. There is no cooking or heating step. No pressure is applied. As in blue cheese, the starter continues to produce acid until it becomes self-inhibited at pH 4.7 to 4.8.

Cheese is removed from the forms and brined or dry salted (salt rubbed or sprinkled on the surface). After the salt is absorbed (1 day), spores of *Penicillium camemberti* are sprayed onto the surface. The cheese is not pierced as in blue cheese so mold does not grow in the interior of the cheese, even though the milk may have been inoculated with mold spores. The cheese is transferred to shelves in rooms of high relative humidity (90%–95%) at 10°C. The cheese is placed on mats or perforated sheets to allow air contact with as much surface area as possible, thus permitting growth of the mold evenly over the entire surface area. The cheese is also turned regularly to expose bottom areas and keep the soft cheese from being imprinted with the perforated mats or sticking to them. After approximately 2 weeks in the ripening rooms, the mold has developed sufficiently, and the cheese is wrapped for sale. The cheese is then stored at a low temperature (4°C–7°C) for further ripening (2–4 weeks).

Slow growth of mold may indicate a too-high salt content or too-low pH. To prevent the latter, water may be added to the milk or whey to remove some of the lactose before the curd is transferred to the forms. The practice has also been applied in the manufacture of blue cheese.

Before growth of *P. camemberti*, the cheese is firm, crumbly, and acid. As the mold grows, it metabolizes lactic acid and hydrolyzes protein. Just beneath the surface growth, the cheese is very soft, creamy, and appears slightly darker and more yellowish than the interior portion of the cheese. As ripening continues, the interior becomes progressively softer and creamier just as at the surface. This progression is referred to as ripening from the outside to the inside. The change in the body of the cheese results, in part, from migration of calcium lactate from the inside to the outside of the cheese. In addition, migration of ammonia from the outside to the inside of the cheese raises its pH (6.5–7.0). This alters the protein and protein-serum phase with the net result of an increase in the fluidity of the cheese. With an increase in pH, the naturally occurring milk proteinase, plasmin (not active at low pH), hydrolyzes the protein, further softening the cheese. Eventually, the entire cheese becomes soft and creamy. Overripening either by poor stock maintenance or by design results in cheese that is very fluid and that "runs" when cut open.

Metabolism of *P. camemberti* results in hydrolysis of milk fat (lipolysis) and subsequent oxidative decarboxylation of free fatty acids to methyl ketones (Molimard and Spinnler, 1996). Of particular importance to the flavor of Camembert is 1-octen-3-ol, 1,5-octadien-3-ol, and 2-methylisoborneol (Karahadian and Lindsay, 1985). In the United States, Camembert and Brie are generally ripened with mold only. However, Karahadian and Lindsay (1985) postulated that growth of *Brevibacterium linens* on cheeses imported from France resulted in development of sulfur compounds: dimethyl disulfide, dimethyl trisulfide, and methional. Other regional differences in flavor may arise from metabolism of particular microflora contaminating the surface of the cheese. Nooitgedagt and Hartog (1988) found yeasts (predominantly *D. hansenii*, *Yarrowia lipolytica*, *Kluyveromyces marxianus*, and *Candida* spp.), and *Geotrichum candidum* and a few cheeses with greater than 10^4 staphylococci, greater than 10^5 *E. coli*, and greater than 10^7 *Enterobacteriaceae*.

E. Cheeses with Eyes

Swiss, baby Swiss, Gouda, and Edam are among the cheeses characterized by development of circular openings called eyes. Eyes develop through formation of CO_2 by metabolism of specific secondary bacteria. In Swiss-type cheeses, gas (CO_2) is formed by *P. fruedenreichii* subsp. *shermanii* through metabolism of lactic acid. In Gouda and Edam cheeses CO_2 is formed from metabolism of citric acid by *Leuconostoc* spp. and *Lc. lactis* subsp. *lactis* biovar *diacetylactis*.

These are the most difficult of cheeses to manufacture because of the strict grading regimen they must pass. Eye development is key and is sometimes the only criterion by which these cheeses are graded. Detailed descriptions on the manufacture of these cheeses have been provided by Reinbold (1972) and Olson (1969). The method of manufacture is similar for all cheeses with eyes.

Starters for Swiss and baby Swiss cheese are predominantly *Streptococcus thermophilus* with small amounts of *Lactobacillus delbrueckii* subsp. *bulgaricus, Lb. helveticus*, and *Lb. lactis*. Depending on the manufacturer, lactococci may also be used to ensure fermentation of all sugar, including residual galactose. The propionibacteria are added with the lactic starter.

Lactococci are used as the starter for Gouda and Edam cheeses. Gouda and Edam are manufactured similarly, but Edam is lower in moisture and fat content.

The main objective in the manufacture of cheeses with eyes is to produce a pliable curd mass. This is necessary for development of round eyes rather than slits or cracks. The pliability or elasticity of the cheese results from both protein density and physicochemistry. As CO_2 is formed, it accumulates at locations where air has been entrapped during processing or at sites where the curd is not tightly fused. Gas exerts pressure on the protein network. If the protein network is elastic, the protein bends or gives but does not break, forming an eye. If the protein network cannot withstand the pressure, it breaks and a slit is formed. Elasticity is a phenomenon related to calcium–phosphate bonding and electrostatic interactions between casein molecules. Thus, the rate and extent of acid development (pH) at whey drainage and pressing must be carefully controlled. To accomplish this, slow acid development is necessary and separation of curd and whey generally occur at a relatively high pH.

In Swiss cheese, after cutting, the curd is heated to 48°C to 53°C and held for 30 to 60 minutes depending on the desired moisture content and pH. In Gouda and baby Swiss varieties, a portion of the whey (25% of the milk weight) is drained and replaced with hot water to raise the temperature of the curds and whey to 38°C to 39°C. Addition of water not only heats the curd but also dilutes the lactose, thereby controlling the pH of the cheese. Alternatively, Swiss cheese manufacturers can control the pH by adding warm water to the milk and cold water to cool the curd before whey drainage (combined water addition is approximately 7%–10% of the weight of the milk). The high heat used in Swiss cheese manufacture inhibits acid development and partially inactivates the coagulant (depends on type of coagulant). The starter is not killed and resumes acid development as the curd cools.

In larger commercial manufacturing plants, regardless of the cheese type, the curd and whey mixture is pumped together into a smaller vessel or rectangular tower that concentrates the curd into a single large mass. Pressure is applied and serum is squeezed from the curd.

In the traditional method of Swiss cheese manufacture, all the curd from the cheese vat (a round kettle) is enclosed in a cheese cloth, lifted into the cheese form, whey is manually pushed out of the curd mass, and the cheese is pressed. Because serum is at a high pH during pressing, less calcium–phosphate is dissolved in the whey as compared to cheeses of a more acidic nature at drain (cheddar or mozzarella). Phosphate acts as a buffer and helps keep the curd from getting too low. Low pH (<5.1) inhibits growth of *Propionibacterium* sp. and is involved in development of a short, inelastic body in cheese. After pressing, the curd mass is brine salted.

To ensure curd fusion and complete sugar metabolism, the cheese is held at 7°C (prewarm room) for several days. The cheese is then placed in a warm room (10°C–13°C for Gouda and 20°C–26°C for Swiss).

The temperature affects both the growth of the eye former and the elasticity of the protein network. The warmer the cheese, the more elastic the protein. Rate of gas development is critical. If gas develops too rapidly and the curd cannot handle the gas pressure, the curd splits. If gas forms too slowly, the cheese maker may leave the cheese in the warm room for too long, resulting in too much proteolysis. When the gas does develop, the curd is no longer elastic, causing splits. After the eyes form, Swiss cheese is cooled and stored (<5°C) to prevent further gas development.

Gouda cheese may be ripened for extended periods at the warm room temperature. Because citric acid is limiting in cheese, there is no fear of excessive gas being formed by the added lactococci or *Leuconostoc*. However, in Swiss cheese, there is excess substrate (L-lactic acid) and potential for continued gas formation unless the cheese is cooled (Fedio et al., 1994; Hettinga et al., 1974). The search is underway to find *Propionibacterium* sp. that does not form gas during storage (Hofherr et al., 1983). Cold cheese is not elastic, so if gas is formed, it expands the existing eyes and they split. As cheese ages, the cheese protein is hydrolyzed (proteolysis) by residual coagulant, nonstarter bacteria, and plasmin (native milk proteinase). Proteolysis eventually destroys elastic properties of protein. Thus, if gas is formed in cheese after much proteolysis has occurred, slits are formed.

In all cheeses with eyes, undesirable gas formation can occur if high numbers of *Clostridium tyrobutyricum* are active. Their growth usually occurs months after the cheese is made and after much proteolysis has occurred. The result is split eyes or newly formed large slits called cracks. Metabolism by *C. tyrobutyricum* also results in rancidity and H_2S formation (Langsrud and Reinbold, 1974). The latter gives rise to the term "stinker cheese" or "stinkers" for short. Metabolism at the surface of cheese by *Pseudomonas* spp. also produce stinkers.

Certain varieties of cheeses with eyes, for example, Gruyere and Danbo, are also surface ripened (see Sec. IV G).

F. Surface-Ripened Cheeses

Limburger and traditional brick cheese are known for their highly malodorous character. For the olfactorally challenged, they literally stink; to the connoisseur, they stink in a pleasant sort of way. The strong odors arise from putrefication of protein, which releases ammonia and sulfide compounds (H_2S and methyl mercaptan).

These cheeses are made from whole milk with *Lactococcus* spp. starters. After cutting the coagulum, a portion of the whey is removed (25%–50% of the milk weight) and replaced with hot water. This raises the temperature and removes lactose from the curd, which prevents the cheese from becoming too acidic, a development that could delay ripening. There are several variations to this procedure, but the net result is the same. After 30 to 60 minutes, the whey is drained while the curd is stirred. If an open-bodied cheese is desired, all or most of the whey is removed. The curd is put into forms and may or may not be pressed. Once pressed, the cheese is brined, "smeared," and placed in a high humidity room (90%–95 relative humidity) at 13°C to 15°C. The cheese is slightly acid after brining (pH 5.2–5.4). The pH depends on the amount of whey removed and water added during manufacture.

Once brined, the cheese is inoculated with the smear. The smear is a mixture of several microorganisms, most importantly yeasts, micrococci, and *B. linens*, which develop as a layer on a cheese surface as it ripens. Smearing is done by scraping the smear layer from an already ripened cheese into a brine solution and then rubbing fresh cheeses with the smear-containing brine. This procedure is repeated every few days until luxurious growth occurs. The microorganisms in the smear may be purchased separately, mixed, and the cheeses inoculated.

B. linens gives the smear a red-orange color. Yeasts (*D. hanssenii, Candida* spp., *G. candidum, Y. lipolytica*) metabolize lactic acid and the pH of the cheese increases (Eliskases-Lechner and Ginzinger, 1995; Iya and Frazier, 1949). *Micrococcus* sp. (*M. varians, M. caseolyticus*, and *M. freudenreichii*) begin to grow, followed by *B. linens* (Lubert and Frazier, 1955). The pH must be greater than 5.5 for *B. linens* to grow (Kelly and Marquardt, 1939). The yeasts also synthesize vitamins (pantothenic acid, niacin, and riboflavin), which may be essential for *B. linens* to grow. A symbiotic relationship thus exists between growth of yeasts and *B. linens* (Purko et al., 1951). The length of time the smear is left on the cheese and the size of the cheese influence its flavor intensity. Limburger cheese is cut into small loaves (6.4 cm × 6.4 cm × 13 cm) before smearing, and the smear is not removed. Traditional brick cheeses are larger pieces and, again, the smear is not removed. In less pungent brick cheese, the smear is washed off after 4 to 10 days. If the smear is left on the cheese, ripening continues. Ripening of the cheese involves extensive proteolysis, with release of ammonia, H_2S, and methyl mercaptan (Grill et al., 1966). These flavor compounds diffuse into the cheese. Metabolism of lactic acid at the surface of the cheese and proteolysis on the inside of Limburger, caused by coagulant and plasmin, eventually lead to a fluid interior. The cheese is runny when cut. Because of the pH increase, microorganisms once held in check by the low pH can then begin to multiply. *L. monocytogenes*, staphylococci, and coliforms are of major concern (see Chapter 11).

G. Colby, Sweet Brick, Muenster, Havarti

Colby, sweet brick, Muenster, and Havarti are consumed with minimum ripening, generally between 1 to 3 months. Inferior products are often sold if the cheese is aged for a longer time. These cheeses are low in acid because lactose is rinsed from the curd, and they have a pH of approximately 5.2 to 5.4. Lactococci are used as the starter in Colby cheese but *S. thermophilus* is sometimes used in brick cheese and is preferred for Muenster. Some manufacturers also use *Lb. delbrueckii* subsp. *bulgaricus* if *S. thermophilus* is used. Havarti cheese is manufactured with lactococci with added *Leuconostoc* sp. and citrate-metabolizing *Lc. lactis* subsp. *lactis* biovar *diacetylactis*. Once the coagulum is cut, the curds are heated to 36°C to 37°C if lactococci are used as the starter, and 39°C to 41°C if thermophilic cultures are used. In Muenster and brick cheeses, there is little or no acid development before putting the curd into forms.

In Colby manufacture, the pH of the curd at whey drainage is approximately 6.1. Whey is drained and curds are continually stirred as cool water (30°C–32°C) is sprinkled on them. This not only cools the curd but also removes lactic acid and lactose. All the whey is then removed and curd is salted, put into forms, pressed, and stored. The result is a cheese with a pH of 5.2 to 5.3 and many mechanical openings. Cool, firm curd does not fuse completely even when pressed. Vacuum packaging closes the openings, and the cheese forms a tight knit texture, but this is not allowed in authentic Wisconsin Colby.

In Muenster cheese, the curd and whey are pumped into rectangular shaped open-ended forms, lightly pressed, brined, and stored. In the manufacture of brick and Havarti, once the coagulum is cut, a portion of the whey is drained and replaced with hot water to heat the curds to 36°C to 37°C. The curds are then handled as with Muenster. The more whey removed before putting curd into forms, the more mechanical openings appear in the cheese. The cheeses are brined and stored. Havarti is ripened at 13°C to 16°C for 2 to 6 weeks; Muenster and sweet brick are stored at 7°C and are ready for consumption within a month.

H. Cheddar Cheese

Cheddar cheese is consumed when it is anywhere from 1 month to several years of age. Pasteurized or heat-treated (67°C–70°C for 20 seconds) whole milk is used. *Lactococcus* spp. is the starter, with *Lc. lactis* subsp. *cremoris* preferred for long-hold cheese. Once the coagulum is cut, the curd is heated to 38°C to 39°C. After proper stir-out, whey is completely drained and the curd is either continuously stirred (stirred curd cheddar) or allowed to mat (cheddared curd, also called milled curd cheddar). Stirred curd is preferred if the cheese will be used for process cheese. Cheddared curd is preferable for table cheese. Cheddared curd cheese is thought to develop a better flavor and has a smoother body than stirred curd cheese. When the curd reaches the desired pH, curd is salted. In cheddared curd, the matted curd is cut into large pieces, which are periodically turned (cheddared) in the vat until the proper pH is reached (pH 5.4–5.5). The slabs of curd are then cut into finger-sized pieces (milled), salted, put into forms and pressed. The cheeses are generally stored at 7°C to 9°C. There is considerable variation in the details of cheddar cheese manufacture resulting from mechanization of the process, size of vats, rate of acid development by the starter, and whims of the manufacturer. The pH of the cheese at 1 week is generally 4.95 to 5.1.

I. Pasta Filata Cheeses: Mozzarella and Provolone

Manufacture of pasta filata cheeses is almost identical to milled curd cheddar cheese. *S. thermophilus* (cocci) and *Lb. delbrueckii* subsp. *bulgaricus* or *Lb. helveticus* (rods) are used as the starter. The ratio of cocci to rods used varies from manufacturer to manufacturer, but ratios of 1:1, 3:1, or 5:1 are commonly used. The cocci are the main acid producers. Lactococci are sometimes added. As with cheddar cheese, considerable variation exists in the manufacture of mozzarella and provolone cheeses. Mozzarella is a common name applied to various cheeses made similarly (Code of Federal Regulations, 1995).

Provolone is lower in moisture but higher in fat than mozzarella cheese. Lipases are added to the milk and lipolysis results in a light piquant or rancid flavor in provolone. After the coagulum is cut, the curd and whey mixture is heated to 42°C to 43°C. At pH 6.1, whey is drained and the curd cheddared. At a pH of 5.15–5.35, the curd is milled and placed in a hot water bath (70°C–88°C) and kneaded. The pulling or stretching of the molten curd mass gives the pasta filata cheeses their name, but also imparts a fibrous body to the cheese. The temperature of the cheese (57°C–63°C) varies with the time of exposure to mixing and water temperature. Coagulating enzymes vary in their heat sensitivity and the residual coagulant can play a major role in the physical characteristics of the cheese (melt, stretch, oiling off, burning, chewiness) (Kindstedt, 1993). The starter may or may not survive the heat treatment of the curd. This has a major impact on the metabolism of residual sugar and consequently Maillard-browning reactions when the cheese is subsequently heated on pizza (Johnson and Olson, 1985). After the cheese is stretched, it is shaped (usually into cylinders), placed in cold water to cool, and eventually brined.

High-moisture fresh mozzarella can be eaten immediately, but the more familiar pizza-type mozzarella (low-moisture, part skim) is aged for a few days. This short ripening period (4–7 days) is essential to allow entrapped water (via the kneading process) to be absorbed. If the water is not absorbed, it (expressible serum) causes matting of shredded cheese. As the cheese ages, proteolysis results in an increase in melt and a decrease in stretchability when used in cooking. This occurs in all cheeses but is most noticeable in mozzarella because of the demands that are placed on the physical characteristics of this cheese when baked or fried.

The physical properties are linked to pH, composition, and proteolysis (Kindstedt, 1993). Thermophilic starter strains do not use the galactose portion of the lactose molecule, and it accumulates in the cheese (Hutkins and Morris, 1987). Use of galactose-fermenting starter strains (Mukherjee and Hutkins, 1994) may reduce the level of galactose. Residual galactose and lactose are responsible, in part, for the browning of the cheese when baked (Johnson and Olson, 1985). Dehydration and scorching of protein during baking also results in browning. Residual sugar is also a prime substrate for heterofermentative lactobacilli and coliforms. Gas formation by these bacteria leads to "blown" cheeses (splits, eyes) and puffy packages.

J. Parmesan and Romano

S. thermophilus (cocci) and *Lb. helveticus* or *Lb. delbrueckii* subsp. *bulgaricus* (rods) are used to manufacture grana cheeses, Parmesan and Romano. The ratio of cocci to rods varies according to the manufacturer but ratios of 1:1 or 3:1 cocci to rod are common. Some manufacturers also add a small amount of *Lc. lactis* to ensure complete sugar metabolism. Reduced-fat milk is used for both. Moisture content is low (32% maximum for Parmesan, 34% maximum for Romano). Parmesan must be aged 10 months and has a minimum FDM of 32%, whereas Romano must be aged 5 months and has a minimum FDM of 36% (Code of Federal Regulations, 1995). Both are manufactured similarly. The coagulum is cut softer and finer than for other cheeses to ensure a drier finished cheese. Fast acid development by the starter is desired. After cutting, the curd and whey mixture is heated to 45°C to 47°C and stirred until the pH of the whey is approximately 6.0. Whey is then drained. Curds are continuously stirred until all the whey is removed. The low pH and high heat during stir-out enhances syneresis. Curds are put into forms (usually 9- 18-kg wheels), pressed overnight, and brine salted for several days to 2 weeks. Some manufacturers brine only 1 day and apply salt to the cheese after it is removed from the brine. This method of salting is called dry salting and may require several days of application to achieve the desired salt level. Alternatively, the curds first may be salted, put into forms, and then pressed. This process produces what is referred to as barrel cheese. The cheese is not brine salted and the process usually requires a longer stir-out and application of less salt. If the salt content is too high it may inhibit fermentation of all the sugar. Residual sugar may participate in Maillard-browning reactions, especially if the cheese is further dried (with heat) to produce grated cheese. After brining, the cheese is stored at 7°C to 10°C. Traditionally, wheels of cheese are coated with an oil to prevent mold growth and coated with wax at a later date. Some modern manufacturers coat the cheese with a polymer containing a mold inhibitor (natamycin).

Although Parmesan and Romano cheese are made similarly, they taste distinctively different. Pregastric esterase or lipase is added to the milk for manufacture of Romano but not to milk for manufacture of Parmesan. Thus, the flavor of Romano is rancid or picante, whereas that of Parmesan is described as sweet and nutty.

K. Reduced-Fat Cheese

Demand by consumers has led to development of reduced-fat versions of popular cheese varieties. Early attempts did not meet with tremendous success because of poor physical and flavor characteristics. Adjustments to the manufacturing protocol, including use of selected starter strains and particular attention to dairy plant hygiene, have greatly improved the quality of these cheeses. Young, mild-flavored cheese with a reduction in fat content of 25% to 33% as compared to the full-fat cheese, have almost, if not actually, duplicated the quality (flavor and body) of the full-fat counterpart. Cheeses with a fat reduction of greater than 50% have yet to achieve similar results. It is more likely that these cheeses have taken on their own unique flavor and are being accepted on their own merits rather than in comparison to other cheeses.

Reduced-fat versions of cheese are similar to their full-fat counterparts in that they are subject to the same microbiologically induced defects and for the same reasons. However, the ecology (variety of bacteria and changes over time) of the cheeses may or may not be the same; this has not yet been studied. Cheese with less fat is firmer than cheese with higher fat content. To overcome this problem, reduced-fat cheeses are manufactured to contain much higher moisture contents. But, higher moisture means higher lactose in the cheese, which, in turn, means that the cheese is high in acid after the starter ferments the sugar. To avoid producing an acidic cheese, many manufacturers or reduced-fat cheese (regardless of the type) use whey dilution or curd rinsing to remove some of the lactose. However, reduced-fat cheddar is also being made commercially without a rinse treatment by using a specific manufacturing protocol to retain the buffering capacity of the cheese (Johnson and Chen, 1995). The cheese contains more acid (up to 2% lactic acid compared with less than 1.6% lactic acid in full-fat cheddar) but both have the same pH. However, compared with full-fat cheeses, most reduced-fat versions are higher in moisture and pH and generally lower in salt (lower S/M) and lactic acid. Thus, the cheese environment and chemistry is not the same between full-fat and reduced-fat cheeses.

Consequences of reduction of fat on the flavor of cheddar cheese are well recognized (i.e., lack of similar flavor intensity at similar age), but the difference in flavor is less evident in reduced-fat versions of cheeses in which the full-fat version is mild-flavored (Muenster, brick, Gouda). Reduced-fat cheddar cheese made with a curd rinse tastes similar to Colby. Reduced-fat cheddar made without a curd rinse has more cheddar flavor than one made with a rinse, but the development of flavor still lags behind that of full-fat cheddar.

Although reduced-fat cheeses have met with some success, there appears to be a universal concern that cheeses not ripened by yeasts or molds lack flavor. As a result, adjunct bacteria, particularly *Lactobacillus* spp. and *B. linens* are being used commercially to enhance cheese flavor.

V. CHEESE RIPENING—THE INFLUENCE OF MICROORGANISMS

Microorganisms found in cheese can be classified into two groups: those that are deliberately added, such as starters and adjuncts, and those that are adventitious contaminants. The primary role of the starter bacteria is to produce acid at a consistent rate, but it would be wrong to assume that their role is limited to this. The starter has a major impact on flavor in cheese consumed fresh. As the cheese matures, the direct contribution to flavor by the starter diminishes as the non-starter flora grow. Although, in most instances, the exact means by which starter bacteria or adjunct microorganisms contribute to development of flavor is controversial, they can both influence cheese maturation. The development of flavor in blue, Camembert, Limburger, Romano, and provolone cheeses is clearly dominated by microorganisms or enzymes deliberately applied to them. With other varieties of cheese, however, development of flavor is not clearly understood. Scores of compounds with the potential to affect flavor have been isolated from a variety of cheeses. But the full duplication of cheese flavor chemically has eluded us.

Olson (1990) described the possible role of starter bacteria in cheese flavor development as follows: fermentation and depletion of fermentable carbohydrates creates an environment that controls growth and composition of adventitious flora. This is accomplished through development of acids, creation of low oxidation-reduction potential during early stages of cheese maturation, and competition for nutrients. In addition, starters can develop flavor compounds directly and indirectly through their metabolic activities (Crow et al., 1993). (See Chapter 7 for more details on the influence of carbohydrate metabolism and proteolysis by bacteria on cheese flavor development).

Autolysis of starters (and adjuncts) releases nutrients that serve as metabolites for other microorganisms in cheese (Thomas, 1987b). Also, activity of released intracellular peptidases can contribute to the increase in the free amino acid pool within cheese (Lane and Fox, 1996). Amino acids can, in turn, be metabolized by other bacteria directly to flavor compounds or can react chemically with other constituents in cheese to produce flavor compounds (Griffith and Hammond, 1989). Any bacteria thriving in cheese can potentially influence the flavor of cheese through synthesis of flavorful metabolites. The characterization of flavor in most cheeses is lacking, thus the direct connection between microbial metabolism and cheese flavor is also limited. Another problem hampering the study of the influence of starter and non-added microflora on flavor in cheese is a lack of consensus on what constitutes cheese flavor, especially in varieties of cheese not ripened by yeasts or molds. There is an element of distrust in that what one person perceives to be true cheese flavor may not be the same as what another might consider to be cheese flavor. Consequently, results of experiments on flavor enhancement or acceleration of flavor development are often met with skepticism.

Starters are the dominant bacteria found in cheese initially. Numbers range from 10^6 to 10^9/g cheese. As the cheese ages, their numbers decrease and the numbers of nonstarter bacteria increase. The rate at which this happens depends on the strain of starter and the initial numbers and type of nonstarter bacteria. Lactobacilli constitute most nonstarter lactic acid bacteria in cheddar cheese (and probably most cheeses), with the dominant species of quality cheese being *Lc. casei* and *Lc. plantarum* (Franklin and Sharpe, 1963; McSweeney et al., 1993; Peterson and Marshall, 1990). Heterofermentative lactobacilli may be present with no visible sign of gas production (Laleye et al., 1987). *Lactobacillus* numbers in raw milk are greatly reduced by pasteurization. The presence of lactobacilli in pasteurized milk generally indicates high (>10,000/mL) numbers in raw milk or postpasteurization contamination. The type and strains of nonstarter bacteria found in cheese is dependent on the initial numbers in the milk (especially if raw milk is used), biofilm formation on equipment and subsequent contamination, and the ability of individual strains to survive and compete in the cheese environment (pH, acidity, temperature, availability of nutrients).

Addition of *Lactobacillus* adjuncts has been suggested as a means of controlling the numbers of adventitious lactobacilli by, at least initially, outcompeting the other microflora in the cheese (Martley and Crow, 1993). However, dependent on the strain, the adjunct culture may die or may not compete well against nonstarter microflora; thus, the ecology of the cheese can change as the cheese matures. The dominance of the cheese microflora by lactobacilli has led to numerous studies advocating addition of defined strains of lactobacilli to milk or cheese to reduce bitterness, enhance flavor, or develop particular textural or physical attributes in the cheese (El-Soda, 1993). Other bacteria, particularly *B. linens*, have also been used commercially to enhance the flavor of cheddar and reduced-fat cheeses. The use of adjunct bacteria to accelerate flavor development has met with some resistance by manufacturers because the flavor developed in the cheese is not the same as the flavor of the cheese without the adjunct. Consistency of flavor quality is a major goal of the cheese maker. Not surprisingly, reduced-fat varieties of cheese more closely mimic the full-fat counterpart if the same adjunct is used in both cheeses.

VI. ASSESSMENT OF MICROBIOLOGICALLY INDUCED DEFECTS IN CHEESE

It is difficult at times to assess quality problems occurring with cheese. A thorough knowledge of all aspects of cheese making is required for the detective work necessary to determine cause and effect relationships that may lead to a cheese quality problem. Foremost is identification of the problem. Is the problem really microbially related or is it the result of mechanistic shortcomings of manufacture? For example, openings in cheese can be a result of either pressing cold curd (mechanical) or gas formation by heterofermentative bacteria.

Second, if the problem is microbially induced, how did the organism gain access to the product and is the problem exacerbated by the manufacturing protocol, handling, storage, pH, or composition of the cheese? For example, residual sugar in cheese as the result of incomplete fermentation by the starter can be fermented by contaminating heterofermentative bacteria leading to gassy cheese. Incomplete fermentation, in turn, can result from a change in the composition of the starter because of improper starter preparation. Perhaps the cheese was cooled prematurely, too much salt was added, or phage killed the starter. Regardless of the circumstances, a contaminating organism must be present and must grow. If the microorganism is present but does not grow, there is no problem. Many legal questions have arisen because of this simple concept.

The problem results from *growth* of a microorganism. Perhaps, had the cheese not been temperature abused, the microorganism would not have grown. Cheese is not made in a sterile environment. It is inevitable that contaminating microorganisms will be present in the cheese. It is not inevitable that they will cause a problem in the cheese.

Prevention of undesirable growth of microorganisms in cheese involves four steps. One, keep the microorganism out of milk or prevent its growth in milk (hygiene on the farm, quickly cooling the milk, short time between milking and cheese making). Two, kill the bacteria (pasteurization). Three, manufacture the cheese to prevent contamination (dairy plant hygiene). Four, create an environment within the cheese such that, if the microorganism is present, its growth will be limited (proper pH, salt, fermentation of all sugar, low storage temperature). However, the most universally accepted (but not always properly practiced) method of preventing defects caused by microorganisms is sanitation. In this regard, development of biofilms is important.

Many bacteria can form biofilms or can become associated with them. Biofilms consist of microorganisms immobilized at a surface, typically embedded in an organic polymer matrix of bacterial origin (Marshall, 1992). Biofilms can develop at almost any wet surface (equipment) (Criado et al., 1994). Microorganisms attach to the surface, or to other organic material already attached to the surface, excrete copious amounts of extracellular polymers, and grow vigorously, creating a biofilm. Bacteria can form biofilms within a few hours of initial attachment to a surface. As the biofilm becomes thicker, the outer layer is sloughed as the result of turbulence (e.g., milk stirring in a vat). Microorganisms within the sloughed-off pieces contaminate the milk. Other organisms can also attach to the biofilm. Sanitizers are less affective against biofilms because the sanitizer reacts only with the outer layer and the extracellular polymers protect the microorganisms. Therefore biofilms must be removed before sanitizers are applied. The cleaning regimen becomes paramount in controlling bacterial contamination.

1. Molds

Airborne mold spores are ubiquitous, but, upon germination, they require oxygen to grow and sporulate. Therefore, mold growth on the surface of cheeses exposed to air is to be expected. Molds are not supposed to grow on cheeses that are vacuum packaged, but they sometimes do. Molds tend to grow on cheese where pockets of air exist between the packaging material and the cheese surface (Hocking and Faedo, 1992). Growth is limited by the amount of residual oxygen. Low oxygen levels may dictate the species of molds found. The most common molds found on vacuum-packaged cheddar cheeses are *Penicillium* spp. (especially *Penicillium commune*, a blue mold), and *Cladosporium* spp. (especially *Cladosporium cladosporioides*, a black mold). Other molds found on different cheeses include *Aspergillus, Fusarium, Mucor, Scopulariopsis*, and *Verticillium*. *Penicillium* spp. appear to be the dominant type of molds that grow on cheeses

(Lund et al., 1995). *P. commune* is the most widespread and frequently occurring species found on all cheese types and in smear of surface-ripened cheeses (Lund et al., 1995). Although *Aspergillus* spp. and *Penicillium* spp. dominate the fungi isolated from air in cheese plants, *Penicillium* spp. dominate the fungi isolated from cheese with very low levels of *Aspergillus* also being present.

Potassium sorbate and natamycin are used to control mold growth. Sorbate-resistant strains of *Penicillium* metabolize sorbic acid to yield 1,3-pentadiene, which has a kerosenelike odor (Marth et al., 1966). If sorbic acid is added to the cheese and the cheese is made into processed cheese, the sorbate is diluted. Sensitive strains are then able to grow and may produce 1,3-pentadiene. The maximum amount of sorbic acid permitted for use in cheese is 0.3% by weight (Code of Federal Regulations, 1995), which is not enough to inhibit all strains of *Penicillium* but adequate to inhibit *Aspergillus* spp. (Liewen and Marth, 1984). In cheeses with a rind (e.g., Gouda, Parmesan), a polymer coating is applied to prevent mold contamination and for appearance. Sorbates or natamycin are incorporated into the coating. Sorbates diffuse into the cheese and may cause off-flavors, but natamycin diffuses very little and does not give the cheese an objectionable flavor (de Ruig and van den Berg, 1985).

2. Yeasts

Although the growth of yeasts is desirable in surface-ripened and some mold-ripened cheeses, it is not desirable in most other varieties. The heterofermentative metabolic activity (alcohol and CO_2) of yeasts makes them particularly easy to identify as spoilage organisms, even though visible colonies are not observed. The cheese tastes yeasty, a taste reminiscent of raw fermented bread dough (Horwood et al., 1987). Lipolytic activity can lead to rancid flavors (free fatty acids), and the combination of alcohols and free fatty acids can lead to fruity flavors. Although yeasts are commonly associated with slimy surface defects, other putrefactive organisms such as *Pseudomonas* spp. contribute greatly to the defect. A major factor contributing to growth of yeast, or any contaminating organism, is a wet cheese surface. This situation can occur for several reasons. As cheese matures, proteolysis results in release of moisture held by the protein network. If the cheese is warmed or if the cheese is ripened at greater than 7 to 8°C, moisture collects at the surface of the cheese, that is, the cheese "sweats." Moisture (serum), laden with potential nutrients (lactic acid, dissolved peptides, amino acids) accumulates between the packaging material and the cheese, setting up an ideal situation for rapid microbial growth. The cheese must first be contaminated. Excellent plant hygiene is a necessary because yeasts are common contaminants in the dairy plant environment (wet surfaces, spilled milk, whey). A major source of yeasts are brines (Kaminarides and Laskos, 1992; Viljoen and Greyling, 1995) and thus brined cheeses tend to be more susceptible to yeast contamination. In addition, the high salt at the surface of the cheese draws moisture, creating an environment that favors yeasts. The most frequently isolated yeasts are *Candida* spp., *Y. lipolytica, K. marxianus, G. candidum, D. hansenii,* and *Pichia* spp. (Fleet, 1990; Hocking and Faebo, 1992; Rohm, 1992; Viljoen and Greyling, 1995).

Yeasts and molds are common on the surface of rind cheeses, a large group of traditional European cheeses. These are cheeses that are not covered or packaged but rather allowed to mature "in air." The low-moisture and high-salt environment at the surface (most are brined cheeses) creates conditions selective for yeasts and molds. However, with these cheeses, the growth of mold and yeasts is expected if not demanded.

3. Gassy Defects in Cheese

In Swiss, Gouda, Havarti, Roquefort, and similar varieties of cheese, the controlled development of gas by bacteria during maturation is desired. The result of gas formation in these cheeses is development of eyes (Swiss, Gouda) and expansion of preexisting mechanical openings deliberately formed during manufacture. In any cheese, however, gas formation can lead to the undesirable development of slits, small round eyes (sweet holes), or blown, "puffy" packages. Whether a slit or a sweet hole develops is determined by the physical properties of the cheese. Eyes are formed if the cheese can be deformed without fracturing. This property is determined by cheese composition, temperature of the cheese, rate of gas formation, and, most importantly, pH and degree of proteolysis (Grappin et al., 1993; Luyten et al., 1991).

In general, a minimum population on the order of 10^6 colony-forming units per gram is necessary before openness from gas production occurs (Martley and Crow, 1996). The nonstarter flora most often associated with slit formation in cheese are obligate heterofermentative lactobacilli, *C. tyrobutyricum,* and facultative lactobacilli. Others encountered but far less often are coliforms, yeasts, "wild" propionibacteria, and *Leuconostoc.* The incidence of slits or blown cheese and the causative organism is reflective of the microbial quality of the milk, overall dairy plant hygiene, heat treatment given the milk, post heat-treatment contamination, rate and extent of acid development, residual sugar, and cheese environment, pH, and redox potential. Pasteurization is very effective at killing all coliforms, leuconostocs, most strains of lactobacilli, and greatly reducing the level of all microorganisms except clostridial spores.

Fermentation of residual sugar (lactose or galactose) is a common source of carbon dioxide in cheese. The level of sugar and the speed at which it is eliminated by the homofermentative starter is critical. Slow starter activity and incomplete fermentation by thermophilic starters are chief causes of residual sugar. *S. thermophilus* and *Lb. delbrueckii* subsp. *bulgaricus* do not ferment the galactose moiety of the lactose molecule and release it into the cheese. Addition of mesophiles or *Lb. helveticus* to the starter can eliminate galactose from the cheese. However, in pasta filata cheeses, the heat treatment given the cheese can greatly reduce the level of starter. The starter must be able to ferment the sugar at the low temperature (<7°C) at which the cheese is stored, an unlikely possibility with thermophilic starter.

Cometabolism of citric and lactic acid by facultative lactobacilli, *Lb. casei*, and *Lb. plantarum* is another source of carbon dioxide (Fryer et al., 1990; Laleye et al., 1987; Lindgren et al., 1990; Thomas, 1987a). Because facultative lactobacilli are ubiquitous in cheese, it is by default (no validation) that their metabolism is regarded as the cause of tiny slits in cheese when no other potential gas-forming bacteria are found.

Lactic acid fermentation by propionibacteria and clostridia is also a major source of gas in cheese. These organisms are regarded as the culprits in late blowing of cheese. As the cheese ages, extensive proteolysis results in an increase in pH and the release of amino acids, which stimulate their growth. Although many strains of clostridia can ferment lactic acid, *C. tyrobutyricum* is probably the only one that is significant in cheese (Klijn et al., 1995).

Other minor contributors to gas formation in cheese are amino acid catabolism (nonstarter lactobacilli, propionibacteria, *Lc. lactis* subsp. *lactis*) and use of urea by streptococci (Martley and Crow, 1996). However, decarboxylation of glutamic acid into carbon dioxide and 4-aminobutyric acid is the main source of eye and split formation in cheese made with a particular thermophilic starter composed of *S. thermophilus* and *Lb. helveticus* (Zoon and Allersma, 1996).

4. Discoloration in Cheese

Color is an important sensory attribute of cheese, and consumers avoid cheese that exhibit discoloration. Annatto-colored cheeses (cheddar, Colby) are susceptible to light-induced, oxidation which turns affected areas pink (Hong et al., 1995). Govindarajan and Morris (1973) reported that hydrogen sulfide produced from amino acid metabolism by nonstarter bacteria in cheese is responsible for formation of a pink precipitate of norbixin, a component of annatto. Cheese color can also be bleached under acid conditions but the color returns as the pH of the cheese increases during maturation. This defect is common when whey is entrapped between curd particles (mechanical openings). Lactose in the whey is fermented, forming localized areas of low pH (<5.0) and consequently bleached color.

Paremsan, Romano, and Swiss cheeses are susceptible to a defect known as pink ring. As the name implies, a pink ring develops around the outside of the cheese and can progressively develop throughout the cheese from the outside to the inside. The pink becomes brown with age. It is most common in air-ripened cheeses (non–vacuum-sealed cheese). Shannon et al. (1977) implicated metabolism of tyrosine by certain strains of *Lb. helveticus* and *Lb. delbruekii* subsp. *bulgaricus* as the cause for the pink ring defect. Presence of oxygen appears necessary for development of the defect. It is more common in stirred-curd direct-salted Parmesan cheese (nonbrined) in which air is incorporated during the lengthy stir-out and is not subsequently removed by fermentation or vacuum packaging. Mallaird browning has also been implicated in some instances in which residual galactose is present in the cheese because of the metabolism of thermophilic starters.

Brown or red spots in Swiss cheese have been traced to growth of certain strains of "wild props," *Propionibacterium thoenii* (Baer and Ryba, 1992) or *Propionibacterium jensenii* (Britz and Riedel, 1994).

5. Calcium Lactate Crystals

White crystalline material on the surface of cheddar and Colby cheese is often confused with mold growth. It is, however, calcium lactate, a racemic mixture of L(+) and D(–)-lactic acid (Severn et al., 1986). Lactose fermentation by *Lactococcus* spp. produces L(+)-lactic acid. Growth of nonstarter lactic acid bacteria, particularly lactobacilli and pediococci racemize L(+)-lactic acid to D(–)-lactic acid (Thomas and Crow, 1983). Crystals can also form in the interior of the cheese but generally form where moisture (serum) can collect (Johnson et al., 1990b). Not all crystalline material is calcium lactate but may be composed of tyrosine (from proteolysis) or calcium phosphate (Conochie and Sutherland, 1965).

REFERENCES

Atlas RM, Bej AK. Polymerase chain reaction. In: Gerhardt P, Murray RGE, Wood WA, Krieg NR, eds. Methods for General Molecular Bacteriology. Washington, DC: American Society for Microbiology, 1994.

Baer A, Ryba I. Serological identification of propionibacteria in milk and cheese samples. Int Dairy J 2:299, 1992.

Bertozzi L, Panari G. Cheeses with appellation d'origine contrôlée (AOC): factors that affect quality. Int Dairy J 3:297, 1993.

Besancon X, Smet C, Chabalier C, Rivemale M, Reverbel JP, Ratomahenina R, Galzy P. Study of surface yeast flora of Roquefort cheese. Int Dairy J 17:9, 1992.

Bodyfelt FW. Sensory and shelf-life characteristics of cottage cheese treated with sorbic acid. Proceedings of the Second Biennial Marschall International Cheese Conference, Madison, WI, 1981, pp 406–425.

Bodyfelt FW, Tobias J, Trout GM. Sensory evaluation of cultured milk products. In: *The Sensory Evaluation of Dairy Products.* New York: Van Nostrand Reinhold, 1988, p 227.

de Boer E, Kuik D. A survey of the microbiological quality of blue-veined cheeses. Neth Milk Dairy J 41:227, 1987.

Britz TJ, Riedel K-HJ. *Propionibacterium* species diversity in Leerdammer cheese. Int J Food Microbiol 22:257, 1994.

Brockelhurst TF, Lund BM. Microbiological changes in cottage cheese varieties during storage at +7°C. Food Microbiol 2:207, 1985.

Carini S. Lysozyme: activity against clostridia and use in cheese production—a review. Microbiol Aliments Nutr 3:299, 1995.

Chen JH, Hotchkiss JH. Effect of dissolved carbon dioxide on the growth of psychrotrophic organisms in cottage cheese. J Dairy Sci 74:2941, 1991.

Chen JH, Hotchkiss JH. Growth of *Listeria monocytogenes* and *Clostridium sporogenes* in cottage cheese in modified atmosphere packaging. J Dairy Sci 76:972, 1993.

Code of Federal Regulations. Title 21. Part 133. Cheeses and related cheese products. Rockville, MD: Office of the Federal Register National Archives and Records Administration, 1995, p 295.

Coghill D. The ripening of blue vein cheese: a review. Aust J Dairy Technol 34:72, 1979.

Collins EB. Resistance of certain bacteria to cottage cheese cooking procedures. J Dairy Sci 44:1989, 1961.

Conochie J, Sutherland BJ. The nature and cause of seaminess in cheddar cheese. J Dairy Res 32:35, 1965.

Cousin MA. Presence and activity of psychrotrophic microorganisms in milk and dairy products: a review. J Food Prot 45:172, 1982.

Crow VL, Coolbear T, Holland R, Pritchard GG, Martley FG. Starters as finishers: starter properties relevant to cheese ripening. Int Dairy J 3:423, 1993.

Criado M, Suárez B, Ferreirós CM. The importance of bacterial adhesion in the dairy industry. Food Technol 48:123, 1994.

Daeschel MA. Antimicrobial substances from lactic acid bacteria for use as food preservatives. Food Technol 43:164, 1989.

Davis PA, Babel FJ. Slime formation on cottage cheese. J Dairy Sci 37:176, 1954.

Drake M, Small CL, Spence KD, Swanson BG. Rapid identification of *Lactobacillus* sp. in dairy products using the polymerase chain reaction. J Food Prot 59:1031, 1996.

Eliskases-Lechner F, Ginzinger W. The yeast flora of surface-ripened cheeses. Milchwissenschaft 50:458, 1995.

El Soda M. Accelerated maturation of cheese. Int Dairy J 3:531, 1993.

Emmons DB, Tuckey SL. Cottage Cheese and Other Cultured Milk Products. New York: Pfizer Inc., 1967.

Fedio WM, Ozimek L, Wolfe FH. Gas production during the storage of Swiss cheese. Milchwissenschaft 49:3, 1994.

Fleet GH. A review: yeasts in dairy products. J Appl Bacteriol 68:199, 1990.

Franklin JG, Sharpe ME. The incidence of bacteria in cheese milk and cheddar cheese and their association with flavor. J Diary Res 30:87, 1963.

Fryer TF, Sharpe ME, Reiter B. Utilization of milk citrate by lactic acid bacteria and "blowing" of film wrapped cheese. J Dairy Res 37:17, 1970.

Godinho M, Fox PF. Effect of NaCl on the germination and growth of *Penicillium roqueforti*. Milchwissenschaft 36:205, 1981.

Golding NS. Gas requirements of mold 1. *Penicillium roqueforti.* J Dairy Sci 20:319, 1937.

Govindarajan S, Morris HA. Pink discoloration in cheddar cheese. J Food Sci 38:675, 1973.

Grappin R, Lefier D, Dasen A, Pochet S. Characterizing ripening of Gruyère de Comté: influence of time × temperature and salting conditions on eye and slit formation. Int Dairy J 3:313, 1993.

Griffith R, Hammond EG. Generation of Swiss cheese flavor compounds by reactions of amino acids with carbonyl compounds. J Dairy Sci 72:604, 1989.

Griffiths MW, Phillips JD, Muir DD. Thermostability of proteases and lipases from a number of species of psychrotrophic bacteria of dairy origin. J Appl Bacteriol 50:289, 1981.

Grill H Jr, Patton S, Cone JF. Aroma significance of sulfur compounds in surface-ripened cheese. J Dairy Sci 49:409, 1966.

Herman LMF, De Ridder HFM, Vlaemynck GMM. A multiplex PCR method for the identification of *Listeria* spp. and *Listeria monocytogenes* in dairy samples. J Food Prot 58:867, 1995.

Hettinga DH, Reinbold GW, and Vedamuthu ER. The split defect of Swiss cheese. 1. Effect of strain of *Propionibacterium* and wrapping material. J Milk Food Technol 37:322, 1974.

Hocking AD, Faebo M. Fungi causing thread mould spoilage of vacuum packaged cheddar cheese during maturation. Int Dairy J 16:123, 1992.

Hofherr LA, Glatz BA, Hammond EG. Mutagenesis of strains of *Propionibacterium* to produce cold-sensitive mutants. J Dairy Sci 66:2482, 1983.

Hong CM, Wendorff WL, Bradley RL. Factors affecting light-induced pink discoloration of annatto-colored cheese. J Food Sci 60:94, 1995.

Horwood JF, Stark W, Hull RR. A fermented, yeasty flavour defect in cheddar cheese. Aust J Dairy Technol 47:91, 1987.

Hutkins RW, Morris HA. Carbohydrate metabolism by *Streptococcus thermophilus*: a review. J Food Prot 50:876, 1987.

Johnson EA, Nelson JH, Johnson ME. Microbiological safety of cheese made from heat-treated milk. Part II. Microbiology. J Food Prot 53:519, 1990a.

Johnson ME, Chen CM. Technology of manufacturing reduced-fat cheddar cheese. In: Malin EL, Tunick MH, eds. Chemistry of Structure–Function Relationships in Cheese. New York: Plenum Press, 1995, p 331.

Johnson ME, Olson NF. Non-enzymatic browning of mozzarella cheese. J Dairy Sci 68:3143, 1985.

Johnson ME, Riesterer BA, Chen C, Tricomi B, Olson NF. Effect of packaging and storage conditions on calcium lactate crystallization on the surface of cheddar cheese. J Dairy Sci 73:3033, 1990b.

Iya KK, Frazier WC. The yeast in the surface smear of brick cheese. J Dairy Sci 32:475, 1949.

Kaminarides SE, Laskos NS. Yeasts in factory brine of feta cheese. Aust J Dairy Technol 47:68, 1992.

Karahadian C, Lindsay RC. Contribution of *Penicillium* sp. to the flavors of Brie and Camembert cheese. J Dairy Sci 68:1865, 1985.

Kelly CD, Marquardt JC. The influence of hydrogen ion concentration and salt on the surface flora of Limburger cheese. J Dairy Sci 23:309, 1939.

Kinsella JE, Hwang D. Biosynthesis of flavors by *Penicillium roqueforti*, Biotechnol Bioeng 18:927, 1976.

Kindstedt PS. Effect of manufacturing factors, composition, and proteolysis on the functional characteristics of mozzarella cheese. Crit Rev Food Sci Nutr 33:167, 1993.

Klijn N, Nieuwendorf FFJ, Hoolwerf JD, van der Waals CB, Weerkamp AH. Identification of *Clostridium tyrobutyricum* as the causative agent of late blowing in cheese by species-species PCR amplification. Appl Environ Microbiol 61:2919, 1995.

Laleye LC, Simard RE, Lee B-H, Holley RA, and Giroux RN. Involvement of heterofermentative lactobacilli in development of open texture in cheeses. J Food Prot 50:1009, 1987.

Lane CN, Fox PF. Contribution of starter and adjunct lactobacilli to proteolysis in cheddar cheese during ripening. Int Dairy J 6:715, 1996.

Langsrud T, Reinbold GW. Flavor development and microbiology of Swiss cheese—a review. J Milk Food Technol 37:26, 1974.

Liewen MB, Marth EH. Inhibition of penicillia and aspergilli by potassium sorbate. J Food Prot 47:554, 1984.

Liewen MB, Marth EH. Growth and inhibition of microorganisms in the presence of sorbic acid: a review. J Food Prot 48:364, 1985.

Lindgren SE, Axelsson LT, McFeeters RG. Anaerobic L-lactate degradation by *Lactobacillus plantarum*. FEMS Microbiol Letters 66:209, 1990.

Lubert DJ, Frazier WC. Microbiology of the surface ripening of brick cheese. J Dairy Sci 38:981, 1955.

Lund F, Filtenborg O, Frisvad JC. Associated mycoflora of cheese. Food Microbiol 12:173, 1995.

Luyten H, van Fleet T, Walstra P. Characterization of the consistency of Gouda cheese: rheological properties. Neth Milk Dairy J 45:33, 1991.

Mans J. Looking for Mr. Longlife. Dairy Foods 96:65, 1995.

Marshall KC. Biofilms: an overview of bacterial adhesion, activity, and control at surfaces. ASM News 58:202, 1992.

Marth EH. Spoilage of cottage cheese by psychrophilic microorganisms. Cultured Dairy Prod J 5:14, 1970.

Marth EH, Capp CM, Hasenzahl L, Jackson HW, Hussong RV. Degradation of potassium sorbate by *Penicillium* sp. J Dairy Sci 49:197, 1966.

Martley FG, Crow VL. Interactions between non-starter microorganisms during cheese manufacture and ripening. Int Dairy J 3:461, 1993.

Martley FG, Crow VL. Open texture in cheese: the contributions of gas production by microorganisms and cheese manufacturing practices. J Dairy Res 63:489, 1996.

Masui K, Yamada T. French Cheeses. New York: DK Publishing, 1996.

McSweeney PLH, Fox PF, Lucey JA, Jordan KN, Cogan TM. Contribution of the indigenous microflora to the maturation of cheddar cheese. Int Dairy J 3:613, 1993.

Molimard P, Spinnler HE. Review: compounds involved in the flavor of surface mold-ripened cheeses: origins and properties. J Dairy Sci 79:169, 1996.

Morris HA. Blue-Veined Cheeses. New York: Pfizer Inc, 1981.

Morris HA, Jezeski JJ, Combs WB, Kuramoto S. Free fatty acids, tyrosine, and pH changes during ripening of blue cheese made from variously treated milks. J Dairy Sci 46:1, 1963.

Mukherjee KK, Hutkins RW. Isolation of galactose-fermenting thermophilic cultures and their use in the manufacture of low browning mozzarella cheese. J Dairy Sci 77:2839, 1994.

Nooitgedagt AJ, Hartog BJ. A survey of the microbiological quality of Brie and Camembert cheese. Neth Milk Dairy J 42:57, 1988.

Olson NF. Ripened Semisoft Cheeses. New York: Pfizer Inc, 1969.

Olson NF. The impact of lactic acid bacteria on cheese flavor. FEMS Microbiol Rev 87:131, 1990.

Peterson SD, Marshall RT. Nonstarter lactobacilli in cheddar cheese: a review. J Dairy Sci 73:1395, 1990.

Purko M, Nelson WO, Wood WA. The associative action between certain yeasts and *Bacterium linens*. J Dairy Sci 34:699, 1951.

Reinbold GW. Swiss Cheese Varieties. New York: Pfizer Inc, 1972.

de Ruig WG, van den Berg G. Influence of the fungicides sorbate and natamycin in cheese coatings on the quality of the cheese. Neth Milk Dairy J 39:165, 1985.

Rohm H, Eliskases-Lechner F, Bräuer M. Diversity of yeasts in selected dairy products. J Appl Bacteriol 72:370, 1992.

Salama MS, Sandine WE, Giovannoni SJ. Isolation of *Lactococcus lactis* subsp. *cremoris* from nature by colony hybridization with rRNA probes. Appl Environ Microbiol 59:3941, 1993.

Sandine WE, Anderson AW, Elliker PR. How to avoid floating curd when making cottage cheese. Food Engin 29:84, 1957.

Seitz EW, Sandine WE, Elliker PR, Day EA. Distribution of diacetyl reductase among bacteria. J Dairy Sci 46:186, 1963.

Severn DJ, Johnson ME, Olson NF. Determination of lactic acid in cheddar cheese and calcium lactate crystals. J Dairy Sci 69:2027, 1986.

Shannon EL, Olson NF, Deibel RH. Oxidative metabolism of lactic acid bacteria associated with pink discoloration in Italian cheese. J Dairy Sci 60:1693, 1977.

Smith DA, Mikolajcik EM, Lindamood JB. Causative organisms and chemical nature of the Swiss cheese rind rot defect. Cultured Dairy Products J 22:9, 1987.

Sofos JN, Busta FF. In: Branen AL, Davidson PM, eds. Sorbates, Antimicrobiols in Foods. New York: Marcel Dekker, 1983, p 141.

Thomas TD. Acetate production from lactate and citrate by non-starter bacteria in cheddar cheese. N Z J Dairy Sci Technol 22:25, 1987a.

Thomas TD. Cannibalism among bacteria found in cheese. N Z J Dairy Sci Technol 22:215, 1987b.

Thomas TD, Crow VL. Mechanism of D(−)-lactic acid formation in cheddar cheese. N Z J Dairy Sci Technol 18:131, 1983.

Viljoen BC, Greyling T. Yeasts associated with cheddar and Gouda making. Int Dairy J 28:79, 1995.
Zoon P, Allersma D. Eye and crack formation in cheese by carbon dioxide from decarboxylation of glutamic acid. Neth Milk Dairy J 50:309, 1996.

10

Fermented By-Products

DAVID R. HENNING
South Dakota State University, Brookings, South Dakota

I. INTRODUCTION

The manufacture of cheese from milk generates approximately 9 pounds of whey for each pound of cheese. This by-product of cheese manufacturing has posed serious disposal problems for most cheese makers. Table 1, from Kosikowski and Mistry (1997), shows that sweet whey, derived from most rennet-coagulated cheeses, is approximately 93% water and 6.35% solids, with about 76% of the solids being lactose. Acid whey, derived from acid-coagulated cheeses, is compositionally similar to sweet whey except it has higher lactic acid and ash contents. Using a ratio of 9 pounds of whey produced for each 1 pound of cheese, it can be calculated that approximately 64 billion pounds of whey was produced in 1994 when U.S. cheese production (including cottage cheese curd) totaled 7.193 billion pounds, according to the National Agricultural Statistical Service, USDA (1996). Approximately 1.153 billion pounds of dry whey was produced for human and animal consumption. This dry whey represents approximately 28% of the whey solids resulting from cheese manufacture during the year. The remaining whey solids were used as concentrated whey for human and animal food (85.4 million pounds), whey protein concentrates (181.8 million pounds), lactose

TABLE 1 Composition of Some Commercial Fluid and Dried Wheys

Component	Fluid Sweet Whey[a]	Fluid Acid Whey[b]	Condensed Acid Whey	Dried Sweet Whey	Dried Acid Whey
Total solids, %	6.35	6.5	64.0	96.5	96.0
Moisture, %	93.70	93.50	33.5	3.5	4.0
Fat, %	0.5	0.04	0.6	0.8	0.6
Protein-total, %	0.8	0.75	7.6	13.1	12.5
Lactose, %	4.85	4.90	34.9	75.0	67.4
Ash, %	0.50	0.80	8.2	7.3	11.8
Lactic acid,[c] %	0.05	0.40	12.0	0.2	4.2

[a]From cheddar cheese
[b]From cottage cheese
[c]Estimated true lactic acid after substituting basic acidity.
Source: Partly derived from Mavropoulou and Kosikowski, 1973.

(253.3 million pounds), partially delactosed and demineralized whey (95.8 million pounds), whey solids in wet blends (99.8 million pounds), fermented whey products (including ethanol production), sewage disposal, land disposal, or animal feed.

As cheese manufacturing plants increase in size, the use of sewage treatment facilities, land disposal, and return of whey to farmers become less viable options. Whey disposal through sewage systems can overload the treatment facilities by its high biochemical oxygen demand (BOD). The BOD for whey is between 30,000 and 60,000 ppm (Sienkiewicz, 1990). A plant discharging a thousand gallons of whey to a sewage system creates a load equivalent to that of more than 1000 domestic users.

The component of whey that poses the greatest disposal problem is lactose. Ultrafiltration of whey in the production of whey protein concentrates results in a deproteinized effluent that retains nearly the same level of BOD as the original whey. To effectively deal with the whey problem, lactose must be removed from the waste stream and converted to nonpolluting products. Several industrial products are derived from whey. Refined food-grade lactose can be used in the pharmaceutical industry for compression with additives into pills, used as carriers for antibiotic powder or used in infant foods. Some industrial products derived from lactose by chemical processes are lactitol, lactitol palmitate, lactosylurea, and galactose/glucose syrups. This chapter describes some of the fermentation options that convert lactose and whey into products that can relieve the burden of whey disposal for cheese manufacturers. Production of methane from lactose and dairy wastes is discussed in Chapter 14. The discussions in this chapter are grouped into human food products and industrial products.

Fermentation of lactose is one form of energy-yielding microbial metabolism in which the organic substance, lactose, is incompletely oxidized and another organic compound acts as the electron acceptor. The microbial fermentation processes do not result in stoichiometric conversion of lactose into the desired products because the microorganisms use lactose as a source of carbon for cellular processes.

Fermentation products from whey were reviewed by Marth (1970). Since that review, the most significant change in the techniques of whey fermentation processes has been membrane processing of whey to customize the feedstock for fermentations. Ultrafiltration allows separation of proteins from the lactose-rich permeate. Reverse osmosis allows concentration of lactose to more useable levels and thereby assists in obtaining desired fermentation end products at levels that make economic recovery possible, for example, to obtain a beer with 4% to 5% ethanol instead of 2% to 2.5%. This increased ethanol level makes the distillation process more cost efficient. The use of electrodialysis can reduce the mineral level in concentrated whey and permeate preparations. High mineral content of the concentrated whey retards microbial growth.

Anaerobic digestion of whey or whey permeate is discussed in Chapter 14. The primary product from the anaerobic digestion is "biogas," a mixture of CO_2 and CH_4 with small amounts of H_2 and H_2S.

II. WHEY FERMENTATION BY-PRODUCTS

A. Food Products

1. Wine

Whey wines can be made from whey or whey permeates with a final alcohol content of approximately 2.25% to 14%. The fermentation of lactose to carbon dioxide and ethanol results in 0.54 g of ethanol for each gram of lactose. The alcohol content of a wine derived from cheese whey would be approximately 2.4% assuming there is an efficiency of approximately 90% conversion of the lactose to ethanol.

Kosikowski and Wzorek (1986) prepared whey wines from a reconstituted acid whey powder. They removed the proteins by ultrafiltration. The permeate had a total solids concentration of 26% to 28% before demineralizing to a mineral level of 1% or less. Any residual whey taints after fermentation were removed with bentonite and charcoal at the levels of 5 g/L and 2 g/L, respectively.

Processes for whey wine production before the advent of ultrafiltration were limited to a low alcohol percentage in the final wine. The lactose level of the fermentable whey had to be increased to about twice the level of the desired alcohol content of the finished product. Without mineral removal techniques, growth of most ethanol-producing yeasts and bacteria would be inhibited. To overcome these limitations, whey was fortified with other fermentable carbohydrates. According to Sienkiewicz and Riedel (1990), the most common carbohydrate additives are sucrose and glucose, although honey has been used. Demineralization of the carbohydrate fortified whey before fermentation was recommended by Larson (1976).

2. Whey Champagne

Serwovit, whey champagne, is produced in Poland (Sienkiewicz and Riedel, 1990). The starting material is deproteinated acid whey, sucrose and caramelized sugar, a dry yeast, fruit flavoring, and water. The yeasts and fruit flavorings are added to the pasteurized base of whey, water, and sugars. The mixture is bottled and the fermentation occurs in the bottle at 18°C in 8 to 12 hours.

3. Whey Beer

Whey beers are produced with and without malt addition (Sienkiewicz nd Riedel, 1990). Whey has a high mineral content similar to wort obtain from malt mash. Also, carmelization of the lactose develops a flavor suggestive of dried malt. The use of β-galactosidase to hydrolyze lactose provides the brewer's yeast with the glucose moiety of lactose, but the galactose is generally not fermented. Use of ultrafiltration to provide a protein-free permeate has improved the flavor by removing insoluble proteins which have a distinctive flavor. The makers of whey beer are particular concerned that all the fat be removed so as to not suppress the foaming ability of the beer. Hops are usually brewed into the beer.

A malted whey beer contains a maximum of 30% whey and is brewed with hops. Bottom-fermenting yeasts are used with this product.

A whey malt beer is a sweeter beverage that contains a maximum of 50% whey and additional fermentable carbohydrates such as starch and glucose syrup. Hops are brewed into the beer and the color is adjusted with roasted malt. A top-fermenting yeast is used. This product is pasteurized and a secondary fermentation with yeast and added sugar proceeds before bottling.

A whey-nutrient beer is weakly alcoholic. It is brewed from deproteinated whey and hops and has a mineral mix added to it before fermentation. The product is pasteurized before bottling.

Elmer and Clark (1982) reported on the use of an 85% hydrolyzed whey permeate as a replacement for 12.2% of the brewing extract in the production of beer. The permeate was produced by reverse osmosis and had approximately 18% total solids. A 50-barrel brew was produced and held for 5 months. Flavor stability and foam character were similar to those of commercial brews from the same brewer.

Supplementing mash with lactose and lactase (0.2% of the weight of the added lactose) gave good results according to Polish workers (Sienkiewicz and Riedel, 1990).

4. Beerlike Beverages

Russian workers developed two beverages in the late 1960s that contained added sugar, flavorings, and raisins (Sajanschsehaukas, 1968). Botschyu is brewed with hops and fermented with yeasts; it contains approximately 3.8% alcohol. The other, Bodrost, is fermented with kefir fungi and lactic acid bacteria.

B. Industrial Products

1. Culture Media

Richardson et al. (1977) pioneered work on use of fresh cheddar or Swiss cheese whey as a low-cost alternative to nonfat dry milk or commercial bacteriophage-inhibitory media for propagation of lactic starter cultures for cheese makers. Incorporation of a phosphate-stimulant blend provided protection for the lactic starter culture organisms against bacteriophage action and provided minerals and vitamins needed for acceptable activity. The recommended use of the phosphated whey medium (PWM) was in a system incorporating pH control. The fortifying phosphate-stimulant blend consists of 44.4% NaH_2PO_4, 42.2% Na_2HPO_4, 8.66% Ardamine "Yes" (a dehydrated yeast autolysate), 4.43% NZ Amine NAK (a protein hydrolysate), 0.22% $MgSO_4$, 0.04% $MnSO_4$, and 0.04% $FeSO_4$. The phosphate-stimulant blend is added to liquid cheddar or Swiss cheese whey at a rate of 1.17%. The PWM is heated to 89°C to 95°C for 45 minutes, cooled to 21°C to 27°C, and inoculated with the desired culture strains. The incubation of the culture in a specially equipped tank allows the monitoring of the pH of the PWM and controlled addition of neutralizing agents such as NH_4OH, anhydrous ammonia, or NaOH to maintain the pH in a range of 6.0 and 6.3.

Starter cultures prepared with the PWM-pH control procedure have greater activity than those grown in conventional bacteriophage-inhibitory or milk media. The amounts of PWM-pH control starter culture needed for Monterey cheese manufacture was 20% to 33% of the amount of milk medium starter culture. The cheese made with PWM-pH control was normal in all parameters, including yield.

Whitehead et al. (1993) reviewed the subject of starter media for cheese making and introduced the concept of internal neutralization of the lactic acid produced during starter culture production. The acid-neutralizing components of the medium can be added as insoluble salts, which dissolve when acid is produced, or they may be encapsulated in compounds that gradually release neutralizing compounds during acid production. Experiments reported by this group showed that neutralized starters maintain their activity for longer periods (up to 10 days) than starter cultures prepared in unneutralized milk or phage-inhibitory media. Trials with the internally neutralized medium in cottage cheese manufacture by Ogden (1981) showed a reduction of 42% in the amount of starter was possible in comparison to skim milk starter. This work also showed an increase in curd yield of 2.8%.

2. Ethanol

Several lactose-fermenting yeasts can produce ethanol during the fermentation of lactose. Strains of *Kluyveromyces marxianus* (formerly *Kluyveromyces fragilis*), *Torula cremoris*, and *Candida kefyr* efficiently convert lactose to ethanol. The strain *K. marxianus* NRRLY 2415 produced up to 12% ethanol for Kosikowski and Wzorek (1982) from a demineralized acid whey permeate with 24% lactose in 7 to 14 days at 30°C. Laboratory studies by Rogosa (1947) indicated a 91% conversion of lactose to alcohol, but Rajagopalan and Kosikowski (1982) were only able to obtain 84.3% of the theoretical maximum yield, or about 0.45 g of ethanol, from each gram of lactose.

Commercial whey-to-ethanol facilities in the United States are based on the integrated Carbery process developed at Express Dairy in West Cork, Irish Republic. The ethanol plant was commissioned in April 1978 and is located on site at a cheese manufacturing plant that has condensing and drying capabilities. The cheddar cheese whey is ultrafiltrated, with the retentate destined for whey protein concentrate (WPC) powders and the permeate as a feedstock for the ethanol fermenters. A bottom-fermenting yeast culture and fresh permeate is pumped to one of six 25,000-gallon fermenter vessels. The vessels are similar in design to those used in English breweries. They have a cone bottom, are 50 feet in height and 12 feet in diameter. Large compressors inject filtered air into the bottom of the fermenter tanks. This injection of air prevents the yeast from settling. The fermentation requires approximately 20 hours. The cheese operation generates about 110,000 gallons of permeate per day, so a battery of six 25,000-gallon fermenters allows one fermenter to be filling, one to be emptied, and four to be in active fermentation at any time. At the end of fermentation, the fermenter contents, which contain approximately 2.8% alcohol are centrifuged and the yeast cream is recovered. The clear liquid is pumped to a balance tank until it is distilled. After distillation, the ethanol is 96.5% by volume (Sandbach, 1981). The residue from the still has a BOD that is reduced to approximately 5% to 10% that of the whey permeate.

The Carbery process has been improved by several of the whey ethanol operations in the United States. One improvement is preconcentration of the whey permeate by reverse osmosis to increase the lactose concentration to obtain a greater percentage of ethanol in the fermenter.

Although food-grade ethanol can be converted into other products such as acetic acid, the use for most ethanol is in gasoline blends. National Chemical Products in South Africa initiated whey fermentation in the mid-1970s for alcohol fuel production. Other countries that produce alcohol fuel from whey fermentation are Canada, France, Ireland, Japan, the Netherlands, New Zealand, and the United States. In the United States, the largest whey-to-alcohol facility is at the Golden Cheese Company in Corona, CA. This integrated operation processes more than 2.25 million pounds of whey per day and makes more than 150,000 pounds of WPC human food ingredients and 50,000 gallons of 200-proof (100%) fuel-grade ethanol per week. The ethanol plant has eight 48,000-gallon fermenters that are used to batch-ferment the lactose to ethanol in 18 to 30 hours. Yeast is recovered from beer (4% to 5% ethanol) and reused in subsequent batches for up to ten times. The beer is collected in two beer wells until it is heated and distilled. The distillation process uses glycol to form an azeotrope (Morris, 1986). After distillation of the azeotrope, 200-proof ethanol is obtained. The residues after distillation are further processed into animal feed proteins. The BOD level of the waste stream is reported to be less than 10,000 ppm (Ahmed, 1991).

Cheese whey or whey permeate has a relative cost advantage as a substrate for ethanol production. If permeate is used, the costs include transportation, ultrafiltration, and concentration. However, the cheese manufacturer avoids the cost of reducing the BOD of the whey at the cheese-making facility. Although corn starch has been the substrate used for most fuel ethanol in the United States, production costs of ethanol-from-corn depend on corn prices and value of by-products associated with the ethanol produced. The 1995 corn-to-ethanol industry included 43 plants in 20 states producing 1.4 billion gallons of ethanol. This ethanol-from-corn used 5.3% of the U.S. corn crop of 1994.

Zhou and coworkers (1992) at the University of Nebraska-Lincoln, developed a cofermentation process to combine whey and corn substrates. The cost of alcohol production from corn is high relative to whey because of the value of the corn. Corn represents about half the cost of the ethanol. A process that could provide approximately 28% of the fermentable carbohydrates from whey can positively affect the economics of the process. Also, a whey ethanol plant is usually limited in capacity by the availability of whey in region.

A cofermentation process could use all the whey available and still have the cost advantages of a large ethanol production facility. Acid or sweet whey or their permeates can be used as the liquid portion of the corn mash. The cofermentation process requires a staggered inoculation procedure. Lactose fermentation requires a high level of *K. marxianus* inoculum followed by the addition of

Saccharomyces cerevisiae for the fermentation of the α-amylase–hydrolyzed corn starch. The *S. cerevisiae* is tolerant of higher alcohol concentrations and can yield a final concentration of more than 9.5% ethanol in 72 hours. The Nebraska process trials achieved an alcohol yield 29% higher than with corn alone, or 3.3 gallons per bushel of corn.

3. Vinegar

Ethanol produced from whey can be used as a substrate for production of vinegar. The alcoholic fermentation is terminated when formation of acetic acid starts. The *Acetobacter* and *Gluconobacter* species used in the conversion of alcohol to acetic acid oxidize ethanol, so there must be a plentiful supply of oxygen. There are at least four industrial methods for conversion of ethanol to acetic acid.

4. Lactic Acid

Lactic acid from cheese whey is produced commercially in Slovakia, Italy, and the United States. Whereas synthetic production from acetaldehyde or lactonitrile is cheaper, the low cost of whey makes a large-scale production facility competitive. The starting materials for lactic acid production include rennet and acid wheys and their permeates. Lactic acid bacteria such as *Lactococcus lactis, Lactobacillus delbrueckii* subsp. *bulgaricus, Lactobacillus acidophilus, Lactobacillus helveticus, Lactobacillus casei,* and mixed cultures of these organisms have been successfully used to convert lactose to lactic acid. Fermentation of whey or whey permeate requires enrichment with approximately 0.5% corn steep liquor, 0.1% glucose, 0.05% β-galactosidase, and several other growth factors (Rosenau, 1986). The pH is maintained in the range of 5.0 to 6.0 by neutralization of the lactic acid with $CaCO_3$, $Ca(OH)_2$, or 27% NH_4OH. After 85% to 90% of the lactose is fermented, usually in 24 to 48 hours, the pH is adjusted to 12 with $Ca(OH)_2$. The liquid is boiled, allowed to settle, and then filtered to remove whey proteins and calcium phosphate. The pH is adjusted to 7.0. The calcium lactate crystallizes and is redissolved by adding H_2SO_4 and $ZnSO_4$ at about 95°C. Calcium sulfate and zinc lactate are formed after stirring for 2 hours. After sedimentation occurs, the calcium sulfate is centrifuged off and the zinc lactate solution is cooled to 10°C and allowed to crystallize for 48 hours. The zinc lactate crystals are harvested and treated with H_2SO_4. After a second filtration, the liquid is electrodialyzed to remove the heavy metals, excess sulfuric acid, and other impurities.

5. Propionate

Production of propionic acid from lactose and lactate by *Propionibacterium freundenrichii* was the basis of a patent by Sherman and Shaw (1923). Whey was fortified with pulverized limestone, which led to formation of calcium salts of propionic and acetic acids, then the salts are recovered. A preferred use of calcium and sodium propionates is in bakery products. They have relatively no effect on yeast growth, but inhibit the growth of spoilage molds and the *Bacillus* species, which causes ropiness in bakery products.

6. Calcium Magnesium Acetate

An innovative product proposed by Yang and coworkers (1992) at Ohio State University is production of calcium magnesium acetate (CMA) for use as a road deicer. A sequential conversion of lactose to lactate and then to acetate with a yield of greater than 90% acetate was developed. The fermentations are carried out in continuous, immobilized cell bioreactors. *Lactococcus lactis* is the culture used for the homolactic fermentation of the lactose to lactate. *Clostridium formicoaceticum* ATCC 27076 is used for fermenting the lactate to acetate. The overall yield of acetate from lactose in this laboratory trial was approximately 95%. This compares with yields of approximately 60% conversion to acetate in the aerobic vinegar process. The acetate concentration obtained from permeate was approximately 4%. Therefore, the recovery and concentration of acetic acid becomes a significant portion of the process. A mixture of 50% Alamine 336, a tertiary amine from the Henkel Corporation, in 2-octanol was the solvent selected to extract the acetic acid from the fermentation process. The acetic acid was completely stripped from the solvent and reacted with CaO/MgO solution to form CMA by vigorous mixing. The CMA produced from lactose had deicing ability similar to commercial CMA.

Calcium magnesium acetate has been identified by the Federal Highway Administration as one of the most promising alternative road deicers to the currently used salt. Salt corrodes bridge metals and concrete on highways. Also, salt is harmful to vegetation and is a threat to ground water quality in some regions. However, salt costs approximately 5% as much as currently available CMA (30 vs. $650/ton). An estimated cost of $215/ton for CMA was calculated when using a 1.5-million pound whey permeate per day feedstock entering a facility with a $7,000,000 capital investment. The output of CMA would be about 40 tons per year.

7. β-Galactosidase

β-galactosidase is an industrially important enzyme that has application in producing dairy products with reduced lactose content. It has been used to prevent sandiness in ice cream, for example. However, the most compelling reason for the use of the lactase enzyme is to provide products for consumers with lactose malabsorption. There are industrial applications requiring lactose hydrolysis, such as the fermentation of lactose by a non-lactose–fermenting yeast.

Myers and Stimpson (1956) patented a process to produce lactase. The process involved the heat coagulation of whey proteins at pH 4.5. The clear supernatant liquid was fortified with 0.1% corn steep liquor and a nitrogen source. After cooling to 30°C, the substrate was inoculated with 10% of a culture of actively growing *K. marxianus*. The patent suggested both aerated and nonaerated incubation. When aerated, one volume of air per volume of medium per minute was the rate of aeration. The yeast cells were washed to improve flavor and then dried. The yeast was stored at 4.4°C until used in a lactose fermentation.

Several commercial companies have isolated β-galactosidase from yeast. Gist-Brocades isolated its Maxilact from *K. marxianus*. Sumitomo of Japan uses *Aspergillus oryzae* as the source of lactase. These companies have bound these lactose-splitting enzymes to support materials to provide immobilized enzymes for commercial applications.

8. Other Fermentation Products from Whey

Table 2 lists fermentation products that have been produced from whey and for which there is technical feasibility of producing these products. In most instances, the economic potential is poor or unknown (Hobman, 1984).

TABLE 2 Fermentation Products from Whey with Limited Potential for Commercialization

Food grade yeast/single cell protein	Amino acids
Bakers' yeast	Vitamins
Acetone/butanol	Riboflavin
Methane	B12
Food acids	Ascorbic acid
Citric	2-keto-L-gulonic acid
Lactobionic	Antibiotics/penicillin
Itaconic	Other biochemicals
Malic	D(-)-3-hydroxy-butyric acid
Enzymes/β-galactosidase	Gibberellic acid
Food gums	2,3-butylene glycol
Xanthan	Hydrogen
Pullulan	Diacetyl
Alginic acid	Calcium gluconate
Indican	Pyruvic acid

Source: Hobman PG (1984).

REFERENCES

Ahmed I, Morris D. Trends in Cheese Whey Production and Utilization. An Alternate Feedstock for Alcohol Fuel Production in the United States. Washington, DC: Institute for Local Self-Reliance, 1991, pp 1–19.

Elmer RA, Clark WS Jr. A new whey of doing it. Modern Brewery Age, December, 1982.

Hobman PG. Review of processes and products for utilization of lactose. J Dairy Sci 67:2630, 1984.

Kosikowski FV, Wzorek W. Whey wine from concentrates of reconstituted acid whey powder. J Dairy Sci 60:1982, 1986.

Kosikowski FV, Mistry VV. Cheese and Fermented Milk Foods. Vol 1. Stamford, CT: FV Kosikowski LLC, 1997, p 427.

Larson PK, Yang HY. Some factors of clarification of whey wine. J Milk Food Technol 39:614, 1976.

Marth EH. Fermentation products from whey. In: Webb BH, Whittier EO, eds. Byproducts from Milk. Westport, CT: AVI Publishing, 1970, pp 43–82.

Mavropoulou IP, Kosikowski FV. Free amino acids and soluble peptides of whey powders. J Dairy Sci 56:1135, 1973.

Myers RP, Stimpson EG. Production of lactase. U.S. patent 2,762,749, 1956.

Morris C. New plant of the year—Golden Cheese Company of California. Food Engineering, March, 1986.

National Agricultural Statistics Service. Dairy Products Summary. Washington, DC: United States Department of Agriculture, 1996.

Ogden LV. Whey starter for commercial cottage cheese (abstr). J Dairy Sci 64(suppl 1):53, 1981.

Rajagopalan K, Kosikowski FV. Alcohol from membrane processed concentrated cheese whey. I&EC Product Research & Development 21:82, 1982.

Richardson GH, Cheng CT, Young R. Lactic bulk culture system utilizing a whey-based bacteriophage inhibitory medium and pH control. I. Applicability to American style cheese. J Dairy Sci 60:378, 1977.

Rogosa M, Browne HH, Whittier EO. Ethyl alcohol from whey. J Dairy Sci 30:263, 1947.

Rosenau JR. Lactic acid production from whey via electrodialysis. In: Le Maguer M, Jelen P, eds. Food Engineering and Process Application. Vol 2. New York: Elsevier Science Publishers, 1986, pp 245–250.

Sajanschshankas PV. Dairy beverage "Bachyu" made from whey. Molčnaya Promyšlennost 29:12, 24–25, 1968. Cited in Dairy Sci Abstr 31:844, 1969.

Sandbach DML. Production of potable grade alcohol from whey. Cultured Dairy Prod J, November, 1981, pp 17–19, 22.

Sherman JM, Shaw RH. Process for the production of propionates and propionic acid. U.S. Patent 2,826,502, 1923.

Sienkiewicz T, Riedel C-L. Whey and Whey Utilization. Gelsenkirchen-Buer, Germany: Verlag Th Mann, 1990.

Whitehead WE, Ayres JW, Sandine WE. Symposium: recent developments in dairy starter cultures: microbiology and physiology. A review of starter media for cheese making. J Dairy Sci 76:2344, 1993.

Yang S-T, Zhu H, Lewis VP, Tang I-C. Calcium magnesium acetate (CMA) production from whey permeate: process and economic analysis. Resources, Conservation and Recycling 7:181, 1992.
Zhou KP, Yang HH, Shahani RK, Whalen PJ, Shahani KM. Producing alcohol fuel from whey and co-products. Proceedings, Whey Products Conference, Chicago, Illinois, April 29–30, 1992, pp 37–42.

11
Public Health Concerns

ELLIOT T. RYSER

University of Vermont, Burlington, Vermont

I. INTRODUCTION

Milk, a highly nutritious food ideally suited for growth of both pathogenic and spoilage organisms, is the basis for an extremely large industry in the United States. In 1991, more than 17 billion gallons of milk were produced by 10 million cows on 182,000 dairy farms with total sales exceeding $18 billion (Anonymous, 1992a). Even though dairy products are consumed daily by most individuals in the United States, milk, ice cream, and cheese are still among the safest foods marketed and have most recently accounted for less than 1.5% of all foodborne illness cases reported annually (Bean et al., 1996). Dairy products manufactured in the United States continue to be safer than those produced in many other countries with 2%, 4%, 6%, and 8% of all foodborne outbreaks in France, Spain, Scotland, and Germany, respectively, traced to milk products during 1987 (Notermans and Hoogenboom-Verdegaal, 1992). However, two outbreaks in 1985—the first involving up to 85 deaths in southern California from *Listeria*-contaminated cheese and the second in the Chicago-area in which more than 16,000 cases of salmonellosis were traced to one particular brand of pasteurized milk—reaffirms the need for continued vigilance by the dairy industry to safeguard public health.)

(Outbreaks of milk-borne illness date from the inception of the dairy industry. Bacterial infections including diphtheria, scarlet fever, tuberculosis, and typhoid fever predominated before World War II and were almost invariably linked to consumption of raw milk, with the greatest public health concerns at that time perceived to be poor sanitation, inadequate milk handling procedures, and animal health issues.) Although reports of experimental milk pasteurization first appeared in the public health literature during the early 1900s, supporters of the certified raw milk movement denounced pasteurization, claiming that this practice led to nutritional deficiencies and flavor defects and allowed the marketing of "sterilized filth." Conversely, the pasteurized milk movement maintained that certified milk was unsafe, despite the adherence to strict sanitary practices. In 1923, the Public Health Service began publishing summaries of gastrointestinal outbreaks attributed to milk. These early surveillance efforts soon led to passage of the first Model Milk Ordinance, which stressed nationwide pasteurization and the eventual reduction in the incidence of milk-borne enteric diseases with no milk-borne cases of diphtheria, scarlet fever, tuberculosis, or typhoid fever reported in more than 40 years. However, interstate shipment of raw milk continued to be legal until 1973. Banning the interstate shipment of all raw milk products, both certified and noncertified, in 1986 helped reduce the number of raw milk–related outbreaks (Bean et al., 1996). Sporadic illnesses continue to be reported, however, particularly among farm families who routinely consume milk from their own dairy herds.

The importance of various etiological agents in milk-borne disease has changed dramatically over time, with routine pasteurization of milk having a significant impact. However, more than 90% of all reported cases of dairy-related illness continue to be of bacterial origin with at least 21 milk-borne or potentially milk-borne diseases currently being recognized (Table 1). Typhoid fever and scarlet fever accounted for most cases of milk-borne illness until the late 1930s. During and shortly after World War II, brucellosis, salmonellosis, and staphylococcal poisoning emerged as major public health concerns, with salmonellosis continuing to be the most important dairy-related illness currently in terms of overall numbers of cases. As reports of staphylococcal poisoning subsided during the 1970s, campylobacteriosis and yersiniosis emerged as major public health concerns for those individuals who still drank raw milk. In 1985, as many as 85 people in California died of cheeseborne listeriosis, a rare and seldom diagnosed disease that was previously only weakly associated with consumption of raw and pasteurized milk. Most recently, *Escherichia coli* O157:H7 has emerged as a serious threat to the dairy industry with several outbreaks of potentially fatal hemolytic uremic syndrome reported in Wisconsin and Oregon. Even though they are able to cause potentially serious health problems, the rickettsiae, parasites, and viruses are each generally responsible for less than 1% of all dairy-related illnesses, with chemical contaminants other than aflatoxin also posing minimal public health concerns. Despite modern-day epidemiological strategies and extensive laboratory testing, a significant number of reasonably large and noteworthy outbreak investigations still fail to identify a specific cause of illness (Anonymous, 1984b; Maguire et al., 1991; Osterholm et al., 1986).

The types of dairy products implicated in outbreaks of disease since 1900 are listed in Table 2. Consumption of raw milk and cream was the leading cause of dairy-related illnesses before 1950 with numerous outbreaks of typhoid and scarlet fever reported. Although the number and size of these outbreaks have decreased in response to increased pasteurization, approximately one-third of all dairy-related illnesses still involve raw milk, with many of these outbreaks confined to small family farms. Except for the unusually large 1985 salmonellosis epidemic in the Chicago area, few additional outbreaks have been positively linked to pasteurized milk in recent years. Nonfat dry milk and butter, which are generally far less supportive of bacterial growth, have posed relatively few public health problems. However, numerous outbreaks have been traced to cheese, particularly cheddar and soft surface-ripened varieties, that support growth or extended survival of such noted milk-borne pathogens as *Salmonella, Listeria monocytogenes, Staphylococcus aureus*, and certain strains of *E. coli*. The number of ice cream–related outbreaks has steadily increased. From 1900 to about 1925, ice cream was most commonly associated with typhoid fever. Thereafter, staphylococcal poisoning emerged and predominated through the 1950s. Recent popularity of homemade ice cream containing eggs has led to a rapid increase in the number of outbreaks involving *Salmonella*, principally *Salmonella enteritidis*. Consequently, many individuals would consider these recent outbreaks to be more closely linked to eggs than to dairy products.

Based on the Food, Drug and Cosmetic Act of 1938, pasteurized milk and dairy products are considered adulterated and therefore unfit for human consumption if they contain potentially hazardous levels of pathogenic microorganisms, toxins, drugs, or other hazardous substances. In accordance with federal law, the Food and Drug Administration requests that firms voluntarily recall such adulterated products from the market. Hence, an examination of dairy product recalls offers another means of assessing the importance of current public health concerns.

Adoption of a "zero tolerance" policy for *L. monocytogenes* in milk and dairy products had a profound economic impact on the dairy industry.

TABLE 1 Percentage of Milk-Borne Illnesses of Various Causes Reported in the United States from 1900 to 1992

Cause	1900–1909	1910–1919	1920–1929	1930–1939	1940–1949	1950–1959	1960–1969	1970–1979	1980–1982	1988–1992
Total bacterial	100	99	96	92	71	83	67	52	92	74
Bacillus cereus poisoning	—	—	—	—	—	—	—	—	1	—
Botulism	—	1	<1	1	—	1	—	—	—	—
Brucellosis	—	—	—	1	8	4	9	1	—	<1
Campylobacteriosis	—	—	—	—	—	—	—	3	40	14
Citrobacter freundii	—	—	—	—	—	—	—	—	—	—
Corynebacterium	—	—	—	—	—	—	—	—	—	—
Diphtheria	8	2	4	1	1	—	—	—	—	—
Escherichia coli diarrhea	—	—	—	—	—	—	—	1	1	3
Escherichia coli O157:H7	—	—	—	—	—	—	—	—	—	1
Haverhill fever	—	—	<1	—	—	—	—	—	—	—
Johne's and Crohn's disease	—	—	—	—	—	—	—	—	—	—
Listeriosis	—	—	—	—	—	—	—	—	—	—
Paratuberculosis	—	—	—	—	—	—	—	—	—	—
Salmonellosis	—	1	3	2	7	21	28	42	50	56
Scarlet fever	14	15	18	27	8	—	—	—	—	—
Shigellosis	—	—	1	2	4	3	—	—	—	—
Staphylococcal poisoning	—	—	—	8	26	50	30	5	—	—
Tuberculosis	<1	<1	<1	<1	<1	<1	—	—	—	—
Typhoid fever	78	80	68	50	17	3	—	—	—	—
Yersiniosis	—	—	—	—	—	—	—	1	—	—

Total rickettsial	0	0	0	0	0	1	0	0	0	0
Q fever	—	—	—	—	—	1	—	—	—	—
Total parasites	0	0	0	0	0	0	0	1	0	0
Cryptosporidiosis	—	—	—	—	—	—	—	—	—	—
Tick-borne encephalitis	—	—	—	—	—	—	—	—	—	—
Toxoplasmosis	—	—	—	—	—	—	—	1	—	—
Total viral	0	<1	<1	0	<2	0	2	1	1	<1
Hepatitis	—	—	—	—	<1	—	2	1	1	<1
Poliomyelitis	—	<1	<1	—	<1	—	—	—	—	—
Total chemical	0	0	0	0	0	0	0	4	<1	6
Aflatoxin and other mycotoxin	—	—	—	—	—	—	—	—	—	—
Antibiotics	—	—	—	—	—	—	—	—	—	—
Chemicals	—	—	—	—	—	—	—	3	<1	6
Histamine poisoning	—	—	—	—	—	—	—	1	—	—
Unknown etiology	<1	1	3	8	26	17	31	41	6	18

Source: Bryan (1983) and Bean et al. (1996).

TABLE 2 Percentage of Reported United States Outbreaks Involving Various Dairy Products: 1900 to 1992

Product	1900–1909	1910–1919	1920–1929	1930–1939	1940–1949	1950–1959	1960–1969	1970–1979	1988–1992
Raw milk and cream	100	86	93	90	54	35	36	23	31
Certified raw milk	—	—	1	—	—	2	—	4	—
Pasteurized milk	—	4	3	3	16	2	—	3	—
Nonfat dry milk	—	—	—	—	2	2	—	4	—
Cheese	—	1	<1	3	9	33	21	17	14
Butter	—	1	—	—	<1	—	—	4	—
Ice cream	<1	8	2	4	17	24	44	44	56
Number of outbreaks	(n = 173)	(n = 333)	(n = 390)	(n = 403)	(n = 301)	(n = 45)	(n = 39)	(n = 94)	(n = 36)

Source: Bryan (1983) and Bean et al. (1996).

From 1985 to 1989, 30 cheese recalls (primarily imported soft, surface-ripened varieties) were issued along with 26 ice cream recalls involving more than 3.1 million gallons of product (Ryser and Marth, 1991). Even though the number of *Listeria*-related recalls has decreased slightly, similar product trends can be seen in the period from 1990 to May 1997, with *L. monocytogenes* still being responsible for 74% of all dairy product recalls (Table 3). Other reasons for recalling dairy products, principally cheese, during this period have included the presence of *E. coli* and potentially toxigenic molds as well as the presence of *Cryptosporidium* and changes in cheese spread formulations that may lead to potential growth of *Clostridium botulinum* and the accompanying threat of botulism.

In this discussion, the various public health concerns affecting the dairy industry have been organized into three arbitrary categories. The first section of this chapter deals with public health concerns primarily of historical interest such as diphtheria, scarlet fever, tuberculosis, and typhoid. Major public health concerns of current interest are discussed in greater detail in the following section and include the common bacterial infections (e.g., campylobacteriosis, listeriosis, salmonellosis) and intoxications (e.g., staphylococcal poisoning) as well as potential health concerns related to presence of aflatoxins and drug residues in the milk supply. Uncommon and suspected milk-borne bacteria, rickettsiae, parasites, viruses, and toxins responsible for infrequent dairy-related illnesses are briefly reviewed in the last section; a few of these etiological agents are likely to increase in significance within the next 10 to 20 years.

TABLE 3 Number (%) of Microbiologically Related FDA Recalls Issued for Dairy Products: 1990–May 1997

| | Product | | | | | |
Agent	Fluid Milk	Butter	Ice Cream and Frozen Yogurt	Dried Milk and Whey	Cheese	Total
Listeria monocytogenes	—	4	17	—	25	46 (74)
Clostridium botulinum	—	—	—	—	7	7 (11)
Salmonella	—	—	1	2	—	3 (5)
Mold	—	—	—	—	2	2 (3)
Escherichia coli	—	—	—	—	2	2 (3)
Aflatoxin	1	—	—	—	—	1 (2)
Cryptosporidium	—	—	—	—	1	1 (2)
Total	1 (2)	4 (7)	14 (25)	2 (4)	34 (62)	62

Source: FDA Enforcement Reports 1990–1997.

II. HISTORICAL CONCERNS

The presence of pathogenic bacteria in milk has been a matter of public health concern since the early days of the dairy industry. From the turn of the century to 1940, numerous health hazards were associated with ingesting raw milk and dairy products prepared from raw milk. Typhoid fever, the primary milk-borne disease during this period, accounted for 50% to 80% of all milk-related illnesses, with scarlet fever being responsible for an additional 14% to 27% of milk-borne infections (Bryan, 1983). Both of these diseases were frequently fatal because suitable treatments such as with antibiotics were unavailable. The high incidence of milk-borne typhoid and scarlet fever, coupled with sporadic dairy-related outbreaks of diphtheria, poliomyelitis, and tuberculosis, soon confirmed the need for increased pasteurization, a process that was used only sporadically during the 1920s and 1930s. After World War II, almost universal adoption of pasteurization, coupled with modernization of milk production practices emphasizing improved farm and dairy factory sanitation, udder health, herd inspection, and cooling, handling, and storage of milk have, for all practical purposes, eliminated the threat of these diseases, with the last cases of milk-borne typhoid fever, scarlet fever, and diphtheria in the United States being reported more than 40 years ago.

A. Diphtheria

Dreaded by mothers of small children for more than 2000 years, diphtheria has come to be one of the best understood and controlled human bacterial diseases, with long-standing immunization programs virtually eliminating diphtheria in the United States, Canada, and most of Europe. *Corynebacterium diphtheriae*, the bacterial pathogen responsible for diphtheria, is an obligate parasite, with humans serving as the natural host and reservoir. Morphologically, *C. diphtheriae* is a gram-positive, nonmotile, non–spore-forming, club-shaped bacterium that grows in characteristic branching Chinese "letter" arrangements and stains irregularly because of intracellular metachromatic (polyphosphate) granules (Barksdale, 1986; Dixon et al., 1990). Diphtheria-producing strains of *C. diphtheriae* secrete diphtheroid toxin, an extremely potent extracellular, simple protein toxin, the production of which is dictated by a prophage carrying the tox^+ gene (Barksdale, 1986).

Diphtheria is normally acquired through contact with asymptomatic carriers harboring the organism in their nasal passages and only rarely by contact with actual clinical cases. In classical infections, *C. diphtheriae* multiplies within epithelial cells of the nasopharynx. Early symptoms include mild fever, sore throat, and prostration, but continued toxin production leads to formation of a tough grayish pseudomembrane composed of dead tissue and fibrin, which adheres to the tonsils and the posterior pharyngeal wall (Dixon et al., 1990;

McCloskey, 1986). Subsequent spreading of this membrane downward into the larynx and trachea produces severe respiratory problems and eventual suffocation. Because all human tissues are vulnerable to this toxin, further complications including degeneration of the heart muscle, nervous system, and most other internal organs, result in almost certain death once the toxin enters the bloodstream (McCloskey, 1986). Hence, neutralization of the toxin with diphtheria antitoxin at the first suspicion of infection, together with administration of penicillin or erythromycin (Dixon et al., 1990) are both vital for full recovery.

Raw milk consumption was epidemiologically linked to 11 diphtheria outbreaks in the United States between 1919 and 1948 (Bryan, 1979), with three similar outbreaks also reported in England (Goldie and Maddock, 1943; Wilson, 1933) and Australia (Bryan, 1979) during this period. In most of these outbreaks, dairy workers who either exhibited active infections or carried *C. diphtheriae* asymptomatically were assumed to have contaminated the milk during milking or subsequent handling. Evidence for direct transmission of *C. diphtheriae* by cows is limited to two related cases of bovine diphtheritic mastitis in a small South African village (Pfeiffer and Viljoen, 1945) and one additional outbreak in which superficial teat and udder infections developed in cows from contact with a human carrier (Henry, 1920). Several early outbreaks also were associated with consumption of ice cream (Bryan, 1979) and butter (Hammer, 1938). More recently, 149 diphtheria cases were reported in the Arab Republic of Yemen from August 1981 to January 1982 (Jones et al., 1985). Twenty-one children died, giving a mortality rate of 14%. Subsequent epidemiological evidence implicated one commercial brand of yogurt as a possible source of infection; however, the product was no longer available for testing. The fact that most children younger than 10 years of age were not fully immunized against diphtheria was a major contributing factor in this outbreak. With the help of well-developed immunization programs, an average of only three annual cases of diphtheria was recorded in the United States from 1984 to 1994 (Anonymous, 1994c). Thus, the rarity of diphtheria cases in highly industrialized countries coupled with routine pasteurization has all but eliminated dairy products as a source of *C. diphtheriae* infections.

B. Scarlet Fever and Septic Sore Throat

Streptococcus pyogenes, a gram-positive, β-hemolytic, group A streptococcus, causes scarlet fever, septic sore throat, pharyngitis, and tonsillitis in humans and mastitis in dairy cattle (Bryan, 1979). Human symptoms of infection include severe sore throat, hoarseness, headache, muscle pain, fever, prostration, weakness, chills, diarrhea, nausea, and vomiting (Decker et al., 1985). Scarlet fever, which manifests with a characteristic rash, results from infection with certain erythrogenic toxin-producing strains of *S. pyogenes*. Prompt treatment with anti-biotics greatly minimizes further complications such as rheumatic fever and nephritis; however, 25% of patients who have received proper treatment can become lingering carriers of *S. pyogenes* (Valkenburg, 1986).

Dairy-related outbreaks of scarlet fever and septic sore throat were common before pasteurization became routine (Eyler, 1986) with at least 40 such outbreaks (13,939 cases and 20 deaths) occurring between 1907 and 1927; 37 of these outbreaks presumably resulted from ingestion of raw milk (Hammer, 1938). The largest of these outbreaks occurred in Chicago, with at least 10,000 cases linked to faulty milk pasteurization (Capps and Miller, 1912). In most instances, the initial source of contamination was traced to dairy farmers with scarlet fever who either infected their cows or the milk during milking (Hammer, 1938) with subsequent growth of *S. pyogenes* also possible between 20°C and 37°C (Davis, 1914). Even though a few additional outbreaks have been traced to ingestion of ice cream (Hammer, 1938) and dried milk (Allen and Baer, 1944; Purvis and Morris, 1946), routine pasteurization has virtually eliminated milk as a vehicle for scarlet fever in the United States.

C. Tuberculosis

Tuberculosis was one of the greatest scourges of humans and animals since antiquity with detailed descriptions of this disease recorded by Hippocrates in 400 BC (Grange, 1990). The turning point finally came in 1882 when Robert Koch isolated and showed *Tuberkelbacillin* (bacilli of tuberculosis) to be the causative agent of tuberculosis (Collins and Grange, 1983). Similar organisms were subsequently isolated from cases of tuberculosis-like disease in various animals, giving rise to three main types of tubercle bacilli now recognized as *Mycobacterium tuberculosis* (human type), *Mycobacterium bovis* (bovine type), and *Mycobacterium avium* (avian type). In 1901, Koch erroneously claimed that "the human subject is immune against infection with bovine bacilli . . ." and that "human tuberculosis differs from bovine, and cannot be transmitted to cattle." In 1911, the Royal Commission on Tuberculosis concluded that cows with bovine tuberculosis indeed posed a hazard to human health (Collins and Grange, 1983). Two years later, cattle vaccinated with supposedly attenuated strains of *M. tuberculosis* were shown to shed viable virulent organisms in their milk (Griffith, 1913). Today, tuberculosis is an infectious granulomatous disease primarily acquired by inhaling *M. tuberculosis*. Dairy herd immunization programs and mandatory pasteurization have virtually eliminated milk-borne *M. bovis* infections in developed countries after 1960 (Habib and Warring, 1966).

1. General Characteristics

M. bovis, the primary organism responsible for milk-borne tuberculosis, is an aerobic, nonmotile, non–spore-forming, nonencapsulated, straight to slightly curved, slender, weakly gram-positive, acid-fast (resistant to decolorization by acidified organic solvents after initial staining) bacillus (Nolte and Metchock, 1995). Most commonly isolated mycobacteria, including *M. bovis*, grow very slowly and may require up to 8 weeks of incubation at 35°C for visible growth to appear on laboratory media. Although the technique is virtually obsolete, inoculation of guinea pigs was historically used to identify raw milk and clinical samples containing *Mycobacterium* spp., particularly *M. bovis* (Wilkins et al., 1987). Biochemically, *M. bovis* fails to produce niacin or reduce nitrate, with *M. tuberculosis* giving the reverse reactions (Nolte and Metchock, 1995). High cell wall lipid levels account for the unique resistance of the organism to drying, chemical disinfection, and other environmental stresses (Mitscherlich and Marth, 1984), which, in turn, makes *M. bovis* difficult to eradicate from farm environments.

2. Clinical Manifestations

Nonpulmonary tuberculosis is an infectious granulomatous disease characterized by development of lesions at the site of penetration, typically the oropharnyx and intestinal tract in milk-borne cases involving *M. bovis*. Spread of the organism to the kidneys and genitourinary tract via the lymphatic system can produce additional lesions in these areas. A condition affecting the bones and joints known as kyphosis (hunchback) frequently occurs in older children and adults, whereas children younger than 5 years of age are most prone to complications of the meninges, including meningitis (Bryan, 1979; Hammer, 1938). Given such widespread organ involvement, prognosis was poor, with 2000 of 4000 childhood cases in Great Britain ending terminally in 1932 (Anonymous, 1932). Development of antituberculosis drugs, including isoniazid, rifampicin, and pyrazinamide has revolutionized modern-day treatment of tuberculosis, making operative intervention and sanatoriums part of a bygone era (Grange, 1990).

3. Outbreaks

According to Park and Krumwiede (1911), *M. bovis* infections were relatively common, with this organism accounting for 7% of all tuberculosis cases observed in New York City and 9% of all such cases reported worldwide. Reports circumstantially linking raw milk consumption to tuberculosis also abound in the early literature (Bryan, 1979; Hammer, 1938); however, only three reports are supported by strong bacteriological evidence. In the first of these outbreaks (Price, 1934), *M. bovis* was recovered from raw milk consumed by 3 of 45 Canadian children in whom nonpulmonary tuberculosis developed. The second outbreak occurred in 1936 and was traced to a small Swedish village (Stahl, 1939). Milk from a cow with active tuberculosis of the udder was reportedly consumed raw by 29 of 32 individuals in whom tuberculosis developed even though the local dairy farm had a rigorous tuberculosis screening program in place at the time of the outbreak. Quinn et al. (1974) reported that a young boy living on a Michigan farm reacted positively to a tuberculin skin test after ingesting raw milk from his parents' herd of 34 dairy cows, several animals of which were heavily infected. The last two cases of bovine tuberculosis within the United States were diagnosed in 1976 (Passes et al., 1978) with both victims reportedly being foreign born and having spent much time in India.

Changing milk consumption habits, mandatory pasteurization, and cattle immunization programs have drastically reduced but probably not totally eliminated milk-borne transmission of *M. bovis* tuberculosis. In the United States, *M. bovis* accounted for only 6 of 2086 culturally confirmed tuberculosis cases at the Mayo Clinic from 1950 to 1958 (Steele and Ranney, 1958). However, another survey conducted in England, Scotland, and Wales showed that 26%, 22%, and 17%, respectively, of nonpulmonary tuberculosis cases were caused by *M. bovis*, with many of these patients giving a history of raw milk consumption. According to Collins and Grange (1983), *M. bovis* is still of concern in nonpulmonary tuberculosis, with 109 cases being reported in southeast England (including London) from 1977 to 1981. Nonetheless, the last three confirmed cases of milk-borne tuberculosis in England and Wales were reported during the 1950s (Galbraith et al., 1982) which, in turn, suggests that potential milk-borne *M. bovis* cases are being incompletely investigated or underreported (Collins and Grange,1983).

4. Occurrence and Survival in Milk and Dairy Products

Factors influencing milk-borne transmission of *M. bovis* include the incidence of infection in cows as well as the incidence and level of contamination in milk. *M. bovis* infections in dairy cattle are long-lasting, with 1% to 2% of cases involving udder lesions and excretion of *M. bovis* in the milk (Stiles, 1989). However, dairy cattle also can shed *M. bovis* in their milk as a result of septicemic and cutaneous infections. In both instances, sufficient levels of mycobacteria can be excreted from a single cow to make 100 gallons of previously noncontaminated milk infectious for infants (Kleeburg, 1975). Although unable to grow in refrigerated milk, *M. bovis* will persist in such milk during extended storage (Hammer, 1938). Before dairy cattle were routinely screened for tuberculosis, *M. bovis* was commonly found in raw milk, with 3% to 15% of the raw and pasteurized milk supplied at the turn of the century to such cities as New York, Washington, DC, Baltimore, Philadelphia, and Chicago containing *M. tuberculosis* or *M. bovis*. During this period, an estimated 43,750 quarts of contaminated milk were being pasteurized daily for the Chicago market (Hammer, 1938). Elsewhere, the situation was similar, with 8.6% of 16,700 milk samples collected worldwide yielding *M. tuberculosis* (Hammer, 1938). A tuberculosis infection rate of 40% for cattle slaughtered in 1932, 0.5% of which had mycobacteria in their milk, likely accounts for only part of this contamination (Savage, 1933), with the remainder coming from direct contact of the milk with dried fecal material or the farm environment. Given the success of tuberculosis screening programs for cattle in the United States, Canada, and western Europe, coupled with the systematic slaughter of all animals testing positive from infected herds, the incidence of *M. bovis* in the raw milk supply is very low. However, raw milk from northern India (Sakhre and Vyas, 1978) and other less developed areas of the world may harbor *M. bovis* with greater frequency.

Although clearly present in raw milk after the turn of the century, *M. tuberculosis* (*M. bovis*) was seldom recovered from dairy products other than butter, cream cheese, and occasionally cottage cheese. Early reports indicated *M. tuberculosis* contamination rates of 9.5 % and 13.2% for butter marketed in Boston (Rosenau et al., 1914) and Europe (Hammer, 1938), respectively, with the organism surviving at least 153 days in butter prepared from naturally contaminated cream (Mitscherlich and Marth, 1984). In addition, *M. tuberculosis* was present in 13.7% of cream cheese and 3.2% of cottage cheese marketed in Washington, DC (Schroeder and Brett, 1918). While *M. bovis* is seldom found in other dairy products, numerous early studies attest to the hardiness of *M. bovis* in other cheeses, reportedly surviving at least 47 days in Camembert and Muenster cheese as well as 62 days in cheddar cheese and 232 days in Tilsit cheese (Mitscherlich and Marth, 1984), all of which were prepared from naturally contaminated milk. In a more recent survey, Obiger et al. (1970) failed to recover *M. tuberculosis* from 51 soft French cheeses, many of which were likely prepared from raw milk. Consequently, these latter findings combined with a complete lack of cheese-associated cases suggest that cheese is an unlikely vehicle for *M. bovis* infections.

5. Prevention

Veterinary inspection of meat and tracing of infected animals back to their farm of origin, coupled with the systematic slaughter of tuberculosis-positive animals in the herd, have greatly reduced the incidence of tuberculosis in dairy cattle. In 1979, only 0.18% of the herds in Great Britain were found positive for tuberculosis (Collins and Grange, 1983). However, because certain wild animals including the badger and opossum are also susceptible to *M. bovis* infections, total eradication of tuberculosis in dairy cattle appears unlikely. Mandatory pasteurization, the other step in preventing milk-borne tuberculosis, has proven highly effective with high-temperature, short-time (71.7°C for 15 seconds) and vat pasteurization (61.7°C for 30 minutes), both inactivating large populations of *M. bovis* and *M. tuberculosis* in milk with a wide margin of safety (Kells and Lear, 1960). However, as discussed earlier (Quinn et al., 1974), bovine tuberculosis may still present an occasional health risk, particularly for individuals consuming raw milk on farms.

D. Typhoid Fever

Raw milk was first suggested as a vehicle for typhoid fever well before the etiological agent was isolated and identified. In 1857, Dr. Michael Taylor of Penrith, England, identified 13 cases of typhoid fever among seven rural families who obtained raw milk from a family farm (Hammer, 1938). Epidemiological evidence suggested that a servant girl suffering from typhoid fever was most likely responsible for contaminating the milk in this outbreak. The causative agent of typhoid fever, later named *Salmonella typhi*, was not observed in human patients until 1880, and the organism was isolated on culture media 4 years later (Bryan et al., 1979; Marth, 1969). During the first half of the 20th century, consumption of contaminated milk, cheese, and butter was responsible for numerous outbreaks of typhoid fever and many fatalities, with the disease accounting for 50% to 80% of all cases of milk-borne illness reported in the United States from 1900 to 1939 (Bryan, 1983; Hammer, 1938). However, adoption of routine pasteurization of milk after World War II and improvements in sanitation standards led to a precipitous decrease in typhoid fever with no milk-borne outbreaks being reported in the United States sine the 1950s. Whereas milk-borne typhoid fever has been eradicated in most industrialized countries, occasional milk-borne outbreaks still occur in developing countries where the disease is endemic because of contaminated water supplies.

1. General Characteristics

S. typhi, an important bacterium in the family Enterobacteriaceae, is short, motile, gram-negative, facultatively anaerobic and rod shaped (Kuesch, 1986). *S. typhi* readily grows on common laboratory media at 15°C to 41°C with optimal growth occurring at 37°C. This serovar of *Salmonella* is biochemically distinct from the more than 2300 other serovars of salmonellae that are responsible for gastroenteritis, the most common form of *Salmonella* infection. Isolates of *S. typhi* do not produce gas from glucose, utilize citrate, decarboxylate ornithine, or ferment rhamnose. In contrast to other salmonellae that are found in the gastrointestinal tract of many animals, humans are the only known reservoir for *S. typhi*. The closely related organism *Salmonella paratyphi*, which causes paratyphoid fever, rarely produces mastitis in dairy cattle (George et al., 1972) and is almost invariably confined to human carriers. However, both organisms are extremely hardy and can survive many weeks in water, ice, feces, and dust particles (Mitscherlich and Marth, 1984), which makes elimination from the environment difficult.

2. Clinical Manifestations

Typhoid fever normally begins with a bacteremia-related fever that develops 1 to 2 weeks after ingesting at least 10^5 *S. typhi* organisms (Kuesch, 1986; Parker, 1990). Various nonspecific symptoms including anorexia, malaise, lethargy, myalgia, and a continuous headache frequently accompany the fever, which peaks at 104°F to 105°F (40°C to 41°C) within 3 to 4 days. Although some patients experience spontaneous remission of these symptoms within 1 week, sustained high fever over the ensuing 2 weeks most often leads to frank prostration, delirium, and abdominal pain caused by spleen or liver enlargement. A bewildering array of typhoid-related complications ranging from encephalitis, Guillain-Barrè syndrome, and psychiatric disorders to myocarditis, hepatitis, nephritis, hemolytic uremic syndrome, osteomyelitis, and septic arthritis has also been reported. Two prominent complications during the third week of infection, namely intestinal hemorrhage and perforation of the bowel, can lead to peritonitis, which is usually fatal without surgical intervention. Prompt antimicrobial therapy has reduced the mortality rate of typhoid fever in the United States and other developed countries to less than 1%, with chloramphenicol being the antibiotic of choice. Despite proper treatment, 90% of patients shed *S. typhi* in their feces for up to 3 months. Furthermore, approximately 3% of patients become long-term (>1 year) excreters of *S. typhi* at levels of at least 10^6 organisms/g, as was true for the infamous "Typhoid Mary," with such individuals serving to spread the organism to other humans and perpetuate the infectious cycle.

3. Outbreaks

Before 1940, *S. typhi* was responsible for 50% to 80% of all milk-borne illnesses reported in the United States (Bryan, 1983), with a total of 848 typhoid fever outbreaks involving more than 30,000 cases and 300 deaths being recorded (Hammer, 1938). Milk-borne epidemics peaked in the United States during 1911 to 1915 with 238 reported outbreaks, most of which involved *S. typhi*. According to Armstrong and Parran (1927), of 479 milk-borne outbreaks of typhoid fever summarized up to 1927, 444 (93%) and 32 (6.5%) were linked to milk (primarily raw) and ice cream, respectively, with the three remaining outbreaks involving butter and cheese. Only 29 outbreaks were traced to pasteurized milk or dairy products, with *S. typhi* entering the product as a postpasteurization contaminant. Symptomatic and asymptomatic carriers of *S. typhi* were identified as the contamination source in 80% of these outbreaks with milk bottles coming from infected households and use of polluted water to clean dairy utensils cited as additional contributing factors.

Many of these early epidemics involving milk, butter, and cheese have been well documented. In one outbreak from 1900, 65 college students became ill after consuming raw milk (Hammer, 1938). Investigators traced the milk to a farm worker who had recently recovered from typhoid fever. Use of badly polluted well water to wash dairy equipment and temperature abuse of the milk were cited as major contributing factors. During the 1920s, one particularly large outbreak of typhoid fever was traced to pasteurized milk in Montreal, Canada. A total of 5014 cases and 488 deaths occurred primarily among institutionalized children and adults, giving a mortality rate of nearly 10% (Lumsden et al., 1927). Faulty pasteurization, intentional addition of raw milk to pasteurized milk, and handling of the milk by an infected worker were cited as probable causes. Butter contaminated by a convalescent carrier of *S. typhi* was responsible for 35 to 40 cases (including six deaths) of typhoid fever in Minnesota during 1913 (Hammer, 1938). The cheese-borne outbreak alluded to earlier occurred during 1923 and involved 51 cases of typhoid fever (including four deaths) in Michigan (Rich and Fellow, 1923). Raw milk used in cheese making was reportedly contaminated by a farm worker shedding *S. typhi* with the cheese consumed shortly after manufacture. During the 1940s, three additional Canadian outbreaks of typhoid fever were traced to cheddar cheese prepared from contaminated raw milk and consumed before 60 days of ripening (Foley and Poisson, 1945; Gauthier and Foley, 1943; Menzies, 1944). Colby and mold-ripened cheeses also have been implicated in outbreaks of typhoid fever, according to Marth (1969).

Routine pasteurization of milk and improved sanitary standards led to a precipitous decrease in the number of typhoid cases and human carriers of *S. typhi* after World War II, with typhoid fever responsible for only 17% and 3% of all dairy-related illnesses reported in the United States during 1940 to 1949 and 1950 to 1959, respectively (Bryan, 1983). As a result of these efforts, the annual number of typhoid fever cases from all causes decreased from 4211 cases in 1945 to 411 cases in 1994 (Anonymous, 1994c) with no dairy-related cases of typhoid fever being reported in the United States since the 1950s. As recently as 1971, however, officials at the Centers for Disease Control traced 132 cases of typhoid fever in Trinidad to commercially produced ice cream with an infected factory worker and absence of pasteurization cited as the most likely causes (Taylor et al., 1974). Hence, milk-borne typhoid fever can still pose a threat in developing countries where substandard sanitation can lead to substantial numbers of *S. typhi* carriers.

4. Occurrence and Survival in Milk and Dairy Products

Contamination of raw milk, pasteurized milk, and other dairy products has been invariably linked to symptomatic or asymptomatic carriers of *S. typhi*. Scott and Minett (1947) produced mastitis in dairy cows by injecting *S. typhi* into the udder, and the organism was subsequently shed in the milk for up to 85 days. Given the absence of any naturally occurring cases of bovine mastitis involving the shedding of *S. typhi* in milk, however, contamination of all dairy products is assumed to be exclusively of human origin. Even without survey data, because there is an absence of recent milk-borne cases of typhoid fever and because fewer than 700 cases of typhoid fever from all causes have been reported in the United States since 1965 (Anonymous, 1994c), it is suggested that *S. typhi* has been eliminated from the milk supply and other dairy products.

Current vat and high-temperature, short-time pasteurization standards are designed to destroy large populations of *S. typhi* in milk with a wide margin of safety (Evans and Litsky, 1968). However, this pathogen will readily grow as a postpasteurization contaminant in pasteurized and sterilized milk that has been temperature abused (Pullinger and Kemp, 1938) and persist in refrigerated milk for 3 to 6 weeks (Mitscherlich and Marth, 1984). According to results from several additional early studies summarized by Mitscherlich and Marth (1984), *S. typhi* can survive 2 to 4 weeks in butter prepared from contaminated cream and also persist 12 to 39 days in ice cream stored at –4°C to –5°C. These findings clearly support milk, butter, and ice cream as vehicles of infection in the afore-mentioned outbreaks.

The fate of *S. typhi* both on and in cheese also has attracted some attention. Early reports indicated that *S. typhi* survived approximately 2 to 4 and 10 to 15 weeks after cheddar, Swiss, and other cheeses were surface inoculated and stored at ambient and refrigeration temperatures, respectively (Mitscherlich and Marth, 1984). When cheddar cheese was prepared from inoculated milk and ripened at 15°C and 5°C, *S. typhi* generally persisted for 3 and at least 10 months, respectively (Campbell and Gibbard, 1944). In addition, Wade and Shere (1928) found that *S. typhi* survived 2 to 3 months in cheddar cheese prepared from naturally contaminated milk that would not clot properly. Such findings led to the present law that requires that cheddar and other hard cheeses legally prepared from raw or subpasteurized milk be held at greater than or equal to 1.7°C for at least 60 days to eliminate *S. typhi* and other pathogenic bacteria.

5. Prevention

With modern sewage treatment plants and dairy processing facilities, milk-borne typhoid fever is no longer a threat in industrialized countries, and no such cases have been reported in the United States or England during the past 40 years

(Galbraith et al., 1982). However, the potential for milk-borne outbreaks still exists in developing countries where typhoid fever is endemic because of high human carriage rates for *S. typhi*. Consequently, travelers to such areas should be appropriately vaccinated and avoid consuming raw milk and dairy products prepared under poor sanitary conditions.

III. CURRENT PUBLIC HEALTH CONCERNS

Public health concerns impacting the dairy industry continue to change in response to advances in sanitation, milk handling, and animal husbandry practices. Common pre-World War II milk-borne illnesses such as diphtheria, scarlet fever, tuberculosis, and typhoid fever no longer pose a threat to consumers and have been replaced by more immediate concerns. Milk-borne staphylococcal poisoning, a major problem during the middle of this century, has been superseded in importance by salmonellosis and campylobacteriosis, which have accounted for most dairy-related illnesses reported since the early 1980s. Although responsible for comparatively few outbreaks of illness, several recently identified milk-borne pathogens, including *E. coli* O157:H7, *L. monocytogenes*, and *Yersinia enterocolitica*, have received widespread attention because of the particularly severe or fatal complications produced by these organisms. The potential impact of aflatoxin—a highly potent human carcinogen sometimes found in milk—has also received recent attention as have the possible public health ramifications of antibiotic and drug residues in the milk supply. Finally, several recent reports suggest that milk-borne brucellosis is of more than passing interest, particularly in the southwestern United States. Although definition of a "major public health concern" is somewhat arbitrary, 12 major public health concerns impacting the dairy industry since World War II are discussed in this section based on their continued high incidence of disease (e.g., campylobacteriosis, salmonellosis, staphylococcal poisoning), recurring sporadic outbreaks (*Bacillus cereus* poisoning, brucellosis, enteropathogenic *E. coli*), severity of illness (botulism, *E. coli* O157:H7, listeriosis, yersiniosis), and potential impact of chronic exposure (aflatoxin, drug residues). These major public health concerns account for more than 95% of all reported milk-borne illnesses.

A. Aflatoxin

The aflatoxins belong to a subset of secondary metabolites termed mycotoxins, which are produced by certain strains of molds, namely *Aspergillus flavus*, *Aspergillus parasiticus*, and *Aspergillus nomius*. First identified in England in 1960 during an outbreak that involved the death of more than 100,000 turkeys from liver disease (Stevens et al., 1960), the aflatoxins have become recognized as extremely potent liver carcinogens for both animals and humans. Four major forms of aflatoxin designated aflatoxin B_1 (AFB$_1$), B_2 (AFB$_2$), G_1 (AFG$_1$), and G_2 (AFG$_2$) are currently recognized, with AFB$_1$ being the most potent (Applebaum et al., 1982). AFB$_1$ is most often found in moldy peanuts (ground nuts) and animal feeds containing corn or other grains. When dairy cattle ingest contaminated feed, AFB$_1$ is metabolized to aflatoxin M_1 (AFM$_1$), some of which is shed in the milk (see Chap. 1). Animal feeding studies indicate that AFM$_1$ is somewhat less carcinogenic than AFB$_1$. Nonetheless, as of 1987, at least 34 countries had active or proposed legislation regarding aflatoxin limits in animal feed, with the United States and many European countries also having legislated maximum acceptable levels of AFM$_1$ in fluid milk and dairy products (van Egmond, 1989b).

1. General Characteristics

Aflatoxins, named because of their production by the mold *A. flavus* (*A. flavus toxin*), are highly substituted coumarins containing a fused dehydrofurofuran moiety. These toxic and carcinogenic secondary metabolites form a unique group of highly oxygenated, heterocyclic, low molecular weight compounds. Four major aflatoxins are recognized, with AFB$_1$ and AFB$_2$ exhibiting intense blue fluorescence and AFG$_1$ and AFG$_2$ exhibiting intense green fluorescence when viewed under ultraviolet light at 425 and 450 nm, respectively (Applebaum et al., 1982). Relatively few *A. flavus* and *A. parasiticus* isolates produce aflatoxin. However, toxigenic strains typically synthesize two or three forms of aflatoxin, one of which is invariably AFB$_1$—the most potent toxin and carcinogen of the group. When dairy cattle ingest aflatoxin-contaminated feed, AFB$_1$ and AFB$_2$ are metabolized to their respective 4-hydroxy derivatives, namely AFM$_1$ and AFM$_2$, and excreted with the milk. AFM$_1$ is of primary concern to the dairy industry, whereas AFM$_2$ is produced in smaller quantities and is far less toxic than AFM$_1$.

Requirements for aflatoxin synthesis are relatively nonspecific, with these secondary metabolites being produced on most foods and laboratory media that support mold growth. *A. flavus* and *A. parasiticus* fail to produce aflatoxins when grown at temperatures below 7.5°C or above 40°C, with 7 to 21 days of incubation at 25°C to 30°C optimal for aflatoxin synthesis (Schindler, 1977). However, aflatoxin production is reportedly enhanced when temperatures fluctuate between 5°C and 25°C (Park and Bullerman, 1981). Present evidence indicates that levels of AFM$_1$ in milk and dairy products are relatively unaffected by pasteurization, sterilization, fermentation, cold storage, freezing, concentrating, or drying (Yousef and Marth, 1989). However, treating milk with hydrogen peroxide, benzoyl peroxide, sulfites, bisulfites, riboflavin, lactoperoxidase, or ultraviolet irradiation has proven effective in experimentally reducing levels of AFM$_1$ in contaminated milk (Applebaum and Marth, 1982a; Yousef and Marth, 1985;

Yousef and Marth, 1986). AFM_1 levels in milk also can be markedly reduced by physically adsorbing the toxin onto certain particulate materials such as bentonite (Applebaum and Marth, 1982b).

2. Detection

Several methods based on thin-layer chromatography (TLC) and high performance liquid chromatography (HPLC) are available for detecting AFM_1 in milk and dairy products (Scott, 1990; Stubblefield and van Egmond, 1989). Regardless of the method used, AFM_1 must first be extracted from the sample either directly using chloroform or indirectly from columns on which AFM_1 has been adsorbed using more polar solvents such as methanol and acetonitrile. After further purification by solvent partition column chromatography or dialysis (Diaz et al., 1993), the extract is analyzed by one- or two-dimensional TLC on silica gel plates with the latter being the Association of Official Analytical Chemists (AOAC)-approved method preferred for samples likely to contain less than 0.1 μg AFM_1/kg. However, an alternative method using reverse-phase HPLC is equally sensitive and has superseded two-dimensional TLC in many laboratories. After ultraviolet detection at 365 nm, positive results can be confirmed by re-chromatographing the sample after trifluroacetic acid derivitization and inspecting the chromatogram for specific end products.

More recently, immunochemical strategies have been developed for detecting AFM_1 in milk and milk products (Fremy and Chu, 1989). Polyclonal and monoclonal antibodies produced by conjugating AFM_1 to a protein before injection into rabbits have been adapted for use in radioimmunoassays (Sun and Chu, 1977) and more recently enzyme-linked immunosorbent assays (Candlish et al., 1985; Fremy and Chu, 1984) following extraction of AFM_1 from the product in question. However, as of 1996, these newer immunochemical methods had not received AOAC approval.

3. Clinical Importance

Concern regarding human exposure to aflatoxin is based on results from animal feeding trials. Acute toxicity of aflatoxin is well documented in laboratory animals with 50% of 1-day-old ducklings dying after receiving single doses of AFB_1, AFG_1, AFB_2, and AFG_2 at levels of 0.73, 1.18, 1.76, and 2.83 mg/kg body weight, respectively (Wogan et al., 1971). Gross liver failure is the normal cause of death in animal studies involving acute oral exposure to AFB_1. Long-term exposure to low levels of AFB_1 (i.e., 1 ppm) in feed usually leads to terminal liver cancer with mutagenic and teratogenic effects also widely recognized.

Studies assessing the toxicity of AFM_1 are far fewer because of limited availability of AFM_1 in pure form. Allcroft and Carnaghan (1963) were first to report that milk from cows that received aflatoxin-contaminated feed produced liver lesions and kidney damage in day-old ducklings. Pathological changes in the liver also were similar to those produced by AFB_1. According to Pong and Wogan (1971), AFM_1 was lethal to laboratory rats at a dose of 1.5 mg/kg body weight with death again resulting from acute liver failure. Overall, the hepatotoxicity of AFM_1 in ducklings and rats appears to be similar or slightly less than that of AFB_1.

Like AFB_1, AFM_1 is also a potent liver carcinogen. In early feeding studies using rainbow trout, Sinnhuber et al. (1974) found that 60% of fish on a continuous diet of 20 μg AFM_1/kg body weight developed liver carcinomas within 12 months. In another study by Cullen et al. (1985), four groups of laboratory rats were maintained on diets containing 0, 0.5, 5, and 50 μg AFM_1/kg with a fifth control group receiving 50 μg AFB_1/kg. Hepatocarcinomas developed in only 5% of rats receiving 50 μg AFM_1/kg, whereas liver cancers were detected in 95% of the AFB_1 control group. Thus, AFM_1 appears to possess only about 5% of the hepatocarcinogenic potential when compared to AFB_1. Based on studies using rhesus monkeys (Seiber et al., 1979), the carcinogenic potential of AFM_1 in humans is likely 10- to 100-fold less than that of AFB_1.

4. Occurrence and Fate in Dairy Products

Ingestion of aflatoxin-contaminated animal feed leads to the excretion of AFM_1 in milk within 12 to 24 hours (van Egmond, 1989a). However, only 0.4% to 2.2% of ingested AFB_1 appeared in milk as AFM_1 (Frobish et al., 1986; Patterson et al., 1980). Levels of AFM_1 normally peak within 3 to 6 days and decrease to undetectable levels 2 to 4 days after exposure to contaminated feed is stopped. Assuming that 6 kg of feed containing 10 μg AFB_1/kg is consumed daily, dairy cows should produce milk containing 0.02 to 0.07 μg AFM_1/L (van Egmund, 1989b). However, AFM_1 levels in milk fluctuate daily and vary between animals (Kiermeier et al., 1977; Patterson et al., 1980; Veldman et al., 1992).

Milk surveillance programs for AFM_1 have been conducted in the United States and elsewhere. During the fall of 1977, 43% to 80% of milk samples collected in Alabama, Georgia, North Carolina, and South Carolina contained a trace to greater than 0.7 μg AFM_1/L with a heavily contaminated corn crop being largely responsible (Stoloff, 1980). While the AFM_1 contamination rate has since subsided, a similar peak in AFM_1-positive milk samples was again observed in late 1988 and early 1989 (van Egmund, 1989b), with a midsummer drought blamed for high levels of AFB_1 in midwestern feed corn. More recently, high levels of aflatoxin also forced a Georgia dairy to recall more than 24,000 gallons of pasteurized dairy products during January of 1991 (Anonymous, 1991c).

In European surveys conducted during the late 1960s and 1970s, 11% to 82% of the milk samples examined contained AFM_1 at levels of 0.2 to 6.5 μg/kg, with fewer positive milk samples recorded during the summer grazing period (van Egmond, 1989b). Legislative action regarding maximum acceptable aflatoxin levels, typically less than 10 µg/kg, has led to marked reductions in AFM$_1$ contamination levels with less than 25% of samples from Austria, Belgium, Finland, France, Ireland, and Spain normally containing AFM$_1$, mostly at very low levels. However, markedly higher contamination rates have been reported in Italy, the Netherlands, and Sweden.

Sporadic contamination of the milk supply has raised concerns regarding the fate and stability of AFM$_1$ during manufacture and storage of both nonfermented and fermented dairy products. Despite some variability in the data reviewed by Yousef and Marth (1989), the level and activity of AFM$_1$ in milk does not appear to change appreciably as a result of pasteurization, sterilization, cold storage, or freezing. Although AFM$_1$ is concentrated in nonfat dry milk, evaporated milk, and freeze-dried milk during manufacture, the stability and activity of AFM$_1$ is again relatively unaffected during concentration and drying. Because AFM$_1$ is primarily water soluble, a natural partitioning of AFM$_1$ also occurs during production of cream and butter. Typically, only approximately 10% and 2% of the AFM$_1$ in milk appears in cream and butter, respectively, with the remainder shed in the two by-products—skim milk and buttermilk.

When cheeses such as cheddar (Brackett and Marth, 1982b), Swiss (Stubblefield and Shannon, 1974), Parmesan (Brackett and Marth, 1982c), and Camembert (Kiermeier and Buchner, 1977) are prepared from AFM$_1$-contaminated milk, the toxin partitions almost equally between the curd and whey, with the higher than expected levels in curd presumably resulting from selective hydrophobic adsorption to casein. The end result is that AFM$_1$ levels in soft and hard cheese are normally 2.5- to 3.3- and 3.9- to 5.8-fold higher, respectively, as compared with original levels in the cheese milk (Yousef and Marth, 1989). Little, if any, change in AFM$_1$ activity has been reported during cheese making or ripening. However, proteolysis of casein during extended storage tends to increase the levels of AFM$_1$ detected (Brackett and Marth, 1982a).

In addition to AFM$_1$-contaminated milk, growth of naturally occurring aflatoxigenic molds on the surface of cheese is also of some public health concern because aflatoxin-positive cheeses have been detected during surveys in the United States, western Europe, and north Africa (Scott, 1989). In several surface mold inoculation studies, AFB$_1$ and AFG$_1$ diffused at least 4 cm into the cheese during storage (Park and Bullerman, 1983; Shih and Marth, 1972). Hence, simply scraping the mold from the cheese surface does not necessarily render the product safe for consumption.

5. Prevention

Minimizing the presence of AFM$_1$ in milk and dairy products is entirely dependent on careful control and monitoring of mold growth and AFB$_1$ levels in animal feed. The United States requirement of less than or equal to 20 ng total aflatoxin per gram of animal feed, if observed, will consistently yield acceptable milk containing less than 0.5 ng (0.5 ppb) AFM_1/g.

B. *Bacillus cereus* Food Poisoning

Aerobic, spore-forming bacteria resembling *B. cereus* have been suspected as agents of foodborne disease for many years with at least six "*B-cereus*-like" outbreaks in Europe described before 1950 (Gilbert, 1979; Kramer and Gilbert, 1989). However, early recognition of pathogenic bacilli other than *Bacillus anthracis*—the causative agent of anthrax—was not possible because of considerable taxonomic confusion in the genus *Bacillus*. Realization that *B. cereus* was a foodborne pathogen came in the early 1950s when the Norwegian investigator Hauge (1950, 1955) described a series of four outbreaks involving 600 cases of diarrheal illness that were traced to consumption of vanilla sauce prepared from corn starch heavily contaminated with *B. cereus*. Even though the role of *B. cereus* in this outbreak was proven in human volunteer feeding studies, conclusive findings concerning the direct involvement of two different toxins, namely diarrheal enterotoxin (Goepfert et al., 1972; Spira and Goepfert, 1972) and emetic enterotoxin (Melling et al., 1976), in two distinct forms of *B. cereus* poisoning were not reported until the 1970s.

A common contaminant of the dairy environment and raw milk supply, *B. cereus* also is one of several spoilage organisms responsible for "sweet curdling" (Overcast and Atmaram, 1974) and "bitty cream" (Cox, 1975). Despite the ability of this spore-forming organism to germinate and grow in refrigerated milk, most recent outbreaks of *B. cereus* food poisoning in the United States (Bean et al., 1996) and elsewhere (Kramer and Gilbert, 1989) have involved an entirely different product, namely Chinese food (primarily fried rice), with fewer than 10 cases traced to dairy products (i.e., nonfat dry milk, cream, ice cream) consumed in the United States (Holmes et al., 1981) and England (Galbraith et al., 1982; Sockett, 1991). However, the ability of *B. cereus* to persist in powdered milk and grow in the reconstituted product as evidenced by a large outbreak in Chile involving newborn infants (Cohen et al., 1984) has led to establishment of rigid international standards for *B. cereus* in infant formula (Becker et al., 1994).

1. General Characteristics

A member of the diverse genus *Bacillus*, which contains more than 60 different species of aerobic and facultatively anaerobic, gram-positive, spore-forming rods, *B. cereus* is larger than most other bacilli and also normally motile by peritrichous flagella (Drobniewski, 1993; Kramer and Gilbert, 1989; Sneath, 1986). Endospores are produced in either a central or near-central position and do not distend the sporangium. Unlike most other organisms discussed in this chapter, *B. cereus* is a psychrotroph and therefore able to grow at temperatures of 50°C to as low as 4°C (Dufrenne et al., 1995; van Netten et al., 1990) with growth also occurring over a pH range of 4.4 to 9.3 and at a water activity (a_w) of 0.92 in the presence of 7% NaCl. Typical isolates grow both aerobically and anaerobically, reduce nitrate to nitrite, liquify gelatin, hydrolyze casein and starch, utilize citrate, and produce a positive Voges-Proskauer reaction, with acid also being produced from glucose, fructose, maltose, sucrose, salicin, trehalose, and glycerol; however, numerous exceptions have been noted. Lecithinase activity, coupled with the inability to utilize mannitol and resistance to polymyxin B, is most often used to identify suspect colonies during primary isolation.

 B. cereus normally produces several types of toxins including hemolysins, proteases, phospholipases, cytotoxins (Christiansson et al., 1989), a heat-labile diarrheal enterotoxin, and a heat-stable emetic toxin, the last two of which are responsible for two distinct forms of food poisoning. Despite these many characteristics, *B. cereus* is closely related to two other prominent bacilli, namely *B. anthracis* (the causative agent of anthrax) and *B. thuringensis* (a well-known insect pathogen) as shown by DNA hybridization studies. It is therefore often difficult to positively identify. Thus far, 23 of 42 flagellar (H) *B. cereus* serotypes have been linked to illnesses, with biotyping (based on biochemical properties), phage typing, pyrolysis mass spectroscopy, and whole-cell fatty acid analysis profiles proving useful in subtyping isolates obtained during epidemiological investigations (Drobniewski, 1993; Valsanen et al., 1991).

2. Analysis of Dairy Products for *Bacillus cereus* and Toxin

Most outbreaks of *B. cereus* poisoning have been traced to foods containing at least 10^6 organisms/g. However, products such as nonfat dry milk frequently contain small numbers of *B. cereus* spores, which can germinate when the milk is reconstituted and grow to dangerous levels during storage.

 When large numbers of *B. cereus* are expected, the suspect food is serially diluted and surface plated on mannitol–egg yolk–polymyxin agar (MYP). Alternatively, a three-tube most probable number (MPN) method using trypticase soy–polymyxin broth can be used for samples suspected of containing low levels of *B. cereus*, with tubes showing growth similarly surface plated to MYP after 48 hours of incubation at 30°C (Harmon et al., 1992; Rhodehamel and Harmon, 1995). Presumptive *B. cereus* isolates on MYP after 24 to 48 hours of incubation at 30°C appear as large, spreading, pink (mannitol-negative) colonies surrounded by an opaque halo indicating lecithinase activity. Selected isolates are then purified and confirmed as *B. cereus* based on anaerobic production of acid from glucose, nitrate reduction, tyrosine decomposition, resistance to lysozyme, and a positive Voges-Proskauer reaction. Tests for motility, rhizoid growth, hemolysin production, and intracellular toxin crystals are also useful in differentiating the various groups of *B. cereus* from *B. thuringensis* and *B. anthracis*. Confirming an outbreak of *B. cereus* poisoning is also dependent on demonstrating that the suspect isolate is toxigenic. Rapid assays for detecting the emetic toxin are not yet available, and current methods rely on monkey feeding trials and in vitro cell culture systems (Drobniewski et al., 1993; Wong et al., 1988). However, a serologically based microslide gel double diffusion assay has been developed for *B. cereus* strains producing the diarrheal form of enterotoxin (Bennett, 1995), with several fluorescent-based immunoblot (Baker and Griffiths, 1995; Granum et al., 1993) and reverse passive latex agglutination assays (Granum et al., 1993; Griffiths, 1990) also available commercially.

3. Clinical Manifestations

B. cereus is responsible for two distinct clinical syndromes of long and short onset, both of which have involved dairy products. The so-called "diarrheal syndrome" resembles *Clostridium perfringens* food poisoning and results from a proteinaceous, heat-labile diarrheal enterotoxin that is presumably produced during growth of *B. cereus* within the small intestine (Donta, 1986; Drobniewski, 1993; Kramer and Gilbert, 1989). However, the precise mode of action of this enterotoxin remains unclear. Within 8 to 16 hours after ingesting a food containing greater than or equal to 10^6 *B. cereus* cells/g, patients typically have abdominal pain and cramps followed by a profuse watery diarrhea devoid of blood or mucus occurring at 15- to 30-minute intervals. Vomiting and fever are normally absent. The entire illness is self-limiting, usually resolving within 12 to 24 hours without complications. The second syndrome caused by *B. cereus*, termed "emetic syndrome," results from ingesting a heat-stable (90 min/126°C) emetic toxin, which is preformed in the food and resistant to proteolysis. This syndrome resembles staphylococcal poisoning in both symptoms and incubation period. After an onset time of 15 minutes to 5 hours, nausea, vomiting, abdominal cramps, and less frequently diarrhea develop, with these symptoms resolving within 1 to 5 hours. Because both syndromes are, by definition, intoxications of short duration, antibiotic therapy is contraindicated and treatment is limited to fluid replacement and, in severe cases, administration of antiemetics.

4. Outbreaks

As discussed, few dairy-related outbreaks of *B. cereus* poisoning have been reported. The largest known of such outbreaks occurred in the Netherlands during the late 1980s when nausea and diarrhea developed in 280 individuals 2 to 14 hours after consuming pasteurized milk containing 4×10^5 enterotoxigenic *B. cereus* colony-forming units per milliliter (cfu/mL) (van Netten et al., 1990). Except for a few scattered cases traced to feta cheese in Canada (Schmitt et al.,

1976), pasteurized cream in England (Galbraith et al., 1982; Gilbert and Parry, 1977), milk in Romania (Gilbert, 1979), and ice cream in both England (Sockett, 1991) and the former Soviet Union (Gilbert, 1979), most of the remaining outbreaks have been small and typically linked to contaminated nonfat dry milk used as an ingredient. Such reported outbreaks have involved Dutch vanilla pudding (Gilbert and Parry, 1977), a Norwegian yellow pudding dessert (Pinegar and Buxton, 1977), Hungarian cream pastries (Pinegar and Buxton, 1977) and English "vanilla slice" pastries (Pinegar and Buxton, 1977). Three additional cases of *B. cereus*-like food poisoning in Canada also have been attributed to the use of nonfat dry milk and malted milk powder as ingredients in unspecified home-prepared foods (Schmitt et al., 1976).

Only one dairy-related outbreak of *B. cereus* poisoning has been reported in the United States since this illness was first discovered in the early 1950s. According to Holmes et al. (1981), the emetic form of *B. cereus* poisoning developed in eight individuals in Alabama after consuming macaroni and cheese at a cafeteria. Investigators found that some of the product not served contained 10^8 to 10^9 *B. cereus* cfu/g with the organism also identified in powdered milk, an ingredient used in preparing the macaroni and cheese. Improper heating and refrigeration were deemed responsible for growth of *B. cereus* in the final product before serving. One additional unusually large outbreak occurred in Chile during May and June of 1981 when 35 neonatal cases of *B. cereus* diarrheal syndrome were traced to infant formula prepared from contaminated powdered milk (Cohen et al., 1984). Growth of *B. cereus* in infant formula during 12 and 24 hours of refrigerated storage was subsequently confirmed. Follow-up studies using suckling mice demonstrated that selected isolates were enterotoxigenic. The fact that virtually all isolates were non-typeable further supports powdered milk as the source of *B. cereus* in this outbreak.

5. Occurrence and Survival in Dairy Products

Psychrotrophic spore-forming organisms belonging to the genus *Bacillus* are common contaminants of raw milk produced in the United States and elsewhere. Spores of *B. cereus* most often enter milk from soil, cattle feed, milking equipment, or the udder during milking (Crielly et al., 1994). However, *B. cereus* also can be shed in cow's milk as a result of mastitis (Horvath et al., 1986; Logan, 1988). One survey of raw milk from Wisconsin demonstrated that 9% of the samples contained *B. cereus* at levels less than or equal to 100 cfu/g (Ahmed et al., 1983a). Working in Scotland, Griffiths and Phillips (1990) found psychro-trophic *Bacillus* spp. in 58% of the raw milk supply. In addition, 39% of the isolates were identified as *B. cereus*, most of which produced diarrheal toxin (Griffiths, 1990). During a 2-year survey in England, Crielly et al. (1994) also noted that *B. cereus* was more commonly recovered from raw milk during the summer months at levels as high as 10^5 cfu/mL with similar observations also made by other investigators (McKinnon and Pettipher, 1983). Because *B. cereus* spores do not germinate in raw milk, rapid growth of vegetative cells during periods of temperature abuse is presumably responsible for the high incidence of this organism in summer milk (Phillips and Griffiths, 1986).

Given the frequency of *B. cereus* in raw milk and the ability of *B. cereus* spores to survive pasteurization and germinate (Stadhouders et al., 1980), it is not surprising that this organism is also a common contaminant of pasteurized milk. According to Ahmed et al. (1983a), 35% of pasteurized milk samples sold in Wisconsin contained *B. cereus* at levels less than or equal to 1000 cfu/mL. Elsewhere, the incidence of *B. cereus* in pasteurized milk is reportedly 2% in China (Wong et al., 1988), 25% in the Netherlands (van Netten et al., 1990), and 33% in Australia (Rangasamy et al., 1993), with levels generally being less than 1000 cfu/mL. Although none of these surveys examined isolates for enterotoxin production, pasteurized milk is an excellent source of enterotoxigenic strains with 59% and 100% of milk isolates from Norway (Granum et al., 1993) and Scotland (Griffiths, 1990), respectively, producing toxins, some isolates of which were also reportedly psychrotrophic.

Growth of enterotoxigenic *B. cereus* strains in pasteurized milk is well documented (Christiansson et al., 1989; Griffiths, 1990; Wong et al., 1988), with this organism exhibiting an average generation time of 17 hours at 6°C (Griffiths and Phillips, 1990). When naturally contaminated retail pasteurized milk was stored at 7°C, van Netten et al. (1990) found that *B. cereus* attained levels of 10^3 to 10^5 cfu/mL in 85% of samples by the "sell by" date. Furthermore, selected *B. cereus* isolates from these samples also grew and produced enterotoxin in pasteurized milk after 24, 12, and 2 days of incubation at 4°C, 7°C, and 17°C, respectively. When Griffiths (1990) inoculated sterile reconstituted skim milk to contain 10^4 *B. cereus* cfu/mL, the organism grew to 10^7 cfu/mL and produced detectable levels of toxin after only 7 days of storage. Nonetheless, enterotoxin is generally confined to pasteurized milk containing greater than 10^7 *B. cereus* cfu/mL, which accounts for the lack of milk-borne cases of *B. cereus* poisoning, and such milks frequently show obvious spoilage.

Presence of *B. cereus* in powdered milk probably poses the greatest public health concern because both pasteurization and spray-drying induce germination and outgrowth of spores in the reconstituted product. According to Rodriguez and Barrett (1986), *B. cereus* was identified in 5 of 8 (62.5%) dried milk samples analyzed in California, with most larger European surveys yielding contamination rates of 27% to 57% (Becker et al., 1994). *B. cereus* was similarly present in 13% to 43% of nonfat dry milk–based infant formula manufactured in West Germany (Becker et al., 1994). However, levels of *B. cereus* in both products seldom exceeded 1000 cfu/g.

Growth of *B. cereus* to hazardous levels in reconstituted nonfat dry milk and infant formula is well documented. Using naturally contaminated reconstituted nonfat dry milk, Rodriguez and Barrett (1986) reported *B. cereus* populations of more than 10^6 cfu/mL following 12 to 22 and 24 to 56 hours of incubation at 30°C and 20°C, respectively, with samples not yet showing signs of spoilage. However, growth of the organism was generally prevented when identical samples were stored at 5°C. Similar growth of *B. cereus* in reconstituted infant formula during ambient storage has been reported (Becker et al., 1994). Because infants are particularly susceptible to *B. cereus* poisoning, a proposal has been introduced in Europe to limit *B. cereus* levels in infant formula to less than 1000 cfu/g (Becker et al., 1994), with even stricter standards likely to be enforced in the future.

B. cereus contamination is not confined to the aforementioned products. According to Ahmed et al. (1983a), *B. cereus* was recovered from 14% of cheddar cheese samples and 48% of ice cream samples tested in Wisconsin, with contamination levels not exceeding 200 cfu/g in cheddar cheese and 3800 cfu/g in ice cream. Spores of *B. cereus* can survive in experimentally produced cheddar cheese for at least 52 weeks (Mikolajcik et al., 1973). However, the pH of properly prepared cheddar cheese (i.e., pH 5.0) is sufficiently low to inhibit spore germination and growth of vegetative cells (van Netten et al., 1990). Consequently, low levels of *B. cereus* in properly fermented dairy products are of minimal public health concern.

6. Prevention

Widespread occurrence of *B. cereus* in the natural environment ensures the continued recovery of this organism from milk and other dairy products during all stages of production. Unlike the other milk-borne pathogens to be discussed, heat-resistant *B. cereus* spores readily germinate as a result of pasteurization with outgrowth and enterotoxin production occurring in products stored at temperatures near refrigeration. However, because *B. cereus* populations greater than 10^5 cfu/g (Langeveld et al., 1996) are invariably needed to induce illness, dairy-related outbreaks of *B. cereus* poisoning are readily prevented by minimizing contamination of raw milk at the farm level and storing both fluid and reconstituted milk at temperatures less than or equal to 4°C. Active starter cultures also minimize growth of this organism during manufacture of fermented dairy products.

C. Botulism

The rarest and most fatal of the milk-borne diseases, botulism results from ingesting minute amounts of a preformed neurotoxin produced by the bacterium

Clostridium botulinum. This toxin, termed botulinal toxin, is 100,000 times stronger than rattlesnake venom with the human lethal dose estimated at 0.1 to 1.0×10^{-6} g (Hobbs, 1986). During the early 1800s, investigators in central Europe traced this illness to liver and blood sausage from which the term "botulism" (from the Latin word *botulus*, meaning sausage), also known as "sausage poisoning," derived its name (Hauschild, 1989). The causative organism was first isolated and named *Bacillus botulinus* by van Ermengem in 1896 after a Belgian outbreak that was traced to home-cured ham. This organism was later reclassified as *C. botulinum.*

Botulism is usually associated with consumption of low acid (pH > 4.6) foods such as home-canned vegetables, canned cured meats, fermented sausage, and cured fish, which are packaged in air-tight containers, with dairy products seldom being implicated (Hauschild, 1989). Only 5 of 971 (0.51%) botulism outbreaks (18 of 2430 cases, or 0.74%) reported in the United States from 1899 to 1994 (Headrick et al., 1996; Solomon et al., 1995) were linked to dairy products (all cheeses), with the last United States case occurring in 1951 (Meyer and Eddie, 1951). Worldwide, only 10 dairy-related outbreaks of botulism involving a total of 127 cases have been documented. Recent work on milk products has focused on different means of preventing *C. botulinum* growth and toxin production in processed cheese spread, a product that was responsible for one fatality in 1951 and a small outbreak in Argentina during 1974 (Briozzo et al., 1983).

1. General Characteristics

C. botulinum is the taxonomic designation given to a group of gram-positive, strictly anaerobic, rod-shaped, spore-forming bacteria that produce a characteristic neurotoxin (Hauschild, 1989). Most strains are motile by peritrichous flagella and produce oval spores either centrally or subterminally, which distend the cell wall (Cato et al., 1986). Although biochemically diverse, all isolates produce gas from glucose and hydrolyze gelatin, with most strains also exhibiting lipase activity. Growth-limiting temperatures are 3.5°C and 50°C; however, considerable variation has been observed between strains (Conner et al., 1989; Hauschild, 1989). Although growth has been demonstrated in laboratory media at pH values as low as 4.0, growth and toxin production do not generally occur in foods having a pH less than 4.6 with some strains failing to grow at less than pH 5.0.

By definition, all *C. botulinum* strains produce at least one of seven antigenically distinct neurotoxins designated types A, B, C, D, E, F, and G, with some strains producing two toxins (e.g., types A and B, A and F, B and F, C and D) (Conner et al., 1989; Hauschild, 1989). Biochemically, these toxins are proteins ranging in molecular weight from 150,000 to 900,000 daltons. Toxin production occurs intracellularly with the toxin released into the external environment during logarithmic growth and subsequent cell lysis. Unlike staphylococcal enterotoxin, all botulinal toxins are heat labile and rapidly destroyed by boiling. Based on proteolytic activity and the type of toxin produced, all *C. botulinum* isolates can be divided into the following four groups: (a) proteolytic types A, B, and F, (b) nonproteolytic types B, E, and F, (c) proteolytic and nonproteolytic types C and D, and (d) proteolytic type G. Thus far, only those strains producing toxin types A and B have been associated with dairy-related outbreaks of botulism with growth and toxin production at refrigeration temperatures confined to nonproteolytic type B strains.

2. Analysis of Dairy Products for *C. botulinum* and Botulinal Toxin

Upon receipt of the sample, one portion is analyzed for viable *C. botulinum* organisms and a different portion is examined for botulinal toxin. Recovery of the organism from dairy products begins with primary enrichment in cooked meat medium (CM) at 26°C to 28°C and trypticase-peptone-glucose-yeast extract broth (TPGYE) at 35°C (Kauter et al., 1992; Solomon et al., 1995). After 7 and, if necessary, up to 17 days of incubation, both broth cultures are microscopically examined for typical tennis racket–shaped, spore-forming bacteria resembling *C. botulinum*. One portion of the positive broth culture is immediately centrifuged and analyzed for botulinal toxin. The remaining portion is ethanol treated or heat treated to eliminate the non–spore-forming background flora and then is surface plated on liver-veal egg yolk agar (LVEY) or anaerobic egg yolk agar (AEY) to obtain isolated colonies. After 48 hours of anaerobic incubation at 35°C, 10 *C. botulinum*-like colonies are recultured in CM or TPGYE and then restreaked to LVEY or AEY for purification.

Toxin analysis begins by macerating and then extracting the sample with an equal volume of gel-phosphate buffer at pH 6.2 (Kauter et al., 1992; Solomon et al., 1995). After centrifugation of the extracted food sample or aforementioned broth culture, one portion of the supernatant liquid is treated with trypsin to activate botulinal toxins produced by nonproteolytic strains. After diluting a portion of the trypsin-treated and untreated supernatant liquids 1:5, 1:10, and 1:100 in gel-phosphate buffer, 0.5 mL of each preparation is injected intraperitoneally into pairs of white mice; a boiled untrypsinized and undiluted preparation serves as the control. During the next 48 hours, the mice are observed for symptoms of botulism, which include ruffled fur, labored breathing, limb weakness, total paralysis, and death resulting from respiratory failure. However, death alone does not provide conclusive evidence that the preparation contained botulinal toxin. Establishing the amount of toxin in the sample is dependent on some of the mice surviving. The type of toxin in the sample can be determined by first injecting the mice with monovalent antitoxins to types A, B, E, and F. Although not yet approved for official use, several enzyme-based immunosorbent assays for toxin detection are available, thus circumventing the problems associated with animal tests.

3. Clinical Manifestations

Dairy-related botulism outbreaks have been confined to products (primarily cheeses) containing toxin types A and B, which, together with type E found in fish, comprise the most lethal of the seven known toxin types. The first symptoms of botulism normally develop within 12 to 36 hours of ingesting the preformed toxin and include diarrhea, nausea, and vomiting followed by persistent constipation (Donta, 1986; Hauschild, 1989; Smith, 1990). Soon after being absorbed by the gastrointestinal tract, the toxin enters the bloodstream and begins to shut down the peripheral nervous system by attaching to the tips of motor nerve endings, which, in turn, prevents the release of acetylcholine at neuromuscular junctions. Neurological symptoms associated with this classic phase of the illness include blurred and double vision, difficulty in speaking and swallowing, a dry mouth, throat, and tongue, fatigue, lack of muscle coordination, and, in extreme cases, total paralysis, with death by respiratory failure within as little as 24 hours from the initial onset of gastroenteritislike symptoms. Because botulism can be confused with Guillain-Barré syndrome, carbon monoxide poisoning, myasthenia gravis, and other types of food poisoning, a quick and accurate diagnosis based on the patient's clinical symptoms and case history is essential for proper treatment and full recovery. Eventual confirmation of suspected cases is dependent on detecting botulinal toxin or viable *C. botulinum* cells in appropriate clinical specimens.

Before antitoxins and modern mechanical respirators were available, at least half of all victims died, making botulism the gravest of the milk-borne diseases. However, the fatality rate has been reduced to 5% to 15% in the United States and most other industrialized countries (Donta, 1986; Hauschild, 1989; Smith, 1990). Initial treatment of botulism is focused on toxin removal or inactivation by neutralizing circulating toxin with antitoxin before the toxin irreversibly binds to the nerve endings. Induced vomiting, gastric lavage, and enemas are also used to help rid the body of toxin. Subsequent treatments are directed toward counteracting paralysis of respiratory muscles with mechanical respirators which are required by 80% of victims in the United States. Chemotherapy is limited to administration of guanidine, which is sometimes helpful in restoring nerve function. However, prolonged use normally leads to serious side effects.

4. Outbreaks

Historically, dairy products have been responsible for less than 1.0% of all food-borne botulism cases, with only 10 outbreaks involving a total of 127 cases reported worldwide since 1899. Six small dairy-related outbreaks affecting 19 people (9 of whom died) have been documented in the United States, with four outbreaks reported in California (1912, 1914, 1935, and 1951) and two outbreaks occurring in New York (1914, 1939) (Meyer and Eddie, 1951). All of these outbreaks were cheeseborne and traced to cheeses such as cottage, Limburger, and Neufchatel prepared at home. In the 1914 New York outbreak, Nevin (1921) reported that three people died after eating homemade cottage cheese stored in sealed tins with *C. botulinum* growth and toxin production demonstrable in inoculated cottage cheese after 72 hours of incubation at 37°C. In another anecdotal account by Meyer and Eddie (1951), three botulism cases and one death were traced to an ethnic-type curd cheese that was ripened below ground in a buried canvas-covered crock; *C. botulinum* undoubtedly entered the product from the soil. The last documented dairy-related case of botulism in the United States occurred in 1951 (Meyer and Eddie, 1951) and involved a 53-year-old man who died within 3 days of eating several ounces of a pasteurized process soft-ripened Limburger cheese spread that reportedly tasted peculiar. Subsequent tests with mice demonstrated type B botulinal toxin in the remaining product, with 4 of 51 additional jars of cheese spread also reportedly toxic.

The four remaining outbreaks occurred outside of the United States during the 1970s and 1980s. During July and August of 1973, ripened Brie cheese was epidemiologically linked to two simultaneous outbreaks of type B botulism, one in Marsaille, France (32 cases) and the other in Switzerland (42 cases) (Gilles et al., 1974; Kauf et al., 1974; Sebald et al., 1974). Surprisingly, no fatalities were reported among the 75 cases. Although the implicated cheese was not available for testing, all cheeses were ripened on the same batch of unclean straw in both Marsailles and Switzerland, thereby providing a plausible means of contamination. Toxin production in the rind of similarly ripened cheeses, was later demonstrated experimentally (Billon et al., 1980).

Early in 1974, Argentinian authorities reported that a commercially produced cheese spread containing onions was responsible for six cases of type A botulism, including three deaths, in Buenos Aires (Briozzo et al., 1983). Investigators eventually blamed the outbreak on several cheese formulation deficiencies including an overly high moisture content and pH, which permitted *C. botulinum* growth and toxin production.

The last reported dairy-related outbreak of botulism occurred in 1989 and involved 27 cases (including one death); it was unusual in several respects (Critchley et al., 1989; O'Mahony et al., 1989). First, the outbreak originated in the United Kingdom, a country that has seen only nine cases of foodborne botulism since 1922 with no cases traced to milk or dairy products. Second, hazelnut yogurt, a very unusual product (i.e., pH < 4.6, refrigerated) never before associated with botulism, was identified as the cause of this epidemic. Third, wide sales distribution of the product led to patients seeking treatment at different

hospitals. Investigators identified type B botulinal toxin and later type B
C. botulinum in opened and unopened cartons of hazelnut yogurt as well as in one
fecal specimen and in a blown can of hazelnut conserve. All implicated lots of
hazelnut yogurt and conserve were recalled and warnings were issued to the
general public. Thermal processing of the hazelnut conserve was later shown to
be inadequate for destruction of *C. botulinum* spores with all *C. botulinum* growth
and toxin production occurring in these cans of product during long-term storage.

5. Occurrence and Survival in Dairy Products

Spores of *C. botulinum* are widespread in the natural environment, with soil
serving as the primary reservoir (Hauschild, 1989; Smith, 1990). Consequently,
vegetation, animal feed, and farm produce are most frequently contaminated. Few
domestic farm animals including dairy cattle are fecal carriers of *C. botulinum*;
this organism also is not known to cause mastitis in ruminant animals.

Spore-forming bacteria, including *C. botulinum*, are frequent contaminants
of raw and pasteurized milk. Although these spores readily survive pasteurization
as evidenced by several spore-related defects, toxin production in raw and pasteurized milk does not occur because of the product's short refrigerated shelf-life
and the inability of this organism to readily compete with the native psychrotrophic background flora. However, Kaufmann and Brillaud (1964) found that
C. botulinum types A and B did grow and produce toxin in cans of sterilized skim
milk after 46 to 56 days of storage at 13°C. Read et al. (1970) also reported
growth and toxin production by *C. botulinum* type E in inoculated cans of commercially sterilized whole milk after 3 to 28 days of storage at 20°C, with one
sample being toxic after 70 days of incubation at 7.2°C. However, *C. botulinum*
type E is typically confined to fish products and has never been associated with
dairy products.

The aforementioned fatalities in 1951 and 1974 involving processed cheese
spread prompted extensive investigations into the safety of these anaerobically
packaged, long shelf-life products. In a series of three studies by Wagenaar and
Dack (1958a, 1958b, 1958c), *C. botulinum* growth and toxin production was
related to the pH, moisture, a_w, and salt content of process cheese spreads
prepared from three different varieties of surface-ripened cheeses, with the toxin
in these cheeses stable for 2 to 4 years (Grecz et al., 1965). Several subsequent
investigators (Briozzo et al., 1983; Kauter et al., 1979) reported toxin production
in inoculated samples of commercially prepared pasteurized process cheese
spread having a pH greater than or equal to 5.70 and an a_w greater than or equal
to 0.936, suggesting that these products should be classified as low acid foods
(pH > 4.6, a_w > 0.85). However, when Tanaka et al. (1979) prepared pasteurized
processed cheese spread according to the United States federal standards of
identity (52% moisture, 2% sodium chloride, and 2.5% disodium phosphate),

inoculated samples remained nontoxic during 11 months of storage at 30°C. Given these parameters, toxin development would not be expected in such cheese spreads having an a_w less than 0.95 (Hauschild, 1989). The fact that the cheese spread implicated in the Argentinian outbreak had an a_w of 0.97 (Brizzo et al., 1983) further supports this conclusion. However, production of nontoxic process cheese spreads containing less salt or up to 60% moisture is also possible by decreasing the pH and increasing the level of various phosphates, which, in turn, lowers the a_w (Eckner et al., 1994; Karahadian et al., 1985; Tanaka et al., 1986). Furthermore, direct addition of nisin, a bacteriocin produced by *Lactococcus lactis* subsp. *lactis* that prevents germination of *C. botulinum* spores, to process cheese spreads at levels up to 250 ppm affords additional protection against *C. botulinum* growth and allows production of reduced sodium and sodium chloride-free spreads (Somers and Taylor, 1987).

6. Prevention

The few reported botulism cases traced to dairy products have primarily involved anaerobically packaged cheeses with process cheese spreads being of greatest concern. Although contamination of such products with spores of *C. botulinum* cannot be prevented, the threat of toxin production can be eliminated by carefully controlling the pH, moisture content, a_w, phosphate level, and nisin content of the finished product. Furthermore, most dairy-related botulism cases have involved proteolytic strains of *C. botulinum* types A and B, with the implicated products showing obvious signs of spoilage. Such products in swollen containers should be immediately discarded and never tasted. Continued Food and Drug Administration (FDA) enforcement of established governmental standards for preventing *C. botulinum* growth and toxin production in high-risk foods also plays an important role in preventing future dairy-related outbreaks of botulism, seven non-complying cheese products were recalled since 1990 without incident.

D. Brucellosis

Human brucellosis, a classic zoonoses presumably prevalent in the Mediterranean countries since antiquity (Anonymous, 1995a; Tarala, 1969), is primarily acquired through direct or indirect contact with infected animals harboring three of six bacterial species belonging to the genus *Brucella*. Two of these species, *Brucella melitensis* and *Brucella abortus*, are pathogenic to goats and sheep and to cattle, respectively, and are consequently of major concern to the dairy industry. The remaining specie, *Brucella suis*, is primarily found in pigs and, as such, has been only rarely associated with milk-borne cases of brucellosis (Horning, 1935). In 1887, while working as a British naval surgeon on the island of Malta, Bruce was first to isolate an organism from four fatal cases of a disease he termed "Malta fever," now commonly known as undulant fever. By 1904, Maltese goat's milk was confirmed as the source of infection (Hammer, 1938; Rammell, 1967; Tarala, 1969) with the causative organism, *B. melitensis*, still recognized as the *Brucella* sp. most pathogenic for goats, sheep, and humans (Hendricks and Meyer, 1975). Although probably present since Biblical times, the second of these two organisms was not identified until 1895 when the Danish veterinarian Bang identified *B. abortus* as the causative agent of contagious abortion (Bang's disease), an economically devastating affliction in dairy cattle. However, the close relationship between *B. abortus* and *B. melitensis* was not recognized until 1918, when Evans linked cow's milk to cases of undulant fever in the United States (Hammer, 1938; Stiles, 1989).

1. General Characteristics

All six *Brucella* spp. are small, nonmotile, coccobacilli or short rod-shaped, gram-negative bacteria that are found singly, in pairs, and in short chains (Moyer and Holcomb, 1995). Both *B. melitensis* and *B. abortus* are intracellular parasites that localize and grow within the rough endoplasmic reticulum of nonphagocytic host cells. Although able to grow aerobically at 10°C to 40°C, with optimal growth occurring at 37°C, some strains grow better in an atmosphere containing 5% to 10% CO_2. Brucellae are nutritionally fastidious and require biotin, panthotenic acid, thiamine, nicotinamide, trace amounts of magnesium, and occasionally bovine serum for growth. Consequently, propagation on ordinary solid media can be difficult. Biochemically, *B. melitensis* and *B. abortus* are catalase-positive, oxidase-positive, and metabolically oxidative (Moyer and Holcomb, 1995). Despite some cultural, biochemical, serological, and host differences among the brucellae, DNA-DNA hybridization and ribotype analyses (Anonymous, 1988a) indicate that all six currently recognized *Brucella* spp. are closely related and comprise only one genospecies, *B. melitensis*.

2. Clinical Manifestations

Human brucellosis, which ranges from a mild flulike illness to a severe disease (undulant fever), defies easy diagnosis because of differences in reported symptoms (Dalrymple-Champneys, 1960; Young, 1983). Even an increasing and decreasing temperature, the symptom for which undulant fever is named, may not always occur. The severity of brucellosis is partially dependent on the species involved, with *B. melitensis* being most pathogenic for humans, followed by *B. suis* and *B. abortus*. Onset of symptoms can be either abrupt or gradual following a normal incubation period of 3 to 21 days. However, incubation periods of 7 to 10 months also have been reported (Moyer and Holcomb, 1995). Brucellosis patients typically have multiple complaints but show few physical abnormalities. Symptoms associated with the sudden onset form of brucellosis have included pyrexia, profuse sweating, chills, weakness, malaise, various aches, chest and joint pain, weight loss, and anorexia with physical findings limited to disturbances of the spleen and lymphatic system (Dalrymple-Champneys, 1960; Young, 1983). Osteomyelitis is the most common complication from *B. melitensis* infection followed by skeletal, genitourinary, cardiovascular, and neurological complaints (Young, 1983). Victims of the gradual onset, chronic form of brucellosis exhibit long histories of recurrent fever and depression, malaise, headaches, sweating, vague pains, impotence, and insomnia, with eventual incapacitation also being reported (Stiles, 1989).

Slow growth of brucellae on laboratory media frequently delays primary isolation of the organism with fewer than 20% of all cases initially confirmed by recovering *Brucella* spp. from blood, bone marrow, or infected tissues (Stiles, 1989). Consequently, preliminary diagnosis is normally based on serological findings (Young, 1991a). Treating brucellosis with antibiotics is also difficult, because the organism is localized intracellularly. Therefore, combined oral administration of several antibiotics with high intracellular activity, such as tetracycline, streptomycin, rifampin, or trimethoprim-sulfamethoxazole (Street, 1975; Young, 1983; Young and Suvannoparrat, 1975), is the prescribed cure for typical *Brucella* infections.

3. Outbreaks

Worldwide, brucellosis remains one of the most widespread and costly diseases afflicting humans and animals, with this disease presently endemic to northern Mexico (Salman and Meyer, 1984; Teclaw et al., 1985; Thapar and Young, 1986; Young, 1991b) as well as many South American (Wallach et al., 1994), Latin American (Wallach et al., 1994), Mediterranean (Anonymous, 1995a), Middle Eastern (Anonymous, 1995a; Nour, 1982; Sabbaghian and Nadim, 1974), and African countries (Anonymous, 1995a; Cherif et al., 1986; Fakuuzi et al., 1993). Consumption of unpasteurized dairy products, including milk (Foley, 1970), cream (Barrow et al., 1968), and cheese (Anonymous, 1995a; Galbraith, 1969; Hammer, 1938; Rammell, 1967; Young and Suvannoparrat, 1975) has traditionally accounted for approximately 10% of all reported brucellosis cases (Anonymous, 1972; Stiles, 1989), with the remainder occurring primarily among male veterinarians, farmers, and meat processors who contract the disease through direct contact with infected livestock.

Mandatory pasteurization of milk and highly effective brucellosis eradication programs for livestock have drastically reduced the number of reported cases in the United States from more than 600 in 1945 to 119 in 1994, giving an annual incidence rate of one case for every 2 million people. Although brucellosis occurs throughout the United States, this disease has a long history in the American southwest (Anonymous, 1994c) and among Hispanic people who become infected after consuming certain types of soft unripened cheese produced in Mexico (Eckman, 1975). El Paso, TX, was the site of three separate brucellosis outbreaks in 1968 (Seyffert and Bernard, 1969), 1973 (Street et al., 1975; Young and Suvannoparrat, 1975), and 1983 (Tharper and Young, 1986), all of which involved consumption of Mexican-produced raw goat's milk cheese. A similar outbreak involving 31 primarily Hispanic patients who consumed fresh goat's milk cheese (queso blanco) illegally imported from Mexico occurred in Houston, TX, during 1983 (Thapar and Young, 1986)—a year in which 84 brucellosis cases were reported to the Texas Department of Health (Thapar and Young, 1986). Three of four Mexican border states, namely Texas, Arizona, and California respectively accounted for 29, 17, and 36 of the 119 (69%) brucellosis cases reported nationally in 1994 (Anonymous, 1994c), with many of these cases presumably linked to consumption of illegally imported raw milk Mexican cheese.

In England and Wales, dairy-related brucellosis outbreaks have been virtually eliminated after instituting similar programs for brucellosis eradication in livestock and mandatory pasteurization of milk (Barrett, 1989). Only 17 cases of milk-borne brucellosis were reported in England and Wales from 1950 to 1989 (Galbraith et al., 1982; Sockett, 1991), of which, 10 cases were linked to raw cow's milk containing *B. abortus* and 7 cases to *B. melitensis* in raw pecorino sheep cheese imported from Italy (Galbraith et al., 1969). However, at least nine additional people also reportedly contracted brucellosis after returning from Spain and the Middle East (Barrett, 1986; Porter and Smith, 1971) with raw sheep's milk, raw goat's milk, and goat's milk cheese being the probable vehicles of infection. Even though the number of brucellosis cases in England and Wales presently appears to be increasing, with 49 cases reported from 1992 to 1995, at least 28 of these cases were acquired abroad during visits to Malta, Spain, Portugal, France, Italy, Greece, Bosnia, Turkey, Egypt, Israel, Jordan, Qatar, Oman, Pakistan, Somalia, and Tanzania (Anonymous, 1995a), with many of these cases presumably being milk-borne or cheeseborne. Two of these cases were linked to a massive outbreak in Malta during the first half of 1995 involving 135 cases of *B. melitensis* brucellosis in which soft cheese prepared from unpasteurized sheep's and goat's milk was identified as the vehicle of infection (Anonymous, 1995a). Consequently, travelers to areas where brucellosis is endemic should consider avoiding raw milk and cheeses prepared from raw milk.

4. Occurrence and Survival in Milk and Dairy Products

Although reliable information concerning the incidence of brucellae in the raw milk supply is lacking, contamination rates are presumably very low in countries with well-developed brucellosis eradication programs. Within infected herds, brucellae can persist in the udders of cows for many years following an abortion and can be intermittently shed at levels up to 15,000 organisms/mL for as long as 5 months (Rammell, 1967). When naturally contaminated raw milk is held at 25°C to 37°C, *Brucella* populations typically decrease to nondetectable levels within 2 to 3 days (Kuzdas and Morse, 1954; Mitscherlich and Marth, 1984). However, brucellae survive at least 42 and 800 days when such milk is stored at 4°C (Mitscherlich and Marth, 1984) and –40°C (Kuzdas and Morse, 1954), respectively.

Cream and butter are unusual sources for *Brucella* spp. with only 5 of 916 cream samples being positive in one outbreak-related survey (Barrow et al., 1968). However, both products can support extended survival of brucellae with *B. melitensis* and *B. abortus* persisting 4 and 6 weeks, respectively, in inoculated cream stored at 4°C (Rammell, 1967). More recently, *Brucella* spp. reportedly survived 94 to 102 days and greater than 140 days in inoculated cream that was stored at 20°C to 25°C and 2°C to 4°C, respectively (Nour, 1982); this further confirms the increased persistence of *Brucella* at refrigeration temperatures. According to several reports referenced by Rammell (1967), brucellae can survive even longer in refrigerated butter (King, 1957), persisting 6 to 13 months in salted and unsalted butter, respectively (Fulton, 1941).

As is true for cream and butter, brucellae have become virtually nonexistent in domestic and imported cheeses sold legally in the United States. This is not true for dairy products sold in Mexico, Argentina, and many of the Mediterranean and Middle Eastern countries as evidenced by the many aforementioned dairy-related brucellosis cases. One recent survey from Turkey (Sancak et al., 1993) indicated that 7 of 40 raw sheep's milk cheeses harbored *B. melitensis* or *B. abortus*.

Long-term survival of *Brucella* in many cheese varieties has been recognized since the 1940s (Rammell, 1967). *B. abortus* survived 6 days in Emmental and Gruyere, 15 days in Tilsit, and 57 days in Camembert cheese prepared from milk inoculated to contain 10,000 brucellae/mL (Kästli and Hausch, 1957). This organism also persisted 90 days in pecorino cheese (Rammell, 1967) and up to 60 days in Roquefort cheese (King, 1957). When cheddar cheese was prepared from milk containing 1000 *B. abortus* cfu/mL and ripened at 4°C, the organism remained viable for 6 months (Gilman et al., 1946). Most recently, traditional manufacturing practices failed to eliminate *B. abortus* from Mexican white soft cheese during 21 days of storage at 5°C (Diaz Cinco et al., 1994). Even though these studies raise serious concerns regarding the safety of raw milk cheeses, standard vat and high-temperature, short-time pasteurization (Bryan, 1979) both inactivate *Brucella* populations in milk with a very large margin of safety.

5. Prevention

Preventing dairy-related cases of brucellosis is based on eliminating this disease in animals through immunization programs, slaughtering of infected animals, mandatory pasteurization of milk, and aging of cheeses that can legally be prepared from raw milk for at least 60 days. Even though this disease has been largely controlled in the United States, the recent upsurge of cases along the United States-Mexican border, combined with increased reports of British citizens acquiring this disease abroad, indicates that travelers to areas where brucellosis is endemic should avoid consuming raw milk as well as fresh cheeses (particularly goat's milk cheese) prepared from unpasteurized milk.

E. Campylobacteriosis

Although recognized since 1909 as an important cause of abortion in cattle and sheep (Stern and Kazmi, 1989), *Campylobacter jejuni*—the causative agent of campylobacteriosis—remained an obscure human enteric bacterial pathogen until the late 1970s. Improved isolation strategies (Dekeyser et al., 1972) leading to recovery of *Campylobacter* from 7.1% of randomly selected patients with diarrhea (Skirrow, 1977) suggested that this organism was of more than passing importance in human gastroenteritis. Although generally considered a sporadic illness with a propensity for children, 45 foodborne campylobacteriosis outbreaks (1308 cases) were reported in the United States between 1978 and 1986, over half of which involved ingestion of raw milk (Anonymous, 1988b). Similar reports linking raw or inadequately pasteurized milk to 13 outbreaks in Great Britain from 1978 to 1980 (Robinson and Jones, 1981) helped to further substantiate *C. jejuni* as an important milk-borne pathogen. With an estimated 2 million annual cases of this non-notifiable disease in the United States (DeMol, 1994) and numerous reports of similar cases in Canada (Lior, 1994a), Europe (Stringer, 1994), and developing countries (Taylor, 1992), campylobacteriosis has come to rival or surpass salmonellosis as the leading form of human gastroenteritis worldwide.

1. General Characteristics

The genus *Campylobacter* (Greek for "curved rod") includes 18 species and subspecies within the family *Campylobacteraceae*, with six of these organisms identified as threats to human health (Hunt and Abeyta, 1995; Nachamkin, 1995). Two seldom differentiated species, *C. jejuni* subsp. *jejuni* (hereafter *C. jejuni*) and *Campylobacter coli*, are well-recognized causes of human foodborne gastroenteritis, with the former accounting for approximately 90% of such cases (Hunt and Abeyta, 1995). Of the four remaining species, *Campylobacter upsaliensis, Campylobacter lari,* and *Campylobacter hypointestinalis* have been only recently linked to sporadic gastrointestinal disorders, and *Campylobacter fetus* subsp. *fetus* is primarily associated with bacteremia and systemic infections in patients with underlying illnesses.

Morphologically, all campylobacters are small, gram-negative, curved, S-shaped or spiral-shaped rods, which are motile by means of a single polar flagellum. All species of the genus are oxidase-positive, urease-negative, and both methyl red- and Voges-Proskauer-negative. Of the six aforementioned species of concern to humans, all produce catalase (Nachamkin, 1995) and all except *C. fetus* grow optimally at 42°C. However, all campylobacters are obligate microaerophiles and, as such, require an atmosphere of 5% O_2, 10% CO_2, and 85% N_2 for optimal growth.

2. Isolation and Identification

High numbers of *Campylobacter* are normally present in human diarrheal specimens, and microaerobic incubation of such samples streaked on commonly used selective media such as Skirrow's medium and charcoal cefoperazone deoxycholate agar allows for relatively simple recovery of the organism (Nachamkin, 1995). However, isolation of *Campylobacter* from raw milk is far more challenging because the organism is likely to be greatly outnumbered by the normal bacterial flora of the milk. Consequently, selective enrichment under microaerobic conditions became a crucial initial step in all early procedures for recovering *Campylobacter* from raw milk (Doyle and Roman, 1982a; Hunt et al., 1985; Lovett et al., 1983). All three methods currently recommended for detecting *Campylobacter* in raw milk (Flowers et al., 1992a; Hunt and Abeyta, 1995; Stern et al., 1992) are complicated and require initial centrifugation of the raw milk sample, selective enrichment of the pellet at 42°C under microaerobic conditions (5% O_2, 10% CO_2, 85% N_2), and subsequent plating on two selective media followed by similar incubation. The FDA procedure (Hunt and Abeyta, 1995) also includes a 4-hour microaerobic preenrichment step at 37°C, followed by 24 hours of incubation at 42°C with continuous shaking. Suspect isolates (round to irregular spreading colonies with smooth edges) obtained using these procedures are then examined microscopically for morphological characteristics and motility and subsequently speciated using a standard series of biochemical tests in addition to resistance to nalidixic acid and cephalothin.

3. Clinical Manifestations

Campylobacter enteritis affects all age groups but is particularly common among children. Most dairy-related cases of campylobacteriosis are presumably acquired indirectly through consumption of raw milk with person-to-person infections infrequent and normally limited to young children with acute diarrhea. Typical attack rates of at least 50% in milk-borne outbreaks suggest that the oral infective dose is relatively low. Robinson (1981) and Block et al. (1978) confirmed that ingesting as few as 500 to 800 total cells of *C. jejuni* can induce illness after the normal 2- to 5-day incubation period with the infective rate, severity of illness, and incubation period remaining unaltered as the oral infective dose increases to as many as 2×10^9 organisms. However, routine consumption of raw milk lowers the attack rate and results in partial immunity to symptomatic infections (Blaser et al., 1987).

Flulike symptoms develop in approximately one-third of patients sufficiently ill to seek medical attention; symptoms include a mild fever that occurs 2 to 3 days before appearance of diarrhea, which likely represents the initial invasive and sometimes septicemic stage of infection (Nachamkin, 1995). A few such individuals may also experience severe prediarrheal appendicitislike abdominal pain, which can lead to unnecessary surgery. Onset of diarrhea is sudden and can be severe, with development of profuse watery stools through the action of a heat-labile choleralike toxin produced by most strains of *C. jejuni* and *C. coli*. Bloody diarrhea sometimes develops, which mimics ulcerative colitis caused by shigellae. Various extraintestinal complications including bacteremia, reactive arthritis, bursitis, urinary tract infections, meningitis, endocarditis, peritonitis, and pancreatitis can occur among elderly and immunocompromised adults as can occasional fatalities. Abortion and neonatal septicemia have been cited as complications affecting pregnant women (Blaser, 1990). Two additional complications, namely Guillain-Barré syndrome (Mishu and Blaser, 1993) and acute motor axonal neuropathy (McKhann et al., 1993), also were reported in Japan and China, respectively, with a single serotype of *C. jejuni* appearing to be responsible.

Presumptive diagnosis of *Campylobacter* enteritis is based on the direct microscopic observation of *Campylobacter*-like organisms in stool specimens with campylobacteriosis being confirmed by isolating the organism on selective plating media. Although most patients spontaneously recover within 3 to 7 days, fluid replacement and oral administration of erythromycin or fluoroquinolones for 7 to 10 days may be needed for more severely ill patients (DeMol, 1994). Patients typically stop shedding the organism after 1 to 3 months. However, a small percentage of individuals remain chronic fecal carriers of *C. jejuni* and *C. coli*, thereby retaining the ability to infect others.

4. Outbreaks

Evidence for *C. jejuni* as a foodborne pathogen dates back to 1938 when milk (possibly raw) was epidemiologically linked to 357 cases of gastroenteritis among inmates of two Illinois prisons (Levey, 1947). Supporting evidence included the isolation of organisms resembling *Vibrio jejuni* (*C. jejuni*) from blood and stool samples, the appearance of symptoms compatible with present-day campylobacteriosis, and the fact that the outbreak ceased after the incriminated milk was boiled. Nevertheless, poor isolation techniques during these early days delayed identifying *C. jejuni* as a prominent foodborne pathogen for more than 40 years.

The foodborne route for *Campylobacter* infection was not suggested again until 1976 when Taylor et al. (1979) identified four Los Angeles residents who presumably acquired campylobacteriosis after consuming certified raw milk. Beginning in 1978, raw milk–related outbreaks were recorded in the United States, with many more accounts being documented up to 1985 (Table 4). Raw milk consumption was implicated in 14 of 23 outbreaks (621, or 83% of 748 cases) reported from 1980 to 1982, with four of these outbreaks involving children (Finch and Blake, 1985). Fourteen of 20 raw milk–related outbreaks documented from 1981 to 1990 (Wood et al., 1992) were traced to children in kindergarten through third grade who became ill after returning from spring and fall field trips to dairy farms.

Campylobacteriosis simultaneously emerged with similar force in Great Britain, with 27 outbreaks from 1978 to 1984 linked to consumption of raw or inadequately pasteurized milk (Hutchinson et al., 1985b; Robinson and Jones, 1981). Two additional outbreaks also were recorded in Switzerland (Stadler et al., 1983) and New Zealand (Brieseman, 1984) during the early 1980s, the former of which involved more than 500 joggers running a race.

Despite the many aforementioned outbreaks, confirmation of raw milk as the source of infection has remained difficult. Only three reports have appeared in which the epidemic strain was recovered from a portion of the lot of milk that was consumed (Bradbury et al., 1984; Patton et al., 1991; Salama et al., 1990); such strains were more commonly identified in fecal samples from incriminated dairy herds (Hutchinson et al., 1985b; Kornblatt et al., 1985; Potter et al., 1983; Stadler et al., 1983; Vogt et al., 1984), thus suggesting that the milk was contaminated during or after milking. However, at least three campylobacteriosis outbreaks in England were traced to glass-bottled pasteurized milk that was pecked by birds (Hudson et al., 1990; Riordan et al., 1993; Southern et al., 1990), with magpies and jackdaws identified as probable carriers of *C. jejuni*.

Milk-borne campylobacteriosis outbreaks have been almost invariably associated with consumption of raw or inadequately pasteurized cow's milk. However, a few cases of *C. jejuni* and *C. coli* enteritis have been traced to ingestion of raw goat's milk in the United States (Harris et al., 1987), Great Britain (Hutchinson et al., 1985a), and Australia (Gilbert et al., 1981), with the epidemic strain identified in fecal samples from the incriminated goats. Other than one additional outbreak in Great Britain in which 37 cases were linked to consumption of milk shakes, no other fluid or fermented dairy products, including yogurt and cheese, have been associated with campylobacteriosis.

TABLE 4 Campylobacteriosis Outbreaks Resulting from Ingestion of Raw Milk in the United States, Canada, and Britain

Location	Year	Number of Cases	Reference
United States			
California	1976	4	Taylor et al. (1979)
Colorado	1978	3	Anonymous (1978)
Oregon	1980–1981	77	Terhune et al. (1981)
Minnesota	1981	25	Korlath et al. (1985)
Arizona	1981	200	Taylor et al. (1982a)
Kansas	1981	264	Kornblatt et al. (1985)
Georgia	1981	50	Potter et al. (1983)
Wisconsin	1982	15	Klein et al. (1986)
Oregon	1982	22	Blaser et al. (1987)
Vermont	1982	15	Vogt et al. (1984)
Pennsylvania	1983	26	Anonymous (1983)
Pennsylvania	1983	57	Anonymous (1983)
Vermont	1983	5	Hudson et al. (1984)
California	1984	12	Anonymous (1984a)
California	1985	23	Anonymous (1986)
Vermont	1986	28	Birkhead et al. (1988)
Canada			
Ontario	1980	14	McNaughton et al. (1982)
Great Britain			
England	1978	63	Robinson et al. (1979)
England	1978	22	Robinson et al. (1979)
England	1979	2500	Jones et al. (1981)
Scotland	1979	148	Porter and Reid (1980)
England	1981	46	Wright and Tillett (1983)
England	1981	22	Wright and Tillett (1983)
England	1985	75	Hutchinson et al. (1985a, 1985b)
England	1992	72	Morgan et al. (1994b)

5. Occurrence and Survival in Dairy Products

Experimentally induced mastitis in dairy cows has led to the excretion of up to 10^5 *C. jejuni* cfu/mL in milk over a period of 7 days (Lander and Gill, 1980). However, evidence supporting such shedding by naturally infected cows is relatively limited (De Boer et al., 1984; Hutchinson et al., 1985b; Logan et al., 1982). The bovine intestinal tract remains the primary reservoir for *C. jejuni* with 19% to 64% of fecal samples being positive (DeBoer, 1984). Consequently, heavy shedding of *Campylobacter* in feces followed by fecal contamination of the milk during or after milking has come to be regarded as the primary route of contamination.

Early surveys in the United States suggested a relatively low incidence of *Campylobacter* spp. in the raw milk supply, with *C. jejuni* recovered from only 0.4% to 1.5% of raw milk bulk tank samples (Doyle and Roman, 1982b; Lovett et al., 1983; McManus and Lanier, 1987). However, after improvements were made in isolation techniques, *C. jejuni* was later detected in 12.3% of farm milk bulk tanks supplying eastern Tennessee (Rohrbach et al., 1992), thus suggesting a markedly higher incidence of *Campylobacter* contamination. In a similar European survey, *C. jejuni* was recovered from 5.9% of raw milk bulk tank samples examined in England (Humphrey and Hart, 1988) with a strong correlation observed between *C. jejuni* and *E. coli* contamination.

The extent to which *Campylobacter* persists in milk is related to the strain of *C. jejuni*, type of milk (i.e., raw, pasteurized, sterilized) and the storage temperature. When raw milk was inoculated to contain 10^7 *C. jejuni* cfu/mL and stored at 4°C, the organism survived 6 to 21 days (Christopher et al., 1982; Doyle and Roman, 1982a; Wyatt and Timm, 1982). However, when this work was repeated using more realistic levels of 1 to 10 *C. jejuni* cfu/mL, the organism survived less than 4 days (DeBoer et al., 1984), which emphasizes the difficulty in recovering *Campylobacter* from raw milk. In similar studies using pasteurized milk, *C. jejuni* persisted somewhat longer because there was less microbial competition and slower acid development (Blaser et al., 1980; Christopher et al., 1982; Doyle and Roman, 1982b). However, *Campylobacter* viability decreased rapidly at higher storage temperatures with this pathogen no longer being detected in pasteurized or sterilized milk after 3 days of storage at 20°C to 25°C. Similar survival has been reported for *C. jejuni* in raw and pasteurized goat's milk (Simms and MacRae, 1989).

Campylobacter is far more sensitive to heat, acid (pH $\leq$ 5.0), oxygen, ambient temperatures, dehydration, chlorine- and iodine-based sanitizers, and the raw milk environment than most other milk-borne pathogens (Koidis and Doyle, 1984; Wyatt and Timm, 1982). Consequently, *C. jejuni* is rapidly inactivated during the cooking step in cottage cheese (Ehlers et al., 1982) and Swiss cheese (Bachmann and Spahr, 1995) manufacture. Furthermore, when cheddar cheese was prepared from pasteurized milk inoculated to contain 10^2 to 10^6 *C. jejuni* cfu/mL, the organism was no longer recoverable from the cheese (pH 5.0) beyond 15 days of curing (Ehlers et al., 1982). Limited survival of *Campylobacter* in cheddar cheese is also supported by an earlier survey in which 127 samples of 60-day-old cheddar cheese were negative for *C. jejuni* (Brodsky, 1984b). The only evidence of cheese contamination comes from Wegmuller et al. (1993) who detected DNA from *C. jejuni* in three raw milk cheeses using a polymerase chain reaction method. However, inability to culture *C. jejuni* from these cheeses suggests that the organisms were no longer viable. Given these findings along with the absence of any reported cheese-associated cases of campylobaceriosis and the lack of any supportive epidemiological evidence (Harris et al., 1986), cheese appears to be a highly improbable vehicle for *Campylobacter* enteritis.

6. Prevention

Proper vat (61.7°C/30 min.) and high-temperature, short-time pasteurization (71.7°C/15 sec) offer complete protection against the spread of milk-borne campylobacteriosis, even if impossibly high populations of *C. jejuni* were present in raw milk to be pasteurized (D'Aoust et al., 1988; Gill et al., 1981; Waterman, 1982). Even though *Campylobacter* spp. are unable to grow in refrigerated raw milk, the organism can persist for several days or more at levels sufficient to induce illness. Because most milk-borne campylobacteriosis outbreaks have been linked to consumption of raw milk, milk-borne *Campylobacter* enteritis can be easily avoided by consuming only pasteurized milk. Because many of the reported campylobacteriosis cases are among children, individuals involved in youth activities and school field trips must be alert to the danger of raw milk if free samples are offered.

F. Drug Residues

Emphasis on increased milk production over the past 50 years has fostered the use of many antibiotics including the β-lactams, tetracyclines, and sulfonamides for treating mastitis and other diseases in dairy cattle. As of May 1992, at least 60 different animal drugs were approved for use (Anonymous, 1992). However, at the same time, 52 non-FDA approved, residue-producing drugs were also suspected of being used illegally. Regardless of the route of administration (i.e., oral, injection, infusion), these antibiotics enter the bloodstream to produce their desired effect at the point of infection and are then metabolized and excreted by the animal at various rates. The FDA has established legally binding limits for at least 16 animal-approved drugs and has also set drug withdrawal periods ranging from a few hours to several weeks, during which time, milk from treated cows must be discarded. Shortened milk withdrawal or discard periods can lead to potentially unsafe drug residue levels in milk. Because milk from various farms is typically commingled, unsafe or illegal animal drug residues can contaminate large volumes of milk, with the FDA estimating that milk from a single sulfamethazine-treated cow can contaminate the milk from 70,000 cows when pooled (Anonymous, 1992). Two widely publicized 1989 surveys published in *The Wall Street Journal* highlighted the scope of this problem with 20% and 38%

of the retail milk samples tested containing animal drug residues and other nonapproved drugs (Place, 1990).

1. General Characteristics

Testing milk for the presence of antibiotic residues in the United States began in 1953 after a revision of the Pasteurized Milk Ordinance to prohibit the sale of milk containing antibiotics (Anonymous, 1990). Since those early days, the β-lactam antibiotics have been the traditional target of state and federally regulated fluid milk testing programs. However, results from a widely publicized 1988 survey raised concerns regarding numerous other drugs and drug residues in the milk supply, with the sulfonamides and tetracyclines also attracting considerable attention.

The β-lactam antibiotics include the penicillins and cephalosporins, both of which consist of a thiazolidine and β-lactam ring with the latter containing various side chains. Penicillin G has traditionally been the most common drug residue found in milk owing to the popularity of the use of this drug on the farm. The level and duration of β-lactam residues in milk are affected by both the route of administration and the number of antibiotics administered (Oliver et al., 1990). When injected, less than 0.3% of the drug appears in milk. However, treatment of mastitis by intramammary infusion leads to almost total excretion in the milk. Most reports suggest that penicillin G and its derivatives are relatively resistant to heat with vat and high-temperature, short-time pasteurization reducing antimicrobial activity in milk less than 10% (Moats, 1988). Penicillin also has the distinction of being the most allergenic drug known, with approximately 10% of the human population reportedly being sensitive (Olson and Sanders, 1975). Because several early reports traced allergic dermatitis to tainted milk (Erskine, 1958), a maximum legal limit of 0.01 ppm has been established for penicillin in fluid milk (Anonymous, 1990).

The sulfonamides, another important group of antimicrobials, have been used to treat systemic and cutaneous infections in farm animals for more than 50 years. All sulfonamides are derivatives of sulfanilamide and ultimately inhibit nucleic acid synthesis. Although available without prescription, the sulfonamides, except sulfadimethoxine, sulfabromomethiazine, and sulfaethoxypyridazine, cannot be used to treat disease in lactating animals (Charm et al., 1988). The latter antimicrobial has a zero tolerance in milk and the former two have a 10 ppb tolerance. Like the penicillins, the sulfonamides are also resistant to most food processing conditions, with activity being retained during prolonged heating (Moats, 1988). Sulfonamides are somewhat less allergenic than penicillin, with approximately 3.4% of the population reportedly being sensitive (Bigby et al., 1986). However, one particular sulfonamide banned for use in lactating dairy cattle, namely sulfamethazine, is a suspected human carcinogen based on animal studies (Anonymous, 1990). Considerable public concern was raised in 1988 when trace levels of sulfamethazine were detected in the United States milk supply (Anonymous, 1990). The estimated maximum allowable level of 1 to 5 ppb sulfamethazine in milk will likely preclude any practical use of this drug in dairy cattle.

In the United States, a highly diverse group of at least 60 FDA-approved and 52 non–FDA-approved drugs were being administered, often illegally, to dairy herds, with 64 of these drugs reportedly leaving residues of concern in milk (Anonymous, 1990). Other antibiotics commonly encountered in the United States milk supply include tetracycline, aminoglycosides, cephalosporins, and chloramphenicol (Brady and Katz, 1988; Kaneene and Miller, 1992). Penicillins remain the drug of choice in treating bovine mastitis followed by cephalosporin, aminoglycosides, novobiocin, and erythromycin, with these five antibiotics accounting for greater than 90% of all drug residues detected in milk. Less frequently encountered antibiotic residues in milk include chlorotetracycline, tetracycline, oxytetracycline, gentamicin, dihydrostreptomycin, and chloramphenicol (Anonymous, 1990). Consumer-safe levels for most of these antibiotics have not yet been established, with the United States generally advocating a policy of "zero tolerance."

2. Detection Methods

Current strategies for detecting antibiotic residues in milk have evolved over the last 50 years; the earliest methods were based on the inability of test bacteria to produce acid, reduce dyes, or grow on solid media in the presence of antibiotics (Bishop and White, 1984). These time-consuming assays, which required overnight incubation, were eventually replaced by the qualitative and quantitative *Bacillus stearothermophilus* disc assays for penicillin and other inhibitors. Both of these AOAC-approved tests are based on measurable inhibition zones that develop around filter paper discs impregnated with the test sample within 3 hours of incubation at 64°C (Bishop et al., 1992; Richardson, 1990). A variation of this assay known as the Delvotest-P (Bishop et al., 1992; Bishop and White, 1984) is even more sensitive for penicillin and uses the pH indicator bromcresol purple to assess acid production by *B. stearothermophilus*. The widely acclaimed and AOAC-approved Charm test, first introduced in 1978, is based on the competitive binding of radioactively labeled penicillin (and later tetracycline, erythromycin, streptomycin, novobiocin, sulfamethazine, and chloramphenicol) to vegetative cells of *B. stearothermophilus*. At least seven different versions of this assay are known, three of which have been simplified for on-farm testing (Bishop et al., 1992). In addition, at least six different enzyme-linked immunosorbent assays covering most other antibiotics of interest are available (Bishop et al., 1992). However, these newly developed rapid methods and several others based on agglutination of antibiotic-coated latex beads, high-pressure liquid chromatography, and reduction of brilliant black dye have not yet received AOAC approval. Consequently, the aforementioned qualitative and quantitative *B. stearothermophilus* disc assay and Charm tests remain the methods of choice for most commonly encountered antibiotics.

3. Risks of Drug Residues

The public health significance of barely detectable levels of animal drug residues in the milk supply is still somewhat controversial. Several international studies have concluded that small amounts of drug residues in milk are not likely to pose a significant human health hazard, with bacterial pathogens clearly constituting a far more serious threat (Anonymous, 1990). However, as previously discussed, ingesting antimicrobials such as penicillin, streptomycin, tetracycline, aminoglycosides, and sulfonamides in food can produce life-threatening allergic reactions including anaphylactic shock in susceptible individuals (Anonymous, 1990). Given the current long life expectancy of humans, increased suppression of the human immune system through long-term exposure to low levels of antibiotics in the milk supply is also a growing concern. At least two drugs used in treating dairy cattle, namely sulfamethazine and nitrofurazone, also can produce cancer in laboratory animals and, as such, are potential human carcinogens (Anonymous, 1990). Other commonly used drugs, including chloramphenicol and ivermectin (an anti-worming agent), have been associated with aplastic anemia (an irreversible and potentially fatal bone marrow disease) and various neurological disorders.

A separate public health issue relates to development of new antibiotic-resistant bacterial pathogens as a result of long-term exposure to low levels of antibiotics in milk. In 1985, the largest known foodborne outbreak involving more than 16,000 cases of salmonellosis in the Chicago area was traced to pasteurized milk that contained a very rare multi-antibiotic-resistant strain of *Salmonella typhimurium* (Ryan et al., 1987). The fact that this organism also contained several plasmids encoding resistance to 14 different antibiotics (Schuman et al., 1989), eight of which are commonly encountered as drug residues in milk (Anonymous, 1992a), highlights the potential danger of antibiotic misuse on the farm.

Contamination of milk with even minute levels of antibiotics also has created several potential safety-related problems for manufacturers of fermented dairy products, including inadequate milk clotting and improper cheese ripening, inadequate acid and flavor development in buttermilk, invalid results from certain quality control tests, and, most importantly, diminished starter culture growth and acid production during cheese making, which can allow pathogens such as *Salmonella* and *Staphylococcus aureus* to grow (Park and Marth, 1972b). Starter culture failure remains a major cause of disease outbreaks involving cheese and other fermented dairy products.

4. Occurrence

Antibiotic residues were relatively common in the United States milk supply as recently as the late 1980s. In a nationwide survey, Collins-Thompson et al. (1988) detected sulfamethazine and tetracycline in 47% and 28%, respectively, of samples tested from 16 states, with penicillin, erythromycin, chloramphenicol, and novobiocin found in less than or equal to 5% of samples. Furthermore, each state yielded samples containing one or more antibiotic residues, with similar findings obtained from 40 retail milk samples collected in four Canadian provinces.

When 64 retail milk samples from eastern Pennsylvania, central New Jersey, and the New York City area were screened, Brady and Katz (1988) found antibiotics in 63% of the samples with 43% and 17% of positive samples containing residues of two and four or more antibiotics, respectively. Sulfonamides and tetracyclines were the most prevalent residues with each present in nearly 40% of the samples tested. Thirty-eight percent of samples contained both sulfonamide and tetracycline with 16% containing both sulfonamides and streptomycin. Chloramphenicol, erythromycin, and the β-lactams were identified in 10%, 5%, and 2% respectively, of the samples.

According to Charm et al. (1988), 71% of retail and tanker truck milk samples tested in the northeast United States were contaminated with sulfonamides at levels of at least 5 ppb. Half of the positive samples contained greater than 25 ppb sulfonamide, with one sample having 15,000 to 20,000 ppb. Sulfamethazine was the dominant sulfonamide detected and was sometimes present at levels as high as 40 ppb, eight times higher than the maximum allowable level suggested by the FDA. In another survey involving retail milk from 10 major United States cities, sulfonamides were detected in 36 of 49 samples, with most positive findings coming from the northwest and northeast (Charm et al., 1988). However, in Prince Edward Island, Canada, where sulfonamides are not sold over the counter, 1000 tanker truck samples tested negative for these drugs.

Much of the controversy concerning the public health significance of antibiotic residues in milk is based on wide disparities between results from regulatory and nonregulatory surveys and a lack of firmly established tolerance levels for many antibiotics. In 1990, the FDA compiled test results from more than 1.4 million bulk tank samples representing 43% of the United States milk supply and reported that only 0.27% and 0.09% of these samples contained unsafe levels of β-lactam antibiotics and sulfamethazine, respectively (Anonymous, 1990). These low contamination rates decreased even further during 1994 and 1995 with illegal levels of β-lactam antibiotics, sulfonamides, sulfamethazine, and tetracycline, present in only 0.15%, 0.13%, 0.007%, and 0.12%, respectively, of the milk supply (Anonymous, 1996). When compared with bacterial pathogens, these low background levels of antibiotic residues clearly pose a negligible public health risk to consumers.

5. Prevention

During the 1980s, faulty dairy herd management practices, including insufficient knowledge concerning milk withdrawal periods, inadequate record keeping on mastitic cows, and inappropriate use of antibiotics, were cited as being primarily responsible for the high incidence of antibiotic residues in the milk supply (Kaneene and Ahl, 1987). In response to these findings, the FDA and dairy industry adopted a joint three-point program to (a) reevaluate antibiotic detection methods for adequacy and efficiency, (b) implement a public awareness program to accurately inform consumers about the safety of the milk supply, and (c) develop a 10-point Hazard Analysis Critical Control Point (HACCP)-based animal drug education program for dairy farmers (Adams, 1994). The latter program focuses on proper use of FDA- approved drugs under a veterinarian's supervision, animal treatment records, employee education, and ongoing drug residue screening programs (Adams, 1994). These efforts appear to have been highly successful given the recent sharp decrease in antibiotic-positive milk samples.

G. Enteropathogenic *Escherichia coli*

Bacterium coli commune, known as *E. coli*, was first described by Theodor Escherich in 1885. Most *E. coli* strains are harmless commensals common to the intestinal tract of humans and animals. Some milk-borne strains of *E. coli* were thought to be responsible for summer diarrhea in children as early as 1900 (James, 1973). However, bacteriological confirmation of such strains did not come until the 1940s when Bray (1945), Bray and Beavan (1948), and later Brown and Bailey (1958) identified several serologically distinct "enteropathogenic" *E. coli* strains responsible for infant diarrhea. Based on distinct virulence properties, different interactions with the intestinal mucosa, distinct clinical symptoms, differences in epidemiology, and variations in O (somatic) and H (flagellar) antigens, more than 60 distinct strains causing different forms of diarrhea in humans have been identified (Hitchens et al., 1995). These strains are grouped into the following five categories: classical enteropathogenic *E. coli* (EPEC), enterotoxigenic *E. coli* (ETEC), enteroinvasive *E. coli* (EIEC), entero-hemorrhagic *E. coli* (EHEC), and, most recently, enteroadherent *E. coli* (EAEC). Both EIEC and ETEC have been linked to major cheese-related outbreaks in the United States and Europe (MacDonald et al., 1985; Marier et al., 1973), and both are discussed in this section as is EPEC, which is a major problem in less developed countries. EAEC is a newly developed and consequently poorly understood category. EHEC, which has recently emerged as a particularly hazardous foodborne pathogen of major public health concern, is discussed separately.

1. General Characteristics

A species in the family *Enterobacteriaceae*, *E. coli* is a short, gram-negative, facultatively anaerobic, rod-shaped bacterium that may be nonmotile or motile by peritrichous flagella. Most isolates grow optimally at or near 37°C, with growth ceasing at a_w values less than 0.95. Identification of *E. coli* is based on fermentation of glucose and other carbohydrates to acid (lactic, acetic, formic) and gas (CO_2, O_2). Whereas most *E. coli* isolates ferment lactose to acid and gas within 48 hours, some strains (particularly those of EIEC) are weakly lactose positive or lactose negative (Hitchins et al., 1995). Important biochemical tests for routine confirmation of *E. coli* include production of indole (usually indole [I] positive), production of stable acid end products from glucose (methyl red [M] positive), production of acetoin from glucose (Voges-Proskauer [Vi] negative), and use of citrate (citrate [C] positive). About 95% of all *E. coli* are IMViC ++– – with the remaining 5% of strains being IMViC –+– – (Doyle and Padhye, 1989). Further characterization and identification of potentially pathogenic strains is partially based on serology with 173 O (somatic), 56 H (flagellar), and 80 K (capsular) antigens yielding an estimated 50,000 to 100,000 serotypes of *E. coli* (Orskov and Orskov, 1992).

ETEC produce a heat-labile enterotoxin (LT), which is immunologically related to cholera toxin and sometimes a second heat-stable enterotoxin (ST) of low molecular weight (Doyle and Padhye,1989; Gyles, 1992). The ETEC strains also produce membrane-bound colonization factors that mediate attachment of the organism to the intestinal wall. Sixteen ETEC serotypes comprising 14 different O serogroups are presently known (Hitchins et al., 1995).

EIEC are *Shigella*-like organisms capable of invading and proliferating in the intestinal epithelium, with such invasive ability reportedly being plasmid mediated (Doyle and Padhye, 1989). Eleven serotypes comprising eight different O serogroups are presently recognized (Hitchins et al., 1995).

EPEC are defined as diarrheagenic strains belonging to serogroups epidemiologically incriminated as pathogens but whose pathogenicity has not been positively linked to production of heat-labile enterotoxins, heat-stable enterotoxins, *Shigella*-like invasiveness (Edelman and Levine, 1983), or verocytotoxin production (Doyle and Padhye, 1989), the last of which is characteristic of EHEC. Twenty-nine serotypes of EPEC comprising 15 different O serogroups are recognized with some EPEC serotypes being mainly associated with infant diarrhea.

2. Isolation and Detection Methods

Procedures for detecting diarrhea-causing strains of *E. coli* in dairy products generally begin with a 3-hour/35°C preenrichment in brain heart infusion broth to resuscitate injured cells (Hitchins et al., 1995). This step is followed by 20 additional hours of incubation at 44°C, which selects for fecal coliforms, including *E. coli*. Thereafter, plates of several standard plating media (i.e., eosin-methylene blue agar and MacConkey agar) are streaked, incubated, and examined for typical and atypical (non-lactose fermenting) *E. coli*. Following standard biochemical confirmation, commercially available antisera that react with many pathogenic serogroups of *E. coli* can be used to screen for the most common, potentially pathogenic strains. However, because pathogenicity cannot be completely correlated with specific O antigens, actual proof of the pathogenicity of the strain is required.

Production of LT and ST toxins by ETEC can be demonstrated using the Y-1 mouse adrenal cell test and the infant mouse test (Doyle and Padhye, 1989; Hutchins et al., 1995), respectively, or by using one of several immunological or DNA probe–based assays (Doyle and Padhye, 1989; Hill et al., 1995). Invasiveness of EIEC isolates is typically shown using HeLa cell cultures or the guinea pig–based Sereny test. Virulent *E. coli* strains not conforming to ETEC, EIEC, or EHEC are likely EPEC. However, because no standard pathogenicity tests for such strains are available, confirmation of most suspect EPEC isolates requires complete serotyping by a qualified *E. coli* reference laboratory.

3. Clinical Manifestations

EPEC is principally responsible for infantile diarrhea, a clinically severe illness in children younger than 2 years of age which is characterized by fever, vomiting, abdominal pain, and a persistent diarrhea that may last for several weeks (Escheverria et al., 1987; Levine, 1987). In adults, EPEC foodborne infections are typically far less severe. Symptoms begin 17 to 72 hours after exposure and include a severe watery diarrhea with mucus, which is frequently accompanied by nausea, vomiting, abdominal cramps, headache, fever, and chills (Doyle and Padhye, 1989). Unlike infantile diarrhea, this illness is of far shorter duration in adults, with spontaneous recovery occurring within 6 to 72 hours.

Widely known as traveler's diarrhea, ETEC gastroenteritis may vary from a mild, 1-day illness consisting of loose stools, abdominal cramps, vomiting, and low-grade fever to a severe choleralike illness lasting several weeks in which profuse rice water-like stools can lead to serious dehydration (Kantor, 1986). Human volunteer studies (DuPont et al., 1971; Levine et al., 1977) demonstrated an unusually high infectious dose with ingestion of 10^8 to 10^{10} ETEC cells required to produce symptoms within 8 to 44 hours. Most individuals stop shedding ETEC 4 to 5 days after cessation of diarrhea. Traveler's diarrhea is typically mild and self-limiting. However, in severe cases of ETEC gastroenteritis resembling cholera, fluids are normally given either orally or intravenously to prevent dehydration (Kantor, 1986). Antibiotic therapy is inappropriate and can even be harmful.

EIEC penetrates and destroys the mucosal tissue of the colon to produce an illness indistinguishable from shigellosis (bacillary dysentary) (Gray, 1995). Symptoms typically develop 8 to 24 hours after receiving a minimum oral infectious dose of at least 10^6 EIEC cells (DuPont et al., 1971) and include severe diarrhea accompanied by chills, fever, headache, muscle pain, and abdominal cramps. Unlike the profuse watery stools observed in ETEC traveler's diarrhea, stools produced by EIEC are less frequent but typically contain blood, mucus, and leukocytes. However, as in traveler's diarrhea, EIEC infections are normally acquired abroad by adults who recover spontaneously without medical intervention.

4. Outbreaks

Global importance of *E. coli* as a cause of diarrheal illness has decreased markedly over the past 50 years following implementation of improved sanitary practices. Although still a major cause of water-borne diarrhea in less-developed countries, dairy-related cases of *E. coli* enteritis are uncommon in industrialized countries, with only two outbreaks and two additional cases thus far reported.

Evidence for possible involvement of EPEC in milk-borne enteritis is limited to one report from England (Anonymous, 1976) in which two 6-month-old infants developed EPEC infantile diarrhea after ingesting raw milk from their father's farm. EPEC O26 was detected in both stool samples and in milk from one mastitic cow supplying the family. After both infants began drinking bottled pasteurized milk, the illness reportedly disappeared.

In the first of two multistate cheese-related outbreaks (Francis and Davis, 1984; Levy, 1983; MacDonald et al., 1985), symptoms of ETEC gastroenteritis developed in 45 individuals in Washington, DC, 1 to 6 days after ingesting imported French Brie cheese. Investigators eventually isolated ETEC O27:H20 (an ST-producing strain) from stool samples and the incriminated cheese. After much publicity, 124 additional cheese-related cases were soon confirmed in Colorado, Georgia, Illinois, and Wisconsin, after which the implicated cheese was recalled nationwide. Identical outbreaks involving the same brand of cheese were simultaneously reported in Denmark, Sweden, and The Netherlands. Although the source of contamination at the cheese factory was never found, illness caused by this epidemic strain was linked to two different cheese lots manufactured 46 days apart, thus suggesting a recurrent contamination problem.

During 1971, imported French Brie cheese was identified as the vehicle of infection in a second multistate outbreak of EIEC gastroenteritis, which also began in Washington, DC (Barnard et al., 1971; Marier et al., 1973; Schnurrenberger and Pate, 1971; Tulloch et al., 1973). A total of 347 cases of diarrhea in Washington, DC, were eventually linked to ingesting imported French Brie, Camembert, and Coulomiers cheese containing high coliform populations and EIEC O124:B17 at a level of 10^5 to 10^7 organisms/g, with growth of the organism during cheese ripening being suspected (Fantasia et al., 1975). Twelve people required hospitalization and were later released. As in the previous outbreak, the epidemic strain was recovered from both stool and cheese samples. The importer subsequently recalled all lots (1200 pounds) of cheese that were distributed. EIEC O124:B17 was later recovered from samples of partially consumed cheese that were originally manufactured over a 13-day period, thus suggesting an ongoing contamination problem that was likely related to inadequate filtration of river water used in factory cleaning operations.

5. Occurrence and Survival in Dairy Products

EPEC, ETEC, and EIEC are classified as fecal coliforms with their presumed primary reservoir being the intestinal tract of humans and animals. However, these organisms are occasionally found in raw milk from normal and mastitic cows. In early eastern European surveys, pathogenic *E. coli* serotypes were identified in less than 2% of the raw milk supply (Bryan, 1983). More recently, 3 of 47 raw milk samples tested in Iowa harbored EPEC serotypes (Glatz and Brudvig, 1980). Pathogenic serotypes of *E. coli* have been seldom identified in pasteurized milk and cream (Jones et al., 1967); however, ETEC can grow in inoculated sterile milk at ambient temperatures and produce small amounts of LT (Olsvik and Kapperud, 1982).

Despite two large cheese-related outbreaks of gastroenteritis, pathogenic serotypes of *E. coli* are rarely found in cheese marketed in the United States with only one documented report involving Mexican-style fresh white cheese contaminated with ETEC, which was detected during a routine surveillance program and recalled without incident (Anonymous, 1991a). According to Frank and Marth (1978), 106 samples of Camembert, Brie, brick, Muenster, and Colby cheese purchased in Wisconsin tested negative for common EPEC serotypes. Whereas 78 cheese samples tested in Iowa were also negative for ETEC (Glatz and Brudvig, 1980), contamination rates are considerably higher in less developed countries such as Iraq, Abbar, and Kaddar (1991), as well as India (Singh and Ranganathan, 1974), where acute *E. coli* gastroenteritis is common in children.

The fate of any organism in cheese is dictated by many interacting factors including the type of cheese, initial populations in the milk, strain differences, amount and type of starter culture, cheese making procedures (e.g., cooking, washing), pH, salt content, and location of the organism in the cheese as well as the temperature and length of ripening and storage. Frank and Marth (1977a, 1977b) examined the fate of ETEC and EIEC in skim milk that was fermented with 0.25% to 2.0% lactic starter culture at 21°C and 32°C for 15 hours and then refrigerated at 7°C. Growth of ETEC and EIEC generally ceased at pH 4.8 to 5.2 with the combination of low incubation temperature and highest starter inoculum most detrimental to growth and survival. ETEC and EIEC survival was also influenced by the type of starter culture with *Lactococcus lactis* subsp. *lactis* being least inhibitory followed by *Lactococcus lactis* subsp. *cremoris* and a mixture of both organisms.

The 1971 outbreak involving Brie cheese prompted several studies that examined the fate of EIEC and ETEC in various cheeses prepared from pasteurized milk inoculated to contain 100 to 1000 organisms/mL. When Camembert cheese was manufactured (Frank et al., 1977), EIEC and ETEC populations increased approximately 100-fold during the first 6 hours of cheese making until the curd attained a pH less than or equal to 5.0. Both types of *E. coli* were slowly inactivated in the cheese during ripening at 12°C and later 7°C with EIEC and ETEC surviving 1 week and 1 to 6 weeks, respectively. However, rapid growth of EIEC and ETEC to levels of 10^5 organisms/g was observed when cheeses were surface-inoculated 5 days after manufacture and similarly ripened, with both organisms persisting well beyond the normal shelf-life of the cheese. In Colby-like cheese (Kornacki and Marth, 1982), *E. coli* populations increased 100- to 1000-fold during the first 4 hours of cheese making, with EIEC and ETEC persisting 4 and greater than 12 weeks, respectively, in finished cheese ripened at 4°C to 10°C. These *E. coli* strains behaved similarly during manufacture of brick cheese (Frank et al., 1978) with ETEC again proving hardier than EIEC in 7-week-old brick cheese.

To simulate postmanufacturing contamination, Sims et al. (1989) inoculated commercially prepared cottage cheese (pH 4.7–4.9) to contain 10^4 ETEC/EIEC cfu/g and incubated the product at 7°C to 25°C. Regardless of storage temperature, *E. coli* levels decreased only slightly during the 14-day shelf-life of the product.

6. Prevention

Human carriers are presumed to be the primary reservoir and source of ETEC, EIEC, and EPEC. Because *E. coli* is readily destroyed by pasteurization with a wide margin of safety, the organism typically enters the product as a postpasteurization contaminant. Dairy products are most often contaminated by infected food handlers who practice poor personal hygiene or by contact with water containing human sewage. Consequently, food workers must be educated in safe food handling techniques and proper personal hygiene practices including hand washing after using the lavatory.

H. Enterohemorrhagic *Escherichia coli* O157:H7

In 1982, outbreaks of hemorrhagic colitis in Oregon and Michigan drew attention to an unusual clinical syndrome of gastroenteritis caused by a little known enteric bacterial pathogen, namely enterohemorrhagic *E. coli* (EHEC) O157:H7—a serotype identified only once 7 years earlier at the Centers for Disease Control from a single case of human diarrheal illness (Riley et al., 1983; Wells et al., 1983). A total of 47 cases of hemorrhagic colitis were identified in these two outbreaks with undercooked hamburgers subsequently identified as the vehicle of infection. At least 15 additional outbreaks of *E. coli* O157:H7 infection have been linked to consumption of undercooked ground beef (Hancock et al., 1994), apple cider (Besser et al., 1993; Steele et al., 1982), and mayonnaise (Borczyk et al., 1987; Neill, 1989) and this pathogen once again gained considerable notoriety in 1993 after more than 500 hamburger-related cases of illness and the deaths of four children were reported (Conner and Kotrola, 1995; Knight, 1994; Wuethrich, 1994) in Washington, Idaho, California, and Nevada (Anonymous, 1993b; Anonymous, 1994a; Hancock et al., 1994). The seriousness of these aforementioned outbreaks, combined with 1400 reported cases in the United States during 1994 (Anonymous, 1995b), 1600 cases in Canada during 1992 (Lior, 1994b), and additional sporadic cases reported worldwide (Griffin and Tauxe, 1991) have raised *E. coli* O157:H7 to a foodborne pathogen of international importance (Knight, 1993). Of concern to the dairy industry are the likely presence of *E. coli* O157:H7 in 2% to 5% of the raw milk supply (D'Aoust, 1989; Wells et al., 1991) and reports of at least 60 cases of raw milk–associated illness.

1. General Characteristics

Verotoxigenic *E. coli*, or EHEC, produces one or two verotoxins, designated VT-1 and VT-2, which are toxic to Vero (African green monkey kidney) and HeLa cells, as first reported by Konowalchuk et al. (1977). VT-1 is a relatively heat stable, high molecular weight, Shiga-like toxin, whereas VT-2 is immunologically distinct (Doyle, 1991).

Six EHEC serogroups, namely O26, O48, O111, O113, O145, and O157, have been linked to human illness (Hitchins et al., 1995; Goldwater and Bettelheim, 1995) with additional verotoxigenic serogroups detected in both healthy cattle (Montenegro et al., 1990; Wells et al., 1991) and cattle with diarrhea (Mohammad et al., 1986). In all, more than 80 serotypes of EHEC are recognized (Griffin and Tauxe, 1991) with O157:H7 being dominant in the United States and Canada (Lior, 1994b) and O11:H⁻ being particularly common in Australia (Goldwater and Bettelheim, 1995). However, because *E. coli* O157:H7 is the best established foodborne pathogen among the EHEC, this discussion focuses on *E. coli* O157:H7.

E. coli O157:H7 is similar to most other *E. coli* with a few important exceptions (Doyle, 1991; Gray, 1995). Whereas *E. coli* O157:H7 grows optimally at 30°C to 42°C, this serotype grows poorly, if at all, at 44°C to 45.5°C (Buchanan and Klawitter, 1992; Doyle and Schoeni, 1984) and is therefore unlikely to be recovered when samples are analyzed for fecal coliforms. Growth is also poor at temperatures less than 10°C; however, addition of 4% sodium lactate to tryptic soy broth reportedly permits growth at 4°C to levels of 10^8 organisms/mL after a lag period of 4 weeks (Conner and Hall, 1996). Biochemically, *E. coli* O157:H7 generally lacks the enzyme glucuronidase, which is possessed by 92% to 96% of all other *E. coli* strains, and, unlike 80% to 93% of other *E. coli* strains, is unable to ferment D-sorbitol within 24 hours. Both of these biochemical differences are of major importance in screening samples for *E. coli* O157:H7.

2. Isolation and Detection Methods

Selective recovery of *E. coli* O157:H7 from food samples is based on the inability of typical isolates to ferment sorbitol and hydrolyze 4-methyl-umbelliferone glucuronide (MUG) to a fluorogenic product. However, there are a few strains that are positive for sorbitol and MUG (Gunzer et al., 1992). Direct plating media commonly used at 37°C include sorbitol MacConkey agar (SMAC) with and without MUG (Hitchins et al., 1995; McCleery and Rowe, 1995) and hemorrhagic colitis agar, which contains both sorbitol and MUG (Hitchins et al., 1995). Several selective enrichment broths containing novobiocin or other antibiotics also can be used to enhance recovery (Padhye and Doyle, 1992). In one recently reported FDA procedure (Hitchins et al., 1995; Weagant et al., 1995), samples are enriched at 37°C for 7 hours in tryptic soy broth containing vancomycin, cefsulodin, and cefixime and then plated on SMAC supplemented with tellurite and cefixime. Presumptive *E. coli* O157:H7 isolates must be serologically confirmed using either commercially available antisera or a serotype-specific DNA probe in a colony hybridization assay (Hill et al., 1995). Verotoxin production by non-*E. coli* O157:H7 strains can be confirmed using traditional cell culture techniques or the newly developed DNA probe and polymerase chain reaction assays for VT-1. However, identification of other verotoxin-producing strains is infrequent because most such isolates are positive for both sorbitol and MUG (Wells et al., 1991).

3. Clinical Manifestations

Compared to most other foodborne illnesses, infections involving *E. coli* O157:H7 or other EHEC strains are particularly serious, with manifestations ranging from a mild, nonbloody diarrhea to hemorrhagic colitis, hemolytic uremic syndrome, and thrombotic thrombocytopenic purpura, all of which are related to adherence of the pathogen to the intestinal tract lining followed by production of one or more verotoxins (Gray, 1995; Griffin et al., 1988; Griffin and Tauxe, 1991; Padhye and Doyle, 1992; Riley et al., 1983). Furthermore, the oral infectious dose may be relatively low, with fewer than 1000 organisms inducing illness.

Hemorrhagic colitis is characterized by sudden onset of severe appendicitis-like abdominal pain followed by watery and eventually grossly bloody diarrhea described as "all blood and no stool." Vomiting may occur but, unlike in EIEC infections, fever is typically mild or absent. The incubation period ranges from 3 to 5 days, with symptoms generally persisting 2 to 9 days. However, fecal shedding of the organism has been reported for up to 4 weeks. This type of infection is typically self-limiting in adults with antibiotic therapy being of limited value in shortening the duration of bloody diarrhea.

The second manifestation, hemolytic uremic syndrome, develops in 2% to 7% of patients with hemorrhagic colitis and is the leading cause of acute renal failure in children (Karmali et al., 1983). This condition is characterized by hemolytic anemia (intravascular coagulation of erythrocytes), thrombocytopenia (low levels of circulating blood platelets), and kidney failure, which occurs in otherwise healthy individuals. Patients frequently require kidney dialysis and blood transfusions and a number of complications may develop including heart failure, seizures, and a prolonged coma, which can be terminal in 3% to 10% of cases (Gray, 1995).

The third manifestation, thrombotic thrombocytopenic purpura, which usually occurs in adults, is similar to hemolytic uremic syndrome except for development of fever. Central nervous system disorders typically dominate, with the development of terminal blood clots in the brain also being reported. Hence, unlike many other foodborne illnesses, infections with *E. coli* O157:H7 can be particularly devastating.

4. Outbreaks

Consumption of undercooked ground beef has been the traditional mode for *E. coli* O157:H7 infections; however, illnesses from ingestion of raw milk have been reported, with the number of such cases continuing to increase. In April 1986, a group of 60 kindergarten children visited a dairy farm in Ontario, Canada, and were given raw milk to drink (Borczyk et al., 1987). Subsequently, *E. coli* O157:H7 infections developed in 46 children, three of whom also contracted hemolytic uremic syndrome. *E. coli* O157:H7 was later isolated from 1 of 67 fecal samples collected from healthy calves and cows on the same farm.

Several months later, consumption of raw milk on two Wisconsin farms was linked to separate cases of hemolytic uremic syndrome involving a 13-month-old boy and a 5-month-old girl (Martin et al., 1986). Follow-up screening of fecal samples from dairy cattle on both farms yielded *E. coli* O157:H7 in both herds.

Two separate raw milk–related outbreaks also occurred in Oregon during 1992 and 1993. In the first of these outbreaks (Bleem, 1994), *E. coli* O157:H7 infections developed in nine individuals aged 9 months to 73 years after consuming raw milk. Testing the entire herd of 132 animals revealed four cattle as positive for *E. coli* O157:H7, including two 15-month-old heifers, one dry cow, and one milking cow. Furthermore, strain-specific typing demonstrated that six of the nine human isolates were identical to the four bovine strains. In the second raw milk–related outbreak (Bleem, 1994), five cases of *E. coli* O157:H7 infection were identified, including two cases of hemolytic uremic syndrome. Subsequent fecal sampling of the entire herd of 60 animals revealed *E. coli* in four postweaned heifers. Subtyping again demonstrated that the cattle and human isolates were identical. In addition to these milk-related cases, evidence also exists for direct transmission of *E. coli* O157:H7 between fecal-positive dairy calves and a 13-month-old Canadian boy in whom hemolytic uremic syndrome developed after he was playing in straw bedding near the calves (Renwick et al., 1993).

The risk of acquiring *E. coli* O157:H7 infections through ingestion of raw milk is well documented. Additional cases of illness have been linked to farm-manufactured yogurt and pasteurized milk involving a verotoxigenic strain of *E. coli* that is closely related to *E. coli* O157:H7. Sixteen cases of *E. coli* O157:H7 infection that occurred in northwest England during 1991 were epidemiologically linked to consumption of farm-produced yogurt (Anonymous, 1991b; Morgan et al., 1993). Eleven of these cases involved children 10 years old and younger; in five of these children hemolytic uremic syndrome developed. Thirteen individuals required hospitalization and all patients eventually recovered. Although the epidemic strain was never isolated from the implicated yogurt or ingredients obtained from the dairy farm, subsequent inspections yielded strong evidence for postpasteurization contamination. During February and March of 1994, acute bloody diarrhea and abdominal cramps developed in 18 individuals from Helena, MT, from infection with a supposedly rare, but closely related (to other EHEC strains) verotoxigenic strain of *E. coli*, namely *E. coli* O104:H21 (Moore et al., 1995). Epidemiological evidence strongly supported one particular brand of pasteurized milk as the source of infection. Furthermore, company records indicated that coliform counts for at least one of the finished milk products sold during the outbreak exceeded the state allowable maximum level of 10 coliforms per 100 mL of milk. *E. coli* O104:H21 was never recovered from the incriminated milk, the factory environment, or its supposed farm source during subsequent investigations. Because the techniques available for identifying non-O157:H7 verotoxigenic strains of *E. coli* are ill defined and not available to most laboratories, these results are not surprising. However, this outbreak does raise serious new public health concerns regarding the possible presence of verotoxigenic strains of *E. coli* in factory environments and their entry into finished products as postprocessing contaminants.

5. Occurrence and Survival in Dairy Products

The environmental niches for *E. coli* O157:H7 have not yet been clearly established; however, dairy cattle appear to be emerging as a major reservoir for this pathogen with 2% to 6% of such animals shedding *E. coli* O157:H7 in feces (Wells et al., 1991) (also see Chap. 1). Although not currently recognized as a cause of mastitis in dairy cattle, *E. coli* O157:H7 can readily contaminate milk on the farm, with contamination rates of 4.2% and 2.0% reported for raw milk produced in the United States and Canada, respectively (D'Aoust, 1989). However, Padhye and Doyle (1991) found somewhat higher contamination rates, with *E. coli* O157:H7 being present in 10% of raw milk bulk tank samples collected from 69 different Wisconsin farms. According to Kasrazadeh and Genigeorgis (1995), *E. coli* O157:H7 is unable to grow in pasteurized milk held at less than 10°C. Nonetheless, substantial growth can occur in temperature-abused milk, with this pathogen exhibiting generation times of 7.2 and 1.5 hours at 12°C and 20°C, respectively.

Current evidence indicates that *E. coli* O157:H7 is not unduly heat resistant and, like most salmonellae, is readily destroyed in milk by minimum pasteurization (71.7°C/15 sec) (D'Aoust et al., 1988). One study (Conner and Kotrola, 1995) showed the ability of *E. coli* O157:H7 populations to remain relatively constant in laboratory media acidified to pH 4.7 with lactic acid during 56 days of storage at 4°C and 10°C with *E. coli* O157:H7 levels increasing nearly 100-fold in the same medium after 7 days of storage at 25°C. Going one step further, Boor (1996) assessed the ability of *E. coli* O157:H7 to compete with commonly used lactic acid bacteria starter cultures. When pasteurized milk samples were inoculated to contain 10^1 to 10^5 *E. coli* cfu/mL and fermented with *Lactococcus lactis* ssp. *lactis* or *Lactobacillus delbrükii* ssp. *bulgaricus*, the pathogen was completely inactivated within 48 hours and was unable to survive a typical yogurt fermentation. However, *E. coli* O157:H7 remained viable for 40 days when similar milks were fermented with *Streptococcus thermophilus* and *Lactococcus lactis* ssp. *cremoris*.

In response to these findings, several studies were also conducted to determine the fate of this pathogen during manufacture and storage of various fermented dairy products. According to Arocha et al. (1992), *E. coli* O157:H7 was completely destroyed during normal cooking of cottage cheese curd at 57°C.

However, when cheddar cheese was prepared from pasteurized milk inoculated to contain 1 *E. coli* O157:H7 cfu/mL, Reitsma and Henning (1996) reported that the pathogen survived cheese making and persisted for 138 days in finished cheese ripened at 6°C to 7°C, well beyond the minimum 60-day curing period at greater than or equal to 2°C required by the FDA for cheddar cheese prepared from raw or heat-treated milk. Looking at *E. coli* O157:H7 as a post manufacturing contaminant, Kasrazadeh and Genigeorgis (1995) found this organism unable to grow in Hispanic cheese (pH 6.6) during 2 months of storage at 8°C. Whereas growth was observed in temperature-abused cheeses with *E. coli* O157:H7 exhibiting generation times of 23 and 2.5 hours at 10°C and 20°C, respectively, such growth could be delayed or prevented by incorporating 0.3% sodium benzoate or potassium sorbate into the cheese. Most recently, Boor (1996) reported that *E. coli* O157:H7 survived sixty days or longer when retail samples of yogurt, buttermilk, sour cream, and cottage cheese were inoculated to contain as few as 10 *E. coli* O157:H7 cfu per gram or milliliter and stored at 4°C. Given these findings, *E. coli* O157:H7 is likely to persist in other cheeses and fermented dairy products for various times, depending on storage conditions. However, in the only survey thus far reported (Bowen and Henning, 1994), 50 retail samples of natural American and non–American-type cheeses purchased in South Dakota failed to yield *E. coli* O157:H7.

6. Prevention

Unlike ETEC, EIEC, and EPEC, which reside in symptomatic and asymptomatic human carriers, *E. coli* O157:H7 is apparently confined to the intestinal tract of cattle and perhaps other animals. Given the probability for contamination of milk during milking, consumption of raw milk should be avoided. If good manufacturing practices are followed, consumption of pasteurized milk poses little risk because *E. coli* O157:H7 is readily inactivated during high-temperature, short-time pasteurization (D'Aoust et al., 1988). However, because this organism is reasonably acid tolerant, raw milk cheeses and soft-ripened cheeses such as Brie and Camembert could conceivably pose public health concerns if prepared or aged improperly.

I. Listeriosis

L. monocytogenes, the causative agent of listeriosis in humans and animals, was first isolated nearly 75 years ago from the blood of infected rabbits exhibiting a typical monocytosis (Murray et al., 1926). However, this bacterium only recently emerged as a serious foodborne pathogen that can cause abortion in pregnant women and meningitis, encephalitis, and septicemia in newborn infants and immunocompromised adults. Unlike most other foodborne illnesses, the outcome of listeric infections can be particularly devastating, with a mortality rate of 20% to 30%. During the 1980s, three major dairy-related outbreaks of listeriosis—two in the United States and one in Switzerland—were linked to consumption of pasteurized milk, Mexican-style cheese, and Vacherin Mont d'Or soft-ripened cheese and resulted in more than 100 deaths (Farber and Peterkin, 1991; Ryser and Marth, 1991). These outbreaks, combined with a presumably low oral infectious dose, prompted the United States to institute a policy of "zero tolerance" for *L. monocytogenes* in all cooked and ready-to-eat foods, including dairy products. Since 1985, more than 100 class I recalls have been issued for *Listeria*-contaminated dairy products, principally ice cream and cheese, with current financial losses in excess of $120 million. *L. monocytogenes* still accounted for 13 of 18 (72%) dairy-related class I recalls issued during 1994 and 1995, thus suggesting that *Listeria* contamination within dairy processing facilities has not yet been fully controlled.

1. General Characteristics

The genus *Listeria*, which is included among the coryneform bacteria, contains six species. Although three of these species, namely *L. monocytogenes, Listeria ivanovii*, and *Listeria seeligeri*, can cause human or animal infections, only *L. monocytogenes* is important as a foodborne pathogen. *L. monocytogenes* is a gram-positive, non–spore-forming, facultatively anaerobic, short diphtheroidlike, rod-shaped bacterium that occurs singly or in short chains. The organism is psychrotrophic, growing in common laboratory media at temperatures between 1°C and 45°C (optimal growth between 30°C and 3°C) with growth enhanced under reduced oxygen conditions (Farber and Peterkin, 1991; Ryser and Marth, 1991; Swaminathan et al., 1995). Characteristic tumbling motility is visible microscopically in broth cultures incubated at room temperature. Colonies on clear media are small, smooth, and blue-gray when examined under obliquely transmitted light. Biochemically, all listeriae produce catalase, ferment glucose to acid without gas, and hydrolyze esculin. Typical *L. monocytogenes* isolates ferment rhamnose but not xylose and are weakly β-hemolytic. *L. monocytogenes* is unusually tolerant of environmental extremes, being able to grow at pH 4.3 to 10.0, grow in the presence of up to 10% NaCl (a_w 0.92), and survive in refrigerated 25.5% NaCl brine solutions for 4 months (Shahamat et al., 1980). Based on somatic (O) and flagellar (H) antigens, 13 different *L. monocytogenes* serotypes have been identified, with most human illnesses being caused by serotypes 1/2a, 1/2b, and 4b.

2. Isolation and Detection Methods

In the standard FDA protocol (Hitchins, 1995), recovery of *Listeria* from dairy products begins with enrichment of the sample in *Listeria* enrichment broth, a buffered medium containing acriflavin, nalidixic acid, and cycloheximide as selective agents. After 24 and 48 hours of incubation at 30°C, the enrichment culture is streaked to two different *Listeria* selective plating media including Oxford medium (OXA) and either PALCAM *Listeria* selective agar, which contains polymyxin B, acriflavin, and ceftazidime, or lithium chloride-phenylethanol-moxalactam medium with or without esculin and ferric ammonium citrate (LPM). After 24 to 48 hours of incubation at 30°C to 35°C, suspect colonies on OXA, PALCAM, and LPM are black with a black halo resulting from esculin hydrolysis, whereas colonies on LPM without esculin appear blue-green under oblique lighting. Presumptive *Listeria* isolates are speciated based on a standard series of biochemical tests that can take up to 7 days to complete. However, the time required for biochemical confirmation can be shortened using commercially available test kits (i.e., API 20 S, API-ZYM, API Listeria, Micro-ID). Typical *L. monocytogenes* isolates are rhamnose-positive, xylose-negative, and CAMP test-positive with β-hemolysis enhanced in the vicinity of *S. aureus*. Alternatively, several DNA hybridization (Accuprobe, GeneTrak) and enzyme-linked immunosorbent assays (VIDAS) can be used to screen enrichment broths for *Listeria* spp., including *L. monocytogenes* (Hill et al., 1995). Positive test results must be culturally confirmed. Complete serotyping is normally confined to selected isolates of epidemiological importance and conducted by a few select reference laboratories.

3. Clinical Manifestations

Three segments of the population, namely pregnant women, newborn infants, and immunocompromised adults, are at primary risk of contracting listeriosis, with the latter group including the elderly and other people with predisposing conditions such as cancer, organ transplants, cirrhosis of the liver and human immunodeficiency virus (HIV) infections or acquired immunodeficiency syndrome (AIDS) (Ryser and Marth, 1991; Swaminathan et al., 1995). In addition to host susceptibility, development of listeriosis in humans is also affected by gastric acidity, inoculum size, the strain of *L. monocytogenes*, and various virulence factors of the organism. Whereas the oral infective dose varies widely with healthy individuals rarely being infected, ingestion of foods containing greater than or equal to 10^3 organisms/g poses a significant health risk for susceptible individuals.

Listeric infections in immunocompromised adults typically lead to meningitis, encephalitis, or septicemia (Ryser and Marth, 1991; Swaminathan et al., 1995). Symptoms that develop suddenly after an initial incubation period of 2 days to 3 months include severe headache, dizziness, stiff neck or back, incoordination, and other disturbances of the central nervous system. Without proper antibiotic therapy, 20% to 30% of those infected will die, with some survivors developing permanent neurologic complications. In pregnant women, *L. monocytogenes* produces a mild flulike illness characterized by sudden chills, fever, sore throat, headache, dizziness, lower back pain, discolored urine, and occasionally diarrhea. Even though expectant mothers almost invariably recover without complications, infection of the fetus can result in abortion, stillbirth, or the premature delivery of an infant with perinatal septicemia—a severe infection of the respiratory, circulatory, and central nervous systems that can either terminate fatally or lead to permanent mental retardation.

Two factors, namely the growth of *L. monocytogenes* as an intracellular pathogen within macrophage cells of the spleen and liver and the inability of many antibiotics to effectively penetrate the blood–brain barrier, complicate treatment of listeric infections (Ryser and Marth, 1991; Swaminathan et al., 1995). Hence, a favorable prognosis depends on rapid diagnosis and appropriate antibiotic therapy, with oral administration of large doses of ampicillin or penicillin together with an aminoglycoside for 2 to 4 weeks being the currently recommended treatment.

4. Outbreaks

Early animal feeding studies supported the likely importance of food in disseminating listeriosis (Murray et al., 1926). However, this disease was not linked to consumption of a food product until the early 1950s (Gray and Killinger, 1966; Ryser and Marth, 1991), with dairy products thus far being responsible for five major outbreaks of listeriosis (Table 5). In the first of these outbreaks, a sharp increase in stillbirths was observed among pregnant women in post-World War II Germany who consumed raw milk, sour milk, cream cheese, and cottage cheese, with approximately 100 *Listeria*-like infections being reported. The eventual isolation of identical *L. monocytogenes* serotypes from a mastitic cow and stillborn twins whose mother consumed the same raw milk before delivery confirmed raw milk as the vehicle of infection (Potel, 1953, 1954). Despite the presence of *L. monocytogenes* in 2% to 4% of the raw milk supply, only two additional listeriosis cases have been linked to ingestion of raw milk (Ryser and Marth, 1991).

During the summer of 1983, the status of *L. monocytogenes* as a foodborne pathogen began to change when consumption of one particular brand of pasteurized milk was epidemiologically linked to 42 adult and seven infant cases of listeriosis in Massachusetts (Fleming et al., 1985; Ryser and Marth, 1991). Fourteen patients died, giving a mortality rate of 29%. Inspection of the milk processing facility failed to uncover any evidence of improper pasteurization or post-pasteurization contamination. Although the dairy factory received milk from several farms on which veterinarians diagnosed listeriosis in dairy cows during the outbreak, *L. monocytogenes* was never recovered from the incriminated milk,

TABLE 5 Major Listeriosis Outbreaks Involving Milk and Dairy Products

Location	Year	Product	Number of Cases	Reference
Halle, East Germany	1949–1957	Raw milk, sour milk, cream, cottage cheese	Approx. 100	Gray and Killinger (1966) Ryser and Marth (1991)
Massachusetts	1983	Pasteurized milk	49	Fleming et al. (1985) Ryser and Marth (1991)
Los Angeles, CA	1985	Mexican-style cheese	Approx. 300	Linnan et al. (1988) Ryser and Marth (1991)
Vaud, Switzerland	1983–1987	Vacherin Mont d'Or, soft-ripened cheese	122	Bula et al. (1995) Ryser and Marth (1991)
France	1995	Brie cheese	20	Goulet et al. (1995)

which, in turn, raises serious questions concerning the role of pasteurized milk in this outbreak. Except for one additional outbreak in which strain-specific typing positively linked consumption of pasteurized chocolate milk to 19 cases of listeriosis in Wisconsin (Proctor et al., 1995), repeated attempts have failed to culturally confirm other nonfermented dairy products, including ice cream and butter, as vehicles of listeric infection.

Ingestion of *Listeria*-contaminated cheese has been more commonly linked to listeriosis, with three major outbreaks and 10 sporadic cases thus far being reported. The first and largest of these outbreaks occurred in the Los Angeles area during the first half of 1985 and involved an estimated 300 cases (Ryser and Marth, 1991). Consumption of California-made Jalisco-brand Mexican-style cheese contaminated with *L. monocytogenes* serotype 4b was linked to 142 listeriosis cases in Los Angeles County, resulting in 48 deaths (mortality rate of 34%) (Linnan et al., 1988). The contaminated cheese was subsequently recalled nationwide. Factory records suggested that raw milk might have been added to pasteurized milk used in cheese making. Although not isolated from the incoming raw milk supply, the epidemic strain was ubiquitous in the factory environment, which suggests ample opportunity for postpasteurization contamination.

In the second of these outbreaks, consumption of Vacherin Mont d'Or—a soft, surface-ripened, cheese—contaminated with *L. monocytogenes* serotype 4b was linked to 122 listeriosis cases in Switzerland from 1983 to 1987 (Bula et al., 1995; Ryser and Marth, 1991). Thirty-four patients died, giving a mortality rate of 28%. Two different epidemic-associated strains of *L. monocytogenes* serotype 4b were isolated from patients, the incriminated cheese, and the wooden shelves and brushes used in 40 different cheese ripening cellars. Detection of the epidemic strain at levels of 10^4 to 10^6 cfu/g in surface samples of cheese supported both contamination and growth of *L. monocytogenes* on the cheese surface during ripening. The outbreak ceased after installation of metal ripening shelves and thorough cleaning and sanitizing of the ripening rooms.

Most recently, 20 cases of listeriosis in France were traced to consumption of Brie cheese prepared from raw milk (Goulet et al., 1995). Eleven of these cases occurred in pregnant women with the remaining nine cases involving elderly or immunocompromised adults. Unlike the previous outbreaks, no geographic clustering was observed, with cases reported in 8 of 22 French regions. The same epidemic strain was recovered from these patients and the cheese, with this organism likely being present in the raw milk used for cheese making. Remaining reports of cheeseborne listeriosis are confined to 10 isolated cases reviewed by Ryser and Marth (1991), only one case of which was well documented and positively linked to consumption of raw goat's milk cheese (McLauchlin et al., 1990).

5. Occurrence and Survival in Dairy Products

Dairy cattle, sheep, and goats can intermittently shed *L. monocytogenes* in their milk at levels up to 10^4 cfu/mL as a result of listeric mastitis, encephalitis, or a *Listeria*-related abortion. Whereas milk from obviously infected cows is unlikely to reach consumers, mildly infected and apparently healthy animals can shed *L. monocytogenes* in their milk for many months and are thus of greater public health concern. Composite results from numerous bulk tank surveys conducted since 1983 indicate that 3.1%, 2.7%, and 4.8% of all raw milk processed in the United States, Canada, and western Europe, respectively, will likely contain low levels (i.e., < 10 cfu/mL) of *L. monocytogenes* at any given time (Ryser and Marth, 1991). However, *L. monocytogenes* populations in naturally contaminated raw milk can increase 1000-fold after 4 and 10 days of storage at 10°C and 4°C, respectively (Farber et al., 1990).

L. monocytogenes is more heat tolerant than most other non–spore-forming pathogens (Doyle et al., 1987). However, current vat and high-temperature, short-time pasteurization practices ensure total destruction of *L. monocytogenes* as long as the raw milk is properly handled and refrigerated at 4°C to minimize growth. Despite the ability of *L. monocytogenes* to attain populations of 10^6 cfu/mL in skim milk, whole milk, chocolate milk, and whipping cream after 8 days of storage at 8°C (a common temperature of home refrigerators), this organism has been rarely detected in pasteurized fluid milk products (Rosenow and Marth, 1987). Although *L. monocytogenes* has been occasionally recovered from commercially produced butter, with survival up to 70 days also being reported in butter prepared from inoculated cream (Olsen et al., 1988), this pathogen is a far more frequent postpasteurization contaminant of ice cream. Since May 1986, 40 of 45 *Listeria*-related class I recalls issued for unfermented dairy products have involved ice cream, ice cream novelties, and related frozen desserts contaminated with very low levels of *L. monocytogenes*. Increased prevalence of this pathogen in frozen rather than fluid dairy products coincides with the relatively complex handling of such products, particularly ice cream novelties, during manufacture and packaging. Given the presumed low levels of contamination, the inability of *Listeria* to grow in frozen dairy products and the recall of more than 3 million gallons of ice cream without incident, consumption of such products does not appear to pose a major public health threat.

As can be surmised from the previous discussion of outbreaks, *L. monocytogenes* is a frequent contaminant of cheese, most notably soft, surface-ripened varieties such as Brie and Camembert, which support growth of the organism during cheese ripening. Since 1986, 32 class I recalls were issued for domestically produced cheese, principally Mexican-style cheese, contaminated with *L. monocytogenes* (Ryser and Marth, 1991). During this same period, 28 imported cheeses, including French Brie, Danish Esrom, and Anari goat's milk cheese from Cyprus, were similarly recalled. According to Ryser and Marth (1991), approximately 4.25% of European-produced cheeses, primarily soft and semisoft varieties, can be expected to harbor *L. monocytogenes.*

Considerable work has been done to define the behavior of *L. monocytogenes* during manufacture and storage of yogurt, buttermilk, and a wide variety of cheeses, with most of these studies describing what happens if the product is prepared from artificially contaminated pasteurized milk (Ryser and Marth, 1991). In one of several studies assessing postpasteurization contamination, *L. monocytogenes* persisted an average of 3 weeks in refrigerated cultured buttermilk and yogurt inoculated to contain 10^3 and 10^4 *L. monocytogenes* cfu/g (Choi et al., 1988). *Listeria* populations generally increase as much as 10-fold when milk is fermented using a 1% inoculum of a traditional mesophilic or thermophilic lactic acid bacteria starter culture, with growth ceasing at pH less than or equal to 5.2 (Schaack and Marth, 1988a, 1988b). However, physical entrapment of *Listeria* in curd during cheese making results in a 10-fold increase in numbers. Growth of *Listeria* in cheese is primarily confined to soft and semi-soft varieties such as blue, brick, Camembert, and goat cheese, with populations increasing to at least 10^6 cfu/g as the cheese attains a pH greater than 6.0 during ripening (Table 6). Although generally unable to grow in fermented dairy products having a pH less than 5.5, *L. monocytogenes* can survive in many such cheeses for weeks or months, with this pathogen even being recovered from 434-day old cheddar cheese (Ryser and Marth, 1987a). These findings raise serious concerns regarding the adequacy of the mandatory 60-day holding period at greater than or equal to 1.7°C for complete inactivation of *L. monocytogenes* (and other pathogens) in cheddar and certain other hard cheese that can be legally prepared from raw milk. However, barring contamination during packaging, cheeses such as cottage and mozzarella, which undergo severe heat treatments during manufacture, should be *Listeria*-free (Buazzi et al., 1992; Ryser et al., 1985).

6. Prevention

Although *L. monocytogenes* is more heat-resistant than most other foodborne pathogens, current vat and high-temperature, short-time pasteurization practices inactivate expected levels of *L. monocytogenes* in raw milk. Thus, barring postpasteurization contamination, pasteurized fluid milk products pose a minimal public health risk. Given the many gallons of ice cream recalled in the United States, low-level contamination in frozen desserts also appears to be of little health concern. However, certain low-acid, soft, and surface-ripened cheeses such as Mexican-style and Brie cheese can support the growth of *L. monocytogenes* to dangerous levels during ripening, as is evidenced by three major outbreaks of

TABLE 6 Fate of *Listeria monocytogenes* in Various Cheeses During Ripening and Storage

| Cheese | pH | | Ripening | *L. monocytogenes* (log 10 cfu/g or mL) | | | Survival | Reference |
	Initial	Final	Temp (°C)	Milk	Cheese Maximum	Final	(Days)	
Blue	4.6	6.3	9–12/4	3.0	4.0–5.0	1.0–2.3	>120	Papageorgiou and Marth (1989a)
Brick	5.3	7.3	15/10	2.5–3.0	4.6–6.7	2.7–6.1	>168	Ryser and Marth (1988b)
Camembert	4.6	7.5	15/6	2.5–3.0	6.7–7.5	6.7–7.5	>65	Ryser and Marth (1987b)
Cheddar	5.1	5.1	13	2.5–3.0	2.6–3.8	<1.0	70–224	Ryser and Marth (1987a)
Cheddar	5.1	5.1	13	2.5–3.0	3.0–3.7	<1.0–1.5	70–>434	Ryser and Marth (1987a)
Colby	5.1	5.1	4	2.5–3.0	3.6–4.6	2.3–4.1	>140	Yousef and Marth (1988)
Cold-pack	5.3	5.1	4	2.4–2.8	2.4–2.8	1.1–2.0	>180	Ryser and Marth (1988a)
Cottage	5.4	5.2	3	4.0–5.0	1.3–2.8	<1.0–2.4	17–28	Ryser et al. (1985)
Feta	4.7	4.4	22/4	3.7	5.7–6.2	2.8–4.6	>90	Papageorgiou and Marth (1989b)
Goat	5.5	6.2	12	5.0–6.0	6.9	6.2	>126	Tham (1988)
Gouda	5.5	5.5	13	2.5	4.2	3.2	>42	Northolt et al. (1988)
Mozzarella	5.2	5.2	5	4.0–5.0	<10	<10	<1	Buazzi et al. (1992)
Parmesan	5.1	5.1	13	4.0–5.0	3.3–4.3	<10	14–112	Yousef and Marth (1990)

listeriosis involving numerous fatalities. Consequently, individuals at highest risk (i.e., pregnant women, elderly people, and immunocompromised adults) may want to refrain from eating such cheeses.

J. Salmonellosis

Nontyphoid salmonellae were first recognized as foodborne pathogens in the 1880s when acute gastroenteritis developed in 57 people after consuming beef that was contaminated with *Bacterium enteritidis*, which was renamed *S. enteritidis* in 1900 in honor of the American bacteriologist D. E. Salmon (Marth, 1969). From the turn of the century to about 1940, typhoid fever was commonly associated with consumption of raw milk, as described earlier. However, the gastroenteritic form of nontyphoid salmonellosis (hereafter salmonellosis) was not clearly linked to raw milk consumption until the mid-1940s. Interest in milk-borne salmonellosis has peaked twice since the 1940s, first in 1966 when several large outbreaks were traced to nonfat dry milk and again in 1985 when one of the largest recorded outbreaks of foodborne salmonellosis involving more than 180,000 cases was traced to consumption of a particular brand of pasteurized milk in the Chicago area (El-Gazzar and Marth, 1992). Three years before the Chicago outbreak, milk and dairy products were responsible for 5 of 55 (9%) outbreaks of foodborne illness in the United States (MacDonald and Griffin, 1983). Today, *Salmonella* and *Campylobacter* are generally recognized as the two leading causes of dairy-related illness in the United States and western Europe, with rates of infection being particularly high in regions of the world where raw milk is neither pasteurized nor boiled.

1. General Characteristics

All salmonellae are of public health concern given their ability to produce infections ranging from a mild self-limiting form of gastroenteritis associated with consumption of contaminated dairy products to septicemia and life-threatening typhoid fever produced by *S. typhi*, as discussed previously. A prominent group of the family *Enterobacteriaceae*, salmonellae are short, gram-negative, facultatively anaerobic, rod-shaped bacteria (El-Gazzar and Marth, 1992; Kantor, 1986). These organisms grow on common laboratory media at temperatures between 5°C and 45°C (optimum 35°C to 37°C) and at a_w values greater than or equal to 0.95 (Bryan et al., 1979). However, a few strains can multiply in both laboratory media and certain foods at refrigeration temperatures (Matches and Liston, 1968). Biochemically, salmonellae produce gas from glucose, reduce nitrate to nitrite, and can utilize citrate as a sole carbon source (Flowers et al., 1992). Most strains are lysine decarboxylase-positive, hydrogen sulfide-positive, and motile by peritrichous flagella, but important exceptions have been noted. Further classification of the genus *Salmonella* is still a source of confusion because there are three different classification schemes. Using the classical and still popular Kauffmann-White scheme, which is based on somatic (O), flagellar (H), and capsular (Vi) antigens, more than 2300 *Salmonella* serovars (distinct antigenic profiles) are recognized (D'Aoust, 1994). These serovars can be grouped into five different subgenera. In this classification scheme, each serovar has a descriptive and geographical "species" or "popular" name such as *Salmonella typhimurium* or *Salmonella heidelberg*, with these names still widely being used. The Edwards and Ewing scheme, another antigenically based classification system currently decreasing in popularity, recognizes three major species or groups—*S. typhi, Salmonella cholerae-suis*, and *S. enteritidis* with the last species comprising nearly all of the 2300 aforementioned *Salmonella* serovars. Based on DNA hybridization studies, the genus *Salmonella* also can be divided into seven widely accepted subgroups (Flowers et al., 1992), each with its own phenotypic characteristics.

2. Isolation and Detection Methods

Examination of dairy products for *Salmonella* (Andrews et al., 1995; Flowers et al., 1992a, 1992b) begins with preenrichment of the sample in a nonselective medium, most often lactose broth, for resuscitation of injured or debilitated cells. Following 18 to 24 hours of incubation at 35°C, two different selective enrichment media—selenite cystine and tetrathionate broth—are inoculated from the preenrichment broth and similarly incubated. Thereafter, plates of Hektoen enteric, xylose lysine desoxycholate, and bismuth sulfite agar are streaked from both selective enrichment broths and incubated 24 to 48 hours at 35°C for selective isolation of salmonellae. Alternatively, several rapid methods using fluorescent antibodies, hydrophobic grid membrane filtration, enzyme immunoassays, DNA hybridization, immunodiffusion, and conductivity are commercially available for detecting salmonellae in enriched samples. All positive findings must be culturally confirmed. Presumptive isolates are confirmed as *Salmonella* using a standard series of 16 biochemical tests in combination with serological screening tests that use polyvalent O antisera and either polyvalent H or Spicer-Edwards antisera. Five commercially available biochemical test kits (i.e., API 20E, Enterotube II, Enterobacteriaceae II, MICRO-ID, and Vitek GNI) are approved alternatives to traditional biochemical confirmation. Biochemically presumptive salmonellae must still be subjected to serological confirmation with complete serotyping of epidemiologically important strains confined to qualified reference laboratories.

3. Clinical Manifestations

Although commonly referred to as *Salmonella* "food poisoning," gastro-enteritis—the first of three clinical manifestations produced by nontyphoid salmonellae—is an infection (not an intoxication or "poisoning") of the small intestine and less commonly the colon, with no involvement of preformed toxins (D'Aoust, 1994; El-Gazzar and Marth, 1992; Kantor, 1986). The first symptoms to appear after an initial incubation period of 12 to 36 hours include nausea and vomiting, both of which subside within a few hours. Development of mild fever, chills, and abdominal pain sometimes resembling acute appendicitis is soon followed by diarrhea, the most prominent symptom, which can range from a few loose stools to overtly bloody and rice-water choleralike stools in more severe cases. During this period, all infected individuals excrete *Salmonella* in their feces with samples from acute cases often containing 10^6 to 10^9 salmonellae/g. Although this self-limiting illness typically subsides within 5 days without intervention, symptoms can persist up to several weeks with 10% of fully recovered patients excreting salmonellae for at least 2 months.

Septicemia, the second manifestation of salmonellosis (D'Aoust, 1994; El-Gazzar and Marth, 1992; Kantor, 1986), occurs as a complication of gastroenteritis in less than 4% of adult patients with fever being the primary symptom. Even though salmonellosis is generally considered to be among the less serious types of blood infections, fatalities have been reported in 13% of individuals with serious underlying illnesses such as cancer and liver disease.

Localized tissue infections, the third manifestation of salmonellosis (D'Aoust, 1994; El-Gazzar and Marth, 1992; Kantor, 1986), occur as a complication in 8% to 25% of patients with prolonged or untreated septicemic infections. Although any part of the body may become infected, lesions and abscesses are most frequently associated with previously damaged or diseased organs and tissues. Infections most commonly include osteomyelitis, meningitis, and pneumonia followed by pyelonephritis, endocarditis, and suppurative arthritis.

Salmonellosis can only be clinically confirmed by isolating salmonellae from stool, blood, or other specimens. Because this disease is most often mild and self-limiting, treatment is usually aimed at preventing dehydration through fluid replacement (Kantor, 1986). As in other types of gastroenteritis, administration of antibiotics is contraindicated and limited to patients who either have septicemic or localized tissue infections or are at high risk of development of such complications. When necessary, the drug of choice is chloramphenicol given only intravenously.

4. Outbreaks

Dairy-related outbreaks of nontyphoid salmonellosis were first recognized in the 1940s, with raw milk being most commonly identified as the source of infection. At least four notable outbreaks involving raw milk consumption occurred in the United States since 1967 with *S. typhimurium* and *Salmonella dublin* identified as the causative serovars (Table 7). Although certified raw milk legally sold in

California was responsible for one of these outbreaks involving *S. dublin*, similar outbreaks have been documented as far back as 1958 (Anonymous, 1981). Findings from one epidemiological case-control study (Richwald et al., 1988) suggest that one-third of all *S. dublin* infections in California are raw milk related, with the incidence of infection being highest among immunocompromised adults.

In England and Wales where records are more complete, raw milk consumption was responsible for 132 of 148 predominantly small, dairy-related outbreaks (2369 of 2466 cases) from 1951 to 1980 (Galbraith et al., 1982). As in the United States, the predominant *Salmonella* serovars again included *S. typhimurium* (88 outbreaks) and *S. dublin* (14 outbreaks), both of which are frequently recovered from raw milk, dairy cattle, and farm environments (Marth, 1969). During the 1980s, raw milk consumption was linked to at least 58 outbreaks involving 1088 cases in England and Wales with the size of these outbreaks increasing because of commercial distribution of raw milk (Barrett, 1986, 1989; Sockett, 1991). This problem is particularly evident in Scotland where 21 outbreaks (1090 cases) were reported between 1980 and 1982 (Reilly et al., 1983). One of these outbreaks sickened 654 people and cost an estimated $120,000 (Cohen et al., 1983).

American interest in milk-borne salmonellosis first peaked in 1966 when 29 cases of gastroenteritis diagnosed in 17 states over a 10-month period were linked to nonfat dry milk containing *Salmonella newbrunswick*, a serovar rarely encountered in the United States (Collins et al., 1968). Of the 29 victims, more than half were infants or children younger than 5 years of age. The contaminated product was recalled from sale and soon traced to a single midwest factory producing approximately 11 million pounds of nonfat dry milk annually (Marth, 1969). Although the source of contamination was never identified, incomplete pasteurization before spray drying was suspected as one likely cause, with post-processing contamination cited as a contributing factor. Two additional noteworthy outbreaks also have been traced to nonfat dry milk produced in other countries. During 1973, more than 3000 cases of gastroenteritis on the island of Trinidad occurred primarily among infants and young children and were traced to consumption of nonfat dry milk contaminated with *Salmonella derby* (Weissman et al., 1977). Although faulty packaging equipment may have contributed to this outbreak, the source of contamination was never identified. During 1985, consumption of one particular brand of an infant dried milk product was also responsible for at least 46 cases of *Salmonella ealing* gastroenteritis in England; the source of infection was traced to a malfunctioning spray dryer (Rowe et al., 1987).

Pasteurized milk can also serve as a vehicle for salmonellosis. Three small pre-1985 outbreaks were linked to ingestion of inadequately pasteurized milk in Louisiana, Arizona, and Kentucky. Such outbreaks did not attract widespread

TABLE 7 Major Salmonellosis Outbreaks Associated with Milk and Milk Products from 1965 to 1994

Location	Year	Product	*Salmonella* Serovar	Number of Cases	Reference
United States					
Nationwide	1965–1966	Nonfat dry milk	*newbrunswick*	29	Collins et al. (1968)
New York	1967	Ice cream	*typhimurium*	1,790	Armstrong et al. (1970)
Washington	1967	Raw milk	*typhimurium*	40	Francis and Allard (1967)
California	1971–1975	Raw milk	*dublin*	44	Werner et al. (1979)
Maine	1973	Egg-nog	*typhimurium*	32	Steere et al. (1975)
Louisiana	1975	Pasteurized milk	*newport*	43	Blouse et al. (1975)
Colorado	1976	Cheddar	*heidelberg*	339	Fontaine et al. (1980)
Arizona	1978	Pasteurized milk	*typhimurium*	23	Dominguez et al. (1979)
Washington	1980–1981	Raw milk	*dublin*	125	Nolan et al. (1981)
Montana	1981	Raw milk	*typhimurium*	59	Day et al. (1981)
Kentucky	1984	Pasteurized milk	*typhimurium*	16	Adams et al. (1984)
Illinois	1985	Pasteurized milk	*typhimurium*	16,000	Ryan et al. (1987)
Nationwide	1989	Mozzarella	*javiana/oranienburg*	164	Hedberg et al. (1992)
Kansas	1992	Ice cream	*enteritidis*	15	Anonymous (1992)
Kansas	1992	Ice cream	*enteritidis*	31	Anonymous (1992)
Florida	1993	Ice cream	*enteritidis*	14	Buckner et al. (1994)
National	1994	Ice cream	*enteritidis*	224,000	Anonymous (1994), Hennessy et al. (1996)

Canada					
Ontario	1980–1983	Cheddar	*muenster*	33	Styliadis and Barnum (1984)
5 Provinces	1984	Cheddar	*typhimurium*	2,000	Bezanson et al. (1985)
Europe					
England	1972–1973	Raw milk	*typhimurium*	316	MacLachlan (1974)
Scotland	1976	Raw milk	*dublin*	700	Small and Sharp (1979)
England	1977	Raw milk	*typhimurium*	334	Anonymous (1977)
Poland	1978	Pasteurized milk	*enteritidis*	890	Suchowiak and Haiat (1980)
Scotland	1981	Raw milk	*typhimurium*	654	Cohen et al. (1993)
Poland	1981	Ice cream	*typhimurium*	881	Polewska-Jeske et al. (1984)
Italy	1981	Mozzarella	*typhimurium*	100	Felip and Toti (1984)
Sweden	1985	Pasteurized milk	*saintpaul*	153	Anderson et al. (1986)
Switzerland	1985	Vacherin Mont d'Or	*typhimurium*	40	Sharp (1987)
England/Wales	1989	Irish soft cheese	*dublin*	42	Maguire et al. (1992)

attention until 1985 when more than 16,000 culture-confirmed cases of *Salmonella* gastroenteritis in the Chicago area were traced to consumption of 2% pasteurized milk contaminated with a rare multiantibiotic-resistant, plasmid-containing strain of *S. typhimurium* (Ryan et al., 1987; Schuman et al., 1989). One follow-up survey placed the number of people affected at nearly 200,000, making this the second largest outbreak of foodborne salmonellosis ever recorded. [The largest outbreak occurred in 1994, involving approximately 240,000 persons, and was associated with nationally distributed ice cream made in Minnesota (Hennessy et al., 1996)]. Although the milk outbreak affected an estimated 3 of every 1000 residents in the Chicago area, with the highest attack rate observed in children, illness was particularly common among individuals who were taking antibiotics to which this particular strain of *S. typhimurium* was resistant, with 2500 such hospitalized cases being reported. Further complications developed in 16 of these patients, including osteomyelitis, brain abscesses, and meningitis, or they had unnecessary appendectomies. Eighteen fatalities were reported with the epidemic strain cited as either the primary or contributing cause of death. The implicated milk containing the epidemic strain was traced to a northern Illinois dairy processing facility and was immediately recalled from the market. Microbiological studies indicated that the outbreak-related strain was heat-sensitive and would not be expected to survive pasteurization, but inspection of the dairy plant revealed a potential cross-connection between several holding tanks that would have allowed raw milk to contaminate pasteurized skim milk, condensed milk, and cream. The factory ceased operations when the outbreak occurred and has not reopened.

During 1992 and 1993, homemade ice cream was linked to three small outbreaks of *S. enteritidis* gastroenteritis in Florida and Kansas with a similar family outbreak also recently reported in England (Morgan et al., 1994a). The source of contamination in all of these outbreaks was traced to raw eggs, an ingredient of homemade ice cream particularly noted for harboring *S. enteritidis*. One year later, commercially produced ice cream was responsible for 2000 (final estimate 240,000 cases) cases of *S. enteritidis* gastroenteritis in Minnesota, Wisconsin, South Dakota, and elsewhere with the tainted product eventually recalled nationwide. Tankers used to haul liquid raw eggs also were used to haul pasteurized ice cream mix to the ice cream factory where the mix was not repasteurized. The tankers were the likely source of *S. enteritidis* (Hennessy et al., 1996).

Salmonellosis outbreaks involving fermented dairy products have been primarily confined to cheese, with six notable outbreaks being reported since 1976. In the first of these outbreaks, 339 cases of *S. heidelberg* gastroenteritis were identified in Colorado and traced to cheddar cheese prepared in Kansas from pasteurized milk (Fontaine et al., 1980). The incriminated cheese contained less than 1 organism per 100 grams, thus suggesting a low oral infectious dose, with prompt recall of the product possibly averting up to 25,000 additional cases. Cheddar cheese was again identified as the vehicle of infection in two Canadian outbreaks reported during the 1980s. In the first outbreak, *S. muenster* was recovered from aged raw milk cheddar cheese and was traced to an infected dairy herd with one mastitic cow shedding 2000 *S. muenster* per milliliter of milk. Few serovars other than *S. muenster* and *S. dublin* can reportedly infect the bovine mammary gland and contaminate milk in this manner. The second and largest cheeseborne salmonellosis outbreak occurred in Ontario and the four maritime provinces with more than 2000 culture-confirmed cases in 1984 linked to cheddar cheese prepared from heat-treated or pasteurized milk (Bezanson et al., 1985; D'Aoust et al., 1985; Ratnam and March, 1986). Two distinct strains of *S. typhimurium* were implicated in this outbreak. Both strains were recovered from cheese at levels of less than 10 organisms per 100 grams, which again suggests a low oral infectious dose.

Four outbreaks have been traced to cheeses other than cheddar, with one of these outbreaks responsible for 164 cases of salmonellosis in Minnesota, Wisconsin, Michigan, and New York during 1989. Mozzarella cheese containing two epidemic strains, *Salmonella javiana* and *Salmonella oranienburg*, was the vehicle of infection. Inadequate factory sanitation practices and contamination of the cheese by infected production workers were suggested as probable causes. The three remaining outbreaks occurred in Europe, the most recent of which involved a soft raw milk Irish-type cheese prepared on a family farm in England. Contamination was traced to four family-owned cows that were asymptomatically excreting *S. dublin*, the epidemic strain in this outbreak. The remaining milk-borne cases of salmonellosis have almost invariably involved milk from dairy cows, with only a few small outbreaks outside the United States being linked to goat's milk (Sharp, 1987).

5. Occurrence and Survival in Dairy Products

Numerous *Salmonella* infections have been reported in dairy cattle and other ruminant animals with symptomatic and asymptomatic shedding of the organism in feces (Marth, 1969; Styliades and Barnum, 1984). Although salmonellae are seldom associated with mastitis, *S. dublin* and *S. muenster* can colonize the udder and be shed in milk at levels up to 2000 organisms/mL (Fontaine et al., 1980). According to McManus and Lanier (1987), raw milk is a good source of salmonellae with 32 of 678 (4.7%) raw milk bulk tank samples testing positive in Wisconsin, Michigan, and Illinois. However, lower contamination rates have been reported elsewhere. Following the 1980 to 1983 cheeseborne outbreak in Ontario, Canada, McEwen et al. (1988) detected salmonellae, including *S. muenster* (the epidemic serovar), in raw milk bulk tanks from 9 of 759 (1.2%) dairy farms participating in this year-long study, with most positive samples being observed during autumn. In England, 2 of 1138 (0.2%) raw milk samples on sale to the public harbored salmonellae (Humphrey and Hart, 1988), thereby reaffirming the potential hazard of raw milk consumption.

Standard vat and high-temperature, short-time pasteurization destroys expected levels of salmonellae (i.e., <100 cfu/mL), including *Salmonella senftenberg* 775W (the most heat-resistant serovar) with a wide margin of safety (D'Aoust et al., 1987). Inadequate pasteurization and postprocessing contamination have occasionally resulted in milk and cream that test positive for *Salmonella* as evidenced from the aforementioned outbreaks. Unlike *Listeria* and *Yersinia*, which can grow during refrigeration, numbers of salmonellae decrease in fluid milk products and butter prepared from inoculated cream during extended storage at less than or equal to 7°C (Kasrazadeh and Genigeorgis, 1994; Sims et al., 1970; Wundt and Schnittenhelm, 1965). However, at 12°C and 20°C, *Salmonella* populations double every 8.8 and 20 hours, respectively, which reinforces the need for constant refrigeration.

Nonfat dried milk also can occasionally harbor salmonellae as demonstrated by the highly publicized 1966 outbreak in the United States. Surveys conducted on nonfat dried milk over the following 2 years showed that 0.2% of all samples contained salmonellae (Marth, 1969), and another study (LiCari and Potter, 1970a) showed that commercial spray drying conditions killed more than 99.9% of salmonellae in inoculated skim milk but did not yield *Salmonella*-free nonfat dry milk at the inoculum levels used. In a follow-up study (Licari and Potter, 1970b), *Salmonella* populations in heavily inoculated nonfat dried milk decreased sharply when the product was held at 25°C to 55°C. However, persistence of salmonellae in some samples for at least 8 weeks indicates that such storage cannot be used as a substitute for good manufacturing practices.

Ice cream and related frozen desserts can become contaminated before freezing and give rise to outbreaks of salmonellosis. Except for the aforementioned 1994 outbreak involving approximately 240,000 cases in the United States, such contamination has been primarily confined to homemade ice cream with raw eggs being the invariable source of *S. enteritidis*. Contamination rates are very low in commercially produced ice cream, with two recent European surveys identifying salmonellae in 0 of 157 (Massa et al., 1989) and 1 of 67 (Rodriguez-Alvarez et al., 1994) samples sold in Italy and Spain, respectively. However, higher *Salmonella* contamination rates have been reported in less developed countries such as Iraq (Al-Rajab et al., 1986) where 12 of 110 (10.9%) locally produced ice cream samples tested positive.

Despite the recent cheese-related salmonellosis outbreaks, salmonellae are rarely isolated from commercially produced cheeses including cheddar. In surveys responding to the 1980 to 1983 outbreak in Ontario, Canada, Brodsky (1984a, 1984b) failed to recover *Salmonella* from 250 samples of freshly prepared cheddar cheese or 127 samples of 60-day-old cheddar cheese prepared from raw milk. According to Mor-Mur et al. (1992), 42 samples of 60-day-old farm-produced goat cheese in Spain were also free of salmonellae. However, the presence of salmonellae in 8 of 142 (5.6%) locally produced Iranian cheeses (Farkhondeh et al., 1974) again raises concerns regarding the safety of dairy products manufactured in less developed countries where salmonellosis is endemic.

The fate of salmonellae has been assessed during the manufacture and ripening of many different cheeses (Table 8). Modest growth of salmonellae occurs during cheddar cheese making, as predicted by Park and Marth (1972a), with populations increasing 10- to 100-fold beyond the expected 10-fold increase, which results from physical entrapment of the organism during curd formation. Furthermore, when cheeses from the same lot were ripened at 0°C to 13°C, salmonellae survived 84 to 300 days, respectively, with the pathogen always persisting longer in cheese ripened at the lower temperature. White and Custer (1976) subsequently reported that 16 of 48 (33%) and 6 of 48 (12.5%) lots of cheddar cheese similarly prepared from milk containing 10^5 salmonellae cfu/mL were still positive after 9 months of ripening at 4.5°C and 10°C, respectively. Most important, when naturally contaminated cheddar cheese from the two Canadian outbreaks was stored at 5°C, salmonellae persisted up to 125 days (Styliades and Barnum, 1984; Wood et al., 1984) and 240 days (D'Aoust, 1985), well beyond the required 60-day holding period at greater than or equal to 1.7°C for such cheeses prepared from raw or heat-treated milk. Using cheddar cheese samples from the outbreak, D'Aoust (1985) estimated the oral infective dose for six different patients at 1 to 6 total cells of *Salmonella*, which suggests that even very low levels of contamination can pose serious health risks.

Additional cheese varieties studied have included mozzarella following the 1985 outbreak in the United States as well as cottage and Brie cheeses. According to Eckner et al. (1990), *Salmonella* was completely inactivated during molding and stretching of mozzarella cheese curd at 60°C. However, contamination of mozzarella cheese during shredding or packaging can lead to extended survival of salmonellae and possible health risks as evidenced from the aforementioned outbreak. Cottage cheese would appear to be of minimal public health concern, with large populations of salmonellae completely inactivated after cooking the curd and whey mixture at 125°F (52°C) for 20 minutes (McDonough et al., 1967). However, salmonellae also can persist in cottage cheese as postpasteurization contaminants throughout the normal shelf-life of the product (Sims et al., 1989). Soft surface-ripened cheeses such as Brie and Vacherin Mont d'Or have been implicated in major outbreaks involving other pathogenic organisms, including *E. coli* and *L. monocytogenes*, which can attain high levels on the surface of these cheeses during ripening. Although growth of salmonellae on the

TABLE 8 Fate of Salmonellae in Various Cheeses During Ripening and Storage

Cheese	pH		Ripening Temp (°C)	Salmonella (log 10 cfu/g or ml)			Survival (Days)	Reference
	Initial	Final		Milk	Maximum	Final		
Cheddar	5.52	5.22	0	3.14	5.78	1.00	180	Hargrove et al. (1969)
Cheddar	5.52	5.15	4	3.14	5.78	0.30	150	Hargrove et al. (1969)
Cheddar	5.70	5.40	7	2.00	4.10	1.30	210	Park et al. (1970a)
Cheddar	5.80	5.60	13	2.00	5.30	1.00	>210	Park et al. (1970a)
Cheddar	5.10	—	7.5	1.90	5.00	<1.00	112	Goepfert et al. (1968)
Cheddar	5.10	—	13	1.90	5.00	<1.00	84	Goepfert et al. (1968)
Mozzarella	—	—	—	6.00	6.00	<1.00	<1	Eckner et al. (1990)
Monterey Jack	—	—	4.5	6.50	9.00	4.50	>183	Eckner and Zottola (1991)
Montasio	5.30	5.60	12	6.80	5.80	0.80	90	Stecchini et al. (1991)
Feta (pasteurized cow's milk)	5.10	5.69	4	3.40	6.20	1.90	>75	Erkmen and Bozoglu (1995)
Feta (raw ewe's milk)	4.80	4.40	4	7.30	8.90	1.00	20	Papadopoulou et al. (1993)
Cold-pack	5.10	4.70	12.8	—	2.70	1.00	>188	Park et al. (1970b)
Cold-pack	5.10	5.05	4.4	—	2.70	1.70	>188	Park et al. (1970b)

surface of such cheeses is prevented during ripening at 4°C to 20°C (Little and Knochel, 1994), continued survival of the organism during ripening may again pose a potential health hazard.

6. Prevention

Historically, salmonellosis has been most commonly traced to raw milk, and consumption of such milk is best avoided. Despite several outbreaks and reports of extended survival of salmonellae in cheese, most cheeses—including those legally prepared from raw or heat-treated milk and then properly aged—appear to pose a minimal health risk. All salmonellae are readily destroyed by pasteurization. Hence, if postpasteurization contamination is prevented, all pasteurized dairy products will be free of salmonellae. However, use of raw eggs (a potential source of *S. enteritidis*) in homemade ice cream is strongly discouraged as evidenced from a series of recent outbreaks.

K. Staphylococcal Poisoning

A classical foodborne intoxication, staphylococcal poisoning results from ingesting a preformed, heat-stable toxin (termed enterotoxin) produced by the bacterium *S. aureus*. Although reports of cases resembling present-day staphylococcal poisoning date back to 1830, the organism was not observed microscopically until the 1870s (Bergdoll, 1979). Ogston coined the term "staphylococcus" (from the Greek words *staphyle*, bunch of grapes, and *coccus*, a grain or berry) to describe this organism in 1881, with Rosenbach proposing the genus *Staphylococcus* and the species *S. aureus* 3 years later. Known to cause skin infections during the 1870s, staphylococci were not associated with foodborne illness until 1884 when Vaughan and Sternberg recovered the organism from cheddar cheese linked to approximately 300 cases of food poisoning in Michigan (Hendricks et al., 1959). In 1914, the relationship between staphylococcal food poisoning and the toxin produced by *S. aureus* was established by Barber using human volunteers during a milk-borne outbreak in the Phillipines. These findings were later confirmed by Dack et al. (1930) using sterile culture filtrates, with the first of 10 known enterotoxins being purified during the late 1950s (Bergdoll et al., 1959).

Dairy products are well-known vehicles of staphylococcal poisoning, with cheese and raw milk linked to outbreaks before the turn of the century (Bergdoll, 1979). Following a marked decrease in the incidence of milk-borne typhoid and scarlet fever, staphylococcal poisoning emerged as the major milk-borne illness by the late 1930s, accounting for 26%, 50%, and 30% of all milk-borne diseases reported in the United States during the 1940s, 1950s, and 1960s, respectively (Bryan, 1983). These cases of staphylococcal poisoning involved various dairy products including raw milk, pasteurized milk, cheese, ice cream, butter, and nonfat dry milk. Staphylococcal poisoning has been most commonly traced to nondairy foods (e.g., ham, cream-filled pastries), with improvements in milk pasteurization and dairy sanitation standards making dairy-related outbreaks rare in the United States (Headrick et al., 1996), England (Galbraith et al., 1982), and most other industrialized countries.

1. General Characteristics

In the family Micrococcaceae, the genus *Staphylococcus* includes 32 species of facultatively anaerobic, nonmotile, small gram-positive cocci, most of which are catalase-positive and oxidase-negative (Kloos and Bannerman, 1995; Kloos and Schleifer, 1986). When viewed microscopically, the staphylococci appear in pairs, short chains, tetrads, and grapelike clusters, with the latter arrangement being most evident in cultures grown on solid media. Although 15 *Staphylococcus* spp. are of varying clinical importance in humans, *S. aureus* clearly dominates as the primary human pathogen, being responsible for a wide range of cutaneous and life-threatening systemic infections in addition to toxic shock syndrome and staphylococcal food poisoning.

On nonselective media more than 90% of *S. aureus* (*aureus*: Latin for golden) strains produce pigmented colonies ranging from cream yellow to orange (Kloos and Bannerman, 1995; Kloos and Schleifer, 1986). All isolates grow in common laboratory media at 10°C to 45°C (optimum: 30°C to 37°C) and at pH 4.2 to 9.3 (optimum: pH 7.0 to 7.5). Although a few strains can grow at temperatures as low as 6.7°C (Angelotti et al., 1961), production of enterotoxin is typically limited to temperatures above 15°C and pH values above 5.0. *S. aureus* growth and toxin production are generally poor in the presence of competing microflora. Unlike most other foodborne pathogens, *S. aureus* grows at a_w 0.84 (Lee et al., 1981) and in the presence of up to 15% NaCl (Bergdoll, 1989), with enterotoxin produced at a_w values greater than 0.86. Production of several key enzymes including coagulase, thermonuclease, and β-hemolysin is used almost universally to differentiate *S. aureus* from other staphylococci, with sensitivity to lysostaphin and anaerobic utilization of glucose and mannitol also being helpful. However, attempts to associate enterotoxin production in *S. aureus* with specific biochemical properties and phage types have generally failed. Consequently, confirmation of the toxin by serological or other means provides the only proof that the particular strain is enterotoxigenic.

Ten serologically distinct, enterotoxigenic proteins known as enterotoxin types A, B, C_1, C_2, C_3, D, E, F, G, and H are recognized in *S. aureus* (Bergdoll, 1989; Pereira et al., 1996; Su and Wong, 1995), with some strains producing two or three types of enterotoxin (Lopes et al., 1993). Classified as relatively low molecular weight, single-chain polypeptides, these plasmid or chromosomally linked extracellular enterotoxins are resistant to most proteolytic enzymes and a pH of 2.0, which allows their passage into the gastrointestinal tract without loss of activity. Although *S. aureus* is readily destroyed in milk during pasteurization, the staphylococcal enterotoxins are relatively heat stable and are not easily inactivated in foods during cooking. Enterotoxin production is not limited to *S. aureus*, with 10 coagulase-negative and two coagulase-positive species of staphylococci (*Staphylococcus hyicus* and *Staphylococcus intermedius*) also known to contain enterotoxigenic strains (Bergdoll, 1989). However, other than one recent butter-related outbreak traced to an enterotoxigenic strain of *S. intermedius*, all remaining reports of staphylococcal food poisoning have been confined to *S. aureus* (Khambaty et al., 1994).

2. Isolation and Detection Methods

The significance of finding *S. aureus* in foods suspected of causing staphylococcal poisoning should be interpreted with caution. Although foods must typically contain at least 10^6 enterotoxigenic *S. aureus* cfu/g to induce illness, small numbers of *S. aureus* present in thermally processed foods may represent the survivors of very large populations. Consequently, staphylococcal poisoning can only be verified by isolating enterotoxigenic staphylococci from the food or demonstrating the presence of enterotoxin in the food.

In dairy-related outbreaks of staphylococcal poisoning, samples are normally surface plated on Baird-Parker agar (BPA) (Bennett and Lancette, 1995; Flowers et al., 1992a; Lancette and Tatini, 1992). Following 48 hours of incubation at 35°C, presumptive *S. aureus* colonies appear gray to black from reduction of tellurite, with lipolytic strains surrounded by an opaque halo from hydrolysis of egg yolk. A three-tube most probable number method using trypticase soy broth containing 10% NaCl is recommended for samples likely to contain either low numbers of *S. aureus* or high levels of competing background flora. After 48 hours of incubation at 35°C, tubes showing growth are streaked to plates of BPA, which are incubated and examined as just described. Presumptive *S. aureus* isolates on BPA are then tested for coagulase activity using either the standard rabbit plasma test or a rapid latex agglutination assay kit. Coagulase-positive strains should be confirmed as *S. aureus* based on results from one of the commercially available rapid test kits or a series of standard biochemical tests, which includes catalase and thermonuclease production, sensitivity to lysostaphin, and anaerobic utilization of glucose and mannitol. Because multiple enterotoxigenic strains of *S. aureus* are frequently encountered in foods, specialized strain-specific typing techniques such as phage typing, plasmid analysis, antibiotic susceptibility pattern, restriction enzyme analysis, and pulsed-field gel electrophoresis are often necessary to clearly identify the source of intoxication (Khambaty et al., 1994).

Identifying enterotoxigenic strains of *S. aureus* in foods has traditionally involved the use of specific monoclonal or polyclonal antibodies, which react with antigenically distinct antigens. Isolates are specially cultured for enterotoxin production using either the membrane-over agar, sac culture, or semisolid agar method, the last of which is AOAC approved and recommended by the FDA (Bennett and Lancette, 1995; Flowers et al., 1992a; Lancette and Tatini, 1992). Two traditional serological methods, namely the AOAC-approved microslide method (the standard method) and the optimum sensitivity plate can be used to detect enterotoxin. However, several highly sensitive and rapid methods including latex agglutination, enzyme-linked immunosorbent assays, radio-immunoassays, and DNA hybridization assays are also available for identifying enterotoxins in culture fluids.

Detecting enterotoxin in suspect foods is complicated by the minute amounts of toxin that may be present (Bennett and Lancette, 1995; Flowers et al., 1992a; Lancette and Tatini, 1992). If the standard microslide method is to be used, the toxin must first be extracted from 100 g of food and then concentrated to 0.2 mL in a long and complicated procedure. However, use of the aforementioned rapid assays, which possess greater enterotoxin sensitivity, greatly shortens and simplifies sample preparation.

3. Clinical Manifestations

Staphylococcal food poisoning is a severe foodborne intoxication of short duration. Symptoms normally develop 1 to 6 hours after ingestion of food containing enterotoxin, with nausea, vomiting, diarrhea, abdominal cramps, and mild leg cramps occurring most commonly (Bergdoll, 1989). During the acute stage of illness, individuals may also experience brief headaches, cold sweats, rapid pulse, slight fluctuations in body temperature, and various degrees of prostration and dehydration, all of which depend on the sensitivity of the individual and the amount of toxin ingested. Early studies with human volunteers and results from a recent outbreak involving chocolate milk have both confirmed that ingesting as little as 1×10^{-7} g of enterotoxin is sufficient to induce the aforementioned symptoms in susceptible individuals (Evenson et al., 1988). Acute symptoms typically last only 1 to 8 hours, with the patient fully recovering within 1 to 2 days. Consequently, most outbreaks are never reported or investigated. Hospitalization is seldom required. However, intravenous therapy and fluid replacement may be necessary in severe cases of dehydration and collapse. Complications from staphylococcal poisoning are seldom encountered and are limited to a few reports of acute gastritis and pseudomembranous enterocolitis. Although highly unusual, several fatalities have been recorded in the early literature.

4. Outbreaks

Milk and dairy products have been associated with staphylococcal poisoning in the United States for more than 100 years, with numerous accounts of illness documented before 1950. According to Stone (1943), at least 23 outbreaks of staphylococcal poisoning ($\geq$1332 cases) were traced to dairy products during the 28-year period from 1914 to 1942. Raw milk was most frequently implicated (7 outbreaks/500 cases) followed by ice cream (5 outbreaks/360 cases), hollandaise sauce (5 outbreaks/90 cases), butter (2 outbreaks/150 cases), evaporated milk (1 outbreak/90 cases), pasteurized milk (1 outbreak/29 cases), and Jack cheese (1 outbreak/5 cases). Such epidemics were particularly common during the 1940s when staphylococcal poisoning was responsible for 22 of 49 (44.9%) milk-borne outbreaks reported during 1945, 1946, and 1947. Although staphylococcal poisoning is not generally considered a fatal illness, several deaths did occur among individuals who had consumed raw goat's milk (Weed et al., 1943) and butter (Fanning, 1935). Most of the raw milk outbreaks were traced to staphylococcal mastitis in dairy cows, with temperature abuse of the milk cited as a contributing factor (Stone, 1943). Postpasteurization contamination, poor product handling, and transmission by human carriers were most often responsible for outbreaks involving ice cream (Geiger et al., 1935), butter (Stone, 1943), and pasteurized milk (Caudil and Meyer, 1943; Hackler, 1939).

Cheese and nonfat dry milk emerged as major vehicles for staphylococcal poisoning after World War II. According to Hendricks et al. (1959), 18 outbreaks involving at least 475 cases of illness were traced to cheese from 1944 to 1958. The three largest outbreaks were linked to cheddar cheese (200 cases), cheese sauce (80 cases), and Colby cheese (60 cases) (Allen and Stovall, 1960). In the latter two outbreaks, the cheese was prepared from raw milk containing *S. aureus* and the identical phage type was identified in raw milk from dairy herds supplying the cheese factory. According to Bryan (1983), nonfat dry milk caused 27 outbreaks of staphylococcal poisoning in 1956; 19 of these outbreaks affected 775 school children in Puerto Rico (Armijo et al., 1957). Although the incriminated milk was free of *S. aureus*, toxin was demonstrated using human volunteers, thereby suggesting that the organism grew and produced enterotoxin in the milk before spray drying.

During the 1960s, staphylococcal poisoning accounted for 30% of all dairy-related illnesses in the United States (Bryan, 1983; Woodward et al., 1970). The largest documented outbreak during this period involved 42 cases and was traced to cheddar and Monterey cheese prepared with a contaminated starter culture (D'Aoust, 1989; Zehren and Zehren, 1968a, 1968b). Using the aforementioned microslide method, which was developed in response to this outbreak, cheese from 59 of 2112 vats was shown to contain an average of 12 μg of enterotoxin A/100 g.

Given increased monitoring programs for mastitis in dairy cattle coupled with routine milk pasteurization and heightened attention to dairy sanitation, enterotoxigenic staphylococci are now responsible for less than or equal to 5% of all milk-borne disease outbreaks (Bryan, 1983; Holmberg and Blake, 1984), with only four major outbreaks reported in the United States since 1970. Even though it is an unusual vehicle for any foodborne illness because of the small amounts typically consumed, butter products were responsible for three of these outbreaks (Table 9) with an enterotoxigenic strain of *S. intermedius* identified as the causative agent in the most recent outbreak. In the remaining epidemic (Evenson et al., 1988), more than 850 school children in Kentucky became ill after consuming half pints of pasteurized 2% chocolate milk containing extremely low levels of enterotoxin A.

Reports of dairy-related staphylococcal poisoning are not limited to the United States. Between 1951 and 1970, a total of 30 dairy-related outbreaks involving raw milk (20 outbreaks/590 cases/2 deaths), dried milk (2 outbreaks/ 1100 cases), canned milk (1 outbreak/70 cases), cream (6 outbreaks/131 cases), and ice cream (1 outbreak/8 cases) were documented in England and Wales, with an additional 23 milk and 5 cheese-related outbreaks identified between 1969 and 1990 (Galbraith et al., 1982; Parry, 1966; Steede and Smith, 1954; Wieneke et al., 1993). According to Maguire (1993), 18 of 31 cheese-related outbreaks of illness reported in England and Wales from 1951 to 1989 were the result of staphylococcal poisoning. Although not completely eliminated, dairy-related outbreaks of staphylococcal poisoning remain an unusual occurrence in the United States, England, and most other modern industrialized countries, with recent outbreaks in less developed countries most often traced to raw milk and dairy products prepared under less than ideal conditions.

5. Occurrence and Survival in Dairy Products

Staphylococci are frequent contaminants of raw milk, with *S. aureus* being widely recognized as a common cause of clinical and subclinical mastitis in dairy cattle, sheep, and goats. The mammary gland represents an important reservoir for *S. aureus*, with up to 15% and 83% of raw milk samples from mastitic dairy cattle (Garcia et al., 1980; Olson et al., 1970) and sheep (Guitierrez et al., 1982), respectively, harboring enterotoxigenic strains. According to surveys conducted in Brazil (dos Santos et al., 1981) and Trinidad (Adesiyun, 1994), *S. aureus* was present in 47% and 94%, respectively, of the raw milk samples at populations typically ranging between 10^5 and 10^6 cfu/mL. In the latter study, 9 of 117 *S. aureus* isolates produced enterotoxins A, B, or D. Growth and enterotoxin production by *S. aureus* in fluid milk are strain dependent and strongly influenced

TABLE 9 Outbreaks of Dairy-Related Staphylococcal Poisoning Reported Worldwide Since 1970

Location	Year	Product	Number of Cases	Toxin Type	Reference
United States					
Alabama	1970	Whipped butter	>26	A	Wolf et al. (1970)
Midwest	1977	Whipped butter	>100	A	Francis et al. (1977)
Kentucky	1985	Chocolate milk	>860	NR	Lecos (1986)
Southwest	1991	Butter-blend spread	>265	A[a]	Khambaty et al. (1994)
Foreign					
Canada	1977	Emmental cheese	15	B	Todd et al. (1981)
England	1983	Unspecified cheese	30	NR	Barrett (1986)
France	1983	Sheep's milk cheese	20	NR	Sharp (1987)
Scotland	1984/1985	Sheep's milk cheese	28	A	Bone et al. (1989)
Czechoslovakia	1986	Ice cream	>16	A	Kristufkova and Simkovicova (1988)
Egypt	1986	Nonfat dry milk	>21	A, B	El-Dairouty (1989)
Israel	1987	Goat's milk	3	B	Gross et al. (1988)
Brazil	1987	Minas-type cheese	NR	A, B, D, E	Sabioni et al. (1988)
Brazil	1994	Minas-type cheese	7	H	Pereira et al. (1996)

[a]*S. intermedius*

Abbreviations: NR, not reported.

by incubation temperature and initial microbial load. Although *S. aureus* is unable to multiply in naturally contaminated raw milk during refrigerated storage, Clark and Nelson (1961) reported that *S. aureus* populations increased as much as 1000-fold when raw milk was held at 10°C for 7 days. Even though it is readily inactivated during high-temperature, short-time and vat pasteurization (Zottola et al., 1969), *S. aureus* can enter such products as a postpasteurization contaminant as evidenced by the aforementioned outbreaks and a survey from Brazil (dos Santos et al., 1981) in which 6% of pasteurized milk samples harbored *S. aureus* at levels of 10^2 to 10^4 cfu/mL. In the absence of a large background flora, *S. aureus* growth is enhanced with enterotoxin detectable in inoculated samples of pasteurized whole milk, skim milk, half and half, and cream after 18 to 24 hours of incubation at 37°C (Halpin-Dohnalek and Marth, 1989b, 1989c; Ikram and Luedecke, 1977; Minor and Marth, 1972; Varadaraj and Nambudripad, 1983). Decreasing the storage temperature to 22°C to 25°C decreased *S. aureus* growth with all four products being nontoxic after 16 to 24 hours. More than 2 days of incubation were required to detect enterotoxin in half and half and in cream.

Large numbers of *S. aureus* are seldom found in ice cream (Massa et al., 1989), nonfat dry milk (Chopin et al., 1978), or butter (Minor and Marth, 1972) because product composition and storage conditions severely limit growth. However, enterotoxin can persist for several years in nonfat dry milk prepared from contaminated fluid milk (Chopin et al., 1978), with staphylococcal enterotoxin also remaining fully active in ice cream during 7 months of frozen storage (Gogov et al., 1984). In butter prepared from inoculated cream (Minor and Marth, 1972) and whey cream (Halpin-Dohnalek and Marth, 1989a), *S. aureus* populations seldom increased more than 100-fold, with numbers more often remaining stable or decreasing during 2 weeks of storage at temperatures ranging from 4°C to 30°C. Whereas enterotoxin production is clearly minimal under these conditions, Minor and Marth (1972) reported that, when cream was inoculated with *S. aureus*, incubated at 37°C for 24 hours, and then churned into butter, the finished product contained at least 1 µg of enterotoxin/100 g or approximately 10% of the enterotoxin originally present in the cream. Because 0.1 µg of enterotoxin can reportedly induce symptoms of staphylococcal poisoning (Eversen et al., 1988), ingesting such butter does pose a potential health risk as demonstrated by the recent butter-related outbreaks.

Enterotoxigenic staphylococci are occasionally found in cheese as evidenced by the aforementioned outbreaks. In several early surveys, 12% to 20% of cheddar cheese sold in the United States contained potential enterotoxigenic strains of *S. aureus*, sometimes at levels exceeding 200,000 cfu/g, with raw milk cheeses being contaminated most often (Donnelly et al., 1964; Mickelsen et al., 1962). However, stricter measures for controlling and preventing staphylococcal mastitis in dairy cattle have sharply reduced these contamination rates over the past 20 years, with less than or equal to 2% of samples tested in the United States (Bowen and Henning, 1994; Khayat et al., 1988) and Canada (Brodsky, 1984a, 1984b; Warburton et al., 1986) containing *S. aureus* populations exceeding the maximum allowable level (Canadian) of 1000 cfu/g. However, this pathogen is still commonly found in certain raw milk cheeses manufactured abroad (Abbar and Mohammed, 1986; Ocando et al., 1991; Sanchez-Rey, 1993).

Starter culture growth and activity have a pronounced inhibitory effect on the proliferation of *S. aureus* during cheese making. When cheddar cheese was prepared from pasteurized milk inoculated to contain less than 1000 enterotoxigenic *S. aureus* cfu/mL, Koenig and Marth (1982) found that populations increased approximately 1000- and 10,000-fold using a 1.0% and 0.5% starter culture inoculum, respectively. An initial 10-fold increase resulted from physical entrapment of *S. aureus* in the curd with subsequent growth generally ceasing within 8 hours at a pH less than or equal to 5.3. During 8 weeks of ripening at 4°C, *S. aureus* levels decreased 100- to 1000-fold in cheese prepared without salt; whereas populations remained relatively stable in cheese containing 1% to 2% NaCl because of the adverse effects of salt on less salt-tolerant background flora. Nevertheless, virtually all 8-week-old cheeses were positive for enterotoxin, with the highest toxin levels being recorded in high-salt cheeses ripened at 10°C. These findings are consistent with those of Ibrahim et al. (1981a, 1981b) who also reported that, in the event of starter culture failure, *S. aureus* growth and enterotoxin production can be minimized by eliminating salt and limiting exposure of the cheese to ambient temperatures during pressing. Similar behavior of *S. aureus* has been reported during manufacture and storage of a wide variety of experimentally produced cheeses including Monterey (Eckner et al., 1991), brick (Tatini et al., 1973), Swiss (Tatini et al., 1973), Gouda (Stadhouders et al., 1978), Brazilian Minas (dos Santos and Genigeorigs, 1981), Sudanese soft-brined cheese (Khalid and Harrigan, 1984), Spanish Burgos (Nunez et al., 1986; Otero et al., 1988), Spanish Manchego (Gomez-Lucia et al., 1986, 1992), Spanish goat (Mor-Mur et al., 1992), Egyptian Ras (Naguib et al., 1979), and Egyptian Domiati cheese (Ahmed et al., 1983b; Helmy et al., 1975). However, in several other studies involving mozzarella (Tatini et al., 1973), blue (Tatini et al., 1973), Italian Montasio (Stecchini et al., 1991), no enterotoxins were detected even though *S. aureus* grew to populations of more than 10^7 cfu/g during cheese making.

6. Prevention

Given that *S. aureus* is ubiquitous within the farm environment and carried by approximately half of the human population, many dairy products contain low levels of enterotoxigenic staphylococci. However, growth and enterotoxin production are both easily prevented by proper refrigeration, with temperature abuse above 10°C and poor starter culture activity during fermentation being most often cited as contributing factors in dairy-related outbreaks of staphylococcal poisoning. Increased recognition of staphylococcal mastitis in dairy cattle, coupled with improvements in milk handling, cooling, and pasteurization practices, has made dairy-related outbreaks of staphylococcal food poisoning an uncommon occurrence in the United States and most other industrialized countries. However, such outbreaks have been observed in less developed countries, with raw milk cheeses being implicated most often. Hence, consumption of raw milk dairy products should be avoided.

L. Yersiniosis

The genus *Yersinia* formed in 1944 and named after the French bacteriologist Yersin who isolated the plague bacillus in 1894, contains 11 different species, three of which are unquestionably human pathogens (Gray, 1995; Scheimann, 1989). *Yersinia pestis*, the causative agent of bubonic plague ("Black Death"), is spread by the bite of infected rat fleas and has ravaged mankind throughout recorded history. First identified in 1883, *Yersinia pseudotuberculosis* is biochemically similar to *Y. pestis*. Most commonly found in rats and birds, *Y. pseudotuberculosis* occasionally infects humans, causing septicemia, acute gastroenteritis and "pseudoappendicitis," with internal lesions resembling those observed during intestinal tuberculosis (Christie and Corbel, 1990). However, supporting evidence for *Y. pseudotuberculosis* as a foodborne pathogen is limited to two reports (Jones et al., 1982; Prober et al., 1979) in which the organism was detected in milk from mastitic goats. *Yersinia enterocolitica*, the primary cause of *Yersinia*-related foodborne gastroenteritis, hereafter termed yersiniosis, was first identified in a human facial lesion in the United States by Schleifstein and Coleman (1939). The name of the organism changed from *Bacterium enterocolitica* to *Pasteurella* X and finally to *Y. enterocolitica* in 1964. However, this pathogen was not widely recognized as a common cause of foodborne gastroenteritis until the 1970s (Schiemann, 1989), with most cases being linked to pork because hogs are the major reservoir for human pathogenic strains. Since 1972, three outbreaks in the United States and one outbreak in Canada were traced to consumption of milk products with more than 500 people being affected. Although readily capable of growing at refrigeration temperatures, *Y. enterocolitica* is generally regarded as an unusual cause of milk-borne illness because of the low incidence of human pathogenic strains in the raw milk supply and the high susceptibility of the organism to pasteurization.

1. General Characteristics

A species in the family Enterobacteriaceae, *Y. enterocolitica* is a gram-negative, non–spore-forming, sometimes encapsulated, facultatively anaerobic, rod-shaped bacterium that is motile at 25°C but not at 37°C by means of peritrichous flagella (Christie and Corbel, 1990; Farrag and Marth, 1992; Schiemann, 1989). This organism grows at 0°C to 45°C with best growth at 30°C to 37°C. However, multiplication in this latter temperature range is slower than for other enteric pathogens, including *E. coli* and *Salmonella*. Like *L. monocytogenes*, the ability of *Y. enterocolitica* to grow in pasteurized whole milk stored at 3°C (Stern et al., 1980) makes this organism a potential health threat in refrigerated dairy products. In addition, *Y. enterocolitica* also grows or survives at pH 4.6 to 9.6, with optimal growth occurring at pH 7.0 to 8.0. Both of these growth characteristics have been used to selectively isolate this organism from food samples. Fermentation of sucrose, cellobiose, and sorbose can be used to biochemically differentiate *Y. enterocolitica* from *Y. pestis* and *Y. pseudotuberculosis*.

Many *Y. enterocolitica* isolates recovered from food samples are avirulent and of no clinical importance. These nonpathogenic strains, which typically lack a virulence-carrying plasmid and two chromosomal genes encoding for cell invasion factors, abound in raw milk and must be differentiated from strains capable of causing disease (Schiemann, 1987). Eight additional biochemical tests can be used to subdivide *Y. enterocolitica* isolates into seven distinct biochemical types (biotypes), with biotypes 1B, 2, 3, 4, and 5 containing human pathogenic strains (Weagant et al., 1995). Alternatively, the presence or absence of somatic (O) antigens can be used to separate *Y. enterocolitica* isolates into 54 serotypes (Weagant et al., 1995), 12 serotypes of which contain human pathogenic strains. Serotypes most frequently encountered in human infections include O:3, O:5,27, O:9, O:8, O:13, and O:21 ("O-Tacoma"), with the last three serotypes predominating in North America (Schiemann, 1987).

2. Isolation and Detection Methods

Several different procedures can be used to recover yersiniae from dairy products (Schiemann and Wauters, 1992), with most of these methods exploiting the ability of the organism to grow at reduced temperatures and survive in an alkaline environment. In the standard procedure for dairy products (Flowers et al., 1992a; Weagant et al., 1995), the sample is enriched in peptone sorbitol bile broth. After 10 days of incubation at 10°C, a portion of the enrichment is treated with 0.5% potassium hydroxide to reduce the background flora and then is surface-plated on two selective plating media—MacConkey agar (MAC) and Cefsulodin-Irgasan-Novobiocin agar (CIN). However, when high levels of yersiniae are expected, it is recommended that samples be plated before beginning the 10-day enrichment step. All plates are examined for suspect colonies after 48 hours of incubation at 22°C to 26°C. Presumptive *Yersinia* colonies on MAC and CIN are confirmed as *Yersinia* spp. based on reactions in lysine arginine iron agar, urea agar, and bile esculin agar. Results from six additional biochemical tests are required to identify *Y. enterocolitica*, with eight further biochemical tests required to separate isolates into six different biotypes. Potentially virulent strains belonging to biotypes 1B, 2, 3, 4, and 5 need to be confirmed as pathogenic through either dye binding, specific gene probe, cell culture, or mouse inoculation assays. Serotyping is normally confined to isolates of epidemiological importance and conducted by only qualified reference laboratories.

3. Clinical Manifestations

Yersiniosis, the disease caused by infection with *Y. enterocolitica*, can assume many different forms depending on the strain and dose of the organism as well as the age and physical condition of the person infected (Christie and Corbel, 1990; Gray, 1995; Schiemann, 1989). The most frequent manifestation of yersiniosis is gastroenteritis, which primarily affects children younger than 7 years of age; infants in their first year of life are most susceptible. Symptoms that develop 12 to 72 hours after ingesting more than 10^9 organisms (D'Aoust, 1989) typically include a low fever, diarrhea, severe abdominal pain, and cramps, with nausea and vomiting being reported less frequently. Although this illness is normally self-limiting and of short duration, with symptoms subsiding after 1 to 3 days, intestinal complications have been reported. Severe abdominal pain in older children is often mistaken for appendicitis, and a normal or only mildly inflamed appendix is sometimes removed. Such individuals also normally exhibit enlarged mesenteric lymph nodes and acute terminal ileitis, sometimes with involvement of the colon. Although relatively rare, intestinal obstruction, gangrene of the small intestine, and peritonitis have been reported as additional complications.

Acute generalized septicemia, the second major manifestation of yersiniosis, occurs far less frequently and is most often seen in elderly patients suffering from severe underlying illnesses such as alcoholism, liver disease, hemolytic anemia, leukemia, and other immunosuppressive disorders (Christie and Corbel, 1990; Schiemann, 1989). However, several reports of septicemic infection also have involved healthy infants. Despite proper antibiotic therapy, the mortality rate for such infections is still more than 50%.

Secondary complications develop in approximately 5% of all yersiniosis patients with reactive arthritis and skin infections being reported most frequently (Christie and Corbel, 1990; Schiemann, 1989). Other complications include endocarditis, thyroid disorders, eye infections, glomerulonephritis, liver disease, respiratory infections, muscle abscesses, and osteomyelitis.

Isolation of *Y. enterocolitica* from stool samples or from normally sterile materials such as blood and various organ tissues provides definitive diagnosis, with serological tests offering another alternative (Christie and Corbel, 1990; Gray, 1995; Schiemann, 1989). Administration of antibiotics is contraindicated for uncomplicated cases of gastroenteritis. However, prompt and sustained antibiotic therapy with chloramphenicol or tetracyclines is essential if patients with septicemia and severe localized infections are to fully recover.

4. Outbreaks

Evidence supporting *Y. enterocolitica* as a potential milk-borne pathogen dates back to 1975 (Table 10) when raw milk was epidemiologically linked to 138 cases of yersiniosis in Canadian school children after a class field trip (deGrace et al., 1976). However, because clinical and milk isolates belonged to serotypes 0:5,27 and O:6,30, respectively, the incriminated raw milk could not be positively identified as the vehicle of infection (Kasatiya, 1976).

Any doubt regarding *Y. enterocolitica* as a milk-borne pathogen ended in autumn of 1975 when chocolate milk from a school cafeteria was linked to 217 cases of yersiniosis in upper New York State (Black et al., 1978). Sixteen of 36 children who required hospitalization had unnecessary appendectomies. Recovery of *Y. enterocolitica* O:8 from numerous patients and unopened containers of the incriminated milk confirmed pasteurized chocolate milk as the vehicle of infection. Subsequent investigation of the dairy factory suggested that the epidemic strain most likely entered the product when chocolate syrup was added to previously pasteurized milk and mixed by hand in an open vat.

Six years later, another large milk-related outbreak occurred at a New York State summer camp with gastroenteritis developing in 239 of 455 campers and staff members (Shayegani et al., 1983). Five of 7 victims requiring hospitalization had appendectomies before this epidemic was diagnosed as yersiniosis. Epidemiological findings suggested a common source for this outbreak, with *Y. enterocolitica* O:8 being recovered from more than half of the patients and eventually found in reconstituted powdered milk, a milk dispenser, and turkey chow mein. These findings and additional results from pathogenicity studies both supported the aforementioned foods as potential vehicles of infection in this outbreak, with isolation of the epidemic strain from 4 of 11 food handlers suggesting contamination during food preparation.

The largest and most unusual outbreak of milk-borne yersiniosis occurred during summer of 1982 in Tennessee, Arkansas, and Mississippi (Tacket et al., 1984). According to the report, 172 cases of *Y. enterocolitica* infection were culturally confirmed, with the total number of cases estimated at several thousand. Unlike the aforementioned outbreaks, infections in 14 individuals were confined to a sore throat and fever with no symptoms of gastroenteritis (Tacket

TABLE 10 Major Outbreaks of Milk-Borne Yersiniosis

Location	Year	Product	Number of Cases	Serotype	Reference
Montreal, Canada	1975	Raw milk	58	O:5,27	de Grace et al. (1976)
New York	1976	Chocolate milk	36	O:8	Black et al. (1978)
New York	1981	Powdered milk	239	O:8	Shayegani et al. (1983)
Tennessee, Arkansas, Mississippi	1982	Pasteurized milk	172	O:13	Tacket et al. (1984)
Vermont, New Hampshire	1995	Pasteurized milk	10	O:8	Ackers (1995)
England	1985	Pasteurized milk	36	O:10K and O:6,30	Greenwood and Hooper (1990)

et al., 1983). In addition, the epidemic strain belonged to serotype O:13, an unusual serotype previously recognized only in monkeys, and was resistant to many antibiotics. No yersiniae were found in the incriminated milk, which was properly pasteurized. However, after learning that the factory delivered unsold milk to a farm for feeding pigs, investigators recovered the epidemic strain from the bottom of several returned milk crates (Auliso et al., 1982). As a result of inadequate crate washing procedures, the tops of the milk bottles likely became contaminated when the crates were stacked. Given that *Y. enterocolitica* can survive on the outside of refrigerated milk cartons for at least 21 days (Stanfield et al., 1985), the organism likely entered the product during consumer handling and then grew to infectious levels during refrigerated storage.

During October 1995, pasteurized milk was also epidemiologically linked to 10 cases of yersiniosis in Vermont and New Hampshire, with the epidemic strain identified as *Y. enterocolitica* O:8 (Ackers, 1995). Investigators traced the milk to a single dairy processing facility in New Hampshire but were unable to recover the organism from pasteurized milk. Although the milk was packaged in bulk containers and glass bottles, only the latter milk was associated with illness. Given factory records indicating proper pasteurization, *Y. enterocolitica* presumably entered the milk during bottling, with the epidemic strain possibly coming from a dairy farm on which pigs were also raised.

Additional outbreaks of milk-borne yersiniosis are limited to three reports from England and Wales, two of which involved a total of five cases and were linked to pasteurized milk (Barrett, 1986; Barrett, 1989). In the remaining outbreak, gastroenteritis developed in 36 hospitalized children after they consumed pasteurized milk contaminated with *Y. enterocolitica* O:10K and O:6,30 (Greenwood and Hooper, 1990). The incriminated milk was delivered to the hospital in glass bottles from a single supplier. Although both epidemic strains were detected in the incoming raw milk supply, the milk was properly pasteurized, with additional thermal inactivation studies demonstrating complete destruction of yersiniae (Greenwood et al., 1990). After finding the bottle washing procedure to be unsatisfactory, investigators concluded that the epidemic strain most likely entered the milk as a postpasteurization contaminant.

5. Occurrence and Survival in Dairy Products

Domestic animals are widely recognized as fecal carriers of yersiniae, with pigs identified as the primary reservoir for pathogenic strains of *Y. enterocolitica* (Schiemann, 1989). According to Davey et al. (1983), 62 of 124 (50%) healthy dairy cows in Scotland were fecal shedders of yersiniae. However, only 3 of 74 (4%) *Y. enterocolitica* isolates belonged to serotypes associated with the aforementioned outbreaks of milk-borne yersiniosis.

Yersiniae are frequent contaminants of raw milk. Because this organism is not known to cause mastitis in dairy cattle, most contamination is thought to occur through contact with feces or polluted water (Schiemann, 1989). In two surveys from Wisconsin, Michigan, and Illinois (McManus and Lanier, 1987; Moustafa et al., 1983b). *Y. enterocolitica* was demonstrated in 12% and 48% of the raw milk supply. However, none of the isolates were virulent. Working in Canada, Schiemann and Toma (1978) detected *Y. enterocolitica* in 29 of 131 (22%) raw milk samples. In contrast to the American surveys, seven different serotypes were recovered, all of which were previously associated with cases of human yersiniosis in Canada. Elsewhere, *Y. enterocolitica* has been identified in 5.5% to 36.6% of raw milk samples analyzed in Brazil (dos Reis Tassinari, 1994; Tibana et al., 1987), Northern Ireland (Walker and Gilmour, 1986), and Morocco (Hamama et al., 1992), with this organism also recovered from 81.4% of raw milk samples tested in France (Vidon and Delmas, 1981). In addition, 5% of Spanish (Tornadijo et al., 1993) and 12.8% of Australian (Hughes and Jensen, 1980) raw goat's milk samples also harbored *Y. enterocolitica*. Even though very few raw milk samples from these surveys contained human pathogenic strains, *Y. enterocolitica* is still one of the most frequent raw milk contaminants of public health concern.

Compared with other milk-borne pathogens, *Y. enterocolitica* is relatively heat sensitive with current minimum high-temperature, short-time and vat pasteurization standards being sufficient to inactivate unusually high populations of clinically important strains in milk (D'Aoust et al., 1988; Francis et al., 1980; Hanna et al., 1977; Lovett et al., 1982; Toora et al., 1992). Consequently, the occasional presence of yersiniae in properly pasteurized dairy products is indicative of postpasteurization contamination. According to Archer (1988), *Yersinia* spp. were recovered from 10 of 351 (2.9%) pasteurized milk, 5 of 80 (6.3%) chocolate milk, and 1 of 232 (0.4%) ice cream samples, with these organisms absent from butter, cottage cheese, and nonfat dry milk. Although similar findings have been reported from Canada (Schiemann, 1978) with only 1 of 265 (0.4%) pasteurized dairy product samples positive for *Y. enterocolitica*, contamination rates as high as 6% have been reported from Northern Ireland (Walker and Gilmour, 1986) and Brazil (Tibana et al., 1987). Many of these isolates presumably were nonpathogenic serotypes.

As was true for *L. monocytogenes, Y. enterocolitica* also can grow in milk during refrigeration and thus pose a potential health hazard. When pasteurized milk was inoculated to contain 10 *Y. enterocolitica* cfu/mL and refrigerated at 4°C, Amin and Draughon (1987) found that the population doubled every 19 hours and reached 10^6 cfu/mL after 14 days of storage. Furthermore, *Y. enterocolitica* was able to readily compete with the natural background flora. Given these findings and an additional report indicating that *Y. enterocolitica* was present in 4.9% to 19.9% of environmental samples collected from dairy factory floors and coolers (Pritchard et al., 1995), special precautions are needed to minimize contamination and subsequent growth of this organism to potentially hazardous levels in fluid dairy products.

Yersiniae are seldom recovered from fermented dairy products. According to Brodsky (1984a), only 1 of 127 (0.8%) 60-day-old samples of Canadian raw milk cheddar cheese harbored *Y. enterocolitica*, with the isolate being nonpathogenic. In addition, Schiemann (1978) failed to recover *Y. enterocolitica* from 49 samples of Canadian-produced cheddar and Italian cheese. These findings, along with a lack of reported outbreaks, suggest that fermented dairy products manufactured under good sanitary conditions are generally safe. However, the risk of yersiniosis may be somewhat higher in less developed countries, with 4% to 5% of traditional Moroccan fermented milk products and raw milk cheeses containing *Y. enterocolitica* (Hamama et al., 1992) as well as 28.8% of feta-type cheeses produced in Turkey (Aytac and Ozbas, 1992) containing *Y. enterocolitica*.

Pathogenic strains of *Y. enterocolitica* can persist in fermented dairy products for varying lengths of time depending on the initial inoculum level, storage temperature, pH, salt content, and environmental conditions. When pasteurized milk for Colby cheese manufacture was inoculated to contain 10^2 to 10^3 *Y. enterocolitica* cfu/mL, populations increased 1000-fold during cheese making, with one strain surviving at least 8 weeks in cheese ripened at 3°C (Moustafa et al., 1983a). This organism also can proliferate on the surface of ripened Brie cheese during storage at 4°C to 20°C (Little and Knochel, 1994). Although *Y. enterocolitica* is unable to survive the cooking step in cottage cheese manufacture (Golden and Hou, 1996), contamination during packaging can lead to substantial growth, with yersiniae persisting throughout the normal shelf-life of the product (Sims et al., 1989). However, survival of *Y. enterocolitica* in yogurt prepared from inoculated milk is limited to 6 days or less depending on the rate of acid development and final pH (Mantis et al., 1982; Williams et al., 1996).

6. Prevention

Pasteurization readily destroys both pathogenic and nonpathogenic strains of yersiniae and, as such, provides the primary means of defense against milk-borne yersiniosis. However, given the high incidence of yersiniae in dairy processing facilities and the ability of *Y. enterocolitica* to reach hazardous levels in fluid dairy products during refrigerated storage, it is imperative that postpasteurization contamination be minimized. Whereas consumption of raw milk should again be avoided, any risks associated with fermented dairy products appear to be minimal.

IV. UNCOMMON AND SUSPECTED CONCERNS

During the early 1900s, reported milk-borne illnesses were confined to a small number of classical diseases, principally diphtheria, scarlet fever, tuberculosis, and typhoid, with the importance of other milk-borne pathogens such as *Salmonella* and *S. aureus* not being fully realized until the late 1940s. Subsequent improvements in microbial isolation and detection techniques coupled with refinements in investigative strategies for foodborne outbreaks later led to the identification of such organisms as *E. coli* O157:H7, *L. monocytogenes*, and *Y. enterocolitica* as prominent milk-borne pathogens during the 1980s. While the major public health concerns discussed in the preceding section easily account for more than 95% of all dairy-related illnesses of known cause, the list of "new" and "emerging" milk-borne pathogens continues to evolve. Consequently, this section briefly discusses 13 additional uncommon or suspected concerns of potential public health significance which represent milk-borne pathogens and toxins (e.g., *Citrobacter freundii, Corynebacterium ulcerans,* Johne's and Crohn's disease, mycotoxins, toxoplasmosis), long-known disease agents of infrequent illness (e.g., Haverhill fever, Q fever, shigellosis), emerging agents of milk-borne disease (e.g., histamine poisoning, *Streptococcus zooepidemicus*), and disease agents for which milk and dairy products can serve as potential vehicles of infection (e.g., cryptosporidiosis, infectious hepatitis, tick-borne encephalitis).

A. *Citrobacter freundii*

Classified among the gram-negative enterics, *C. freundii* is a well-recognized opportunistic pathogen that normally inhabits the gastrointestinal tract of both humans and animals (Stiles, 1989). However, gastroenteritis caused by *C. freundii* is typically confined to those strains that have acquired plasmids for enterotoxin and verotoxin production (Tschape et al., 1995), colonization factors, or other pathogenic mechanisms. Following an incubation period of 12 to 48 hours, symptoms of *C. freundii* gastroenteritis, in descending order of frequency, include diarrhea, abdominal pain, fever, chills, headache, vomiting, and nausea (Bryan, 1979), with spontaneous recovery occurring with 7 days (Stiles, 1989). Sporadic cases of milk-borne *C. freundii* gastroenteritis were first suspected during the 1940s (Edwards et al., 1948) with a subsequent outbreak involving 14 adults eventually being traced to milk (Sedlak, 1973). In 1983, three separate outbreaks of gastroenteritis affecting 45 people in Washington, DC, were linked to one particular brand of imported French Brie cheese (Levy et al., 1983). Despite extensive testing for routine foodborne pathogens, *C. freundii* was the only organism common to both the cheese and three victims, thus supporting possible involvement of *C. freundii* in this outbreak.

B. *Corynebacterium ulcerans*

First isolated from human throat lesions in 1927 (Hart, 1984), *C. ulcerans* is considered a variant of *C. diphtheriae* and is able to produce several toxins associated with *C. diphtheriae* and *Corynebacterium pseudotuberculosis* (Stiles, 1989). Cases of human illness have been reported only occasionally, with *C. ulcerans* producing pharyngitis of varying severity and, in a few instances, an illness that mimics diphtheria (Hart, 1984). Even though it is considered a highly unusual cause of mastitis in dairy cattle, *C. ulcerans* has been recovered from raw milk (Wilson and Richards, 1980), with ingestion of such milk accounting for most human infections (Hart, 1984; Meers, 1979). Hence, this illness can be classified as a zoonosis.

C. Cryptosporidiosis

Protozoan parasites in the genus *Cryptosporidium* are responsible for one of the most common, acute, self-limiting gastrointestinal infections in healthy individuals, with 30% to 35% of the United States population being seropositive for this organism (Smith, 1993). Wild and domestic animals including cows, sheep, and goats are also highly susceptible to such infections (Tzipori, 1983). The entire life cycle of *Cryptosporidium* occurs within a single host with greater than 10^8 infectious oocysts eventually shed in the feces and deposited in the environment to infect the next host by inhalation or ingestion. Infectivity of *Cryptosporidium* oocysts is best maintained under cool moist conditions (Smith, 1993). Exposure to temperatures less than 0°C or greater than 65°C inactivates the organism. Even though milk pasteurization also presumably results in lost infectivity, an extremely thick outer wall makes cryptosporidia oocysts highly resistant to most commonly used sanitizers, including chlorine.

After ingesting as few as 10 cryptosporidia oocysts, a short-term gastrointestinal illness characterized by profuse watery diarrhea, abdominal cramps, vomiting, mild fever, and headaches typically develops in infants and immunocompetent adults and symptoms resolve spontaneously within 1 to 2 weeks (Jokippi and Jokippi, 1986; Smith, 1993). However, a persistent choleralike diarrhea and other potentially life-threatening complications frequently develop in elderly and immunocompromised patients.

Evidence for cryptosporidiosis as a milk-borne disease is growing, with at least three outbreaks (43 cases) outside of the United States being epidemiologically linked to consumption of raw milk from cows (Casemore et al., 1986; Elsser et al., 1986) and goats (Anonymous, 1984c). More recently, ingestion of kefir (a fluid milk product prepared using a mixed lactic acid and alcoholic fermentation) was responsible for 13 cases of infant cryptosporidiosis, with *Cryptosporidium* oocysts detected in filtered milk samples from the factory

(Romonova et al., 1992). Thus far, dairy products have not been positively linked to any cases of cryptosporidiosis in the United States. However, in conjunction with a massive outbreak involving approximately 500,000 waterborne cases of cryptosporidiosis in Milwaukee, WI, several precautionary recalls were issued for cottage and other cheeses that may have been prepared using *Cryptosporidium*-contaminated water (Anonymous, 1993a).

D. Haverhill Fever

Streptobacillus moniliformis, the etiologic agent of both Haverhill and rat-bite fever, is a gram-negative, facultatively anaerobic, highly pleomorphic, rod-shaped bacterium (Ryan, 1986). This organism and the disease were first described in 1926 after 89 cases of febrile illness in Haverhill, MA, were attributed to raw milk consumption (Parker and Hudson, 1926). However, food-borne infections involving *S. moniliformis* remain rare, with most cases acquired from the bite of infected rats and termed rat-bite fever rather than Haverhill fever, which is foodborne (Stiles, 1989).

The onset of Haverhill fever is abrupt, with chills, headache, rash, and severe back and joint pain occurring 2 to 10 days after initial exposure (Ryan, 1986; Stiles, 1989). Various complications including arthritis in 50% of patients as well as endocarditis, pneumonia, brain abscesses, anemia, severe dehydration, and severe weight loss have been reported, particularly in children. Whereas administration of penicillin generally leads to full recovery, some of the afore-mentioned complications have been fatal.

Only one additional epidemic of milk-borne Haverhill fever has been reported since 1926. In this outbreak, as many as 130 children attending an English boarding school became ill in February 1983 after consuming raw milk from a local farm (Shanson et al., 1983). However, as in 1926, investigators were again unable to recover *S. moniliformis* from the incriminated milk. Because *S. moniliformis* grows poorly in milk (Parker and Hudson, 1926) and is readily inactivated during pasteurization (Stiles, 1989), milk-borne cases of Haverhill fever are likely to remain rare.

E. Histamine Poisoning

Certain strains of lactobacilli and lactococci found in raw milk and many cheeses possess the enzyme histidine carboxylase, which can convert unbound histidine to potentially toxic levels of histamine (Stratton et al.,1991). Whereas levels of free histidine are usually very low in fresh milk, histidine concentrations in aged cheeses such as cheddar and Swiss are often much higher from proteolysis of milk proteins during ripening (Hinz et al., 1956). Cheeses in which free histidine has been converted by certain lactic acid bacteria to greater than or equal to

100 mg histamine/100 g of cheese have been most frequently associated with histamine poisoning. However, histamine levels as low as 30 mg/100 g also have induced illness, with histamine toxicity being enhanced by several biogenic amines (e.g., tyramine, tryptamine) that potentiate histamine activity (Edwards and Sandine, 1981) and certain drugs (e.g., antihistamines, isoniazid) that inhibit histamine-metabolizing enzymes (Hui and Taylor, 1985; Stratton et al., 1991).

Biologically, histamine acts to contract smooth muscle within the intestine (Taylor, 1986) and dilate blood vessels (Stratton et al., 1991). Symptoms of histamine poisoning generally develop 30 minutes to 2 hours after ingesting cheese containing greater than or equal to 100 mg histamine/100 g and include abdominal cramps, diarrhea, nausea, and vomiting as well as hypotension, headache, palpitations, tingling, flushing, and burning sensations in the mouth (Stratton et al., 1991). Medical intervention is usually unnecessary, with most symptoms disappearing a few hours after onset.

Dairy-related outbreaks of histamine poisoning have been confined to aged cheeses, with a total of 52 cases thus far reported worldwide (Stratton et al., 1991). The first of these cases occurred in 1967 in The Netherlands and was traced to 2-year-old Gouda cheese containing 85 mg of histamine/100 g (Doeglas et al., 1967). In the United States, three separate outbreaks in Washington (38 cases) (Taylor, 1985), California (1 case) (Taylor, 1985), and New Hampshire (6 cases) (Taylor et al., 1982b) were documented from 1976 to 1980, with all cases linked to Swiss cheese containing more than 100 mg of histamine/100 g. Sumner et al. (1985) subsequently recovered a histamine-producing strain of *Lactobacillus buchneri* from Swiss cheese implicated in the New Hampshire outbreak. The remaining six cases of histamine poisoning involved Canadian cheddar (Kahana and Todd, 1981), French Cheshire (Uragoda and Lodha, 1979), and French Gruyere cheese (Taylor, 1985) consumed in the country of origin.

Lactic acid bacteria responsible for histamine production include various strains of *Lactobacillus acidophilus, Lactobacillus delbrueckii* ssp. *bulgaricus, Lactobacillus helveticus, Lactococcus lactis* ssp. *lactis* and propionibacteria (Stratton et al., 1991), all of which could potentially be used as cheese starter cultures, with such use being of obvious public health concern. These strains are assumed to be present in milk at the time of cheese making, with few such organisms thought to be postprocessing contaminants (Sumner et al., 1990). Consequently, aged cheeses prepared from raw or heat-treated milk typically contain higher levels of histamine and pose a greater public health threat than cheeses prepared from pasteurized milk.

F. Infectious Hepatitis

Hepatitis A, or infectious hepatitis, is a common infectious disease worldwide and the best known of the milk-related viral diseases, with sporadic outbreaks recorded in the United States since the 1940s (Bryan, 1983). Common source outbreaks are most often recognized in industrialized countries where this illness is rare because of natural immunity. Typical symptoms appearing 15 to 50 days after exposure via the fecal-oral route include jaundice, anorexia, and extreme malaise, with some individuals also experiencing abdominal pain, nausea, fever, and chills (Hirschmann, 1986). Infectious hepatitis is usually a mild illness with bed rest leading to complete recovery within a few weeks. Despite the general lack of serious complications, some individuals with more pronounced cases may complain of fatigue for several months. According to Cliver (1979), milk and dairy products were implicated in five outbreaks (599 cases) of infectious hepatitis, with one of these reports traced to the use of fecally contaminated water in a Czechoslovakian dairy processing facility (Raska et al., 1966). Two additional outbreaks involving ice cream (MacDonald and Griffin, 1983) and cheese (Bean et al., 1996) were also recorded in the United States during 1982 and 1990, respectively. Whereas the virus is only partially inactivated by pasteurization, complete destruction of the virus is ensured by normal chlorination (Hirschman, 1986).

G. Johne's and Crohn's Disease

M. paratuberculosis, a gram-positive, acid-fast bacillus, is the etiological agent of Johne's disease in dairy cattle, goats, and other ruminant animals (Collins et al., 1984; van den Heever, 1984). This economically devastating disease is characterized by a chronic granulomatous ileocolitis that eventually leads to diarrhea, weight loss, debilitation, and death (Benedictus et al., 1987; Chiodini et al., 1984). Fecal shedding of the organism at levels approaching 10^8 cfu/g leads to heavy contamination of the environment, which, in turn, helps perpetuate this disease. Although control programs have traditionally focused on minimizing consumption of contaminated feed by young calves, *M. pseudotuberculosis* is also shed in body fluids, including milk and colostrum, with as many as 35% of clinically infected cattle (Taylor et al., 1981) and 12% of asymptomatic carriers (Sweeny et al., 1992) yielding positive milk samples. Consequently, transmission of Johne's disease to calves via contaminated milk cannot be ignored.

Considerable interest has been generated concerning the possible association between Johne's disease in ruminant animals and human Crohn's disease, a nearly identical form of granulomatous ileocolitis often requiring surgical intervention (Graadal and Nygaard, 1994; Pounder, 1994). Strains of *M. paratuberculosis* similar to those identified in dairy cattle have been isolated from 20% to 38% of Crohn's disease patients (Chiodini and Hermon-Taylor, 1993) with the DNA of the organism also detected in 65% of 12.5% of tissue samples obtained from patients with and without confirmed Crohn's disease, respectively

(Sanderson et al., 1992). Several laboratory studies prompted by possible milk-borne transmission of Crohn's disease have concluded that heat treatments simulating vat and high-temperature, short-time pasteurization do not completely inactivate *M. paratuberculosis* in milk inoculated to contain 10^3 to 10^4 cfu/mL (Chiodini and Herman-Taylor, 1993; Grant et al., 1996). However, the public health significance of these findings and an additional report attesting to the presence of *M. paratuberculosis* DNA in 6% of retail pasteurized milk samples collected in England and Wales (Grant et al., 1996) will not be completely understood until the relationship between *M. paratuberculosis* and certain genetic and environmental factors is fully clarified.

H. Mycotoxins

Aflatoxin, a highly potent carcinogen produced by certain strains of *A. flavus, A. parasiticus* and *A. nemius*, is the primary mycotoxin of public health concern, as discussed previously. However, mycotoxin production is not limited to afla-toxigenic molds, with certain strains of *Alternaria, Aspergillus, Cladosporium, Fusarium, Geotrichum, Mucor,* and *Penicillum* isolated from cheese also being capable of synthesizing toxins (Scott, 1989). Several early studies demonstrated that 30% and 35% of *Penicillium* isolates from cheddar (Bullerman and Olivigni, 1974) and Swiss cheese (Bullerman, 1976), respectively, were toxic to chicken embryos, with such strains now known to produce a wide range of mycotoxins, including cyclopiazonic acid, citrinin, ochratoxin A, patulin, penicillic acid, and penitrem A, all of which are either nephrotoxic, neurotoxic, teratogenic, or car-cinogenic to laboratory animals (Scott, 1989). Cyclopiazonic acid is normally produced by *Penicillium camemberti* during ripening of Camembert cheese, and patulin, penicillic acid, mycophenolic acid, and roquefortine are synthesized by certain strains of *Penicillium roqueforti* used in manufacturing Roquefort cheese (Engel and Teuber, 1989). Although Bullerman and Olivigni (1974) identified only 6.6% of cheddar cheese molds as *Aspergillus* spp., nearly half of these strains were reportedly toxic to chick embryos. Certain cheese isolates of *Asper-gillus* have come to be recognized producers of aflatoxin as well as cyclopiazonic acid, β-nitropropionic acid, kojic acid, and sterigmatocystin, the last of which is carcinogenic and structurally related to aflatoxin (Scott, 1989). In addition, *Fusarium* spp. are well known producers of trichothecenes, zearalenone, and moniliformin (Ueno, 1985), with a few cheese strains of *Geotrichum* also produc-ing ergot alkaloids (El-Refai et al., 1970).

As was true for aflatoxins, the direct impact of these remaining toxins on human health is unknown. However, because many of these mycotoxins are marginally toxic and relatively unstable in cheese (Scott, 1989), any potential public health impact is presumed to be minimal.

I. Q Fever

Coxiella burnetti, a rickettsia-like obligate intracellular parasite that localizes and proliferates within cell vacuoles, is the etiological agent of Q ("Query") fever in humans (Baca and Paretsky, 1983). First observed in Australia in 1935, Q fever is now known to occur worldwide. Ticks and ruminant animals, including dairy cattle, sheep, and goats, are common asymptomatic carriers of *C. burnetti,* with most human cases being traced to dairy workers, farmers, and meat factory employees who work in close contact with animals (Wisniewski and Krumbiegel, 1970b).

Clinical symptoms of Q fever, which mimic viral influenza, generally occur 2 to 4 weeks after ingesting or inhaling *C. burnetti* and include an abrupt fever followed by malaise, anorexia, muscle pain, and intense headache (Turck, 1986). Even though many serious complications affecting the central nervous system, lungs, liver, and other internal organs have been reported, most patients fully recover in 2 to 4 weeks when given tetracycline or chloramphenicol.

Concern regarding Q fever as a milk-borne illness comes from the demonstrable presence of *C. burnetti* in milk from infected cows (Biberstein et al., 1974; Evans, 1956; Huebner and Bell, 1951; Wisniewski and Krumbiegel, 1979a) and goats (Fishbein and Raoult, 1992), with regular consumers of raw milk often displaying high antibody titers to *C. burnetti* (Stiles, 1989). Ingestion of raw milk has been directly linked to Q fever in the United States (Bryan, 1983) and England (Brown et al., 1968), with the latter outbreak involving 23 cases at a detention center. However, given the volume of raw milk consumed worldwide, reports of milk-borne Q fever are far fewer than would be expected. Furthermore, in one study in which contaminated raw milk was ingested by human volunteers, illness did not occur (Krumbiegel and Wisniewski, 1970). Several early studies also attest to the high thermal resistance of *C. burnetti* in milk, with the organism surviving 30 minutes of heating at 61.7°C (Huebner et al., 1949; Lennette et al., 1952). However, heating raw milk at 62.8°C for 30 minutes or 71.1°C for 15 seconds is sufficient to completely destroy *C. burnetti,* with these time-temperature pasteurization standards currently being required by law to prevent milk-borne Q fever.

J. Shigellosis

Outbreaks of bacillary dysentery, which resemble present-day shigellosis, date back to the time of Hippocrates (Wachsmuth and Morris, 1989). However, *Shigella* spp. were not recognized as the cause of this disease until the late 1800s. In the family Enterobacteriaceae, the genus *Shigella* contains four species— *Shigella dysenteriae, Shigella flexnori, Shigella boydii,* and *Shigella sonnei*—all of which are highly infectious and closely related to enterohemorrhagic strains of

E. coli. These organisms, which also produce a vero cytotoxin that is immunologically indistinguishable from *E. coli* O157:H7 Shiga-like toxin, are host-adapted to humans and other primates. However, shigellae are relatively fragile and unable to readily compete with other enteric flora. This disease is usually transmitted either person to person or by the fecal-oral route, with food and water often serving as vectors.

Shigellosis is normally an acute, self-limiting infection of the intestinal tract. Symptoms appearing 1 to 7 days after ingesting up to 100 organisms (D'Aoust, 1989) typically include fever, abdominal pain, and watery diarrhea, which can develop into a fulminating dysentery characterized by grossly bloody diarrhea, dehydration, chills, and toxemia (Kantor, 1986; Wachsmuth and Morris, 1989). Children and elderly patients may go into shock from excessive dehydration and further complications including seizures, pneumonia, hemolytic uremic syndrome, bacteremia, peripheral neuropathy, and Reiter's syndrome (urethritis, conjunctivitis and arthritis) may develop. Most individuals recover spontaneously within 2 weeks. Medical intervention is usually confined to replacement of fluids, with antibiotic therapy reserved for severe cases.

A few small sporadic outbreaks of milk-borne shigellosis were documented in the United States between 1920 and 1960, with this disease accounting for less than or equal to 4% of all reported milk-borne illnesses (Bryan, 1979, 1983). These outbreaks generally involved raw milk that was contaminated with *S. dysenteriae* by a human carrier and then held without refrigeration for several hours. However, in 1952, improperly pasteurized milk containing *S. sonnei* was linked to one particularly large outbreak in Tennessee involving 639 school children (Tucker et al., 1954). No additional cases of milk-borne shigellosis have been reported in the United States since the 1950s (Bryan, 1983), but one recent dairy-related outbreak in the former Soviet Union was traced to a milk processor's water supply that was contaminated with *S. sonnei* (Solodovnikov and Aleksandrovskaia, 1992). Other dairy products, including sour milk and white cheese, have been only rarely implicated in shigellosis. One notable outbreak did occur in 1982 in which French cheeses purchased at a Paris airport were responsible for at least 50 cases of *S. sonnei* infection subsequently reported in Scandinavia (Sharp, 1987); however, improved personal hygiene standards, pasteurization practices, and cold storage conditions serve to keep dairy-related shigellosis outbreaks as relatively rare.

K. *Streptococcus zooepidemicus*

Human infections caused by *S. zooepidemicus*, a β-hemolytic streptococcus belonging to Lancefield Group C, are generally uncommon (Stiles, 1989), with this pathogen a more frequent cause of animal infections and subacute or chronic mastitis in dairy cattle. Because most cases of human illness have been acquired through consumption of raw milk or contact with horses (Francis et al., 1993), *S. zooepidemicus* infections can be classified as another zoonosis. In humans, *S. zooepidemicus* produces mild flulike upper respiratory symptoms as well as more serious manifestations including glomerulonephritis, cervical lymphadenitis, pneumonia, septicemia, endocarditis, meningitis, septic arthritis, and cellulitis (Francis et al., 1993). Even though such infections are usually treatable with penicillin, some fatalities have been reported, particularly among elderly patients.

Since the 1960s, five *S. zooepidemicus* outbreaks involving more than 100 cases of illness have been linked to raw milk. In the first and largest of these outbreaks, 85 individuals in a small Romanian town became ill after ingesting improperly pasteurized milk, with *S. zooepidemicus* eventually being isolated from the incriminated milk and several asymptomatic workers at the dairy processing facility (Duca et al., 1969). Three subsequent outbreaks attributed to raw milk were reported in England. Two of these outbreaks were small and confined to family farms (Barnham et al., 1983; Ghoneim and Cook, 1980). The remaining outbreak, which involved 12 cases, including eight fatalities, was directly linked to retail raw milk, with *S. zooepidemicus* eventually traced to subclinical mastitis in the incriminated dairy herd (Edwards et al., 1988). Most recently, Francis et al. (1993) reported that three family members in Australia became ill shortly after ingesting milk from their own dairy herd. *S. zooepidemicus* isolates from family members and cow's milk were later proven identical by molecular subtyping, thereby confirming raw milk as the vehicle of infection. In 1983, 16 cases of *S. zooepidemicus* infection, including one fatality, also occurred among primarily elderly Hispanics living in New Mexico (Espinosa et al., 1983). However, unlike the previous outbreaks, illness was directly linked to fresh "queso blanco" cheese, which was illegally prepared from raw cow's milk on a small family farm and consumed without aging.

L. Tick-Borne Encephalitis

Tick-borne encephalitis is the primary zoonotic viral disease acquired through milk. Dairy animals in central and eastern Europe can become infected through tick bites and later shed the virus in their milk (Cliver, 1979). Although readily destroyed by pasteurization, the tick-borne encephalitis virus can remain infectious for many months in heat-treated milk and fermented dairy products, including cheese. In humans, a moderate fever and symptoms of encephalitis typically develop 7 to 14 days after ingesting the virus. During the mid-1970s, at least 17 cases of tick-borne encephalitis, including one fatality, were traced to raw milk consumed in the former Soviet Union (Vasenin et al., 1975) and Poland (Jezyna et al., 1976), with fresh sheep's milk cheese responsible for one additional cluster of cases in Slovakia (Gresikova et al., 1975).

M. Toxoplasmosis

A worldwide disease of humans and livestock, toxoplasmosis results from infection with the intracellular protozoan *Toxoplasma gondii* (Remington and McLeod, 1986). Ingesting *T. gondii* cysts or oocysts gives rise to rapid multiplication, with the organism eventually transported via the lymphatic and blood system to all body organs and tissues. Major sites of infection include the lymph nodes (lymphadenopathy), eyes (choreoretinitis), central nervous system (meningoencephalitis), lungs (pneumonia), heart (myocarditis), and kidneys (nephritis). Complications including mental retardation, blindness, and deafness have been reported, particularly in infants. The duration of treatment is determined by clinical severity of the illness, with 4 to 6 weeks of drug therapy typically being required. Given that milk from infected animals can harbor *T. gondii* and transmit this disease to their offspring, ingestion of raw milk also can potentially spread toxoplasmosis to humans, as evidenced by two incidents traced to raw goat's milk (Riemann et al., 1975; Sacks et al., 1982), with the latter involving a family cluster of 10 cases in northern California.

REFERENCES

Abbar F, Kaddar HKh. Bacteriological studies on Iraqi milk products. J Appl Bacteriol 71:497, 1991.

Abbar FM, Mohammed MT. Identification of some enterotoxigenic strains of staphylococci from locally processed cheese. Food Microbiol 3:33, 1986.

Ackers M. Summary report: outbreak of *Yersinia enterocolitica* infections in the Upper Valley of Vermont and New Hampshire (letter). Atlanta, GA: Foodborne and Diarrheal Diseases Branch, Centers for Disease Control and Prevention, 1995.

Adams D, Well S, Brown EF, Gregorio S, Townsend L, Skaggs JW, Hinds MW. Salmonellosis from inadequately pasteurized milk—Kentucky. Morbid Mortal Weekly Rep 33:505, 1984.

Adams JB. Symposium: impact of drug control measures—results of drug screening from a producer's view. J Dairy Sci 77:1933, 1994.

Adesiyun AA. Bacteriological quality and associated public health risk of pre-processed bovine milk in Trinidad. Intern J Food Microbiol 21:253, 1994.

Ahmed AA-H, Moustafa MK, Marth EH. Incidence of *Bacillus cereus* in milk and some milk products. J Food Prot 46:126, 1983a.

Ahmed AA-H, Moustafa MK, Marth EH. Growth and survival of *Staphylococcus aureus* in Egyptian Domiati cheese. J Food Prot 46:412, 1983b.

Allcroft R, Carnaghan RBA. Groundnut toxicity: an examination for toxin in human food products from animals fed toxic groundnut meal. Vet Rec 75:259, 1963.

Allen RF, Baer LS. Outbreak of septic sore throat due to reconstituted powdered milk. JAMA 124:1191, 1944.

Allen VD, Stovall WD. Laboratory aspects of staphylococcal food poisoning from Colby cheese. J Milk Food Technol 23:271, 1960.

Al-Rajab WA, Al-Dabbagh WY, Al-Zahawi SM. Occurrence of *Salmonella* in Iraqi milk products. J Food Prot 49:282, 1986.

Amin MK, Draughon FA. Growth characteristics of *Yersinia enterocolitica* in pasteurized milk. J Food Prot 50:849, 1987.

Andersson Y, Genback H, Lindh A. A milkborne outbreak in Sweden in the summer of 1985. In: Proceedings of the 2nd World Congress Foodborne Infections and Intoxications. Berlin, 1986, pp 290–293.

Andrews WH, June GA, Sherrod PA, Hammack TS, Amaguana RM. *Salmonella.* In: FDA Bacteriological Analytical Manual. 8th ed. Gaithersburg, MD: AOAC International, 1995, p 24.01.

Angelotti R, Foster MJ, Lewis KH. Time-temperature effects on salmonellae and staphylococci in foods. Am J Pub Health 51:76, 1961.

Anonymous. Tuberculosis of bovine origin in Great Britain. BMJ 1:618, 1932.

Anonymous. Brucellosis 1971 annual summary. Atlanta, GA: Centers for Disease Control, 1972.

Anonymous. *E. coli* gastroenteritis from food. BMJ 1:911, 1976.

Anonymous. Milk-borne outbreaks of *Salmonella typhimurium.* BMJ 1:1606, 1977.

Anonymous. *Campylobacter* enteritis—Colorado. Morbid Mortal Weekly Rep 27:226, 1978.

Anonymous. Raw-milk-associated illness-Oregon, California. Morbid Mortal Weekly Rep 30:90, 1981.

Anonymous. Campylobacteriosis associated with raw milk consumption—Pennsylvania. Morbid Mortal Weekly Rep 32:337, 1983.

Anonymous. *Campylobacter* outbreak associated with certified raw milk products—California. Morbid Mortal Weekly Rep 33:562, 1984a.

Anonymous. Chronic diarrhea associated with raw milk consumption—Minnesota. Morbid Mortal Weekly Rep 33:521, 1984b.

Anonymous. Cryptosporidiosis surveillance. Weekly Epidemiol Rec 50(10):72, 1984c.

Anonymous. *Campylobacter* outbreak associated with raw milk provided on a dairy tour—California. Morbid Mortal Weekly Rep 35:311, 1986.

Anonymous. International Committee on Systematic Bacteriology: Subcommittee on Taxonomy of *Brucella.* Minutes of the meeting—September 5, 1986. Int J Syst Bacteriol 38:450, 1988a.

Anonymous. *Campylobacter* isolates in the United States, 1982–1986. Morbid Mortal Weekly Rep 37(SS-2):1, 1988b.

Anonymous. FDA's Regulation of Animal Drug Residues in Milk—Hearing Before the Human Resources and Intergovernmental Relations Subcommittee of the Committee on Government Operations, House of Representatives 101st Congress, 2nd Session, February 6, 1990. Washington, DC: U.S. Government Printing Office, 1990.

Anonymous. Class II recall issued for Queso Fresco Margarita Mexican style fresh white cheese. FDA Enforcement Rep Nov 13, 1991a.

Anonymous. *Escherichia coli* O157:H7 illness associated with consumption of farm-manufactured yogurt. Commun Dis Rep 1:47, 1991b.

Anonymous. Milk and milk products recalled. FDA Enforcement Rep Feb 20, 1991c.

Anonymous. Food Safety and Quality; FDA Strategy Needed to Address Animal Drug Residues in Milk. Washington, DC: General Accounting Office, Aug 5, 1992a.

Anonymous. Salmonellosis associated with homemade ice cream in Kansas, 1992. Kansas Med 90:287, 1992b.

Anonymous. Cottage and other cheeses recalled due to *Cryptosporidium*. FDA Enforcement Rep June 9, 1993a.

Anonymous. Update: multistate outbreak of *Escherichia coli* O157:H7 infections from hamburgers—western United States, 1992–1993. Morbid Mortal Weekly Rep 42:258, 1993b.

Anonymous. *Escherichia coli* O157:H7 outbreak linked to home-cooked hamburger—California, July 1993. Morbid Mortal Weekly Rep 43:213, 1994a.

Anonymous. Outbreak of *Salmonella* enteritis associated with nationally distributed ice cream products—Minnesota, South Dakota and Wisconsin, 1994. Morbid Mortal Weekly Rep 43:740, 1994b.

Anonymous. Summary of notifiable diseases, United States 1994. Morbid Mortal Weekly Rep 43(53):1, 1994c.

Anonymous. Brucellosis associated with unpasteurized milk products abroad. Commun Dis Rep 5(32):1, 1995a.

Anonymous. Summary of notifiable diseases, United States 1994. Morbid Mortal Weekly Rep 43:69, 1995b.

Anonymous. National milk drug residue data base fiscal year 1995 annual report: October 1, 1994–September 30, 1995. Arlington, VA: GLH, 1996.

Applebaum RS, Marth EH. Inactivation of aflatoxin M_1 in milk using hydrogen peroxide and hydrogen peroxide plus riboflavin or lactoperoxidase. J Food Prot 45:557, 1982a.

Applebaum RS, Marth EH. Use of sulfite or bentonite to eliminate aflatoxin M_1 from naturally contaminated raw whole milk. Z Lebensm Unters Forsch 174:303, 1982b.

Applebaum RS, Brackett RE, Wiseman DW, Marth EH. Aflatoxin: toxicity to dairy cattle and occurrence in milk and milk products—a review. J Food Prot 45:752, 1982.

Archer DL. Review of the Latest FDA Information on the Presence of *Listeria* in Foods. Geneva, Switzerland: WHO Working Group on Foodborne Listeriosis, Feb 15–19, 1988.

Armijo R, Henderson DA, Tinothee R, Robinson HB. Food poisoning outbreaks associated with spray-dried milk—an epidemiologic study. Am J Pub Health 47:1093, 1957.

Armstrong C, Parran T. Milk-borne typhoid fever—a review of cases. Public Health Rep Suppl 62, 1927.

Armstrong RW, Fodor T, Curlin GT, Cohen AB, Morris GK, Martin WT, Feldman J. Epidemic *Salmonella* gastroenteritis due to contaminated imitation ice cream. Am J Epidemiol 91:300, 1970.

Arocha MM, McVey M, Loder SD, Rupnow JH, Bullerman L. Behavior of hemorrhagic *Escherichia coli* O157:H7 during the manufacture of cottage cheese. J Food Prot 55:379, 1992.

Aulisio CCG, Lanier JM, Chappel MA. *Yersinia enterocolitica* associated with outbreaks in three southern states. J Food Prot 45:1263, 1982.

Aytac SA, Ozbas ZY. Isolation of *Yersinia enterocolitica* from Turkish pickled white cheese. Austral J Dairy Technol 47:60, 1992.

Baca OG, Paretsky D. Q fever and *Coxiella burnetti*: a model for host–parasite interactions. Microbiol Rev 47:127, 1983.

Bachmann HP, Spahr U. The fate of potentially pathogenic bacteria in Swiss hard and semihard cheeses made from raw milk. J Dairy Sci 78:476, 1995.

Baker JM, Griffiths MW. Evidence for increased thermostability of *Bacillus cereus* enterotoxin in milk. J Food Prot 58:443, 1995.

Barksdale L. Diphtheria bacilli and other corynebacteria. In: Braude AI, Davis CE, Fierer J, eds. Infectious Diseases and Medical Microbiology. Philadelphia: WB Saunders, 1986, p 254.

Barnard R, Callahan W. Follow-up on gastroenteritis attributed to imported French cheese—United States. MMWR Morbid Mortal Weekly Rep 20:445, 1971.

Barnham M, Thornton TJ, Lange K. Nephritis caused by *Streptococcus zooepidemicus* (Lancefield Group C). Lancet i:945, 1983.

Barrett NJ. Communicable disease associated with milk and dairy products in England and Wales: 1983–1984. J Infect 12:265, 1986.

Barrett NJ. Milkborne disease in England and Wales in the 1980s. J Soc Dairy Technol 42:4, 1989.

Barrow GI, Miller DC, Johnson DL, Hingston CWJ. *Brucella abortus* in fresh cream and cream products. BMJ 2:596, 1968.

Bean NH, Goulding JS, Lao C, Angulo FJ. Surveillance of foodborne disease outbreaks—United States, 1988–1992. Morbid Mortal Weekly Rep 45(SS-5):1, 1996.

Becker H, Schaller G, von Wiese W, Terplan G. *Bacillus cereus* in infant foods and dried milk products. Intern J Food Microbiol 23:1, 1994.

Benedictus G, Dijkhuizen AA, Stelwagen J. Economic losses due to paratuberculosis in dairy cattle. Vet Rec 121:142, 1987.

Bennett RW. *Bacillus cereus* diarrheagenic enterotoxin. In: FDA Bacteriological Analytical Manual. 8th ed. Gaithersburg, MD: AOAC International, 1995. p 15.01.

Bennett RW, Lancette GA. *Staphylococcus aureus*. In: FDA Bacteriological Analytical Manual. 8th ed. Gaithersburg, MD: AOAC International, 1995. p 12.01.

Bergdoll MS. Staphylococcal intoxications. In: Riemann H, Bryan FL, eds. Food-Borne Infections and Intoxications. 2nd ed. New York: Academic Press, 1979, p. 443.

Bergdoll MS. *Staphylococcus aureus*. In: Doyle MP, ed. Foodborne Bacterial Pathogens. New York: Marcel Dekker, 1989, p 463.

Bergdoll MS, Sugiyama H, Dack GM. Staphylococcal enterotoxin. I. Purification. Arch Biochem Biophys 85:62, 1959.

Besser RE, Lett SM, Weber JT, Doyle MP, Barrett TJ, Wells JG, Griffin PM. An outbreak of diarrhea and hemolytic uremic syndrome from *Escherichia coli* O157:H7 in fresh-pressed apple cider. JAMA 269:2217, 1993.

Bezanson GS, Khakhria R, Duck D, Lior H. Molecular analysis confirms food source and simultaneous involvement of two distinct but related subgroups of *Salmonella typhimurium* bacteriophage type 10 in major interprovincial *Salmonella* outbreak. Appl Environ Microbiol 50:1279, 1985.

Biberstein EL, Behymer DE, Bushnell R, Crenshaw G, Riemann HP, Franti CE. A survey of Q fever (*Coxiella burnetti*) in California dairy cows. Am J Vet Res 35:1577, 1974.

Bigby M, Jick S, Arndt K. Drug-induced cutaneous lesion reactions. A report from the Boston Collaborative Drug Surveillance Program on 15,438 consecutive inpatients 1975–1982. JAMA 256:3358, 1986.

Billon J, Guerin J, Sebald M. Studies on toxinogenesis of *Clostridium botulinum* type B during maturation of soft cheese. Le Lait 60:343, 1980.

Birkhead G, Vogt RL, Heun E, Evelti CM, Patton CM. A multiple-strain outbreak of *Campylobacter* enteritis due to consumption of inadequately pasteurized milk. J Infect Dis 157:1095, 1988.

Bishop JR, White CH. Antibiotic residue detection in milk—a review. J Food Prot 47:647, 1984.

Bishop JR, Senyk GF, Duncan SE, Case RA. Detection of antibiotic/drug residues in milk and dairy products. In: Marshall RT, ed. Standard Methods for the Examination of Dairy Products. 16th ed. Washington, DC: American Public Health Association, 1992, p 347.

Black RE, Jackson RJ, Tsai T, Medvesky M, Shayegani M, Feeley JC, MacLeod KIE, Wakelee AM. Epidemic *Yersinia enterocolitica* infection due to contaminated chocolate milk. N Engl J Med 298:76, 1978.

Blaser MJ. *Campylobacter* species. In: Mandell GL, Douglas RG, Bennet JE, eds. Principles and Practice of Infectious Diseases. New York: Churchill Livingstone, 1990, p 1649.

Blaser MJ, Sazie E, Williams LP Jr. The influence of immunity on raw milk-associated *Campylobacter* infection. JAMA 257:43, 1987.

Blaser MJ, Hardesty HL, Powers B, Wang W-LL. Survival of *Campylobacter fetus* subsp. *jejuni* in biological milieus. J Clin Microbiol 11:309, 1980.

Blaser MJ, Cravens J, Powers BW, Laforce FM, Wang W-LL. *Campylobacter* enteritis associated with unpasteurized milk. Am J Med 67:715, 1979.

Bleem A. *Escherichia coli* O157:H7 in raw milk—a review. In: Animal Health Insight, Spring/Summer 1994. Fort Collins, CO: USDA:APHIS:VS Centers for Epidemiology and Animal Health, 1994.

Block RE, Levine MM, Clements ML, Hughes TP, Blaser MJ. Experimental *Campylobacter jejuni* infection in humans. J Infect Dis 157:142, 1988.

Blouse J, Genglei LE, Lathrop GD, Hodder RA, Nowosiwski T, Caraway CT. A common-source outbreak of *Salmonella newport*—Louisiana. Morbid Mortal Weekly Rep 24:413, 1975.

Bone FJ, Bogie D, Morgan-Jones SC. Staphylococcal poisoning from sheep milk cheese. Epidemiol Infect 103:449, 1989.

Boor KJ. Survival of *Escherichia coli* O157:H7 in fermented dairy products. Seattle, WA: Abstracts Annual Meeting of the International Association of Milk, Food, and Environmental Sanitarians, July 2, 1996.

Borcyzk AA, Karmali MA, Lior H, Duncan LMC. Bovine reservoir for verotoxin-producing *Escherichia coli* O157:H7. Lancet i:98, 1987.

Bowen DA, Henning DR. Coliform bacteria and *Staphylococcus aureus* in retail natural cheeses. J Food Prot 57:253, 1994.

Brackett RE, Marth EH. Association of aflatoxin M$_1$ with casein. Z Lebensm Unters Forsch 174:439, 1982a.

Brackett RE, Marth EH. Fate of aflatoxin M$_1$ in cheddar cheese and in process cheese spread. J Food Prot 45:549, 1982b.

Brackett RE, Marth EH. Fate of aflatoxin M$_1$ in Parmesan and mozzarella cheese. J Food Prot 45:597, 1982c.

Bradbury WC, Pearson AD, Marko MA, Congi RV, Penner JL. Investigation of a *Campylobacter jejuni* outbreak by serotyping and chromosomal restriction endonuclease analysis. J Clin Microbiol 19:342, 1984.

Brady MS, Katz SE. Antibiotic/antimicrobial residues in milk. J Food Prot 51:8, 1988.

Bray J. Isolation of antigenically homogeneous strains of *Bact. coli neapolitanum* in summer diarrhea. J Pathol Bacteriol 57:239, 1945.

Bray J, Beavan TED. Slide agglutination of *Bacterium coli* var. *neapolitanum* in summer diarrhea. J Pathol Bacteriol 60:395, 1948.

Brieseman MA. Raw milk consumption as a probable cause of two outbreaks of *Campylobacter* infection. N Z Med J 97:411, 1984.

Briozzo J, de Lagarde EM, Chirife J, Parada JL. *Clostridium botulinum* type A growth and toxin production in media and process cheese spread. Appl Environ Microbiol 45:1150, 1983.

Brodsky MH. Evaluation of the bacteriological health risk of 60-day aged raw milk cheddar cheese. J Food Prot 47:530, 1984a.

Brodsky MH. Bacteriological survey of freshly formed cheddar cheese. J Food Prot 47:546, 1984b.

Brown EH, Bailey EH. Acute hemolytic anaemia in a case of infantile gastro-enteritis treated with oxytetracycline. BMJ 2:1204, 1958.

Brown GL, Colwell DC, Hooper WL. An outbreak of Q fever in Staffordshire. J Hyg Camb 66:649, 1968.

Bryan FL. Infections and intoxications caused by other bacteria. In: Riemann H, Bryan FL, eds. Food-Borne Infections and Intoxications. 2nd ed. New York: Academic Press, 1979, p 211.

Bryan FL. Epidemiology of milk-borne disease. J Food Prot 46:637, 1983.

Bryan FL, Fanelli MJ, Riemann H. *Salmonella* infections. In: Riemann H, Bryan FL, eds. Food-borne Infections and Intoxications. 2nd ed. New York: Academic Press, 1979, p 73.

Buazzi MM, Johnson ME, Marth EH. Fate of *Listeria monocytogenes* during the manufacture of mozzarella cheese. J Food Prot 55:80, 1992.

Buchanan RL, Klawitter LA. The effect of incubation temperature, initial pH and sodium chloride on the growth kinetics of *Escherichia coli* O157:H7. Food Microbiol 9:185, 1992.

Buckner P, Ferguson D, Anzalone F, Anzalone D, Taylor J, Hlady WG, Hopkins RS. Outbreak of *Salmonella enteritidis* associated with homemade ice cream—Florida, 1993. Morbid Mortal Weekly Rep 43:669, 1994.

Bula CJ, Bille J, Glauser MP. An epidemic of food-borne listeriosis in western Switzerland: description of 57 cases involving adults. Clin Infect Dis 20:66, 1995.

Bullerman LB. Examination of Swiss cheese for incidence of mycotoxin producing molds. J Food Sci 41:26, 1976.

Bullerman LB, Olivigni FJ. Mycotoxin producing-potential of molds isolated from cheddar cheese. J Food Sci 39:1166, 1974.

Campbell AG, Gibbard J. The survival of *E. typhosa* in cheddar cheese manufactured from infected raw milk. Can J Public Health 35:158, 1944.

Candlish AAG, Stimson WH, Smith JE. A monoclonal antibody to aflatoxin B_1: detection of the mycotoxin by enzyme immunoassay. Lett Appl Microbiol 1:57, 1985.

Capps JA, Miller JL. The Chicago epidemic of streptococcus sore throat and its relation to the milk supply. JAMA 58:1848, 1912.

Casemore DP, Jessop EC, Douce D, Jackson FB. *Cryptosporidium* plus *Campylobacter*: an outbreak in a semi-rural population. J Hyg Camb 96:95, 1986.

Cato EP, George WL, Finegold SM. Genus *Clostridium*. In: Sneath PH, Mair NS, Sharpe ME, Holt JG, eds. Bergy's Manual of Systematic Bacteriology. Vol 2. Baltimore: Williams & Wilkins, 1986, p 1141.

Caudill FW, Meyer MA. An epidemic of food poisoning due to pasteurized milk. J Milk Technol 6:73, 1943.

Charm SE, Zomer E, Salter R. Confirmation of widespread sulfonamide contamination in northeast U.S. milk market. J Food Prot 51:920, 1988.

Cherif A, Benelmouffok A, Doudou A. Consumption of goat cheese and human brucellosis at Ghardaia (Algeria). Arch Inst Pasteur Algerie 55:9, 1986.

Chiodini RJ, Hermon-Taylor J. The thermal resistance of *Mycobacterium paratuberculosis* in raw milk under conditions simulating pasteurization. J Vet Diagn Invest 5:629, 1993.

Chiodini RJ, Kruiningen HJV, Merkal RS. Ruminant paratuberculosis (Johne's disease): the current status and future prospects. Cornell Vet 74:218, 1984.

Choi HK, Schaack MM, Marth EH. Survival of *Listeria monocytogenes* in cultured buttermilk and yogurt. Milchwissenschaft 43:790, 1988.

Chopin A, Tesone S, Vila J-P, Le Graet Y, Mocquot G. Survie de *Staphylococcus aureus* au cours de la preparation et de la conservation de lait ecreme en poudre. Problemes poses par le denombrement des survivants. Can J Microbiol 24:1371, 1978.

Christiansson A, Naidu AS, Nilsson I, Wadstrom T, Pettersson H-E. Toxin production by *Bacillus cereus* dairy isolates in milk at low temperatures. Appl Environ Microbiol 55:2595, 1989.

Christie AB, Corbel MJ. Plague and other yersinial diseases. In: Smith GR, Easmon CSF, eds. Topley and Wilson's Principles of Bacteriology, Virology and Immunity. 8th ed. Vol 3—Bacterial Diseases. Philadelphia: BC Decker, 1990 p 399.

Christopher FM, Smith GC, Vanderzant C. Effect of temperature and pH on the survival of *Campylobacter fetus*. J Food Prot 45:253, 1982.

Clark WS, Nelson FE. Multiplication of coagulase-positive staphylococci in grade A raw milk samples. J Dairy Sci 44:232, 1961.

Cliver DO. Viral infections. In: Riemann H, Bryan FL, eds. Food-Borne Infections and Intoxications. 2nd ed. New York: Academic Press, 1979, p 299.

Cohen J, Marambio E, Lynch B, Moreno A. *Bacillus cereus* food poisoning amid newborns. Rev Chilean Pediat 55:20, 1984.

Cohen DR, Porter IA, Reid TMS, Sharp JCM, Forbes GI, Paterson GM. A cost benefit study of milk-borne salmonellosis. J Hyg Camb 91:17, 1983.

Collins CH, Grange JM. A review—the bovine tubercle bacillus. J Appl Bacteriol 55:13, 1983.

Collins P, Davies DC, Matthews PRJ. Mycobacterial infection in goats: diagnosis and pathogenicity of the organism. BVJ 140:196, 1984.

Collins RN, Treger MD, Goldsby JB Boring JR III, Coohon DB, Barr RN. Interstate outbreak of *Salmonella newbrunswick* infection traced to powdered milk. JAMA 203:838, 1968.

Collins-Thompson DL, Wood DS, Thomson IQ. Detection of antibiotic residues in consumer milk supplies in North America using the Charm II test procedure. J Food Prot 51:632, 1988.

Conner DE, Hall GS. Temperature and food additives affect growth and survival of *Escherichia coli* in poultry meat. Dairy Food Environ Sanitat 16:150, 1996.

Conner DE, Kotrola JS. Growth and survival of *Escherichia coli* under acidic conditions. Appl Environ Microbiol 61:382, 1995.

Conner DE, Scott VN, Bernard DT. Potential *Clostridium botulinum* hazards associated with extended shelf-life refrigerated foods: a review. J Food Safety 10:131, 1989.

Cox WA. Problems associated with bacterial spores in heat-treated milk and milk products. J Soc Dairy Technol 28:59, 1975.

Crielly EM, Logan NA, Anderton A. Studies on the *Bacillus* flora of milk and milk products. J Appl Bacteriol 77:256, 1994.

Critchley EMR, Hayes PJ, Isaacs PET. Outbreak of botulism in northwest England and Wales, June, 1989. Lancet 2:849, 1989.

Cullen JM, Ruebner BH, Hsieh DSP. Comparative hepatocarcinogenicity of aflatoxins B_1 and M_1 in the rat. Food Chem Toxicol 22:1027, 1985.

Dack GM, Cary WE, Woolport WE, Wiggers H. Outbreak of food poisoning proved to be due to yellow hemolytic *Staphylococcus*. J Prevent Med 4:167, 1930.

Dalrymple-Champneys W. *Brucella* Infection and Undulant Fever in Man. London: Oxford University Press, 1960.

D'Aoust J-Y. Infective dose of *Salmonella typhimurium* in cheddar cheese. Am J Epidemiol 122:717, 1985.

D'Aoust J-Y. Manufacture of dairy products from unpasteurized milk: a safety assessment, J Food Prot 52:906, 1989.

D'Aoust J-Y. *Salmonella* and the international food trade. Int J Food Microbiol 24:11, 1994.

D'Aoust J-Y, Warburton DW, Sewell AM. *Salmonella typhimurium* phage-type 10 from cheddar cheese implicated in a major Canadian foodborne outbreak. J Food Prot 48:1062, 1985.

D'Aoust J-Y, Park CE, Szabo RA, Todd ECD, Emmons DB, McKellar RC. Thermal inactivation of *Campylobacter* species, *Yersinia enterocolitica*, and hemorrhagic *Escherichia coli* O157:H7 in fluid milk. J Dairy Sci 71:3230, 1988.

D'Aoust J-Y, Emmons DB, McKellar R, Timbers GE, Todd ECD, Sewell AM, Warburton DW. Thermal inactivation of *Salmonella* species in fluid milk. J Food Prot 50:494, 1987.

Davey GM, Bruce J, Drysdale EM. Isolation of *Yersinia enterocolitica* and related species from the faeces of cows. J Appl Bacteriol 55:439, 1983.

Davis J. The growth and viability of streptococci of bovine and human origin in milk and milk products. J Infect Dis 15:378, 1914.

Day C, DeCov WR, Feffer D, Glosser J, Desonia B, Abbott B, Skinner M, Andersen JS. Salmonellosis associated with raw milk—Montana. Morbid Mortal Weekly Rep 30:211, 1981.

DeBoer E, Hartog BJ, Borst GHA. Milk as a source of *Campylobacter jejuni*. Neth Milk Dairy J 38:183, 1984.

Decker MD Lavely GB, Hutcheson RH, Schaffner W. Food-borne streptococcal pharyngitis in a hospital pediatrics clinic. JAMA 253:679, 1985.

DeGrace M, Laurin MF, Belanger C, Rolland PE, Blais R, Breton JP, Martineau G. *Yersinia enterocolitica* gastroenteritis outbreak—Montreal. Can Dis Weekly Rep 2:41, 1976.

Dekeyser P, Gossuin-Detrain M, Butzler JP, Sternon J. Acute enteritis due to related *Vibrio*: first positive stool cultures. J Infect Dis 125:390, 1972.

DeMol P. Human campylobacteriosis: clinical and epidemiological aspects. Dairy Food Environ Sanitat 14:314–316, 1994.

Diaz Cinco ME, Acedo F, Evelia LD, Alma B. Survival of *Brucella abortus* in the Mexican white cheese processing. Dairy Food Environ Sanitat 14:608, 1994.

Diaz S, Moreno MA, Dominguez L, Suarez G, Blanco JL. Application of a diphasic dialysis technique to the extraction of aflatoxins in dairy products. J Dairy Sci 76:1845, 1993.

Dixon JMS, Noble WC, Smith GR. Diphtheria; other corynebacterial and coryneform infections. In: Smith GR, Easmon CSF, eds. Topley and Wilson's Principles of Bacteriology, Virology and Immunity. 8th ed. Vol. 3—Bacterial Diseases. Philadelphia: BC Decker, 1990, p 55.

Doeglas HMG, Huisman J, Nater JP. Histamine intoxication after cheese. Lancet 2:1361, 1967.

Dominguez LB, Relter A, Marks FJ, Press WB, Starko KM. *Salmonella* gastroenteritis associated with milk—Arizona. Morbid Mortal Weekly Rep 28:117, 1979.

Donnelly CB, Black LA, Lewis KH. Occurrence of coagulase-positive staphylococci in cheddar cheese. Appl Microbiol 12:311, 1964.

Donta ST. Food poisoning. In: Braude AI, Davis CE, Fierer J, eds. Infectious Diseases and Medical Microbiology, 2nd ed. Philadelphia: WB Saunders, 1986, p 892.

dos Reis Tassinari A, de Melo Franco BDG, Landgraf M. Incidence of *Yersinia* spp. in food in Sao Paulo, Brazil. Int J Food Microbiol 21:263, 1994.

dos Santos EC, Genigeorgis C. Survival and growth of *Staphylococcus aureus* in commercially manufactured Brazilian Minas cheese. J Food Prot 44:177, 1981.

dos Santos EC, Genigeorgis C, Farver TB. Prevalence of *Staphylococcus aureus* in raw and pasteurized milk used for commercial manufacturing of Brazilian Minas cheese. J Food Prot 44:172, 1981.

Doyle MP. *Escherichia coli* O157:H7 and its significance in foods. Int J Food Microbiol 12:289, 1991.

Doyle MP, Padhye NV. *Escherichia coli.* In: Doyle MP, ed. Foodborne Bacterial Pathogens. New York: Marcel Dekker, 1989, p 71.

Doyle MP, Roman DJ. Prevalence and survival of *Campylobacter jejuni* in unpasteurized milk. Appl Environ Microbiol 44:1154, 1982a.

Doyle MP, Roman DJ. Recovery of *Campylobacter jejuni* and *Campylobacter coli* from inoculated foods by selective enrichment. Appl Environ Microbiol 43:1343, 1982b.

Doyle MP, Schoeni JL. Survival and growth characteristics of *Escherichia coli* associated with hemorrhagic colitis. Appl Environ Microbiol 48:855, 1984.

Doyle MP, Glass KA, Berry JT, Garcia GA, Pollard DJ, Schultz RD. Survival of *Listeria monocytogenes* in milk during high-temperature, short-time pasteurization. Appl Environ Microbiol 53:1433, 1987.

Drobniewski FA. *Bacillus cereus* and related species. Clin Microbiol Rev 6:324, 1993.

Duca E, Teodorovici G, Radu C, Vita A, Talasman-Niculescu P, Bernescu E, Feldi C, Rosca V. A new nephritogenic streptococci. J Hyg Camb 67:691, 1969.

Dufrenne J, Bijwaard M, te Giffel M, Beumer R, Notermans S. Characteristics of some psychrotrophic *Bacillus cereus* isolates. Intern J Food Microbiol 27:175, 1995.

DuPont HL, Formal SB, Hornick RB, Snyder MJ, Libonati JP, Sheahan DG, LaBrec EH, Kalas JP. Pathogenesis of *Escherichia coli* diarrhea. N Engl J Med 285:1, 1971.

Echeverria P, Taylor DN, Bettelheim KA, Chatkaeomorakot A, Changchawalit S, Thongcharoen A, Leksomboon U. HeLa cell-adherent enteropathogenic *Escherichia coli* in children under 1 year of age in Thailand. J Clin Microbiol 25:1472, 1987.

Eckman MR. Brucellosis linked to Mexican cheese. JAMA 232:636, 1975.

Eckner KF, Zottola EA. The behavior of selected microorganisms during manufacture of high moisture Jack cheeses from ultrafiltered milk. J Dairy Sci 74:2820, 1991.

Eckner KF, Dustman WA, Rys-Rodriguez AA. Contribution of composition, physico-chemical characteristics and polyphosphates to the microbial safety of pasteurized cheese spreads. J Food Prot 57:295, 1994.

Eckner KF, Roberts RF, Strantz AA, Zottola EA. Characterization and behavior of *Salmonella javiana* during manufacture of mozzarella-type cheese. J Food Prot 53:461, 1990.

Edelman R, Levine MM. From the National Institute of Allergy and Infectious Diseases—Summary of a workshop on enteropathogenic *Escherichia coli.* J Infect Dis 147:1108, 1983.

Edwards AT, Roulson M, Ironside MJ. A milk-borne outbreak of serious infection due to *Streptococcus zooepidemicus* (Lancefield Group C). Epidem Infect 101:43, 1988.

Edwards PR, West MG, Bruner DW. Antigenic studies of a group of paracolon bacteria (Bethesda group). J Bacteriol 55:711, 1948.

Edwards ST, Sandine WE. Public health significance of amines in cheese. J Dairy Sci 64:2431, 1981.

Ehlers JG, Chapparo-Serrano M, Richter RL, Vanderzant C. Survival of *Campylobacter fetus* subsp. *jejuni* in cheddar and cottage cheese. J Food Prot 45:1018, 1982.

El-Dairouty KF. Staphylococcal intoxication traced to non-fat dried milk. J Food Prot 12:901, 1989.

El-Gazzar FE, Marth EH. Samonellae, salmonellosis, and dairy foods: a review. J Dairy Sci 75:2327, 1992.

El-Refai A-MH, Sallam LAR, Naim N. The alkaloids of fungi. I. The formation of ergoline alkaloids by representative mold fungi. Jpn J Microbiol 14:91, 1970.

Elsser KAA, Moricz M, Procter EM. *Cryptosporidium* infections: a laboratory survey. Can Med Assoc J 135:211, 1986.

Engel G, Teuber M. Toxic metabolites from fungal cheese starter cultures (*Penicillium camemberti* and *Penicillium roqueforti*). In: van Egmond HP, ed. Mycotoxins in Dairy Products. London: Elsevier Applied Science, 1989, p 163.

Erkmen O, Bozoglu TF. Behaviour of *Salmonella typhimurium* in feta cheese during manufacture and ripening. Food Sci Technol 32:480, 1995.

Erksine D. Dermatitis caused by penicillin in milk. Lancet 1:431, 1958.

Espinosa FH, Ryan WM, Vigil PL, Gregory DF, Hilley RB, Romig DA, Stamm RB, Suhre ED, Taulbee PS, Zucal LH, Honsinger RW Jr, Lindberg PJ, Barcheck M, Miller JA, Mitzelfelt R, Montes JM, Nims LJ, Rollag OJ, Weber N, Mann JM. Group C streptococcal infections associated with eating homemade cheese—New Mexico. Morbid Mortal Weekly Rep 32:510, 1983.

Evans AD. "Q" fever in South Wales. Mon Bull Minist Health 21:215, 1956.

Evans DA, Litsky W. Thermal inactivation studies on certain pathogenic organisms in milk. Bacteriol Proc 19:10, 1968.

Evenson ML, Hinds MW, Bernstein RS, Bergdoll MS. Estimation of human dose of staphylococcal enterotoxin A from a large outbreak of staphylococcal food poisoning involving chocolate milk. Int J Food Microbiol 7:311, 1988.

Eyler JM. The epidemiology of milk-borne scarlet fever: the case of Edwardian Brighton. Am J Public Health 76:573, 1986.

Fakuuzi FR, Zyambo GCN, Morris SD. A serological survey of bovine brucellosis in the Melopo District of Bophuthatswanta. J South Afr Vet Assoc 64:154, 1993.

Fanning J. An unusual outbreak of gastroenteritis. BMJ 1:583, 1935.

Fantasia LD Mestrandrea L, Schrade JP, Yager J. Detection and growth of enteropathogenic *Escherichia coli* in soft ripened cheese. Appl Microbiol 29:179, 1975.

Farber JM, Peterkin PI. *Listeria monocytogenes*, a food-borne pathogen. Microbiol Rev 55:476, 1991.

Farber JM, Sanders GW, Speirs JI. Growth of *Listeria monocytogenes* in naturally-contaminated raw milk. Lebensm Wiss Technol 23:252, 1990.

Farkhondeh A, Ghazvinian R, Lavhal Ph. Contamination par les *Salmonella* du fromage iranien frais, non sale, mis en vente dans la region de Teheran. Le Lait 535/536:302, 1974.

Farrag SA, Marth EH. *Escherichia coli* O157:H7, *Yersinia enterocolitica* and their control in milk by the lactoperoxidase system: a review Lebensm Wiss Technol 25:201, 1992.

Felip G, Toti L. Foodborne diseases caused by gram-negative microorganisms in milk products. Microbiol Aliments Nutr 2:251, 1984.

Finch MJ, Blake PA. Foodborne outbreaks of campylobacteriosis: the United States experience, 1980–1982. Am J Epidemiol 122:262, 1985.

Fishbein DB, Raoult D. A cluster of *Coxiella burnetii* infections associated with exposure to vaccinated goats and their unpasteurized dairy products. Am J Trop Med Hyg 47:35, 1992.

Fleming DW, Cochi SL, MacDonald KL, Brondum J, Hayes PS, Plikaytis BD, Homes MB, Audurier A, Broome CV, Reingold AL. Pasteurized milk as vehicle of infection in an outbreak of listeriosis. N Engl J Med 312:404, 1985.

Flowers RS, Andrews W, Donnelly CW, Koenig E. Pathogens in milk and milk products. In: Marshall RT, ed. Standard Methods for the Examination of Dairy Products. 16th ed. Washington, DC: American Public Health Association, 1992, p 103.

Flowers RS, D'Aoust J-Y, Andrews WH, Bailey JS. *Salmonella.* In: Vanderzant C, Splittsoesser DF, eds. Compendium of Methods for the Microbiological Examination of Foods, 3rd ed. Washington, DC: American Public Health Association, 1992b, p 371.

Foley AR, Poisson E. A cheese-borne outbreak of typhoid fever. Can J Public Health 36:116, 1945.

Foley BV, Clay MM, O'Sullivan DJ. A study of a brucellosis epidemic. Irish J Med Sci 3:457, 1970.

Fontaine RE, Cohen ML, Martin WT, Vernon TM. Epidemic salmonellosis from cheddar cheese—surveillance and prevention. Am J Epidemiol 111:247, 1980.

Francis AJ, Nimmo GR, Efstratiou A, Galanis V, Nuttall N. Investigation of milk-borne *Streptococcus zooepidemicus* infection associated glomerulonephritis in Australia. J Infect 27:317, 1993.

Francis BJ, Allard J. *Salmonella typhimurium*—Yakima County, Washington. Morbid Mortal Weekly Rep 16:178, 1967.

Francis BJ, Davis JP. Update: gastrointestinal illness associated with imported semi-soft cheese. Morbid Mortal Weekly Rep 33:16, 1984.

Francis BJ, Langkop C, Mitchel H, Donnell HD Jr, Hernandez C, Telle RD, Halpin TJ. Presumed staphylococcal food poisoning associated with whipped butter. Morbid Mortal Weekly Rep 26:268, 1977.

Francis DW, Spaulding PL, Lovett J. Enterotoxin production and thermal resistance of *Yersinia enterocolitica* in milk. Appl Environ Microbiol 40:174, 1980.

Frank JF, Marth EH. Inhibition of enteropathogenic *Escherichia coli* by homofermentative lactic acid bacteria in milk—I. Comparison of strains of *Escherichia coli*. J Food Prot 40:749, 1977a.

Frank JF, Marth EH. Inhibition of enteropathogenic *Escherichia coli* by homofermentative lactic acid bacteria in milk—II. Comparison of lactic acid bacteria and enumeration methods. J Food Prot 40:754, 1977b.

Frank JF, Marth EH. Survey of soft and semisoft cheese for presence of fecal coliforms and serotypes of enteropathogenic *Escherichia coli*. J Food Prot 41:198, 1978.

Frank JF, Marth EH, Olson NF. Survival of enteropathogenic and non-enteropathogenic *Escherichia coli* during the manufacture of Camembert cheese. J Food Prot 40:835, 1977.

Frank JF, Marth EH, Olson NF. Behavior of enteropathogenic *Escherichia coli* during manufacture and ripening of brick cheese. J Food Prot 41:111, 1978.

Fremy JM, Chu FS. Direct ELISA for determining AFM_1 at picogram levels in dairy products. JAOAC 67:1098, 1984.

Fremy JM, Chu FS. Immunochemical methods of analysis for aflatoxin M_1, In: van Egmond HP, ed. Mycotoxins in Dairy Products. London: Elsevier Applied Science, 1989, p 97.

Frobish RA, Bradley BD, Wagner DD, Long-Bradley PE, Hairston H. Aflatoxin residues in milk of dairy cows after ingestion of naturally contaminated grain. J Food Prot 49:781, 1986.

Fulton JS. Contagious abortion of cattle and undulant fever in man. Med Officer 65:201, 1941.

Galbraith NS, Forbes P, Clifford C. Communicable disease associated with milk and dairy products in England and Wales 1951–1980. BMJ 284:1761, 1982.

Galbraith NS, Ross MS, de Mowbray RR, Payne DJH. Outbreak of *Brucella melitensis* type 2 infection in London. BMJ 1:612, 1969.

Garcia ML, Moreno B, Bergdoll MS. Characterization of staphylococci isolated from mastitic cows in Spain. Appl Environ Microbiol 39:548, 1980.

Gauthier J, Foley AR. A cheese-borne outbreak of typhoid fever. Can J Public Health 34:543, 1943.

Geiger JC, Crowley AB, Gray JP. Food poisoning from ice cream on ships. JAMA 105:1980, 1935.

George JTA, Wallace JG, Morrison HR, Harbourne JF. Paratyphoid in man and cattle. BMJ 3:208, 1972.

Ghoneim A, Cooke EM. Serious infection caused by group C streptococci. J Clin Pathol 33:188, 1980.

Gilbert GL, Davoren RA, Cole ME. Mid-trimester abortion associated with septicaemia caused by *Campylobacter jejuni*. Med J Aust 1:585, 1981.

Gilbert RJ. *Bacillus cereus* gastroenteritis. In: Riemann H, Bryan FL, eds. Food-borne Infections and Intoxications. 2nd ed. New York: Academic Press, 1979, p 495.

Gilbert RJ, Parry JM. Serotypes of *Bacillus cereus* from outbreaks of food poisoning and from routine foods. J Hyg Camb 78:69, 1977.

Gill KPW, Bates PG, Lander KP. The effect of pasteurization on the survival of *Campylobacter* species in milk. BVJ 137:578, 1981.

Gilles G, Jouglard J, Sebald M, Jacquet MJ. Sur une epidemie de botulisme humain enquete et hyptheses etiologiques: role possible du fromage et des paillons. Rec Med Vet 47:165, 1974.

Gilman HL, Dahlberg AC, Marquardt JC. The occurrence and survival of *Brucella abortus* in cheddar and Limburger cheese. J Dairy Sci 29:71, 1946.

Glatz BA, Brudvig SA. Survey of commercially available cheese for enterotoxigenic *Escherichia coli*. J Food Prot 43:395, 1980.

Goepfert JM, Olson NF, Marth EH. Behavior of *Salmonella typhimurium* during manufacture and curing of cheddar cheese. Appl Microbiol 16:862, 1968.

Goepfert JM, Spira WM, Kim HU. *Bacillus cereus*: food poisoning organism. A review. J Milk Food Technol 35:213, 1972.

Gogov Y, Slavchev G, Peeva T. Cold resistance of *S. aureus* and A and C_2 staphylocccus enterotoxins in ice cream. Vet Sci (Sofia) 21:46, 1984.

Golden DA, Hou JJ. Survival of *Yersinia enterocolitica* during manufacture and storage of short- and long-set cottage cheese (abstr). New Orleans, LA: Annual Meeting of the Institute of Food Technologists. June 22–25, 1996.

Goldie W, Maddock ECG. A milk-borne outbreak of diphtheria. Lancet 1:285, 1943.

Goldwater PN, Bettelheim KA. Hemolytic uremic syndrome due to Shiga-like toxin producing *Escherichia coli* O48:H21 in South Australia. Emerg Infect Dis 1:132, 1995.

Gomez-Lucia E, Blanco JL, Goyache J, de La Fuente R, Vazquez JA, Ferri EFR, Suarez G. Growth and enterotoxin A production by *Staphylococcus aureus* S6 in Manchego-type cheese. J Appl Bacteriol 61:499, 1986.

Gomez-Lucia E, Goyache J, Orden JA, Domenech A, Hernandez FJ, Quiteria JAR-S, Lopez B, Blanco JL, Suarez G. Growth of *Staphylococcus aureus* and synthesis of enterotoxin during ripening of experimental Manchego-type cheese. J Dairy Sci 75:19, 1992.

Goulet V, Jacquet C, Vaillant V, Rebiere I, Mouret E, Lorente C, Maillot E, Stainer F, Rocourt J. Listeriosis from consumption of raw-milk cheese. Lancet 345:1581, 1995.

Graadl O, Nygaard K. Crohn's disease: long term effects of surgical treatment. Tidssk Norske Laegeforen 114:1603, 1994.

Grange JM. Tuberculosis. In: Smith GR, Easmon CSF, eds. Topley and Wilson's Principles of Bacteriology, Virology and Immunity. 8th ed. Vol. 3—Bacterial Diseases. Philadelphia: BC Decker, 1990, p 55.

Grant IR, Ball HJ, Neill SD, Rowe MT. Inactivation of *Mycobacterium paratuberculosis* in cows' milk at pasteurization temperatures. Appl Environ Microbiol 62:631, 1996.

Granum PE, Brynestad S, Kramer JM. Analysis of enterotoxin production by *Bacillus cereus* from dairy products, food poisoning incidents and non-gastrointestinal infections. Intern J Food Microbiol 17:269, 1993.

Gray LD. *Escherichia, Salmonella, Shigella* and *Yersinia*. In: Murray PR, Baron EJ, Pfaller MA, Tenover FC, Yolken RH, eds. Manual of Clinical Microbiology. 6th ed. Washington, DC: ASM Press, 1995, p 450.

Gray ML, Killinger AH. *Listeria monocytogenes* and listeric infections. Bacteriol Rev 30:309, 1966.

Grecz N, Wagenaar RO, Dack GM. Storage stability of *Clostridium botulinum* toxin and spores in processed cheese. Appl Microbiol 13:1014, 1965.

Greenwood MH, Hooper WL. Excretion of *Yersinia* spp. associated with consumption of pasteurized milk. Epidemiol Infect 104:345, 1990.

Greenwood MH, Hooper WL, Rodhouse JC. The source of *Yersinia* spp. in pasteurized milk: an investigation at a dairy. Epidemiol Infect 104:351, 1990.

Gresikova M, Sekeyova M, Stupalova S, Necas S. Sheep milk-borne epidemic of tick-borne encephalitis in Slovakia. Intervirol 5:57, 1975.

Griffin PM, Tauxe RV. The epidemiology of infections caused by *Escherichia coli* O157:H7, other enterohemorrhagic *E. coli* and the associated hemolytic uremic syndrome. Epidemiol Rev 13:60, 1991.

Griffin PM, Ostroff SM, Tauxe RV, Greene KD, Wells JG, Lewis JH, Blake PA. Illness associated with *Escherichia coli* O157:H7 infections. Ann Intern Med 109:705, 1988.

Griffith AS. Human tubercle bacilli in the milk of a vaccinated cow. J Pathol Bacteriol 17:323, 1913.

Griffiths MW. Toxin production by psychrotrophic *Bacillus* spp. present in milk. J Food Prot 53:790, 1990.

Griffiths MW, Phillips JD. Incidence, source and some properties of psychrotrophic *Bacillus* found in raw and pasteurized milk. J Soc Dairy Technol 43:62, 1990.

Gross EM, Weizman Z, Picard E, Mates A, Sheinman R, Platzner N, Wolff A. Milkborne gastroenteritis due to *Staphylococcus aureus* enterotoxin B from a goat with mastitis. Am J Trop Med Hyg 39:103, 1988.

Guitierrez LM, Menes I, Garcia ML, Moreno B, Bergdoll MS. Characterization and enterotoxigenicity of staphylococci isolated from mastitic ovine milk in Spain. J Food Prot 45:1282, 1982.

Gunzer F, Bohm H, Russman H, Bitzan M, Aleksic S, Karch H. Molecular detection of sorbitol fermenting *Escherichia coli* O157 in patients with hemolytic uremic syndrome. J Clin Microbiol 30:1807, 1992.

Gyles CL. *Escherichia coli* cytotoxins and enterotoxins. Can J Microbiol 38:734, 1992.

Habib N, Warring FC. A fatal case of infection due to *Mycobacterium bovis*. Am Rev Respir Dis 93:804, 1966.

Hackler JF. Outbreak of *Staphylococcus* milk poisoning in pasteurized milk. Am J Public Health 29:1247, 1939.

Halpin-Dohnalek MI, Marth EH. Fate of *Staphylococcus aureus* in whey, whey cream, and whey cream butter. J Dairy Sci 72:3149, 1989a.

Halpin-Dohnalek MI, Marth EH. Growth and production of enterotoxin A by *Staphylococcus aureus* in cream. J Dairy Sci 72:2266, 1989b.

Halpin-Dohnalek MI, Marth EH. Growth of *Staphylococcus aureus* in milks and creams with various amounts of milk fat. J Food Prot 52:540, 1989c.

Hamama A, El Marrakchi A, El Othmani F. Occurrence of *Yersinia enterocolitica* in milk and dairy products in Morocco. Int J Food Microbiol 16:69, 1992.

Hammer BW. Dairy Bacteriology. 2nd ed. New York: John Wiley & Sons, 1938, p 150.

Hancock DD, Besser TE, Kinsel ML, Tarr PI, Rice DH, Paros MG. The prevalence of *Escherichia coli* O157:H7 in dairy and beef cattle in Washington State. Epidemiol Infect 113:199, 1994.

Hanna MO, Stewart JC, Carpenter ZL, Vanderzant C. Heat resistance of *Yersinia enterocolitica* in skim milk. J Food Prot 40:689, 1977.

Hargrove RE, McDonough FE, Mattingly WA. Factors affecting survival of *Salmonella* in cheddar and Colby cheese. J Milk Food Technol 32:480, 1969.

Harmon SM, Goepfert JM, Bennett RW. *Bacillus cereus*. In: Marshall RT, ed. Standard Methods for the Examination of Dairy Products. Washington, DC: Am Public Health Assoc. 1992, p 593.

Harris NV, Kimball T, Weiss NS, Nolan C. Dairy products and other non-meat foods as possible sources of *Campylobacter jejuni* and *Campylobacter coli* enteritis. J Food Prot 49:347, 1986.

Harris NV, Kimball TJ, Bennett P, Johnson Y, Wakely D, Nolan CM. *Campylobacter jejuni* enteritis associated with raw goat's milk. Am J Epidemiol 126:179, 1987.

Hart RJC. *Corynebacterium ulcerans* in humans and cattle in North Devon. J Hyg Camb 92:161, 1984.

Hauge S. Food poisoning due to *Bacillus cereus*: a preliminary report. Nord Hyg Tigskr 31:189, 1950.

Hauge S. Food poisoning caused by aerobic sporeforming bacteria. J Appl Bacteriol 18:591, 1955.

Hauschild AHW. *Clostridium botulinum.* In: Doyle MP, ed. Foodborne Bacterial Pathogens. New York: Marcel Dekker, 1989, p 111.

Headrick ML, Korangy S, Bean NH, Klontz KC, Angulo FJ, Altekruse SF, Potter ME. Epidemiology of Milk-Associated Foodborne Disease Outbreaks Reported in the United States, 1973–1993. The Continued Risk of Raw Milk Consumption. Atlanta, GA: Centers for Disease Control and Prevention, 1996.

Hedberg CW, Korlath JA, D'Aoust J-Y, White KE, Schell WL, Miller MR, Cameron DN, MacDonald KL, Osterholm MT. A multistate outbreak of *Salmonella javiana* and *Salmonella oranienburg* infections due to consumption of contaminated cheese. JAMA 268:3203, 1992.

Helmy ZA, Abd-el-Malek Y, Mahmoud AA. Growth of *Staphylococcus aureus*, experimentally inoculated in Damietta cheese. Zbl Bakteriol Abt II, 130:468, 1975.

Hendricks SL, Meyer ME. Brucellosis. In: Hubbert WT, McCulloch WF, Schurrenburger PR, eds. Diseases Transmitted from Animals to Man. 6th ed. Springfield, IL: Charles C Thomas, 1975, p 10.

Hendricks SL, Belknap RA, Hausler WJ Jr. Staphylococcal food intoxication due to cheddar cheese. I. Epidemiology. J Milk Food Technol 22:313, 1959.

Hennessy TW, Hedberg CW, Slutsker L, White KE, Besset-Wiek JM, Moen ME, Feldman J, Coleman WW, Edmonson LM, MacDonald KL, Osterholm MT, the Investigation Team. A national outbreak of *Salmonella enteritidis* infections from ice cream. N Engl J Med 334:1281, 1996.

Henry JE. Milk-borne diphtheria. An outbreak traced to infection of a milk handler's finger with *B. diphtheriae.* JAMA 75:1715, 1920.

Hill WE, Datta AR, Feng P, Lampel KA, Payne WL. Identification of foodborne bacterial pathogens by gene probes. In: FDA Bacteriological Analytical Manual. 8th ed. Gaithersburg, MD: AOAC International, 1995, p 24.01.

Hinz PC, Slatter WL, Harper WJ. A survey of various free amino and fatty acids in domestic Swiss cheese. J Dairy Sci 39:235, 1956.

Hirschmann SZ. Viral hepatitis. In: Braude AI, Davis CE, Fierer J, eds. Infectious Diseases and Medical Microbiology. 2nd ed. Philadelphia: WB Saunders, 1986, p 989.

Hitchins AD. *Listeria monocytogenes.* Identification of foodborne bacterial pathogens by gene probes. In: FDA Bacteriological Analytical Manual. 8th ed. Gaithersburg, MD: AOAC International, 1995, p 10.01.

Hitchins AD, Feng P, Watkins WD, Rippey SR, Chandler LA. *Escherichia coli* and the coliform bacteria. In: FDA Bacteriological Analytical Manual. 8th ed. Gaithersburg, MD: AOAC International, 1995, p 4.01.

Hobbs BC. The clostridia. In: Braude AI, Davis CE, Fierer J, eds. Infectious Diseases and Medical Microbiology. 2nd ed. Philadelphia: WB Saunders, 1986, p 407.

Holmberg SD, Blake PA. Staphylococcal food poisoning in the United States: new facts and old misconceptions. JAMA 251:487, 1984.

Holmes JR, Plunkett T, Pate P, Roper WL, Alexander WJ. Emetic food poisoning caused by *Bacillus cereus*. Arch Intern Med 141:766, 1981.

Horning BG. Outbreak of undulant fever due to *Brucella suis*. JAMA 24:1978, 1935.

Horvath G, Toth-Marton E, Meszaros JM, Quarini L. Experimental *Bacillus cereus* mastitis in cows. Acta Vet Hung 34:29, 1986.

Hudson PJ, Vogt RL, Brondum J, Patton CM. Isolation of *Campylobacter jejuni* from milk during an outbreak of campylobacteriosis. J Infect Dis 150:789, 1984.

Hudson SJ, Sobo AO, Russel K, Lightfoot NF. Jackdaws as potential source of milk-borne *Campylobacter jejuni* infection. Lancet 336:1160, 1990.

Huebner RJ, Bell JA. "Q" fever studies in southern California—summary of current results and a discussion of possible control measures. JAMA 145:301, 1951.

Huebner RJ, Jellison WL, Beck MD, Wilcox FP. Q fever studies in southern California—effects of pasteurization on survival of *C. burnetii* in naturally infected milk. Public Health Rep 64:499, 1949.

Hughes D, Jensen N. *Yersinia enterocolitica* in raw goat's milk. Appl Environ Microbiol 41:309, 1981.

Hui JY, Taylor SL. Inhibition of in vivo histamine metabolism in rats by foodborne and pharmacologic inhibitors of diamine oxidase, histamine N-methyl transferase, and monoamine oxidase. Toxicol Appl Pharmacol 8:241, 1985.

Humphrey TJ, Hart RJC. *Campylobacter* and *Salmonella* contamination of unpasteurized cow's milk on sale to the public. J Appl Bacteriol 65:463, 1988.

Hunt J, Abeyta C. *Campylobacter*. In: FDA Bacteriological Analytical Manual. 8th ed. Gaithersburg, MD: AOAC International, 1995, p 7.01.

Hunt, JM, Francis DW, Peeler JT, Lovett J. Comparison of methods for isolating *Campylobacter jejuni* from raw milk. Appl Environ Microbiol 50:535, 1985.

Hutchinson DN, Bolton FJ, Jelley WCN, Mathews WG, Telford DR, Counter DE, Jessop EG, Horsley SD. *Campylobacter* enteritis associated with consumption of raw goat's milk. Lancet 1:1037, 1985a.

Hutchinson DN, Bolton FJ, Hinchliffe PM, Dawkins HC, Horsley SD, Jessop EG, Robertshaw PA, Counter DE. Evidence of udder excretion of *Campylobacter jejuni* as the cause of a milk-borne *Campylobacter* outbreak. J Hyg Camb 94:205, 1985b.

Ibrahim GF, Radford DR, Baldock AK, Ireland LB. Inhibition of growth of *Staphylococcus aureus* and enterotoxin-A production in cheddar cheese produced by induced starter failure. J Food Prot 44:189, 1981.

Ibrahim GF, Radford DR, Baldock AK, Ireland LB. Inhibition of *Staphylococcus aureus* growth and enterotoxin-A production in cheddar cheese produced with variable starter activity. J Food Prot 44:263, 1981b.

Ikram M, Luedecke LO. Growth and enterotoxin A production by *Staphylococcus aureus* in fluid dairy products. J Food Prot 40:769, 1977.

James T. The story of the enteropathogenic *Escherichia coli*. S Afr Med J 47:1476, 1973.

Jezyna Cz, Weglinska T, Nawrocka E, Falecka W, Wieliczko-gebska L, Rodkiewicz T, Piesiak Z, Ciesielski T. A milk-borne outbreak of tick-borne encephalitis in the Olsztyn Province. Przeg Epidemiol 30:480, 1976.

Jokippi L, Jokippi AMM. Cryptosporidiosis. In: Braude AI, Davis CE, Fierer J, eds. Infectious Diseases and Medical Microbiology. 2nd ed. Philadelphia: WB Saunders 1986, p 931.

Jones EE, Kim-Farley RJ, Algunaid M, Parvez MA, Ballad YA, Hightower AW, Orenstein WA, Broome CV. Diphtheria: a possible foodborne outbreak in Hodeida, Yemen Arab Republic. Bull WHO 63:287, 1985.

Jones GA, Gibson DL, Cheng K-J. Coliform bacteria in Canadian pasteurized dairy products. Can J Public Health 58:257, 1967.

Jones PH, Willis AT, Robinson DA, Skirrow MB, Josephs DS. *Campylobacter* enteritis associated with the consumption of free school milk. J Hyg Camb 87:155, 1981.

Jones TO. Caprine mastitis associated with *Yersinia pseudotuberculosis* infection. Vet Rec 110:231, 1982.

Kahana LM, Todd E. Histamine intoxication in a tuberculosis patient on isoniazid. Can Dis Weekly Rep 7:79, 1981.

Kaneene JB, Ahl AS. Drug residues in dairy cattle industry: epidemiological evaluation of factors influencing their occurrence. J Dairy Sci 70:2176, 1987.

Kaneene JB, Miller R. Description and evaluation of the influence of veterinary presence on the use of antibiotics and sulfonamides in dairy herds. J Am Vet Med Assoc 201:68, 1992.

Kantor HS. Bacterial enteritis. In: Braude AI, Davis CE, Fierer J, eds. Infectious Diseases and Medical Microbiology. 2nd ed. Philadelphia: WB Saunders, 1986, p 897.

Karahadian C, Lindsay RC, Dillman LL, Deibel RH. Evaluation of the potential for botulinal toxigenesis in reduced-sodium processed American cheese foods and spreads. J Food Prot 48:63, 1985.

Karmali MA, Steele BT, Petric M, Lim C. Sporadic cases of haemolytic uremic syndrome associated with faecal cytotoxin and cytotoxin-producing *Escherichia coli* in stools. Lancet 1:619, 1983.

Kasatiya SS. *Yersinia enterocolitica* gastroenteritis outbreak—Montreal. Can Dis Weekly Rep 2:73, 1976.

Kasrazadeh M, Genigeorgis, C. Potential growth and control of *Salmonella* in Hispanic type soft cheese. Int J Food Microbiol 22:127, 1994.

Kasrazadeh M, Genigeorgis C. Potential growth and control of *Escherichia coli* O157:H7 in soft Hispanic type cheese. Int J Food Microbiol 25:289, 1995.

Kästli P, Hausch R. The viability of *Br. abortus* in different cheese varieties. Schweiz Arch Tierheilk 99:638, 1957.

Kauf C, Lorent JP, Mosimann J, Schlatter I, Somaini B, Velvart J. Botulismusepidemie vom Typ B. Schweiz med Wschr 104:677, 1974.

Kaufmann OW, Brillaud AR. Development of *Clostridium botulinum* in experimentally inoculated sterile skim milk held at low temperatures. Am J Public Health 54:1514, 1964.

Kauter DA, Lilly T Jr, Lynt RK, Solomon HM. Toxin production by *Clostridium botulinum* in shelf-stable pasteurized process cheese spreads. J Food Prot 42:784, 1979.

Kauter DA, Solomon HM, Lake DE, Bernard DT, Mills DC. *Clostridium botulinum* and its toxins. In: Murray PR, Baron EJ, Pfaller MA, Tenover FC, Yolken RH, eds. Manual of Clinical Microbiology. 6th ed. Washington, DC: ASM Press, 1992, p 605.

Kells HR, Lear SA. Thermal death time curve of *Mycobacterium tuberculosis* var. *bovis* in artificially infected milk. Appl Microbiol 8:234, 1960.

Khalid AS, Harrigan WF. Investigation into the effect of salt level on growth of, and toxin production by, *Staphylococcus aureus* in Sudanese cheese. Lebensm Wiss Technol 17:99, 1984.

Khambaty FM, Bennett RW, Shah DB. Application of pulsed-field gel electrophoresis to the epidemiological characterization of *Staphylococcus intermedius* implicated in a food-related outbreak. Epidemiol Infect 113:75, 1994.

Khayat FA, Bruhn JC, Richardson GH. A survey of coliforms and *Staphylococcus aureus* in cheese using impedimetric and plate count methods. J Food Prot 51:53, 1988.

Kiermeier F, Buchner M. Distribution of aflatoxin M_1 in whey and curd during cheese processing. Z Lebensm Unters Forsch 164:82, 1977.

Kiermeier F, Weiss G, Behringer G, Miller M, Ranfft K. Presence and content of aflatoxin M_1 in milk supplied to a dairy. Z Lebensm Unters Forsch 163:71, 1977.

King NB. The survival of *Brucella abortus* (USDA strain 2308) in manure. J Am Vet Med Assoc 31:349, 1957.

Kleeburg HH, Tuberculosis and other mycobacteriosis. In: Hubbert WT, McCullock WF, Schnurrenberger PR, eds. Diseases Transmitted from Animals to Man. 6th ed. Springfield, IL: Charles C Thomas, 1975, p 315.

Klein BS, Vergeront JM, Blaser MJ, Edmonds P, Brenner DJ, Janssen D, Davis JP. *Campylobacter* infection associated with raw milk—an outbreak of gastroenteritis due to *Campylobacter jejuni* and thermotolerant *Campylobacter fetus* subsp. *fetus*. JAMA 255:361, 1986.

Kloos WE, Bannerman TL. *Staphylococcus* and *Micrococcus*. In: Murray PR, Baron EJ, Pfaller MA, Tenover FC, Yolken RH, eds. Manual of Clinical Microbiology. 6th ed. Washington, DC: ASM Press, 1995, p 282.

Kloos WE, Schliefer KH. Genus IV *Staphylococcus*. In: Sneath PH, Mair NS, Sharpe ME, Holt JG, eds. Bergy's Manual of Systematic Bacteriology. Vol 2. Baltimore: Williams & Wilkins, 1986, p 1013.

Knight P. Hemorrhagic *E. coli*: the danger increases—disease outbreaks and several deaths from contaminated meat products have officials scrambling to change outmoded rules. ASM News, 59:247, 1993.

Koenig S, Marth EH. Behavior of *Staphylococcus aureus* in cheddar cheese made with sodium chloride or a mixture of sodium chloride and potassium chloride. J Food Prot 45:996, 1982.

Koidis P, Doyle MP. Procedure for recovery of *Campylobacter jejuni* from inoculated unpasteurized milk. Appl Environ Microbiol 47:455, 1984.

Konowalchuk J, Speirs JI, Stavric S. Vero response to a cytotoxin in *Escherichia coli*. Infect Immunol 18:775, 1977.

Korlath JA, Osterholm MT, Judy LA, Forfang JC, Robinson RA. A point-source outbreak of campylobacteriosis associated with consumption of raw milk. J Infect Dis 152:592, 1985.

Kornacki JL, Marth EH. Fate of nonpathogenic and enteropathogenic *Escherichia coli* during the manufacture of Colby-like cheese. J Food Prot 45:310, 1982.

Kornblatt AN, Barrett T, Morris GK, Tosh FE. Epidemiologic and laboratory investigation of an outbreak of *Campylobacter* enteritis associated with raw milk. Am J Epidemiol 122:884, 1985.

Kramer JM, Gilbert RJ. *Bacillus cereus* and other *Bacillus* species. In: Doyle MP, ed. Foodborne Bacterial Pathogens. New York: Marcel Dekker, 1989, p 22.

Kristufkova Z, Simkovicova M. Epidemic of enterotoxicosis from ice cream caused by staphylococcal TSST-1. Czech Epidem Mikrobiol Imunol 37:36, 1988.

Krumbiegel ER, Wisniewski HJ. Q fever in the Milwaukee area. II. Consumption of infected raw milk by human volunteers. Arch Environ Health 21:63, 1970.

Kuesch GT. Typhoid fever. In: Braude AI, Davis CE, Fierer C, eds. Infectious Diseases and Medical Microbiology. 2nd ed. Philadelphia: WB Saunders, 1986, p 1189.

Kuzdas CD, Morse EV. The survival of *Brucella abortus*, USDA strain 2308, under controlled conditions in nature. Cornell Vet 44:216, 1954.

Lancette GA, Tatini SR. *Staphylococcus aureus*. In: Vanderzant C, Splittstoesser DF, eds. Compendium of Methods for the Microbiological Examination of Foods. Washington, DC: American Public Health Association, 1992, p 533.

Lander KP, Gill KPW. Experimental infection of the bovine udder with *Campylobacter coli/jejuni*. J Hyg Camb 84:421, 1980.

Langeveld LPM, van Spronsen WA, van Beresteijn CH, Notermans SHW. Consumption by healthy adults of pasteurized milk with a high concentration of *Bacillus cereus*: a double blind study. J Food Prot 59:723, 1996.

Lecos CW. A closer look at dairy safety. Dairy Food Environ Sanitat 6:240, 1986.

Lee RY, Silverman GJ, Munsey DT. Growth and enterotoxin A production by *Staphylococcus aureus* in precooked bacon in the intermediate moisture range. J Food Sci 46:1687, 1981.

Lennette EH, Clark WH, Abinanti MA, Brunetti O, Covent JM. Q fever studies—effect of pasteurization on *Coxiella burnetti* in naturally infected milk. Am J Hyg 55:246, 1952.

Levine MM. *Escherichia coli* that cause diarrhea: enterotoxigenic, enteropathogenic, enteroinvasive, enterohemorrhagic, and enteroadherent. J Infect Dis 155:377, 1987.

Levine MM, Caplan ES, Waterman D, Cash RA, Hornick RB, Snyder MJ. Diarrhea caused by *Escherichia coli* that produce only heat stable enterotoxin. Infect Immun 17:78, 1977.

Levy AJ. A gastro-enteritis outbreak probably due to a bovine strain of *Vibrio*. Yale J Biol Med 20:243, 1947.

Levy ME. Gastrointestinal illness associated with imported Brie cheese—District of Columbia. Morbid Mortal Weekly Rep 32:533, 1983.

LiCari JJ, Potter NN. *Salmonella* survival during spray drying and subsequent handling of skim milk powder. II. Effects of drying conditions. J Dairy Sci 52:871, 1970a.

LiCari JJ, Potter NN. *Salmonella* survival during spray drying and subsequent handling of skim milk powder. III. Effects of storage temperature on *Salmonella* and dried milk products. J Dairy Sci 53:877, 1970b.

Linnan MJ, Mascola L, Lou XD, Goulet V, May S, Salminen C, Hird DW, Yonkura ML, Hayes P, Weaver R, Audurier A, Plikaytis BD, Fannin SL, Kleks A, Broome CV. Epidemic listeriosis associated with Mexican-style cheese. N Engl J Med 319:823, 1988.

Lior H. Campylobacters—epidemiological markers. Dairy Food Environ Sanitat 14:317, 1994a.

Lior H. *Escherichia coli* O157:H7 and verotoxigenic *Escherichia coli* (VTEC). Dairy Food Environ Sanitat 14:378, 1994b.

Little CL, Knochel S. Growth and survival of *Yersinia enterocolitica, Samonella* and *Bacillus cereus* in Brie stored at 4, 8 and 20°C. Int J Food Microbiol 24:137, 1994.

Logan EF, Neill SD, Mackie DP. Mastitis in dairy cows associated with an aerotolerant campylobacter. Vet Rec 110:229, 1982.

Logan NA. *Bacillus* species of medical and veterinary importance. J Med Microbiol 25:157, 1988.

Lopes HR, Solino Noleto AR, de Las Heras MD, Bergdoll MS. Selective enterotoxin production in foods by *Staphylococcus aureus* strains that produce more than one enterotoxin. J Food Prot 56:538, 1993.

Lovett J, Francis DW, Hunt JM. Isolation of *Campylobacter jejuni* from raw milk. Appl Environ Microbiol 46:459, 1983.

Lovett J, Bradshaw JG, Peeler JT. Thermal inactivation of *Yersinia enterocolitica* in milk. Appl Environ Microbiol 44:517, 1982.

Lumsden LL, Leahe JP, Crohurst HR, Waller CE. The Montreal typhoid fever situation— report of a special board of the U.S. Public Health Service. Am J Public Health 17:783, 1927.

MacDonald KL, Griffin PM. Food-borne disease outbreaks, annual summary, 1982. Morbid Mortal Weekly Rep 35:7ss, 1983.

MacDonald LK, Eidson M, Strohmeyer C, Levy ME, Wells JG, Puhr ND, Wachsmuth K, Hargett NT, Cohen ML. A multistate outbreak of gastrointestinal illness caused by enterotoxigenic *Escherichia coli* in imported semisoft cheese. J Infect Dis 151:716, 1985.

MacLachlan J. Salmonellosis in Midlothian and Peeblesshire. Public Health Lond 88:79, 1974.

Maguire H. Continuing hazards from unpasteurized milk products in England and Wales. Int Food Safety News 2(2):22, 1993.

Maguire H, Cowden J, Jacob M, Rowe B, Roberts D, Bruce J, Mitchell E. An outbreak of *Salmonella dublin* infection in England and Wales associated with a soft unpasteurized cow's milk cheese. Epidemiol Infect 109:389, 1992.

Maguire HCF, Boyle M, Lewis MJ, Pankhurst J, Wieneke AA, Jacob M, Bruce J, O'Mahony M. A large outbreak of food poisoning of unknown aetiology associated with Stilton cheese. Epidemiol Infect 106:497, 1991.

Mantis A, Koidis P, Karaioannoglou P. Survival of *Yersinia enterocolitica* in yogurt. Milchwissenschaft 37:654, 1982.

Marier R, Wells JG, Swanson RC, Callahan W, Mehlman IJ. An outbreak of entero-pathogenic *Escherichia coli* foodborne disease traced to imported French cheese. Lancet 2:1376, 1973.

Marth EH. Salmonellae and salmonellosis associated with milk and milk products: a review. J Dairy Sci 52:283, 1969.

Martin ML, Shipman LD, Wells JG, Potter ME, Hedberg K, Wachsmuth IK, Tauxe RV, Davis JP, Arnoldi J, Tilleli J. Isolation of *Escherichia coli* O157:H7 from dairy cattle associated with two cases of haemolytic uremic syndrome. Lancet 2:1043, 1986.

Massa S, Poda G, Cesaroni D, Trovatelli LD. A bacteriological survey of retail ice cream. Food Microbiol 6:129, 1989.

Matches JR, Liston J. Low temperature growth of *Salmonella*. J Food Sci 33:641, 1968.

McCleery DR, Rowe MT. Development of a selective plating technique for recovery of *Escherichia coli* O157:H7 after heat stress. Lett Appl Microbiol 21:252, 1995.

McCloskey RV. Diphtheria. In: Braude AI, Davis CE, Fierer J, eds. Infectious Diseases and Medical Microbiology, Philadelphia: WB Saunders, p 718, 1986.

McDonough FE, Hargrove RE, Tittsler RP. The fate of salmonellae in the manufacture of cottage cheese. J Milk Food Technol 30:354, 1967.

McEwen SA, Martin SW, Clarke RC, Tamblyn SE. A prevalence survey of *Salmonella* in raw milk in Ontario, 1986–1987. J Food Prot 51:963, 1988.

McKhann GM, Cornblath DR, Griffin JW, Ho TW, Li CY, Jiang Z, Wu HS, Zhaori G, Liu Y, Jou LP, Liu TC, Gao CY, Mao JY, Blaser MJ, Mishu B, Asbury AK. Acute motor axonal neuropathy: a frequent cause of acute flaccid paralysis in China. Ann Neurol 33:333, 1993.

McKinnon CH, Pettipher GL. A survey of sources of heat-resistant bacteria in milk with particular reference to psychrotrophic spore-forming bacteria. J Dairy Res 50:163, 1983.

McLauchlin JM, Greenwood H, Pini PN. The occurrence of *Listeria monocytogenes* in cheese from a manufacturer associated with a case of listeriosis. Int J Food Microbiol 3/4:255, 1990.

McManus C, Lanier JM. *Salmonella, Campylobacter jejuni*, and *Yersinia enterocolitica* in raw milk. J Food Prot 50:51, 1987.

McNaughton RD, Leyland R, Mueller L. Outbreak of *Campylobacter* enteritis due to consumption of raw milk. Can Med Assoc J 126:657, 1982.

Meers PD. A case of classical diphtheria and other infections due to *Corynebacterium ulcerans*. J Infect 1:139, 1979.

Melling J, Capel BJ, Turnbull PCB, Gilbert RJ. Identification of a novel enterotoxigenic activity associated with *Bacillus cereus*. J Clin Pathol 29:938, 1976.

Menzies DB. An outbreak of typhoid fever in Alberta traceable to infected cheddar cheese. Can J Public Health 35:431, 1944.

Meyer KF, Eddie B. Perspectives concerning botulism. Z Hyg Infektionskr 133:255, 1951.

Mickelson R, Foltz VD, Martin WH, Hunter CA. The incidence of potentially pathogenic staphylococci in dairy products at the consumer level. II. Cheese. J Milk Food Technol 24:342, 1962.

Mikolajcik EM, Kearney JW, Kristoffersen T. Fate of *Bacillus cereus* in cultured and direct acidified skim milk and cheddar cheese. J Milk Food Technol 36:317, 1973.

Minor TE, Marth EH. *Staphylococcus aureus* and enterotoxin A in cream and butter. J Dairy Sci 55:1410, 1972.

Mishu B, Blaser MJ. Role of infection due to *Campylobacter jejuni* in the initiation of Guillain-Barré syndrome. Clin Infect Dis 17:104, 1993.

Mitscherlich E, Marth EH. Microbial Survival in the Environment—Bacteria and Rickettsiae Important in Human and Animal Health. Berlin: Springer-Verlag, 1984.

Moats WA. Inactivation of antibiotics by heating in foods and other substrates—a review. J Food Prot 51:491, 1988.

Mohammad A, Peiris JSM, Wijewanta EA. Serotypes of verocytotoxigenic *Escherichia coli* isolated from cattle and buffalo calf diarrhea. FEMS Microbiol Lett 35:261, 1986.

Montenegro MA, Bulte M, Trumpf T, Aleksic S, Reuter G, Bulling E, Helmuth R. Detection and characterization of fecal verotoxin-producing *Escherichia coli* from healthy cattle. J Clin Microbiol 28:1417, 1990.

Moore K, Damrow T, Abott DO, Jankowski S. Outbreak of acute gastroenteritis attributable to *Escherichia coli* serotype O104:H21—Helena, Montana, 1994. JAMA 274:529, 1995.

Morgan D, Mawer SL, Harman PL. The role of home-made ice cream as a vehicle of *Salmonella enteritidis* phage type 4 infection from fresh shell eggs. Epidemiol Infect 113:21, 1994a

Morgan D, Newman CP, Hutchinson DN, Walker AM, Rowe B, Majid F. Verotoxin producing *Escherichia coli* O157 infections associated with consumption, yoghurt. Epidemiol Infect 111:181, 1993.

Morgan D, Gunneberg C, Gunnell D, Healing TD, Lamerton S, Soltanpoor N, Lewis DA, White DG. An outbreak of *Campylobacter* infection associated with consumption of unpasteurized milk at a large festival in England. Eur J Epidemiol 10:581, 1994b.

Mor-Mur M, Carretero C, Pla R, Guamis B. A survey on the microbiological quality of a semi-soft on-farm manufactured goat cheese. Food Microbiol 9:345, 1992.

Moustafa MK, Ahmed AA-H, Marth EH. Behavior of virulent *Yersinia enterocolitica* during manufacture and storage of Colby-like cheese. J Food Prot 46:318, 1983a.

Moustafa MK, Ahmed AA-H, Marth EH. Occurrence of *Yersinia enterocolitica* in raw and pasteurized milk. J Food Prot 46:276, 1983b.

Moyer NP, Holcomb LA. *Brucella*. In: Murray PR, Baron EJ, Pfaller MA, Tenover FC, Yolken RH, eds. Manual of Clinical Microbiology. 6th ed. Washington, DC: ASM Press, 1995, p 549.

Murray EGD, Webb RA, Swann MBR. A disease of rabbits characterized by a large mononuclear leucocytosis caused by a hitherto undescribed bacillus *Bacterium monocytogenes* (n. sp.). J Pathol Bacteriol 29:407, 1926.

Nachamkin I. *Campylobacter* and *Arcobacter*. In: Murray PR, Baron EJ, Pfaller MA, Tenover FC, Yolken RH, eds. Manual of Clinical Microbiology. 6th ed. Washington, DC: ASM Press, 1995, p 483.

Naguib MN, Nour MA, Noaman AA. Survival of *Staphylococcus aureus* in Ras cheese. Arch Lebensmittelhyg 30:197, 1979.

Neill MA. *E. coli* O157:H7—current concepts and future prospects. J Food Safety 10:99, 1989.

Nevin M. Botulism from cheese. J Infect Dis 28:226, 1921.

Nolan CM, Anderson HW, Hoyt DET. *Salmonella* surveillance pays off. Epilog July/August 1981. Seattle, WA: Seattle King County Health Department of Public Health, 1981.

Nolte FS, Metchock B. *Mycobacterium*. In: Murray PR, Baron EJ, Pfaller MA, Tenover FC, Yolken RH, eds. Manual of Clinical Microbiology. 6th ed. Washington, DC: ASM Press, 1995, p 400.

Northolt MD, Beckers HJ, Vecht U, Toepoel L, Soentoro PSS, Wisselink HJ. *Listeria monocytogenes*: heat resistance and behavior during storage of milk and whey and making of Dutch types of cheese. Neth Milk Dairy J 42:207, 1988.

Notermans S, Hoogenboom-Verdegaal A. Existing and emerging foodborne diseases. Int J Food Microbiol 15:197, 1992.

Nour MA. Survival of *Brucella* organisms in artificially contaminated cream. Egypt J Food Sci 10:67, 1982.

Nunez M, Chavarri FJ, Garcia BE, Gaytan LE. The effect of lactic starter inoculation and storage temperature on the behaviour of *Staphylococcus aureus* and *Enterobacter cloacae* in Burgos cheese. Food Microbiol 3:235, 1986.

Obiger G von, Neuschulz J, Schonberg A. Tests made on French cream cheese for possible exciters of brucellosis and tuberculosis. Ber Münch tierärzt. Wochenschr 16:318, 1970.

Ocando AJF, Gutierrez DU, Apalmo ZR, Salas LTC, Basanta Y. Microflora isolated from Venezuelan "Palmita-type" cheese. J Food Prot 54:856, 1991.

Oliver SP, Maki JL, Dowlen HH. Antibiotic residues in milk following antimicrobial therapy during lactation. J Food Prot 53:693, 1990.

Olsen JA, Yousef AE, Marth EH. Growth and survival of *Listeria monocytogenes* during making and storage of butter. Milchwissenschaft 43:487, 1988.

Olson JC Jr, Sanders AC. Penicillin in milk and milk products: some regulatory and public health considerations. J Milk Food Technol 38:630, 1975.

Olson JC Jr, Casman EP, Baer EF, Stone JE. Enterotoxigenicity of *Staphylococcus aureus* cultures isolated from acute cases of bovine mastitis. Appl Microbiol 20:605, 1970.

Olsvik O, Kapperud G. Enterotoxin production in milk at 22 and 4°C by *Escherichia coli* and *Yersinia enterocolitica*. Appl Environ Microbiol 43:997, 1982.

O'Mahony M, Mitchell E, Gilbert RJ, Hutchinson DN, Begg NT, Roadhouse JC, Morris JE. An outbreak of foodborne botulism associated with contaminated hazelnut yogurt. Epidemiol Infect 104:389, 1990.

Orskov F, Orskov I. *Escherichia coli* serotyping and disease in man and animals. Can J Microbiol 38:699, 1992.

Osterholm MT, MacDonald KL, White KE, Wells JG, Spika JS, Potter ME, Forfang JC, Sorenson RM, Milloy PT, Blake PA. An outbreak of a newly recognized chronic diarrhea syndrome associated with raw milk consumption. JAMA 256:484, 1986.

Otero A, Garcia MC, Garcia ML, Prieto M, Moreno B. Behavior of *Staphylococcus aureus* strains, producers of enterotoxins C_1 or C_2 during the manufacture and storage of Burgos cheese. J Appl Microbiol 64:117, 1988.

Overcast WW, Atmaram K. The role of *Bacillus cereus* in sweet curdling of fluid milk. J Milk Food Technol 37:233, 1974.

Padhye NV, Doyle MP. Rapid procedure for detecting enterohemorrhagic *Escherichia coli* O157:H7 in food. Appl Environ Microbiol 57:2693, 1991.

Padhye NV, Doyle MP. *Escherichia coli* O157:H7 epidemiology, pathogenesis and methods for detection in food. J Food Prot 55:555, 1992.

Papadopoulou C, Maipa V, Dimitriou D, Pappas C, Voutsinas L, Malatou H. Behavior of *Salmonella enteritidis* during the manufacture, ripening, and storage of feta cheese made from unpasteurized ewe's milk. J Food Prot 56:25, 1993.

Papageorgiou DK, Marth EH. Fate of *Listeria monocytogenes* during the manufacture and ripening of blue cheese. J Food Prot 52:459, 1989a.

Papageorgiou DK, Marth EH. Fate of *Listeria monocytogenes* during the manufacture, ripening and storage of feta cheese. J Food Prot 52:82, 1989b.

Park, HS and Marth EH. Behavior of *Salmonella typhimurium* in skimmilk during fermentation by lactic acid bacteria. J Milk Food Technol 35:482, 1972a.

Park HS, Marth EH. Growth of salmonellae in skimmilk which contains antibiotics. J Milk Food Technol 35:7, 1972b.

Park HS, Marth EH, Goepfert JM, Olson NF. The fate of *Salmonella typhimurium* in the manufacture and ripening of low-acid cheddar cheese. J Milk Food Technol 33:280, 1970a.

Park HS, Marth EH, Olson NF. Survival of *Salmonella typhimurium* in cold-pack cheese food during refrigerated storage. J Milk Food Technol 33:383, 1970b.

Park K-Y, Bullerman LB. Increased aflatoxin production by *Aspergillus parasiticus* under conditions of cycling temperature. J Food Sci 46:1147, 1981.

Park K-Y, Bullerman LB. Effects of substrate and temperature on aflatoxin production by *Aspergillus parasiticus* and *Aspergillus flavus*. J Food Prot 46:178, 1983.

Park WH, Krumwiede C Jr. The relative importance of the bovine and human types of tubercle bacilli in the different forms of human tuberculosis. J Med Res 25:313, 1911.

Parker F, Hudson NP. The etiology of Haverhill fever (erythema arthriticum epidemicum). Am J Pathol ii:357, 1926.

Parker MT. Enteric infections: typhoid and paratyphoid fever. In: Smith GR, Easmon CSF, eds. Topley and Wilson's Principles of Bacteriology, Virology and Immunity. 8th ed. Vol 3—Bacterial Diseases. Philadelphia: BC Decker, 1990, p 423.

Parry WH. Milk-borne diseases: an epidemiological review. Lancet i:216, 1966.

Passes H, Glasser D, Flynn J, Acree KH. Bovine tuberculosis—Maryland. Morbid Mortal Weekly Rep 27:108, 1978.

Patterson DSP, Glancy EM, Roberts BA. The carry-over of aflatoxin M_1 into the milk of cows fed rations containing a low concentration of aflatoxin B_1. Food Cosmet Toxicol 18:35, 1980.

Patton CM, Wachsmuth IK, Evins GM, Kiehlbauch JA, Plikaytis BD, Troup N, Tompkins L, Lior H. Evaluation of 10 methods to distinguish epidemic-associated *Campylobacter* strains. J Clin Microbiol 29:680, 1991.

Pereira ML, Do Carmo LS, Dos Santos EJ, Pereira JL, Bergdoll MS Enterotoxin H in staphylococcal food poisoning. J Food Prot 59:559, 1996.

Pfeiffer DH, Viljoen NF. Diphtheria and mastitis. J South Afr Vet Med Assoc 16:148, 1945.

Phillips JD, Griffiths MW. Factors contributing to the seasonal variation of *Bacillus* spp. in pasteurized dairy products. J Appl Bacteriol 61:275, 1986.

Pinegar JA, Buxton JD. An investigation of the bacteriological quality of retail vanilla slices. J Hyg Camb 78:387, 1977.

Place AR. Animal drug residues in milk—the problem and the response. Dairy Food Environ Sanitat 10:662, 1990.

Polewska-Jeske A, Szulc M, Waluszkiewicz H. An epidemic of alimentary poisoning with ice cream in Szczecin caused by *Salmonella typhimurium*. Przeg Epidemiol 38:61, 1984.

Pong RS, Wogan GN. Toxicity and biochemical and fine structural effects of synthetic aflatoxins M_1 and B_1 in rat liver. J Nat Cancer Inst 47:585, 1971.

Porter AMW, Smith EL. Brucellosis and goat's cheese? BMJ 3:580, 1971.

Porter IA, Reid TMS. A milk-borne outbreak of *Campylobacter* infection. J Hyg Camb 84:415, 1980.

Potel J. Aetilogie der Granulomatosis Infantiseptica. Wiss Z Martin Luther Univ 2:15, 1953/1954.

Potter ME, Blaser MJ, Sikes RK, Kaufmann AF, Wells JG. Human *Campylobacter* infection associated with certified raw milk. Am J Epidemiol 117:475, 1983.

Pounder RE. The pathogenesis of Crohn's disease. Gastroenterology 29(suppl 7):11, 1994.

Price RM. Milk and its relation to tuberculosis. Can Public Health J 25:13, 1934.

Pritchard TJ, Beliveau CM, Flanders KJ, Donnelly CW. Environmental surveillance of dairy processing plants for the presence of *Yersinia* species. J Food Prot 58:395, 1995.

Prober CG, Tune B, Hoder L. *Yersinia pseudotuberculosis* septicemia. Am J Dis Child 133:623, 1979.

Procter ME, Brosch R, Mellen JW, Garrett LA, Kasper CW, Luchansky JB. Use of pulsed-field gel electrophoresis to link sporadic cases of invasive listeriosis with recalled chocolate milk. Appl Environ Microbiol 61:3177, 1995.

Pullinger EJ, Kemp AE. Growth of *Salmonella typhi* and certain other members of the *Salmonella* group in milk and butter stored at atmospheric temperatures. J Hyg 38:587, 1938.

Purvis JD, Morris GC. Report of a food-borne streptococcus outbreak. US Naval Bull 46:613, 1946.

Quinn J, Keon N, Coohon DB, Scott RM. Bovine tuberculosis—Michigan. Morbid Mortal Weekly Rep 23:166, 1974.

Rammel CG. *Brucella* in dairy products: a review. Austral J Dairy Technol 22:40, 1967.

Rangasamy PN, Iyer M, Roginski H. Isolation and characterization of *Bacillus cereus* in milk and dairy products manufactured in Victoria. Austral. J Dairy Technol 48:93, 1993.

Raska K, Helcl J, Jezek J, Kubelka Z, Litov M, Novak K, Radkovsky J, Sery V, Zejdl J, Zikmund V. A milk-borne infectious hepatitis epidemic. J Hyg Epidemiol Microbiol Immunol 10:413, 1966.

Ratnam S, March SB. Laboratory studies on *Salmonella*-contaminated cheese involved in a major outbreak of gastroenteritis. J Appl Bacteriol 61:51, 1986.

Read RB Jr, Bradshaw JG, Francis DW. Growth and toxin production of *Clostridium botulinum* type E in milk. J Dairy Sci 53:1183, 1970.

Reilly WJ, Sharp JCM, Forbes GI, Paterson GM. Milkborne salmonellosis in Scotland 1980 to 1982. Vet Rec 112:578, 1983.

Reitsma CJ, Henning DR. Survival of enteropathogenic *Escherichia coli* O157:H7 during the manufacture and curing of cheddar cheese. J Food Prot 59:460, 1996.

Remington JS, McLeod R. Toxoplasmosis. In: Braude AI, Davis CE, Fierer J, eds. Infectious Diseases and Medical Microbiology. 2nd ed. Philadelphia: WB Saunders, 1986, p 1521.

Renwick SA, Wilson JB, Clark RC, Lior H, Borczyk AA, Spika J, Rahn JK, McFadden K, Bouwer A, Copps A, Anderson NG, Alves D, Karmali MA. Evidence for direct transmission of *Escherichia coli* O157:H7 infection between calves and a human. J Infect Dis 168:792, 1993.

Rhodehamel EJ, Harmon SM. *Bacillus cereus*. In: FDA Bacteriological Analytical Manual. 8th ed. Gaithersburg, MD: AOAC International, 1995, p 14.01.

Rich ED, Fellow CE. A typhoid epidemic traced to cheese. Am J Public Health 13:210, 1923.

Richardson GH. Dairy products. In: Helrich K, ed. Official Methods of Analysis of the Association of Official Analytical Chemists. 15th ed. Arlington, VA: AOAC, 1990, p 802.

Richwald GA, Greenland S, Johnson BJ, Fruedland JM, Goldstein EJC, Plichta DT. Assessment of the excess risk of *Salmonella dublin* infection associated with the use of certified raw milk. Public Health Rep 103:489, 1988.

Riemann HP, Meyer ME, Theis JH, Kelso G, Beymer DE. Toxoplasmosis in an infant fed unpasteurized goat milk. J Pediatr 87:573, 1975.

Riley LW, Remis RS, Helgerson SD, McGee HB, Wells JG, Davis BR, Herbert RJ, Olcott ES, Johnson LM, Hargrett NT, Blake PA, Cohen ML. Hemorrhagic colitis associated with a rare *Escherichia coli* serotype. N Engl J Med 308:681, 1983.

Riordan T, Humphrey TJ, Fowles A. A point source outbreak of *Campylobacter* infection related to bird-pecked milk. Epidemiol Infect 110:261, 1993.

Robinson DA. Infectious dose of *Campylobacter jejuni* in milk. BMJ 282:1584, 1981.

Robinson DA, Jones DM. Milk-borne *Campylobacter* infection. BMJ 282:1374, 1981.

Robinson DA, Edgar WM, Gibson GL, Matchett AA, Robertson L. *Campylobacter* enteritis associated with consumption of unpasteurized milk. Br J Med 1:1171, 1979.

Rodriguez MH, Barrett EL. Changes in microbial population and growth of *Bacillus cereus* during storage of reconstituted dry milk. J Food Prot 49:680, 1986.

Rodriguez-Alvarez C, Hardisson A, Alvarez R, Arias A, Sierra A. Enterobacteriaceae in ice cream sold at retail in Tenerife, Spain. J Food Quality 17:423, 1994.

Rohrbach BW, Draughon FA, Davidson MP, Oliver SP. Prevalence of *Listeria monocytogenes, Campylobacter jejuni, Yersinia enterocolitica*, and *Salmonella* in bulk tank milk: risk factors and risk of human exposure. J Food Prot 55:93, 1992.

Romanova TV, Shkarin VV, Khazenson LB. Group cryptosporidiosis morbidity in children. Meditsink Parazit Parazitar Bolezni 3:50, 1992.

Rosenau MJ, Frost WD, Bryant R. A study of the butter market of Boston. J Med Res 25:69, 1914.

Rosenow EM, Marth EH. Growth of *Listeria monocytogenes* in skim, whole and chocolate milk, and in whipping cream during incubation at 4, 8, 13, 21 and 35°C. J Food Prot 50:452, 1987.

Rowe B, Begg NT, Hutchinson DN, Dawkins HC, Gilbert RJ, Jacob M, Hales BH, Rae FA, Jepson M. *Salmonella ealing* infections associated with consumption of infant dried milk. Lancet 2:900, 1987.

Ryan CA, Nickels MK, Hargrett-Bean NT, Potter ME, Endo T, Mayer L, Langkop CW, Gibson C, McDonald RC, Kenney RT, Puhr ND, McDonnell PJ, Martin RJ, Cohen ML, Blake PA. Massive outbreak of antimicrobial resistant salmonellosis traced to pasteurized milk. JAMA 258:3269, 1987.

Ryan JL. Bites: *P. multocida*, DF-2, *S. moniliformis*, and *S. minor*. In: Braude AI, Davis CE, Fierer J, eds. Infectious Diseases and Medical Microbiology. 2nd ed. Philadelphia: WB Saunders, 1986, p 1499.

Ryser ET, Marth EH. Behavior of *Listeria monocytogenes* during the manufacture and ripening of cheddar cheese. J Food Prot 50:7, 1987a.

Ryser ET, Marth EH. Fate of *Listeria monocytogenes* during manufacture and ripening of Camembert cheese. J Food Prot 50:372, 1987b.

Ryser ET, Marth EH. Survival of *Listeria monocytogenes* in cold-pack cheese food during refrigerated storage. J Food Prot 51:615, 1988a.

Ryser ET, Marth EH. Behavior of *Listeria monocytogenes* during manufacture and ripening of brick cheese. J Dairy Sci 72:838, 1988b.

Ryser ET, Marth EH. *Listeria*, listeriosis and food safety. New York: Marcel Dekker, 1991.

Ryser ET, Doyle MP, Marth EH. Survival of *Listeria monocytogenes* during manufacture and storage of cottage cheese. J Food Prot 48:746, 1985.

Sabbaghian H, Nadim M. Epidemiology of human brucellosis in Isfahan, Iran. J Hyg 73:221, 1974.

Sabioni JG, Hirooka EY, de Souza MLR. Food poisoning from Minas-type cheese contaminated with *Staphylococcus aureus*. Rev Saude Publ S Paulo 22:458, 1988.

Sacks JJ, Roberto RR, Brooks NF. Toxoplasmosis infection associated with raw goat's milk. JAMA 128:1728, 1982.

Sakhre PG, Vyas SH. The incidence of *Mycobacterium tuberculosis* in milk from clinically healthy udders of tuberculin positive Kankrej cows. Food Farming Agric 11:260, 1978.

Salama SM, Bolton FJ, Hutchinson DN. Application of a new phage typing scheme to campylobacters isolated during outbreaks. Epidemiol Infect 104:405, 1990.

Salman MD, Meyer ME. Epidemiology of bovine brucellosis in the Mexicalli Valley, Mexico: literature review of disease-associated factors. Am J Vet Res 45:1557, 1984.

Sancak YC, Boynukara B, Yardimci H. The occurrence and survival of *Brucella* spp. in Van Herby cheese. Veterinarium 4:1, 1993.

Sanchez-Rey R, Poulett B, Caceres P, Larriba G. Microbiological quality and incidence of some pathogenic microorganisms in La Serena cheese throughout ripening. J Food Prot 56:879, 1993.

Sanderson JD, Moss MT, Tizard ML, Hermon-Taylor J. *Mycobacterium paratuberculosis* DNA in Crohn's disease tissue. Gut 33:890, 1992.

Savage WG. Mitchel lecture on human tuberculosis of bovine origin. BMJ 2:905, 1933.

Schaack MM, Marth EH. Behavior of *Listeria monocytogenes* in skim milk during fermentation with mesophilic lactic acid bacteria starter cultures. J Food Prot 51:600, 1988a.

Schaack MM, Marth EH. Behavior of *Listeria monocytogenes* in skim milk and in yogurt mix during fermentation with thermophilic lactic acid bacteria starter cultures. J Food Prot 51:607, 1988b.

Schiemann DA. Association of *Yersinia enterocolitica* with the manufacture of cheese and occurrence in pasteurized milk. Appl Environ Microbiol 36:274, 1978.

Schiemann DA. *Yersinia enterocolitica* in milk and dairy products. J Dairy Sci 70:383, 1987.

Schiemann DA. *Yersinia enterocolitica* and *Yersinia pseudotuberculosis*. In: Doyle MP, ed. Foodborne Bacterial Pathogens. New York: Marcel Dekker, 1989, p 601.

Schiemann DA, Toma S. Isolation of *Yersinia enterocolitica* from raw milk. Appl Environ Microbiol 35:54, 1978.

Schiemann DA, Wauters G. *Yersinia.* In: Vanderzant C, Splittstoesser DF, eds. Compendium of Methods for the Microbiological Examination of Foods. Washington, DC: American Public Health Association, 1992, p 433.

Schindler AF. Temperature limits for production of aflatoxin by twenty-five isolates of *Aspergillus flavus* and *Aspergillus parasiticus.* J Food Prot 40:39, 1977.

Schleifstein JI, Coleman MB An unidentified organism resembling *B. ligineri* and *Past. pseudotuberculosis,* and pathogenic for man. NY State J Med 39:1749, 1939.

Schmitt N, Bowmer EJ, Willoughby BA. Food poisoning outbreak attributed to *Bacillus cereus.* Can J Public Health 67:418, 1976.

Schnurrenberger LW, Pate J. Gastroenteritis attributed to imported French cheese—United States. Morbid Mortal Weekly Rep 20:425, 1971.

Schroeder EC, Brett GW. Public health studies concerning cheese. J Am Vet Med Assoc 5:674, 1918.

Schuman JD, Zottola EA, Harlander SK. Preliminary characterization of a food-borne multiple-antibiotic-resistant *Salmonella typhimurium* strain. Appl Environ Microbiol 55:2344, 1989.

Scott PM. Mycotoxigenic fungal contaminants of cheese and other dairy products. In: van Egmond HP, ed. Mycotoxins in Dairy Products. London: Elsevier Applied Science, 1989, p 193.

Scott PM. Natural poisons—Mycotoxins. In: Helrich K, ed. Official Methods of Analysis of the Association of Official Analytical Chemists. 15th ed. Arlington, VA: AOAC, 1990, p 1184.

Scott WM, Minett FC. Experiments on infection of cows with typhoid bacilli. J Hyg 45:159, 1947.

Sebald M, Jouglard J, Gilles G. Type B botulism in man due to cheese. Ann Microbiol Inst Pasteur 125A:349, 1974.

Sedlak J, Present knowledge and aspects of *Citrobacter.* Curr Topics Microbiol Immunol 62:41, 1973.

Seiber SM, Correa P, Dalgard DW, Adamson RH. Induction of osteogenic sarcomas and tumors of the hepatobiliary system in non-human primates with aflatoxin B_1. Cancer Res 39:4545, 1979.

Seyffert WA, Bernard JA. Brucellosis: report of six cases. Texas Med 65:47, 1969.

Shahamat M, Seaman A, Woodbine M. Survival of *Listeria monocytogenes* in high salt concentrations. Zbl Bakteriol Hyg I Abt Orig A, 246:506, 1980.

Shanson DC, Gazzard BG, Midgley J, Dixey J, Gibson GL, Stevenson J, Finch RG, Cheesbrough J. *Streptococcus moniliformis* isolated from blood in four cases of Haverhill fever. Lancet i:92, 1983.

Sharp JCM. Infections associated with milk and dairy products in Europe and North America, 1980–85. Bull WHO 65:397, 1987.

Shayegani M, Morse M, DeForge I, Root T, Malmberg Parsons L, Maupin PS. Microbiology of a major foodborne outbreak of gastroenteritis caused by *Yersinia enterocolitica* serogroup O:8. J Clin Microbiol 17:35, 1983.

Shih CN, Marth EH. Experimental production of aflatoxin on brick cheese. J Milk Food Technol 35:585, 1972.

Simms L, MacRae IC. Survival of *Campylobacter jejuni* in raw, pasteurized and ultra-heat-treated goat's milk stored at different temperatures. Lett Appl Microbiol 8:177, 1989.

Sims GR, Glenister DA, Brocklehurst TF, Lund BM. Survival and growth of food poisoning bacteria following inoculation into cottage cheese varieties. Int J Food Microbiol 9:173, 1989.

Sims JE, Kelley DC, Foltz VD. Effects of time and temperature on salmonellae in inoculated butter. J Milk Food Technol 32:485, 1970.

Singh RS, Ranganathan B. Occurrence of enteropathogenic *Escherichia coli* serotypes in milk and milk products. Milchwissenschaft 29:529, 1974.

Sinnhuber RO, Lee DJ, Wales JH, Landers MK, Keyl AC. Hepatic carcinogenesis of aflatoxin M_1 in rainbow trout (*Salmo gairdneri*). J Nat Cancer Inst 53:1285, 1974.

Skirrow MB. Campylobacteritis: a "new" disease. BMJ 2:9, 1977.

Small RG, Sharp JCM. A milk-borne outbreak due to *Salmonella dublin*. J Hyg Camb 82:95, 1979.

Smith GR. Botulism. In: Smith GR, Easmon CSF, eds. Topley and Wilson's Principles of Bacteriology, Virology and Immunity. 8th ed. Vol 3—Bacterial Diseases. Philadelphia: BC Decker, 1990, p 513.

Smith JL. *Cryptosporidium* and *Giardia* as agents of foodborne disease. J Food Prot 56:451, 1993.

Sneath PHA. Endospore-forming gram-positive rods and cocci. In: Sneath PH, Mair NS, Sharpe ME, Holt JG, eds. Bergy's Manual of Systematic Bacteriology. Vol 2. Baltimore: Williams & Wilkins, 1986, p 1104.

Sockett PN. Communicable disease associated with milk and dairy products: England and Wales 1987–1989. Commun Dis Rep 1:R-9, 1991.

Sockett PN, Cowden JM, Le Baigue S, Ross D, Adak GK, Evans H. Foodborne disease surveillance in England and Wales: 1989–1991. Comm Dis Rep 3(12):R159, Nov. 5, 1993.

Solodovnikov IP, Aleksandrovskaia IM. Multiyear observations of the dynamic epidemic process in sonne dysentery in a small district of Vladimir Province. Zh Mikrobiol Epidemiol Immunobiol 9–10:41, 1992.

Solomon HM, Rhodehamel EJ, Kautter DA. *Clostridium botulinum*. In: FDA Bacteriological Analytical Manual. 8th ed. Gaithersburg, MD: AOAC International, 1995, p 17.01.

Somers EB, Taylor SL. Antibotulinal effectiveness of nisin in pasteurized process cheese spreads. J Food Prot 50:842, 1987.

Southern JP, Smith RMM, Palmer SR. Bird attack on milk bottles: Possible mode of transmission of *Campylobacter jejuni* to man. Lancet 336:1425, 1990.

Spira WM, Goepfert JM. *Bacillus cereus*—induced fluid accumulation in rabbit illeal loops. Appl Microbiol 24:341, 1972.

Stadhouders J, Cordes MM, van Schouwenburg-van Foeken AWJ. The effect of manufacturing conditions on the development of staphylococci in cheese. Their inhibition by starter bacteria. Neth Milk Dairy J 32:193, 1978.

Stadhouders J, Hup G, Langeveld LPM. Some observations on the germination, heat resistance and outgrowth of fast-germinating and slow-germinating spores of *Bacillus cereus* in pasteurized milk. Neth Milk Dairy J 34:215, 1980.

Stadler H, Isler R, Stutz W, Salfinger M, Lauwers S, Vischer W. Beitrag zur Epidemiologie von *Campylobacter jejuni* von der asymptomatischen Ausscheidung im Stall zur Erkrankung bei über 500 Personen. Schweiz med Wschr 7:245, 1983.

Stahl S. Tuberculosis infection due to milk. Am J Public Health 29:1154, 1939.

Stanfield JT, Jackson GJ, Aulisio CCG. *Yersinia enterocolitica*: survival of a pathogenic strain on milk containers. J Food Prot 48:947, 1985.

Stecchini ML, Sarais I, de Bertoldi M. The influence of *Lactobacillus plantarum* culture inoculation on the fate of *Staphylococcus aureus* and *Salmonella typhimurium* in Montasio cheese. Int J Food Microbiol 14:99, 1991.

Steede FDF, Smith HW. Staphylococcal food-poisoning due to infected cow's milk. BMJ 2:576, 1954.

Steele BT, Murphy N, Arbus GS, Rance CP. An outbreak of hemolytic uremic syndrome associated with ingestion of fresh apple juice. J Pediatr 101:963, 1982.

Steele JH, Ranney AF. Animal tuberculosis. Am Rev Tuberc 77:908, 1958.

Steere AC, Craven PJ, Hall WJ III, Leotsakis N, Wells JG, Farmer JJ III, Gangarosa EJ. Person-to-person spread of *Salmonella typhimurium* after a hospital common-source outbreak. Lancet i:319, 1975.

Stern NJ, Pierson MD Kotula AW. Growth and competitive nature of *Yersinia enterocolitica* in whole milk. J Food Sci 45:972, 1980.

Stern NJ, Kazmi SU. Less recognized and presumptive foodborne pathogenic bacteria. In: Doyle MP, ed. Foodborne Bacterial Pathogens. New York: Marcel Dekker, 1989, p 71.

Stern NJ, Patton CM, Doyle MP, Park CE, McCardell BA. *Campylobacter*. In: Vanderzant C, Splittstoesser DF, eds. Compendium of Methods for the Microbiological Examination of Foods. Washington, DC: American Public Health Association, 1992, p 475.

Stevens AJ, Saunders CN, Spence JB, Newnham AG. Investigation into "disease" of turkey poults. Vet Rec 72:627, 1960.

Stiles ME. Less recognized or presumptive foodborne pathogenic bacteria. In: Doyle MP, ed. Foodborne Bacterial Pathogens. New York: Marcel Dekker, 1989, p 673.

Stoloff L. Aflatoxin M_1 in perspective. J Food Prot 43:226, 1980.

Stone RV. Staphylococci food-poisoning and dairy products. J Milk Technol 6:7, 1943.

Stratton JE, Hutkins RW, Taylor SL. Biogenic amines in cheese and other fermented foods: a review. J Food Prot 54:460, 1991.

Street L Jr, Grant WW, Alva JD. Brucellosis in childhood. Pediatr 55:416, 1975.

Stubblefield RD, Shannon GM. Aflatoxin M_1: Analysis in dairy products and distribution in dairy foods made from artificially contaminated milk. JAOAC 57:847, 1974.

Stubblefield RD, van Egmond HP. Chromatographic methods of analysis for aflatoxin M_1. In: van Egmond HP, ed. Mycotoxins in Dairy Products. London: Elsevier Applied Science, 1989, p 57.

Styliadis S, Barnum D. *Salmonella* Infection in Man and Animals in the Province of Ontario. In: Proceedings of the International Symposium of *Salmonella*, New Orleans, LA, 1984, p 200.

Su, Y-C, Wong ACL. Identification and purification of a new staphylococcal enterotoxin H. Appl Environ Microbiol 61:1438, 1995.

Suchowiak T, Haiat Z. Milk-spread *Salmonella* enteritis epidemic in the Wroclaw province in June 1978. Przeg Epidemiol 34:325, 1980.

Sumner SS, Roche F, Taylor SL. Factors controlling histamine production in Swiss cheese inoculated with *Lactobacillus buchneri*. J Dairy Sci 73:3050, 1990.

Sumner SS, Speckhard MW, Somers EB, Taylor SL. Isolation of histamine-producing *Lactobacillus buchneri* from Swiss cheese implicated in a food poisoning outbreak. Appl Environ Microbiol 50:1094, 1985.

Sun P, Chu FS. A simple solid-phase radioimmunoassay for aflatoxin B_1. J Food Safety 1:67, 1977.

Swaminathan B, Rocourt J, Bille J. *Listeria*, identification of foodborne bacterial pathogens by gene probes. In: Murray PR, Baron EJ, Pfaller MA, Tenover FC, Yolken RH, eds. Manual of Clinical Microbiology 6th ed. Washington, DC: ASM Press, 1995, p 341.

Sweeney RW, Whitlock RH, Rosenberger AE. *Mycobacterium paratuberculosis* cultured from milk and supramammary lymph nodes of infected asymptomatic cows. J Clin Microbiol 30:166, 1992.

Tacket CO, Davis BR, Carter GP, Randolph JF, Cohen ML. *Yersinia enterocolitica* pharyngitis. Ann Intern Med 99:40, 1983.

Tacket CO, Narain JP, Sattin R, Lofgren JP, Konigsberg C, Rendtorff RC, Rausa A, Davis BR, Cohen ML. A multistate outbreak of infections caused by *Yersinia enterocolitica* transmitted by pasteurized milk. JAMA 251:483, 1984.

Tanaka N, Goepfert JM, Traisman E, Hoffbeck WM. A challenge of pasteurized process cheese spread with *Clostridium botulinum* spores. J Food Prot 42:787, 1979.

Tanaka N, Traisman E, Plantinga P, Finn L, Flom W, Meske L, Guggisberg J. Evaluation of factors involved in antibotulinal properties of pasteurized process cheese spreads. J Food Prot 49:526, 1986.

Tarala J. Some observations on bovine brucellosis, its eradication and its relation to undulant fever. State Vet J 24:220, 1969.

Tatini SR, Wesala WD, Jezeski JJ, Morris HA. Production of staphylococcal enterotoxin A in blue, brick, mozzarella, and Swiss cheese. J Dairy Sci 56:429, 1973.

Taylor A Jr, Santiago A, Gonzalez-Cortes A, Gangarosa EJ. Outbreak of typhoid fever in Trinidad in 1971 traced to a commercial ice cream product. Am J Epidemiol 100:150, 1974.

Taylor DN. *Campylobacter* infections in developing countries. In: Nachamkin I, Blaser NJ, Tompkins LS, eds. *Campylobacter jejuni*: Current Status and Future Trends. Washington, DC: American Society of Microbiology, 1992 p 20.

Taylor DN, Porter BW, Williams CA, Miller HG, Bopp CA, Blake PA. *Campylobacter* enteritis: a large outbreak traced to commercial raw milk. West J Med 137:365, 1982a.

Taylor PR, Weinstein WM, Bryner JH. *Campylobacter fetus* infection in human subjects: association with raw milk. Am J Med 66:779, 1979.

Taylor SL. Histamine poisoning associated with fish, cheese and other foods. Monograph WHO, 1985.

Taylor SL. Histamine food poisoning: toxicology and clinical aspects. Crit Rev Toxicol 17:91, 1986.

Taylor SL, Keefe TJ, Windham ES, Howell JF. Outbreak of histamine poisoning associated with the consumption of Swiss cheese. J Food Prot 45:455, 1982b.

Taylor TK, Wilks CR, McQueen DS. Isolation of *Mycobacterium paratuberculosis* from the milk of a cow with Johne's disease. Vet Rec 109:532, 1981.

Teclaw RF, Heck H, Wagner GG, Romo S, Garcia Z. Prevalence of brucellosis infection in cattle in the Mexican state of Nuevo Leon, Tamaulipas and Coahvila as determined by the ELISA. Prev Vet Med 3:445, 1985.

Terhune C, Sazi E, Kalishman N, Bobst J, Bonnlander B, Googins JA, Williams P. Raw-milk-associated illness—Oregon, California. Morbid Mortal Weekly Rep 30:90, 1981.

Tham W. Survival of *Listeria monocytogenes* in cheese made of unpasteurized goat milk. Acta Vet Scand 29:165, 1988.

Thapar MK, Young EJ. Urban outbreak of goat cheese brucellosis. Pediatr Infect Dis 5:640, 1986.

Tibana A, Barbosa Warnken M, Paiva Nunes M, Delgado Ricciardi I, Solino Noleto AL. Occurrence of *Yersinia* species in raw and pasteurized milk in Rio de Janeiro, Brazil. J Food Prot 50:580, 1987.

Todd E, Szabo R, Robern H, Gleeson T, Park C, Clark DS. Variation in counts, enterotoxin levels and Tnase in Swiss-type cheese contaminated with *Staphylococcus aureus*. J Food Prot 44:939, 1981.

Toora S, Budu-Amoako E, Ablett RF, Smith J. Effect of high-temperature short-time pasteurization, freezing and thawing and constant freezing, on the survival of *Yersinia enterocolitica* in milk. J Food Prot 55:803, 1992.

Tornadijo E, Fresno JM, Carballo J, Martin-Sarmiento R. Study of Enterobacteriaceae throughout the manufacturing and ripening of hard goat's cheese. J Appl Bacteriol 75:240, 1993.

Tschape H, Prager R, Steckel W, Fruth A, Tietze E, Bohme G. Verotoxigenic *Citrobacter freundii* associated with severe gastroenteritis and cases of haemolytic uraemic syndrome in a nursery school: green butter as the infection source. Epidemiol Infect 114:441, 1995.

Tucker CB, Fulkerson GC, Neudecker RM. A milk-borne outbreak of shigellosis in Madison County, Tenn. Public Health Rep 69:432, 1954.

Tulloch FE, Ryan KJ, Formal SB, Franklin FA. Invasive enteropathogenic *Escherichia coli*—an outbreak in 28 adults. Ann Intern Med 79:13, 1973.

Turck WPG. Q fever. In: Braude AI, Davis CE, Fierer J, eds. Infectious Diseases and Medical Microbiology. 2nd ed. Philadelphia: WB Saunders, 1986, p 810.

Tzipori S. Cryptosporidiosis in animals and humans. Microbiol Rev 47:84, 1983.

Ueno Y. The toxicology of mycotoxins. CRC Crit Rev Toxicol 14:99, 1985.

Uragoda CG, Lodha SC. Histamine intoxication in a tuberculosis patient after ingestion of cheese. Tubercle 60:56, 1979.

Valkenburg HA. Streptococcal pharyngitis and tonsillitis. In: Braude AI, Davis CE, Fierer J, eds. Infectious Diseases and Medical Microbiology. 2nd ed. Philadelphia: WB Saunders, 1986, p 715.

Valsanen OM, Mwaisumo NJ, Salkinoja-Salonen MS Differentiation of dairy strains of the *Bacillus cereus* group by phage typing, minimum growth temperature, and fatty acid analysis. J Appl Bacteriol 70:315, 1991.

Van den Heever LW. Tuberculosis in milch goats. J South Afr Vet Assoc 55:219, 1984.

van Egmond HP. Introduction. In: van Egmond HP, ed. Mycotoxins in Dairy Products. London: Elsevier Applied Science, 1989a, p 1.

van Egmond HP. Aflatoxin M_1: Occurrence, toxicity, regulation. In: van Egmond HP, ed. Mycotoxins in Dairy Products. London: Elsevier Applied Science, 1989b, p 11.

van Netten P, van de Moosdijk A, van Hoensel P. Psychrotrophic strains of *Bacillus cereus* producing enterotoxin. J Appl Bacteriol 69:73, 1990.

Varadaraj MC, Nambudripad VKN. Growth and production of thermostable deoxyribonuclease and enterotoxin by *Staphylococcus aureus* in milk. Milchwissenschaft 38:23, 1983.

Vasenin AA, Kogan VM, Nekipelova GA, Leonov VA, Gerasimova LF. Epidemiological characteristics of tick-borne encephalitis foci in traditional landscapes of Pribaikalia. Med Parazit Paravit Bol 44:56, 1975.

Veldman A, Meijs JAC, Borggreve GJ, Heeres-van-der-Tol JJ. Carry-over of aflatoxin from cow's food to milk. Animal Prod 55:163, 1992.

Vidon DJM, Delmas CL. Incidence of *Yersinia enterocolitica* in raw milk in eastern France. Appl Environ Microbiol 41:355, 1981.

Vogt RL, Little AA, Patton CM, Barrett TJ, Orciari LA. Serotyping and serology studies of campylobacteriosis associated with consumption of raw milk. J Clin Microbiol 20:998, 1984.

Wachsmuth K, Morris GK. *Shigella*. In: Doyle MP, ed. Foodborne Bacterial Pathogens. New York: Marcel Dekker, 1989, p 447.

Wade EM, Shere L. Longevity of typhoid bacilli in cheddar cheese—a study following an outbreak of typhoid fever traced to cheese. Am J Public Health 18:1480, 1928.

Wagenaar RO, Dack GM. Factors influencing growth and toxin production in cheese inoculated with spores of *Clostridium botulinum* types A and B. I. Studies with surface-ripened cheese type I. J Dairy Sci 41:1182, 1958a.

Wagenaar RO, Dack GM. Factors influencing growth and toxin production in cheese inoculated with spores of *Clostridium botulinum* types A and B. II. Studies with surface-ripened cheese type II. J Dairy Sci 41:1191, 1958b.

Wagenaar RO, Dack GM. Factors influencing growth and toxin production in cheese inoculated with spores of *Clostridium botulinum* types A and B. III. Studies with surface-ripened cheese type III. J Dairy Sci 41:1196, 1958c.

Walker SJ, Gilmour A. The incidence of *Yersinia enterocolitica* and *Yersinia enterocolitica*-like organisms in raw and pasteurized milk in Northern Ireland. J Appl Bacteriol 61:133, 1986.

Wallach JC, Miguel SE, Baldi PC, Guarnera E, Goldbaum FA, Fossati CA. Urban outbreak of a *Brucella melintensis* infection in an Argentine family: clinical and diagnostic aspects. Immun Med Microbiol 8:49, 1994.

Warburton DW, Peterkin PI, Weiss KF. A survey of the microbiological quality of processed cheese products. J Food Prot 49:229, 1986.

Waterman SC. The heat sensitivity of *Campylobacter jejuni* in milk. J Hyg Camb 88:529, 1982.

Weagant SD, Bryant JL, Jinneman KG. An improved rapid technique for isolation of *Escherichia coli* O157:H7 from foods. J Food Prot 58:7, 1995.

Weagant SD, Feng P, Stanfield JT. *Yersinia enterocolitica* and *Yersinia pseudotuberculosis*. In: FDA Bacteriological Analytical Manual. 8th ed. Gaithersburg, MD: AOAC International, 1995, p 8.01.

Weed LA, Michael AC, Harger RN. Fatal *Staphylococcus* intoxication from goat milk. Am J Public Health 33:1314, 1943.

Wegmuller B, Luthy J, Candrain U. Direct polymerase chain reaction detection of *Campylobacter jejuni* and *Campylobacter coli* in raw milk and dairy products. Appl Environ Microbiol 59:2161, 1993.

Weissman JB, Deen RMAD, Williams M, Swanston N, Ali S. An island-wide epidemic of salmonellosis traced to contaminated powdered milk. West Indian Med J 26:135, 1977.

Wells JG, Davis BR, Wachsmuth IK, Riley LW, Remis RS, Sokolow R, Morris GK. Laboratory investigation of hemorrhagic colitis outbreaks associated with a rare *Escherichia coli* serotype. J Clin Microbiol 18:512, 1983.

Wells JG, Shipman LD, Greene KD, Sowers EG, Green JH, Cameron DN, Downes FP, Martin ML, Griffin PM, Ostroff SM, Potter ME, Tauxe RV, Wachsmuth IK. Isolation of *Escherichia coli* serotype O157:H7 and other shiga-like-toxin-producing *E. coli* from dairy cattle. J Clin Microbiol 29:985, 1991.

Werner SB, Humphrey BL, Kamei I. Association between raw milk and human *Salmonella dublin* infection. BMJ 2:238, 1979.

White CH, Custer EW. Survival of *Salmonella* in cheddar cheese. J Milk Food Technol 39:328, 1976.

Wieneke AA, Roberts D, Gilbert RJ. Staphylococcal food poisoning in the United Kingdom, 1969–1990. Epidemiol Infect 110:519, 1993.

Wilkins EGL, Griffiths R, Roberts C. Guinea-pig inoculation for *Mycobacterium bovis*—Is it still necessary? Tubercle 68:311, 1987.

Williams RC, Bodnaurk PW, Golden DA. Survival of *Yersinia enterocolitica* during fermentation and storage of yogurt (abstr). Annual Meeting of the Institute of Food Technologists, New Orleans, LA, June 22–25, 1996.

Wilson CD, Richards MS. A survey of mastitis in the British dairy herd. Vet Rec 107:431, 1980.

Wilson GS. The necessity for a safe milk supply. Lancet 2:829, 1933.

Wisniewski HJ, Krumbiegel ER. Q fever in the Milwaukee area—I. Q fever in Milwaukee area cattle. Arch Environ Health 21:58, 1970a.

Wisniewski HJ, Krumbiegel ER. Q fever in Milwaukee—III. Epidemiological studies of Q fever in humans. Arch Environ Health 21:66, 1970b.

Wogan GN, Edwards GS, Newberne PM. Structure—activity relationship in toxicity and carcinogenicity of aflatoxins and analogs. Cancer Res 31:1936, 1971.

Wolf FS, Floyd C, Carden W, Brown L, Armes WH. Staphylococcal food poisoning traced to butter—Alabama. Morbid Mortal Weekly Rep 19:271, 1970.

Wong H-C, Chang MH, Fan JY. Incidence and characterization of *Bacillus cereus* isolates contaminating dairy products. Appl Environ Microbiol 54:699, 1988.

Wood DS, Collins-Thompson DL, Irvine DM, Myhr AN. Source and persistence of *Salmonella muenster* in naturally contaminated cheddar cheese. J Food Prot 47:20, 1984.

Wood RC, MacDonald KL, Osterholm MT. *Campylobacter* enteritis outbreaks associated with drinking raw milk during youth activities: a 10-year review of outbreaks in the United States. JAMA 268:3228, 1992.

Woodward WE, Gangarosa EJ, Brachman PS, Curlin GT. Foodborne disease surveillance in the United States, 1966 and 1967. Am J Public Health 60:130, 1970.

Wright EP, Tillhet HE. Milk-borne *Campylobacter* enteritis in a rural area. J Hyg Camb 91:227, 1983.

Wuethrich B. Back on the farm: stopping *E. coli* O157:H7 at its source. ASM News 60:409, 1994.

Wundt W, Schnittenhelm P. The behavior of salmonellae and staphylococci in cooled foods. Arch Hyg Bakteriol 149:567, 1965.

Wyatt CJ, Timm EM. Occurrence and survival of *Campylobacter jejuni* in milk and turkey. J Food Prot 45:1218, 1982.

Young EJ. Human brucellosis. Rev Infect Dis 5:821, 1983.

Young EJ. Health issues at the US-Mexican border. JAMA 265:2066, 1991a.

Young EJ. Serologic diagnosis of human brucellosis: analysis of 214 cases by agglutination tests and review of the literature. Rev Infect Dis 13:359, 1991b.

Young EJ, Suvannoparrat U. Brucellosis outbreak attributed to ingestion of unpasteurized goat cheese. Arch Intern Med 135:240, 1975.

Yousef AE, Marth EH. Degradation of aflatoxin M_1 in milk by ultraviolet energy. J Food Prot 48:697, 1985.

Yousef AE, Marth EH. Use of ultraviolet energy to degrade aflatoxin M_1 in raw or heated milk with or without added peroxide. J Dairy Sci 69:2243, 1986.

Yousef AE, Marth EH. Behavior of *Listeria monocytogenes* during the manufacture and storage of Colby cheese. J Food Prot 51:12, 1988.

Yousef AE, Marth EH. Stability and degradation of aflatoxin M_1. In: van Egmond HP, ed. Mycotoxins in Dairy Products. London: Elsevier Applied Science, 1989, p 127.

Yousef AE, Marth EH. Fate of *Listeria monocytogenes* during the manufacture and ripening of Parmesan cheese. J Dairy Sci 73:3351, 1990.

Zehren VL, Zehren VF. Examination of large quantities of cheese for staphylococcal enterotoxin A. J Dairy Sci 51:635, 1968a.

Zehren VL, Zehren VF. Relation of acid development during the cheesemaking process and development of staphylococcal enterotoxin. J Dairy Sci 51:645, 1968b.

Zottola EA, Jezeski JJ, Al-Dulaimi AN. Effect of short-time subpasteurization treatments on the destruction of *Staphylococcus aureus* in milk for cheese manufacture. J Dairy Sci 52:1707, 1969.

12

Control of Microorganisms in Dairy Processing: Dairy Product Safety Systems

ROBERT D. BYRNE

International Dairy Foods Association, Washington, D.C.

J. RUSSELL BISHOP

Center for Dairy Research, University of Wisconsin—Madison, Madison, Wisconsin

I. INTRODUCTION

Control of microorganisms in dairy processing is necessary to produce a safe product of the highest quality. The focus of this chapter is production of safe dairy products. To accomplish this, pathogenic microorganisms need to be controlled. Whereas the techniques described result in a high-quality product, the intent of a dairy product safety system is to ensure that a safe product reaches the consumer. One of the most effective ways to control microorganisms is through use of the Hazard Analysis and Critical Control Point (HACCP) program (Anonymous, 1996a; Pierson and Corlett, 1992). However, a complete dairy processing system encompasses more than just HACCP. To ensure that all hazards are addressed and a safe product is produced, prerequisite programs must be in place before HACCP controls are addressed. A sound prerequisite program also simplifies the HACCP program and minimizes the number of critical control points that need to be monitored. This chapter focuses on those areas that are defined as prerequisites and how to effectively control them, describes the implementation of a HACCP program, and provides a model HACCP program as a guide to developing an effective safety system in a dairy plant.

II. PREREQUISITES/GOOD MANUFACTURING PRACTICES

Before developing HACCP plans under the Dairy Products Safety System (Anonymous, 1996b), it is necessary for dairy plants to have developed, documented, and implemented programs to control factors that may not be directly related to manufacturing controls but support the HACCP plans. These programs are called prerequisite programs and need to be effectively monitored and controlled before HACCP plans are developed. Prerequisite programs are defined as universal steps or procedures that control the operational conditions within a dairy plant, allowing for environmental conditions that are favorable to the production of safe dairy products. Prerequisite areas include premises, receiving and storage, equipment performance and maintenance, personnel training, sanitation, and recalls (Anonymous, 1995, 1996a).

When implementing HACCP in a dairy plant, the first step is to review existing programs to verify whether all the prerequisite requirements are being met and whether all the necessary controls and documentation (e.g., program description, individual responsible, and monitoring records) are in place. Prerequisite programs are evaluated for their conformance to the minimum requirements. The effectiveness of the programs is monitored and the required records are properly maintained.

The importance of the prerequisite programs cannot be overstated. Prerequisite programs are the foundation of the HACCP plans and must be adequate and effective. If any portion of a prerequisite program is not adequately controlled, then additional critical control points would have to be identified, monitored, and maintained under the HACCP plans. In summary, comprehensive, effective prerequisite programs simplify HACCP plans and ensure that the integrity of the HACCP plan is maintained and that the manufactured product is safe.

A. Premises

Building and surroundings must be designed, constructed, and maintained to prevent conditions that may result in contamination of dairy products. Dairy plants must have an adequate program in place to monitor and control all elements in this section and maintain the appropriate records. The premises include all elements of the building and building surroundings: the outside property, roadways, drainage, building design and construction, product flow, sanitary facilities, and water quality. Adherence to the requirements is verified through the written program of the plant, which outlines the procedures that ensure satisfactory conditions are maintained (e.g., areas to be inspected, tasks to be performed, persons responsible, inspection frequencies, and records to be kept).

Land must be free of debris and refuse and must not be in close proximity to any source of pollution (e.g., objectionable odors, smoke, dust, or other contaminants). Roadways must be properly graded, compacted, dust proof, and drained. Premises and shipping and receiving areas must provide or permit good drainage.

The building and facilities must be designed to readily permit cleaning, prevent entrance and harboring of pests, and prevent entry of environmental contaminants. Buildings need to be of sound construction, maintained in good repair, and not present any microbiological, chemical, or physical hazards to the dairy food. The building must be designed to provide suitable environmental conditions, permit adequate cleaning and sanitation, minimize contamination by extraneous materials, prevent access by pests, and provide adequate space for satisfactory performance of all operations. Construction and layout should reflect the approved blueprints, where applicable.

Floors, walls, and ceiling materials, as well as various coating and joint sealants, must be approved materials that are durable, smooth, cleanable, and suitable for production conditions conducted in the area. Walls must be light-colored and well joined. Floors must be sufficiently sloped for liquids to drain into trapped outlets. Windows, if opened, must be equipped with close fitting screens. Doors must have smooth, nonabsorbent surfaces that are close fitting. Stairs, elevators, and other structures must be situated and constructed so that there is no contamination of dairy food and packaging materials. Overhead structures must be designed and installed in a manner that prevents contamination of dairy food and packaging materials and not hamper cleaning operations.

Adequate lighting must be provided throughout the establishment. For operational purposes, the lighting should not alter food colors. Light bulbs and fixtures suspended over exposed dairy food or packaging materials at any stage of production or storage must be of a safety type or be protected to prevent contamination of food if breakage occurs. Ventilation must be provided to prevent a buildup of heat, steam, condensation, or dust and to remove contaminated air. In microbiologically sensitive areas, positive air pressure needs to be maintained. Ventilation openings must be equipped with close-fitting screens or otherwise protected with noncorrodible material. Air intakes must be located so as to prevent an intake of contaminated air.

Drainage and sewage systems must be equipped with appropriate traps and vents. Plants must be designed and constructed so that there is no cross-connection between the effluent of human wastes and any other wastes in the plant. Facilities must be provided for storage of waste and inedible material before removal from the plant. These facilities must be designed to prevent contamination. Containers used for waste must be clearly identified and leak proof.

The traffic pattern of employees and equipment must avoid cross-contamination of the product. Product flow must prevent contamination of the dairy food through physical or operational separation. Plants must provide physical and operational separation of incompatible operations. The facilities must be adequate for the maximum production volume encountered. Living quarters and areas where animals are kept must be completely separated from and not open directly into areas where dairy foods or packaging materials are handled or stored.

Washrooms with self-closing doors must be provided. Washrooms, lunchrooms, and change rooms must be separate from and not lead directly into food processing areas and must also be correctly ventilated and maintained. Washrooms must have hand-washing facilities with a sufficient number of well-maintained sinks with properly trapped waste pipes connected to drains. Hand-washing facilities must have hot and cold potable running water, soap, sanitary hand drying supplies or devices, and, where required, a cleanable waste receptacle.

Processing areas must contain a sufficient number of conveniently located hand-washing stations with properly trapped waste pipes connected to drains. In the processing areas, remote controlled (e.g., foot, knee, timed) hand-washing stations are preferable. Sanitizing facilities (e.g., hand dips) must be in areas where plant employees are in direct contact with microbiologically sensitive dairy foods. Notices must be posted for employees to wash hands.

Plants must provide adequate facilities and means for cleaning and sanitizing equipment. Separate means must be provided for cleaning and sanitizing equipment used for inedible materials.

The water control program evaluates the microbiological, chemical, and physical quality of source and in-plant water (from various points of usage). This water includes the steam supply, cooling medium, process waters, and ice supply. The program establishes frequency of testing, procedures for testing, person responsible, and records to be kept. The plan has procedures in place to deal with water that does not meet specific standards. Records of water potability (laboratory test results) and water treatments applied must be maintained.

Potable hot and cold water is used in dairy food processing, handling, packaging, and storage areas and must be provided at adequate temperatures and pressures and in quantities sufficient for all operational and cleanup needs. Where required, facilities that protect against contamination must be provided for storage and distribution of water. Bacteriological testing of water is done on a semiannual basis for municipal water and on a monthly basis for water from other sources. Records of water potability testing must be maintained.

When chlorination of water occurs on premises, a metering device for adding the correct concentration of chlorine, which is designed to readily indicate a malfunction, must be used. Also, twice daily checks to determine total available chlorine must be done or an automatic analyzer equipped with a recorder and an alarm must be used.

No cross-connections can exist between potable and nonpotable water supply systems. Nonpotable water is never used in dairy food processing, handling, packaging, or storage areas. All hoses, taps, cross-connections, or similar sources of possible contamination must be equipped with antibackflow devices.

Water treatment chemicals used must be appropriate for their intended purpose. The treatment process and recirculated water and process waters must be treated and maintained in a condition so that no health hazard results from their use. Recirculated water must be a separate distribution system, which is readily identified. Records of treatment must be maintained. Microbiological testing needs to be conducted to monitor effectiveness.

Ice must be made from potable water and manufactured, handled, and stored to protect it from contamination. Bacteriological testing of ice must be done on a semiannual basis for plants using municipal water supplies and on a monthly basis for plants using other sources. Records of ice potability testing must be maintained.

Steam coming into direct contact with dairy food or food contact surfaces must be generated from potable water with no harmful substances added. The steam supply must be adequate to meet operational requirements. Boiler treatment chemicals used must be appropriate for their intended use. Records of treatments must be maintained.

B. Receiving and Storage

Plants must receive, inspect, and store ingredients, packaging material, and incoming materials so as to prevent conditions that may result in contamination of dairy foods. Plants must have an adequate program in place to monitor and control all elements in this section and maintain the appropriate records.

Raw materials, ingredients, and packaging material (i.e., incoming materials) must be inspected on receipt and stored and handled in a sanitary manner (i.e., to prevent microbiological, chemical, or physical contamination). Effective measures must be taken to prevent contamination of raw materials, ingredients, and packaging materials by direct or indirect contact with contaminating material. Certification of some incoming materials by letters of guarantee, certificates of analysis, or other satisfactory means may be required and then should be in accordance with the HACCP plan.

Incoming materials must be received into an area separate from the processing area. All food additives must be food grade [i.e., they meet Code of Federal Regulations (CFR) (Anonymous, 1996) specifications or equivalent]. All ingredients must be safe and not impact negatively on the safety of the dairy food. Plants must use packaging materials that are appropriate for their intended use. Incoming raw materials, ingredients, and packaging materials must be monitored on receipt for acceptability for use in dairy foods, and records of this monitoring need to be maintained.

Where applicable, plants must have adequate means of establishing, maintaining, and monitoring the temperature and the humidity of rooms where raw materials, ingredients, packaging materials, and dairy foods are stored. Records of monitoring must be maintained.

Raw materials, ingredients, and packaging materials must be handled and stored in ways to prevent damage and contamination and must be held to avoid growth of microorganisms. Conditions of storage and transport must be such that the safety of the dairy food is not affected.

Returned or damaged goods must be clearly identified and stored in a designated area for appropriate disposition. Conditions of storage must not affect the safety of the finished product. Detergents, sanitizers, or other chemical agents in a dairy plant must be properly labeled, stored, and used in ways that prevent contamination of dairy foods, packaging materials, and food contact surfaces. Chemicals must be stored and handled in an area that is kept dry and well ventilated and is separate from all food handling areas. Chemicals must be mixed and stored in clean, labeled containers and dispensed and handled only by authorized and properly trained personnel.

C. Equipment Performance and Maintenance

Dairy plants must use equipment that is designed for the production of dairy foods and must install and maintain equipment in ways to prevent conditions that may result in contamination of food. Plants must have an adequate program in place to monitor and control all elements in this section and maintain the appropriate records.

Equipment and utensils must be designed and maintained in ways that prevent contamination of dairy foods and be constructed of corrosion-resistant material. Food contact surfaces must be nonabsorbent, nontoxic, smooth, free from pitting, unaffected by food, and able to withstand repeated cleaning and sanitizing. All chemicals, lubricants, coatings, and paints used on equipment in contact with food must be appropriate for their intended use.

Equipment and utensils must be installed in a way that prevents contamination of food with adequate space within and around equipment. Equipment must be accessible for cleaning, sanitizing, maintenance, and inspection. Where required, equipment must be properly vented. Equipment must be maintained in a clean and sanitary manner in accordance with the sanitation program. Equipment and utensils used to handle inedible material must not be used to handle edible material. Containers for inedible and waste material must be clearly identified and be leak proof.

Monitoring devices and any equipment that could have an impact on dairy food safety must be listed together with their intended use. Protocols and calibration methods must be established for those equipment and monitoring devices. This may include thermometers, pH meters, a_w meters, refrigeration unit controls, scales, recording thermometers, recording hygrometers, and other equipment.

The frequency of calibration, responsible person, monitoring and verification procedures, appropriate corrective actions, and record keeping must be specified. If reagents are used for monitoring or verification activities, procedures for keeping and calibrating the reagents must be documented. Required information on the calibration of reagents includes frequency of testing for all reagents, responsible person, dating system, storage conditions, and records to be kept.

A preventive maintenance program must be in place that lists equipment and utensils together with preventive maintenance procedures. The program specifies necessary servicing of the equipment and frequency, including replacement of parts, responsible person, method of monitoring, verification activities, and records to be kept.

D. Personnel Training

Dairy plants must have an adequate program in place to monitor and control training programs and maintain appropriate documentation. The objective of the personnel training program must be to ensure safe food handling practices. The personnel training program must provide, on an ongoing basis, the necessary training for production personnel. A procedure must be developed to verify the effectiveness of the training program.

Production personnel must be trained to understand the critical elements for which they are responsible, what the critical limits are, the importance of monitoring the limits, and actions they must take if the limits are not met. Ongoing training in personal hygiene and hygienic handling of food must be provided to every food handler, and training in personal hygiene and hygienic handling of food must be provided to all persons entering the food handling areas.

Plants must demonstrate that personal hygiene is carried out and controlled. No person, while known to be suffering from or to be a carrier of a disease likely to be transmitted through food, or afflicted with infected wounds, skin infections, sores, or diarrhea, is permitted to work in any food handling area in any capacity in which there is any likelihood of such a person contaminating food with pathogenic microorganisms. All persons having open cuts or wounds may not handle food or food contact surfaces unless the injury is completely protected by a secure, waterproof covering.

All persons entering a dairy food production area must wash their hands thoroughly with soap under warm-running potable water. Hands must be washed after handling contaminated materials and after using toilet facilities. Where required, employees must use disinfectant hand dips.

All persons working in dairy food handling areas must maintain personal cleanliness while on duty. Protective clothing, hair covering, and footwear, functional to the operation in which the employee is engaged, must be worn and maintained in a sanitary manner. Gloves, if worn, must be clean and sanitary. All persons entering dairy food handling areas must remove objects from their person that may fall into, or otherwise contaminate, food. Tobacco, gum, and other food are not permitted in dairy food handling areas. Jewelry must be removed before entering food handling areas. Jewelry, including medic alerts that cannot be removed, must be covered. Personal effects and street clothing must not be kept in food handling areas and must be stored in a manner to prevent the contamination of dairy foods.

Access of personnel and visitors must be controlled to prevent contamination. All necessary precautions must be taken to prevent contamination, including the use of foot baths and hand dips, where required.

E. Sanitation

Plants must have an adequate sanitation program in place and maintain appropriate records. The sanitation program outlines the parameters that need to be controlled to ensure the safety of the dairy food product. Sanitation procedures must be developed for equipment, utensils, overhead structures, floors, walls, ceilings, drains, lighting devices, refrigeration units, and anything else impacting the safety of the dairy food. Equipment and facilities must be cleaned and sanitized as defined in a written schedule. Equipment cleaned must be visually inspected on a routine basis. Equipment must be free of any residue and foreign material before being used.

For each area and each piece of equipment and utensil, the written cleaning and sanitizing program specifies the name of the person responsible, chemicals used, procedures used, and frequency of cleaning and sanitizing.

Chemicals must be used in accordance with the manufacturer's recommendations. The sanitation program must be carried out in a manner that does not contaminate food packaging materials during or after cleaning and sanitizing. Equipment for cleaning and sanitizing food processing equipment must be designed for its intended use and properly maintained.

Hand cleaned or cleaned out-of-place (COP) equipment must be disassembled for each cleaning and inspection, whereas equipment cleaned by an accepted clean in-place (CIP) system must be inspected as prescribed in the CIP program. The general housekeeping and the special sanitation procedures carried out during the operations must be specified (e.g., mid-shift cleanup, responsible person, procedure).

Examples of information to be included in the written sanitation program are (a) area/line, equipment to be cleaned, the frequency, and the responsible person, (b) special instructions for cleaning specific equipment and the responsible person, (c) cleaning equipment that is to be used, along with instructions for its proper operation (e.g., pressure, volume), (d) detergent/sanitizer to be used (including commercial and generic names, dilution factor, temperature), (e) method of application of the solution, contact time, foam consistency, scrubbing (if necessary), high/low pressure, (f) rinsing instructions, water temperature, (g) sanitizing instructions, commercial and generic names, dilution factor, pH, temperature, contact time, (h) final rinsing instructions (if applicable), and (i) safety instructions for products.

Adherence to the written sanitation program must be monitored and recorded (e.g., temperature, concentration, contact parameters). Effectiveness of the sanitation program must be monitored on a routine basis by a company representative (e.g., using microbiological swab tests, visual inspection of areas/equipment, or direct observation of sanitation procedures performed by designated personnel). Operations should begin only after all sanitation requirements are met. Records of all monitoring results need to be maintained. Deviations and corrective actions taken must be recorded.

Dairy plants must have an adequate, effective, safe, and written pest control program in place and must maintain appropriate records. Birds and animals must be excluded from dairy plants. The written pest control program should include the name of a contact person at the establishment for pest control, the name of any applicable extermination company or the name of the person responsible for the program, the list of chemicals and methods used, a map of bait and trap locations, frequency of treatment and inspection, and pest survey and control reports. Chemicals must be used according to manufacturer's instructions, appropriate for their intended use and used in a manner to prevent contamination.

Adherence to the written pest control program must be monitored and recorded. Effectiveness of the pest control program must be verified by on-site inspection of areas for presence of insect and rodent activity. Records of all monitoring results, recommendations, and action taken must be maintained.

F. Recalls

The recall programs outlines procedures that the company would implement in the event of a product recall. The objective of the written recall procedure of the dairy plant is to ensure that an identified dairy food is removed from the market as efficiently, rapidly, and completely as possible via a plan that can be put into operation at any time. The program must be tested to validate its effectiveness.

Each manufacturer of a dairy food product must maintain a system of control that permits a complete and rapid recall of any lot of food product. The written recall procedure includes the following:

1. Documentation pertaining to the product coding system. All products must be identified with a production date or code identifying each lot. Sufficient coding of dairy products is used and explained in the written recall program to permit positive identification and to facilitate an effective recall.
2. Finished product distribution records must be maintained for a time that exceeds the shelf-life of the product. Records must be adequately designed and maintained to facilitate location of the product if it is recalled.
3. A complaint file must be maintained. Records documenting all related complaints and action taken must be included.
4. Responsible individuals who are part of the recall team, along with their respective business and home telephone numbers, must be listed. For each individual, an alternate is designated to act on his or her behalf in case of absence. Roles and responsibilities for every member on the recall team must be clearly defined.
5. The step-by-step procedures to follow for a recall must be described. These procedures should include the extent and the depth of the recall (i.e., consumer, retailer, or wholesaler level) according to the recall classification.
6. Means of notifying the affected customers in a manner appropriate to the type of hazard must be defined. The channels of communication (FAX, telephone, radio, letter, or other means) to be used for trace-back and recovery of all affected products must be identified. Typical messages directed to consumers, retailers, or wholesalers according to the severity of the hazards must be included.
7. Control measures for the returned recalled dairy food must be planned. This includes both returned product and product still in stock on the premises. Control measures and disposal of the affected product must be described according to the type of hazard involved.

8. Means of assessing progress and efficacy of the recall must be stated. A method of checking the effectiveness of the recall needs to be defined.

Any manufacturer who initiates a food safety–related recall of a food immediately notifies the regulatory agency that has jurisdiction with information, including (a) reason for the recall, (b) recalled product identification (e.g., name, code marks or lot numbers, plant number, date of production, date of importation or exportation, if applicable), (c) the amount of recalled product involved, subdivided to include the original quantity of product, the distributed quantity, and the quantity remaining in the possession of the company, (d) areas of distribution of the recalled food, by areas, cities, states, and, if exported, by country, along with names and addresses of retailers and wholesalers, and (e) information on any other product that could be affected by the same hazard.

III. HAZARD ANALYSIS AND CRITICAL CONTROL POINT—AN OVERVIEW

After prerequisite programs have been completed and documented, a HACCP program can be implemented. The use of the HACCP system is not new to the dairy food industry. HACCP is a logical, simple, effective, but highly structured system of food safety control. It is a system designed to identify "hazards and/or critical situations" and to produce a plan to control these situations.

The HACCP system was introduced to the food industry as a "spinoff" of the space program during the 1960s. The National Aeronautics and Space Administration (NASA) used HACCP to provide assurance of the highest quality available for components of space vehicles. This program, to develop assurance of product reliability, was carried over into the development of foods for astronauts.

The U.S. Army Natick Laboratories, in conjunction with NASA, began to develop the foods needed for manned space exploration. They contracted with the Pillsbury Company to design and produce the first foods used in space. While researchers at Pillsbury struggled with certain problems, such as how to keep food from crumbling in zero gravity, they also undertook the task to come as close as possible to 100% assurance that the foods they produced would be free of bacterial or viral pathogens. A foodborne illness that causes severe diarrhea in the confines of a space suit combined with zero gravity could be just as catastrophic to astronauts as a failure of the rockets.

Use of standard quality control methods common to the food industry was soon proven to be unworkable for the task Pillsbury had undertaken. Either the degree of safety desired was not provided or product sampling would have been prohibitive to commercialization of space foods. Pillsbury researchers discarded the standard quality control methods and began an extensive evaluation, in conjunction with NASA and Natick Laboratories, to evaluate food safety. They soon realized that to be successful they would need control over their process, raw materials, environment, and their people. In 1971, they introduced HACCP as a preventive system that provided manufacturers a high degree of assurance that foods were produced safely. If the HACCP system is correctly implemented, there is little requirement for the testing of final product other than for verification purposes.

HACCP is a management tool that provides a more structured approach to the control of identified hazards than that achievable by traditional inspection and quality control procedures. It starts with product design and provides a means to identify potential areas of concern, where failure has not yet been experienced, and is, therefore, particularly useful for new operations. HACCP is a logical basis for better decision-making with respected to product safety. It provides dairy food manufacturers with greater security of control over product safety than is possible with end-product testing. HACCP has international recognition as the most effective means of controlling foodborne disease and is endorsed as such by the joint Food and Agriculture Organization/World Health Organization (FAO/WHO) Codex Alimentarius Commission.

One of the key advantages of the HACCP concept is that it enables a dairy food manufacturing company to move away from a philosophy of control based on testing (i.e., testing for failure) to a preventive approach, whereby potential hazards are identified and controlled in the manufacturing environment (i.e., prevention of product failure).

HACCP has many other benefits as well: it ensures dairy product safety, is science-based, focuses appropriate technical resources on critical processes, lessens emphasis on end-product testing, focuses on prevention, uses resources effectively, and meets customer expectations.

HACCP has been recognized internationally as a logical tool toward a more modern, scientifically based inspection system. The most important element of a HACCP-based system is its preventive nature and the exercising of control throughout the manufacturing process at critical steps. By doing so, defects that could impact on the safety of the dairy food being processed can be readily detected and corrected at these points before the product is completely processed and packaged.

IV. PRINCIPLES OF HAZARD ANALYSIS AND CRITICAL CONTROL POINT

The following seven principles of HACCP were adopted by the National Advisory Committee on Microbiological Criteria for Foods (Pierson and Corlett, 1992). These principles allow for a systematic approach to dairy product safety:

1. Conduct a hazard analysis associated with growing, harvesting, raw materials and ingredients, processing, manufacture, distribution, marketing, preparation, and consumption of the dairy food.
2. Identify the critical control points (CCPs) required to control the identified hazards in the process.
3. Establish the critical limits for preventive measures associated with each identified CCP.
4. Establish CCP monitoring requirements. Establish procedures for using the results of monitoring to adjust the process and maintain control.
5. Establish corrective actions to be taken when monitoring indicates that there is a deviation from an established critical limit.
6. Establish effective record-keeping systems that document the HACCP plan.
7. Establish procedures for verification that the HACCP system is working correctly.

V. HAZARD COMPONENTS

To effectively produce a safe dairy product, all hazards that might occur must be controlled, reduced to an acceptable level, or eliminated. An effective prerequisite program should control many of the environmental hazards. The HACCP system controls any remaining hazards inherent to the food or that may result from processing.

The three hazards that must be controlled are microbiological, chemical, and physical hazards (Anonymous, 1996b; Pierson and Corlett, 1992). There are three types of microbiological hazards: severe, moderate with potentially extensive spread, and moderate with limited spread.

The severe microbiological hazards include *Brucella, Clostridium botulinum, Listeria monocytogenes, Salmonella typhi, Salmonella paratyphi, Salmonella dublin, Shigella dysenteriae,* and hepatitis A and E. Microbiological hazards with potentially extensive spread include: *Salmonella* spp., enterotoxigenic *Escherichia coli,* enteroinvasive *E. coli, E. coli* O157:H7, *Shigella* spp., viruses, and *Cryptosporidium.* The microbiological hazards that are moderate with limited spread include *Bacillus cereus, Campylobacter jejuni* and other species, *Clostridium perfringens, Staphylococcus aureus, Aeromonas* spp., *Yersinia enterocolitica,* and parasites.

Physical hazards that could potentially occur include entry into the food of metal, glass, insect/pest parts, dirt, wood fragments, personal effects, plastic, and any other physical object that may render the food unsafe. Chemical hazards that may occur include presence of natural toxins, metals, drug residues, sanitizer residues, pesticides, food additives, and inadvertent chemicals. The natural toxins include mycotoxins and other natural thyrotoxicoses. Mycotoxins are divided into those causing acute and chronic mycotoxicoses. The acute mycotoxins include ochratoxin, trichothecene, zearalenone, and aflatoxin, whereas the chronic mycotoxins include aflatoxin, sterigmatocystin, and patulin. Metal hazards include presence of copper, cadmium, and mercury. Drug residues are beta-lactams, sulfonamides, tetracyclines, and others. Examples of sanitizer residues are chlorinated, fatty acid, and idophors. Inadvertent chemicals include lubricants and boiler additives, among others.

VI. HAZARD ANALYSIS AND CRITICAL CONTROL POINT IMPLEMENTATION

Implementation of HACCP involves a 12-step process, which, when complete and maintained, ensures a safe dairy product is being produced.

A. Step One: Gain Management Commitment and Assemble the HACCP Team

Before proceeding to the HACCP team selection, it is extremely important to get full commitment from all levels of management to the HACCP initiative. Without a firm commitment of time, personnel, and resources, the HACCP plan may be difficult, if not impossible, to implement effectively. The first step in developing a HACCP plan is to assemble a HACCP team consisting of individuals who have specific knowledge and expertise appropriate to the dairy product and process. It is the team's responsibility to develop each step of the HACCP plan. The team should be multidisciplinary and should include all personnel who are directly involved in the daily process activities, because they are most familiar with the operation.

It is recommended that experts who are knowledgeable about the dairy food and its process should either participate in or verify the completeness of the hazard analysis and the HACCP plan. These individuals should have the knowledge and experience to correctly identify potential hazards; assign levels of severity and risk; recommended controls, criteria, and procedures for monitoring and verification; recommend appropriate corrective actions when a deviation occurs; and recommend research related to the HACCP plan if important information is not known.

B. Step Two: Describe Dairy Food and Method of Distribution

A separate HACCP plan must be developed for each dairy food product that is being processed in a facility. The HACCP team must first fully describe the dairy food product, or intermediate dairy product if only part of the process is studied. The dairy product should be defined in terms of composition, structure, processing, packaging system, storage, required shelf-life, and instructions for use.

The method of distribution should be described along with information on whether the dairy food is to be distributed frozen, refrigerated, or shelf-stable. Consideration should be given to the potential for abuse in the distribution channel and by consumers, but the question, "Is this a hazard or a quality issue?" must be asked.

C. Step Three: Identify Intended Use and Potential Consumers

The intended use of the dairy food should be based on its normal use by end-users, consumers, and consumer target groups. The intended consumers or users may be the general public, a particular segment of the population, another food (dairy or nondairy), or nonfood product. The use of dairy foods as an intermediate or nontraditional product represents a growing market and must be considered more so than in the past.

Intermediate or nontraditional products include dairy foods that serve as ingredients (e.g., cheeses used in processed foods, whey products used in infant formulas, canned cheese, modified atmosphere-packaged dairy foods). Particular attention should be given to lower fat dairy products because the reduction of fat within the product alters its composition as related to water activity, pH, and other characteristics important to the microbiological safety of the product.

D. Step Four: Develop and Verify a Flow Diagram

The purpose of the flow diagram is to provide a clear, simple description of the steps involved in production of the dairy food. The scope of the diagram must cover all steps in the process that are directly under the control of the facility. The flow diagram should consist of words in blocks, not engineering drawings. When developing a flow diagram, certain types of information must be considered: prerequisites/good manufacturing practices already established, all raw materials/ ingredients and packaging used (microbiological, chemical, and physical data), sequence of all process steps (including raw material addition), time/temperature considerations, product recycle/rework loops, and storage and distribution conditions.

The HACCP team should inspect the operation to verify the accuracy and completeness of the flow diagram by taking the diagram to the production floor and walking through the steps to ensure the accuracy of the diagram. The flow diagram should be modified as necessary.

E. Step Five: Conduct a Hazard Analysis (Principle 1)

A hazard is any microbiological, chemical, or physical property that may cause a dairy food to be unsafe for human consumption. The HACCP team conducts a hazard analysis and identifies the steps in the process where hazards of potential significance can occur. The hazards must be of such a nature that their prevention, elimination, or reduction to acceptable levels is essential to production of a safe dairy food. The team must consider what preventive measures, if any, can be applied for each hazard.

The hazard analysis and identification of associated preventive measures allow identification of those hazards of significance and associated preventive measures, modification of a process or product to further assure or improve safety, and determination of CCPs in Principle 2.

During the hazard analysis, the potential significance of each hazard should be assessed by considering its risk and severity. The estimate of risk is usually based on a combination of experience, epidemiological data, and information in the technical literature. Safety concerns must be differentiated from quality concerns. The term "hazard" is limited to safety.

Upon completion of the hazard analysis, the significant hazards associated with each step in the flow diagram should be listed along with any preventive measures to control the hazards. For example, if the HACCP team were to conduct a hazard analysis for the manufacture of yogurt, possible pathogens in raw milk would be identified as a potential hazard. Thus, pasteurization would be listed along with the hazard as the preventive measure. Hazards should be only those that will result in an unsafe product. This same approach may be used for quality or economic issues, but HACCP is limited to product safety only.

F. Step Six: Critical Control Points (Principle 2)

A CCP is any point, step, or procedure at which control can be applied and a dairy food safety hazard can be prevented, eliminated, or reduced to an acceptable level. The hazard analysis conducted in step 5 has identified areas that are necessary to control. The prerequisite/good manufacturing practices program may be used to control many of the identified hazards. Any hazards not controlled through prerequisite programs must be identified as CCPs.

Examples of CCPs include temperature of incoming raw milk, animal drug residue monitoring in raw milk, storage temperature of raw milk or cream, pasteurization temperature and time, and use of metal detectors.

Information developed during the hazard analysis should enable the HACCP team to identify which steps in the process are CCPs. Identification of each CCP can be facilitated by use of the CCP decision tree. All hazards that could reasonably be expected should be considered. Application of the CCP decision

tree can help determine whether a particular step is a CCP for previously identified hazard.

Different facilities preparing the same dairy product can differ in the risk of
hazards and the points, steps, or procedures that are CCPs. This can result from
differences in each facility layout, equipment, selection of ingredients (including
raw versus pasteurized milk), or the process that is used. This is why HACCP
plans must be developed by each individual plant for every product it produces.

G. Step Seven: Critical Limits (Principle 3)

A critical limit is a criterion that must be met for each preventive measure
associated with a CCP. Therefore, there is a direct relationship between the CCP
and its critical limits that serve as boundaries of safety. Critical limits must be
met to ensure the safety of the dairy product. Exceeding a critical limit means a
health hazard may exist or develop or the product was not produced under
conditions assuring safety. Critical limits may be derived from sources such as
regulatory standards and guidelines, literature searches, experimental studies,
and experts. Critical limits may be established for preventive measures such as
temperature, time, a_w, pH, titratable acidity, drug residues, and microbiological
numbers and kinds.

H. Step Eight: Monitoring/Inspection (Principle 4)

Monitoring is a planned sequence of observations or measurements to assess
whether a CCP is under control and to produce an accurate record for future use
in verification. Monitoring serves to track the system operation, determine when
there is a loss of control and a deviation occurs at a CCP (i.e., exceeding the
critical limit), and provide written documentation for use in verification of the
HACCP plan.

Because of the potentially serious consequences of a critical defect, monitoring procedures must be effective. Ideally, monitoring should be at the 100% level.
Continuous monitoring is possible with many types of physical and chemical
methods (e.g., the time and temperature of pasteurization). The person responsible for monitoring also must report a dairy process or product that does not meet
critical limits so that immediate corrective action can be taken.

When it is not possible to monitor a critical limit on a continuous basis, it is
necessary to establish the monitoring interval that is reliable enough to indicate
that the hazard is under control. Statistically designed data collection or sampling
systems lend themselves to this purpose. When using statistical process control,
it is important to recognize that critical limits must not be exceeded.

Most monitoring procedures for CCPs need to be done rapidly because they
relate to an on-line process and there is no time for lengthy analytical testing.

Microbiological testing is seldom, if ever, effective for monitoring CCPs because of the time required to conduct tests. Therefore, physical and chemical measurements are preferred because they may be done rapidly and can indicate the conditions of microbiological control in the process.

The following areas must be addressed when considering monitoring/ inspection: monitoring/inspection controls, procedures, frequency, responsibility, customized contingency plans, monitoring activities, and exceeding the limit. The design of HACCP systems is the most important feature of developing effective monitoring systems. Judgment and discretion are of key importance in designing the CCP and the monitoring system.

I. Step Nine: Corrective Actions (Principle 5)

Corrective actions are procedures to be followed when a deviation occurs. Because of variations in CCPs for different dairy products and the diversity of possible deviations, specific corrective action plans must be developed for each CCP. The actions must demonstrate that the CCP has been brought under control. Individuals who have a thorough understanding of the dairy process, product, and HACCP plan should be assigned responsibility for taking corrective action. Corrective action procedures must be documented in the HACCP plan.

Actions taken should eliminate the actual or potential hazards created by deviation, develop specific corrective actions for each CCP, assure safe disposition of the dairy product involved, and demonstrate that the CCP has been brought under control.

Responsibilities include placing the dairy product on "hold" pending completion of corrective action. If difficult to establish, the effect of deviation on safety, testing and final disposition of the dairy product must be agreed to by the appropriate individuals; records that identify deviant lots and corrective action must be part of records and records must be kept for a reasonable period after the expected end of shelf-life of the dairy product.

J. Step Ten: Records (Principle 6)

The requirement for records is similar to the low-acid canned food requirements. Generally, the records used in the total HACCP system include a listing of the HACCP team and assigned responsibilities, description of the dairy product and its intended use, flow diagram for the entire dairy manufacturing process indicating CCPs, hazards associated with each CCP, and preventive measures, critical limits, monitoring system, corrective action plans for deviations from critical limits, records for all CCPs, and procedures for verification of HACCP system.

K. Step Eleven: Verification (Principle 7)

Verification consists of the use of methods, procedures, or tests in addition to those used in monitoring to determine that the HACCP system is in compliance with the HACCP plan and/or whether the HACCP plan needs modification and revalidation. Verification involves the following:

1. The scientific or technical process to verify that critical limits at CCPs are satisfactory. This consists of a review of the critical limits to verify that they are adequate to control the hazards that are likely to occur.
2. Ensuring that the HACCP plan is functioning effectively. Rather than relying on end-product sampling, firms must rely on frequent reviews of their HACCP plan, verification that the HACCP plan is being correctly followed, review of CCP records, and determinations that appropriate risk management decisions and dairy product dispositions are made when process deviations occur.
3. Documented periodic revalidations, independent of quality audits or other verification procedures, that must be done to ensure the accuracy of the HACCP plan.

Verification inspections should be conducted routinely, or on an unannounced basis, (a) to assure selected CCPs are under control, (b) when intensive coverage of a specific commodity is needed because of new information concerning dairy food safety, (c) when dairy foods produced have been implicated as a vehicle of foodborne disease, (d) when requested on a consultive basis or established criteria have not been met, and (e) to verify that changes have been implemented correctly after the HACCP plan has been modified.

Verification reports should include the following: (a) information about the existence of a HACCP plan and the persons responsible for administering and updating the HACCP plan, (b) the status of records associated with CCP monitoring, (c) direct monitoring data of the CCP while in operation, (d) certification that monitoring equipment is properly calibrated and in working order, (e) deviations and corrective actions, (f) any samples analyzed to verify that CCPs are under control (analyses may involve microbiological, chemical, physical, or organoleptic methods), (g) modifications to the HACCP plan, and (h) training and knowledge of individuals responsible for monitoring CCPs.

L. Step Twelve: Evaluating and Revising HACCP Systems

A full review should take place at least annually and should include validation and assessment of CCP. Other situations that trigger evaluation include (a) new potential hazards for that dairy food such as newly recognized pathogens and new

CCPs, (b) when an existing HACCP is out of date, (c) when records are not available, (d) if changes in production occur and problems are discovered. Another situation that may trigger evaluation is the response to new dairy product development such as raw material change; preparation and processing change; formulation change; packaging change; distribution, storage, or display system change; or new use of dairy product by consumers. The response to a manufacturing change may also trigger evaluation if there are changes in dairy product flow in a plant, equipment change, shift changes, especially if they affect cleaning, and changes in storage or distribution.

VII. EMPLOYEE EDUCATION AND TRAINING

Product safety systems are people programs. Training people is an essential part of safety systems. Employees must develop an awareness of safety and create a proactive environment for dairy product safety. The successful introduction of a safety system needs to be accompanied by both education and training. The information and training needs of staff vary and should be an ongoing process, not a one-time event.

As stated previously, any safety system must have the full support of top-level management who will need to be briefed about the positive benefits of using this approach to assure product safety. This briefing should include resource implications, especially in terms of time input, person-power, and staff training requirements, during the setting up and subsequent operating of the system. Other managers and staff, whether or not they are involved directly in the safety system program, need to be briefed in general terms about the reasons for this approach and their likely role in the resulting safety system. At the very least, managers and staff should be made aware of the reasons for such a new approach. All personnel need to be kept informed of progress during development of safety systems that involve their work, and this may be done via information sheets, meetings, and workshops, among other modes.

Team members need training in (a) the principles of HACCP, (b) approaching the analysis logically, systematically, and thoroughly, (c) benefits of the HACCP system, and (d) role that the team plays in dairy product safety. Production staff managers, supervisors, engineers, and operators need training on two levels to enable them to carry out their parts in changes that result from a safety system program. The first level involves how the application of the safety system program will affect an individual's work. For example, staff who monitor the CCPs need to know what corrective actions to take when a control measure fails (exceeds the specified tolerances) or moves toward failure. Training may also be needed for interpretation of the data produced when monitoring is done. The second level involves specific training in technical skills, for example, taking an accurate and relevant temperature measurement.

Both team members and production staff need to understand that team meetings, verification audits, and changes arising from the findings of these audits all form part of the safety system and are all aimed at achieving the objective of the program in the most effective way. It is suggested that dairy plant personnel be trained in four distinct groups: (a) senior management, (b) HACCP coordinator, (c) HACCP team member, (d) and on-line personnel. Senior management should have general knowledge of HACCP principles and the safety system plan. The HACCP coordinator should have a broad and detailed understanding of HACCP principles and the safety system plan. The HACCP team members need a broad and detailed understanding of HACCP principles and the safety system plan. On-line employees need to know the importance of specific CCPs, and new personnel need to be made familiar with the safety system and be equipped with the necessary skills to carry out their role within it. This information should be conveyed during induction training.

VIII. MODEL HACCP PROGRAMS

Generic HACCP plans can serve as useful guidelines; however, it is essential that the unique conditions within each dairy facility be considered during development of an HACCP plan. Subtle differences in product raw materials and manufacturing require managers to examine CCPs line-by-line and plant-by-plant.

The following model/generic HACCP plan has been developed to serve as a guideline upon which individuals can build their HACCP programs. A hazard analysis chart (Table 1), flow diagram (Fig. 1), and description chart (Table 2) are included. Simple and straightforward are the keys to a successful HACCP plan. If modifications are necessary, only safety issues should be considered if new CCPs are added.

The fluid milk model program is based on a typical high-temperature, short-time system and includes CCPs that were developed to address raw milk receiving, storage, pasteurization, and vitamin addition. A hazard analysis should be conducted if any changes are made to the program to determine whether the change creates a hazard. This model program could also be used for flavored milk products by including additional elements for nondairy ingredient receiving and storage to the flow chart. Other fluid products such as half and half or cream could follow a similar flow chart.

TABLE 1 Hazard Analysis Chart: Fluid Milk

Process Step	Identified Hazard	Preventive Measures	CCP
Raw milk receiving	Microbiological (M)—Pathogens, *Staphylococcus* toxin Chemical (C) — Animal drug residues	Pathogens are eliminated by pasteurization. Temperature control is necessary to prevent *Staphylococcus* toxin production. Testing is necessary to prevent presence of drug residues.	Yes (M, C)
Filter	Physical—presence of any foreign object that may remain in finished product	System prevents passage of a foreign object large enough to be a hazard.	No
Raw milk storage	Microbiological—Pathogens, *Staphylococcus* toxin	Pathogens are eliminated by pasteurization. Temperature control is necessary to prevent *Staphylococcus* toxin production.	Yes (M)
Clarifier/ separator	Microbiological—Pathogens, *Staphylococcus* toxin	Pathogens are eliminated by pasteurization. Resident time not adequate for *Staphylococcus* toxin production.	No
Raw cream storage	Microbiological—Pathogens, *Staphylococcus* toxin	Pathogens are eliminated by pasteurization. Temperature control is necessary to prevent *Staphylococcus* toxin production.	Yes (M)

Homogenation	Microbiological—Pathogens, *Staphylococcus* toxin	Pathogens are eliminated by pasteurization. Resident time not adequate for *Staphylococcus* toxin production.	No
Vitamin addition	Microbiological—Pathogens, *Staphylococcus* toxin Chemical—Toxic levels of vitamin A and/or D	Prerequisite programs are in place for ingredient receiving. Usage records and proper pump calibration ensure proper addition.	Yes (C)
Pasteurization	Microbiological—Pathogens	Pathogens are eliminated by pasteurization.	Yes (M)
Pasteurized storage	Introduction of pathogen hazards after pasteurization	Prerequisite programs are in place to prevent post-pasteurization contamination.	No
Packaging material	Introduction of pathogens, chemicals, or physical hazards after pasteurization	Prerequisite programs are in place to prevent post-pasteurization contamination.	No
Filler	Introduction of pathogens, chemicals, or physical hazards after pasteurization	Prerequisite programs are in place to prevent post-pasteurization contamination.	No
Cold storage	Properly pasteurized, packaged product contains no hazards	Not applicable.	No
Distribution	Properly pasteurized, packaged product contains no hazards	Not applicable.	No

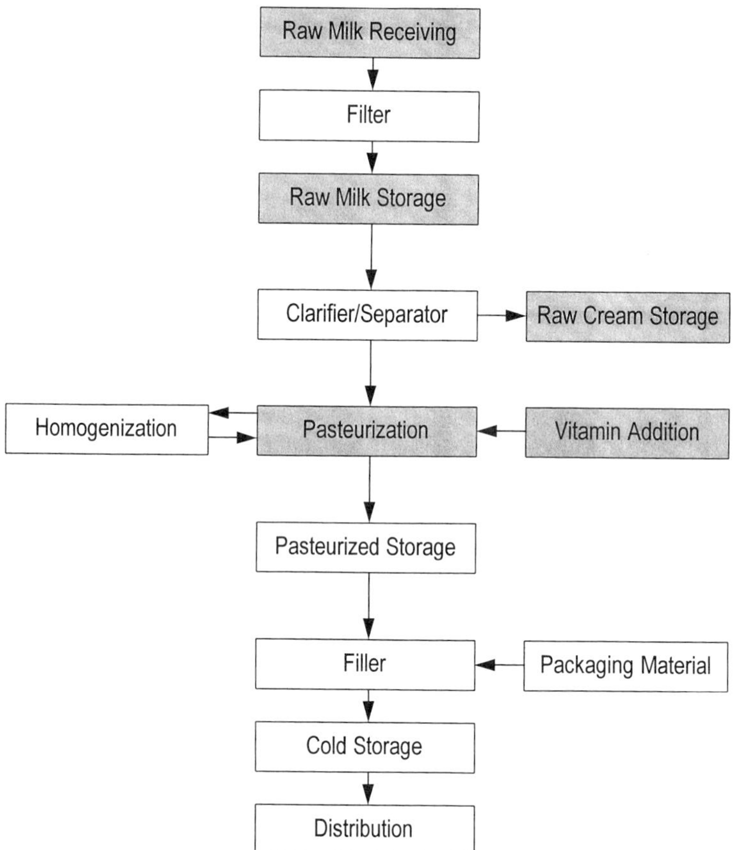

FIGURE 1 Fluid milk flow diagram.

TABLE 2 Hazard Analysis Critical Control Point Description Chart for Fluid Milk

CCP/ Process Step	Hazard/ Concern	Control Point	Critical Limit	Monitoring/ Frequency	Records/ Location	Respon- sibility	Corrective Action	Verification
CCP1: Raw milk receiving	Micro- biological	Temperature	≤45°F (7°C)	Every tanker	Load ticket QA/QC office	Intake operator	Hold and evaluate product	Indicating thermometer
	Chemical— Drug residues (raw milk)	β-Lactam screening	No positives	Every tanker	Receiving log; QA/QC office	Intake operator	Reject	Calibrate test kit
CCP2: Raw milk storage	Micro- biological	Temperature Time	≤45°F (7°C) ≤72 hr	Continuous but checked four times daily	Recording chart; QA/QC office	QA technician	Hold product, investigate cause and adjust	Recording vs. indicating thermometer
CCP3: Raw cream storage	Micro- biological	Temperature Time	≤45°F (7°C) ≤72 hr	Continuous but checked four times daily	Recording chart; QA/QC office	QA technician	Hold product, investigate cause and adjust	Recording vs. indicating thermometer
CCP4: Vitamin addition	Chemical	Proper con- centrations	<300% of label claim	Daily	Vitamin log; QA/QC office	Pasteurizer operator	Hold product, investigate cause and adjust	Pump calibra- tion, usage records
CCP5: Pasteuri- zation	Micro- biological	Temperature Time	≥161°F (72°C) ≥15 sec	Continuous	Recording chart; pro- duction office	Pasteurizer operator	Flow divert, recirculate, and heat	Cut-in/cut-out checks; indicating thermometer calibration

REFERENCES

Anonymous. *Pre-requisites Assessment Plan Manual*. Cornwall, Ontario, Canada: Cibus Consulting, 1995.

Anonymous. *Dairy Product Safety System Manual*. Washington, DC: International Dairy Foods Association, 1996a.

Anonymous. *Title 21, Code of Federal Regulations*. Washington, DC: U.S. Government Printing Office, 1996b.

Pierson MD, Corlett DA Jr, ed. HACCP Principles and Applications, New York: Van Nostrand Reinhold, 1992.

13

Testing Milk and Milk Products

CHARLES H. WHITE

Mississippi State University, Mississippi State, Mississippi

I. INTRODUCTION

Microbiological testing in the dairy plant is critical to ensure that raw milk, other ingredients, and finished products are of high quality. Such testing also serves to verify the adequacy of Hazard Analysis Critical Control Point (HACCP) procedures. Testing for pathogens is not done in the dairy plant but samples are sent to a laboratory located far enough from the plant to preclude introduction of unwanted microorganisms through manipulations in the laboratory (see Chaps. 11 and 12).

This chapter lists the chemical, microbiological, and physical tests that might be done on incoming raw milk and considers the specific microbial aspects of raw milk quality. Also discussed are testing of raw milk and raw ingredients, line sampling and tests for predicting shelf-life of products, testing of various types of dairy products, and the future of testing of milk and milk products.

II. RAW MILK QUALITY

A. General

There are many ways to measure the quality of raw milk. Some of the tests that are done by dairy processing plants either before or after unloading a tanker of milk include the following:

 1. Standard plate count (SPC)
 2. Direct microscopic count (DMC)
 3. Freezing point determination (Cryoscope)
 4. Presence of inhibitory substances (antibiotic screening test)
 5. Sensory evaluation
 6. Preliminary incubation (PI)—SPC
 7. Direct microscopic somatic cell count (DMSCC)
 8. Acid degree value (ADV)
 9. Laboratory pasteurization count (LPC)
10. Thermoduric spore count
11. Fat content
12. Total solids content (can also include protein content)
13. Sediment test

In addition, the weight (total quantity of milk) of the tanker is obtained to ensure proper payment to dairy farmers and to ensure that the processing plant is receiving all the milk for which it is making payment. However, the compositional and chemical quality factors are always important.

Some of the aforementioned tests should be done before unloading the tanker. There is a definite time restraint involved with receiving and unloading a tank load of milk; however, the processor, not the producer, is the customer and should take a reasonable amount of time to obtain satisfactory results from the tests selected. It is recommended that the following tests be done on each tanker load of raw milk before unloading: DMC (until a more definitive test can be done in the same amount of time—bioluminescence may be this test), antibiotic screening test, cryoscope for added water, and sensory evaluation, which should involve checking the odor of the tanker followed by heating the milk and rapid cooling to taste the sample.

Compositional tests (e.g., tests for fat and total solids) should be done on every tanker of milk, although not necessarily before unloading. If the tanker load of milk is from independent producers, tests for abnormal milk, such as DMSCC, are also needed. Most other tests can be used as troubleshooting tests if there is a shelf-life problem.

The number of dairy farms has been decreasing steadily to the point where most of the dairy farmers in business (just as with the processors) take their jobs very seriously. As a result, the quality of raw milk is very good. This is not to imply that all raw milk is of excellent quality and cannot be improved. In 1982, Zall et al. summarized results of the SPC, psychrotropic bacteria count (PBC), and ADV tests of raw milk held at 6.7°C for 0, 3, or 6 days. A summary of their results follows:

	Day of Storage at 6.7°C[a]		
Test (Mean)	0	3	6
SPC	4.92[b]	7.36	8.39
PBC	4.45	6.77	8.46
ADV	0.80	1.38	4.89

[a]Summarized from Zall et al. (1982).
[b]Counts expressed in log numbers.

There was a substantial increase in bacterial numbers regardless of type, whether mesophilic or psychrotrophic. The legal SPC standard for raw milk is 100,000 colony-forming units (cfu)/mL (individual producer) or 300,000 cfu/mL (commingled) milk. Individual raw milk can consistently be produced with less than 10,000 cfu/mL. Counts in tanker loads of milk vary from less than 10,000 to greater than 1,000,000 cfu/mL. The count in most raw milk (tanker loads) currently being received at fluid milk plants in the United States ranges from 30,000 to 70,000 cfu/mL.

The changes in the standard for the DMSCC from 1,000,000 to 750,000 cells per milliliter indicates an improvement in raw milk quality. Although there is no rule about increased bacterial numbers with increased somatic cell counts, this correlation does appear to exist. Within the next few years, it is likely that this standard will be reduced even further, for example, to 500,000/mL.

B. Raw Milk Microflora

According to a recent study (Celestino et al., 1996), gram-positive bacteria are present in raw milk in much smaller numbers than gram-negative species. These workers reported on numbers of *Pseudomonas* as well as other gram-negative and -positive bacteria in both farm bulk tanks and in creamery and plant silos. In farm bulk tanks, regardless of temperature, pseudomonads represented more than 80% of all bacterial isolates. The gram-positive bacteria in milk at the farm bulk tank in this study represented no more than 1% of the total. When the milk was commingled in creamery silos, the pseudomonads represented approximately 70% of the microflora. The gram-positive bacteria increased to 9.0% to 14.1%, depending on temperature. Members of the family Enterobacteriaceae represented up to 15% of the total microflora of milk in the creamery silos.

Celestino et al. (1996) made a most significant conclusion, "As the quality of pasteurized milk improves because of reduction in levels of postpasteurization contamination, the presence of a heat-resistant psychrotrophic bacteria in the milk supply will assume greater importance." Of these, spore-forming microorganisms such as *Bacillus* are the most important. Work in Griffiths' laboratory was reported by these researchers, which indicated that higher heat treatments applied to the milk (70°C rather than 60°C) tended to decrease spore counts, presumably because of the activation of spores, which could subsequently germinate and divide. Using this as evidence, they cautioned that an increase in pasteurization temperature does not necessarily result in an increased shelf-life. This has been the tendency of many processors over the past 10 years (since *Listeria* and *Salmonella* became known to the dairy industry).

C. Spore Formers

It might be concluded that the higher the quality of raw milk, the higher will be the incidence of gram-positive spore-forming bacteria. According to Martin (1974), *Bacillus* species account for 95% of the total spore-forming bacteria in milk, with *Clostridium* species comprising the remaining 5%. He indicated that, in the United States, 43% of *Bacillus* organisms are *Bacillus licheniformis* and 37% are *B. cereus*; however, in other countries, *B. cereus* is predominant. The data in Table 1 (Martin, 1981) indicate that spore-forming bacteria are expected to be present in almost all raw milk supplies. As the dairy processing industry becomes more involved with extended shelf-life (ESL) products, the problem with spore-forming bacilli will probably increase. Thus, an aerobic spore count (80°C for 12 minutes followed by rapid cooling and plating on plate count agar [PCA] with incubation of plates at 32°C for 48 hours) will become a vital microbiological test for raw milk.

D. Psychrotrophic Bacteria

A simple definition of psychrotrophic bacteria is those bacteria that can grow fairly rapidly at refrigeration temperatures. A psychrotroph is unlike a true psychrophile, which is a bacterium whose optimal growth temperature is 10°C or less. There are not many psychrophiles encountered in the dairy industry. In raw milk, the larger the percentage of psychrotrophic bacteria, the greater the number of problems encountered by the dairy processor using that raw milk. A typical psychrotroph (e.g., a pseudomonad) could conservatively have a generation time (the length of time a bacterial population requires to double in numbers) of 9 hours or less at 7°C. Thus, if a load of milk contains 100,000 cfu/mL with 70% of the microflora

TABLE 1 Standard Plate Counts and Aerobic Spore Counts of Raw Milk[a]

Class	SPC Range (per mL)	No. of Samples Analyzed	Average SPC (per mL)	Mesophilic Spore Counts	Thermophilic Spore Counts	
				Average (per mL)	No. of Positive Samples	Average (per mL)
I	<50,000	19	32,000	400	16	46
II	<50,000 to <200,000	36	98,000	400	35	45
II	>200,000 to <1,000,000	48	580,000	710	40	55
IV	>1,000,000 to <5,000,000	73	2,300,000	760	60	41

[a]Spore counts were determined after heating milk at 80°C for 10 min. Mesophilic counts were determined by a pour plate procedure; thermophilic counts by a most probable number dilution tube technique.
Source: 16th International Dairy Congress Proceedings, 1962, pp. 295–304, as reported by Martin (1981).

being psychrotrophic, then, within 36 hours at 7°C, the counts could exceed 1,000,000 cfu/mL. This large number can produce large amounts of proteases and lipases, which can cause serious quality problems for processed products.

In a dated but excellent review of psychrophilic bacteria, Witter (1961) indicated that the choice of the word "psychrophile" was unfortunate because the root name implied "cold-loving." Many people still use the term psychrophile when psychrotroph is what is intended. The key to recognizing the difference is in the optimal growth temperature range. Psychrotrophs have an optimal growth temperature in the range of 21°C to 28°C whereas, as previously discussed, a true psychrophile has a much lower optimal growth temperature. Most of the bacteria that cause problems to the dairy processor are of the psychrotrophic type, which means that, as the temperature is allowed to increase, the generation time is reduced and more psychrotrophs are produced (see Chap. 2).

Witter (1961) indicated that the natural sources of the predominant psychrophilic (psychrotrophic) bacteria are water and soil. Because water and soil are both present in abundance on dairy farms, it is not surprising to find that these psychrotrophs work their way into the milk supply. Hence, it is incumbent upon all segments of the dairy industry to work at keeping equipment clean (as a means of reducing the number of psychrotrophs gaining entrance into the milk) and temperatures as low as possible to retard growth of the psychrotrophs that do get into milk. Witter (1961) also indicated that, at the lower temperatures, from 7°C to 0°C (their minimum growth temperature), the decrease in growth rate was dramatic. Thus, even though the legal limit for holding milk is 7°C, the closer to 0°C that the milk can be held, the higher will be its quality from the standpoint of growth of psychrotrophic bacteria.

For the reasons just outlined, there is a need to monitor the psychrotrophic population of incoming raw milk. Most measurements are by SPC or DMC, both of which measure total bacterial numbers capable of survival (at least with regard to the SPC) at 32°C. The PI-SPC (milk is held at 13°C for 18 hours before it is plated) is one way of estimating the psychrotrophic nature of the microflora. The milk could be incubated for 24 to 48 hours and then plated (SPC). Regardless, it is very important for the dairy processor to have an idea of the psychrotrophic quality of the raw milk, particularly in cheese making. White and Marshall (1973) found that flavor scores were significantly lower for Cheddar cheese made from milk containing a protease from a pseudomonad when compared with control cheese. Witter (1961) indicated that pseudomonads (the primary psychrotrophic/ psychrophilic group) possess certain characteristics that make them important to milk and other foods. Some of these characteristics are (a) ability to use a wide variety of carbon compounds for energy and inability to use most carbohydrates, (b) ability to produce a variety of products that affect flavor, (c) ability to use simple nitrogenous foods, (d) ability to synthesize their own growth factors or vitamins, and (e) proteolytic and lipolytic activity.

Because a high psychrotrophic load can adversely affect the quality of various dairy products, especially cheese and extended shelf-life products, it behooves the processor to routinely monitor the psychrotrophic population of incoming loads of raw milk.

E. Proteases

Because psychrotrophic bacteria can produce both lipases and proteases, it is important to understand the activity of the various enzymes that can be liberated into the milk. Many of the proteases tend to be extremely heat stable, which can result in defects during extended refrigerated storage of milk. Adams et al. (1975) studied heat-resistant proteases produced in milk by psychrotrophic bacteria. They found all of the psychrotrophs obtained from raw milk produced proteases that survived at 149°C for 10 seconds. They reported that 70% to 90% of the raw milk samples contained psychrotrophs capable of producing these heat-resistant proteases. White and Marshall (1973) reported on a heat-stable protease that retained 71% of its original activity after being heated at 71.4°C for 60 minutes. Also, the enzyme hydrolyzed milk protein at 4°C.

In another study, Adams et al. (1976) isolated 10 gram-negative psychro-trophs from raw milk that readily attacked raw milk proteins. They reported that κ- and β-casein were most susceptible to attack by these psychrotrophs, although they indicated that some of the isolates also attacked whey proteins. They further stated that the proteolysis did not require large populations of psychrotrophs; 10% to 20% decrease in κ-casein during 2 days at 5°C accompanied growth of one isolate to a population of only 10,000/mL. Guinot-Thomas et al. (1995) studied proteolysis of raw milk during storage at 4°C. They specifically looked at the effect of plasmin and microbial proteinases. Their study demonstrated the greater importance of microbial proteinases than of plasmin at this temperature. Also, they reported that hydrolysis of caseins by microbial proteinases affected mainly the κ-casein fraction, colloidal calcium, and consequently casein micelles. They concluded that this effect will be noted even more as the number of psychrotrophs become higher. Rollema et al. (1989) compared different methods for detecting these bacterial proteolytic enzymes in milk. This was a study in which two fluorescamine assays, a trinitrobenzene sulfonic acid (TNBS) assay, an azocoll assay, a hide powder azure (HPA) assay, and an enzyme-linked immunosorbent assay (ELISA) were tested for their effectiveness in detection of proteolytic enzymes from six strains of psychrotrophic bacteria. These workers concluded that the TCA-soluble tyrosine and the thin-layer caseinate diffusion assay are too insensitive to be used for quality control of dairy products. They stressed that a good correlation between the proteolytic activity determined with an assay and the keeping quality of the product is a prerequisite for applicability of the assay for quality control of dairy products. Their preliminary study indicated that this requirement could be reasonably satisfied by the fluorescamine, TNBS, and azocoll assays.

III. MICROBIOLOGICAL TESTING OF RAW MILK AND RAW INGREDIENTS

A. Raw Milk

Because the microbiological quality of raw milk does not improve during storage, it is critical that the processor evaluate the raw milk to ensure that only high-quality milk is accepted. With regard to microorganisms, the following information must be known:

1. Total count or aerobic plate count—Classically, this is determined by the use of the SPC procedure. In legal matters concerning acceptability of an incoming tanker of milk or milk from an individual producer, the SPC is the standard to which other screening tests are compared.

2. DMC—In this procedure, as outlined in *Standard Methods for the Examination of Dairy Products* (Marshall, 1992), results can be obtained within 15 minutes by a trained laboratory technician. Dead as well as living cells are counted, so the DMC should result in a slightly higher count than the SPC. The

big advantage is that results may be obtained before milk is unloaded into the processing facility. This allows for much better microbiological control over incoming raw fluid dairy products. The problem that many people encounter when initially using the DMC is that they try to be too "fine" with the results, for example, they may try to distinguish between a count of 40,000 and 45,000 instead of just using the DMC to detect the very high count loads. The DMC was not designed to reflect minor differences in numbers of bacteria; rather, in this instance, the test is strictly used to determine whether a tanker load of milk, cream, or condensed skim milk is of sufficiently high microbiological quality to be unloaded into the plant.

3. Psychrotrophic estimates—There are many types of bacteria in raw milk. It is critical to know what percentage of the population is of a psychrotrophic nature. The standard psychrotropic plate count (PPC) requires incubation of the plate for 10 days at 7°C (Marshall, 1992). This length of time is commercially unacceptable to determine the psychrotrophic population of raw milk. Various elevated incubation temperatures (e.g., incubation of plates at 18°C or 21°C using PCA) give an estimate of the psychrotrophic population. Incubating raw milk (cream or condensed skim milk) for 24 to 36 hours at 7°C followed by SPC incubation also gives some idea as to the number of psychrotrophs present.

4. PI-SPC—Johns (1960) first described this method for evaluating raw milk quality. His method involved incubating raw milk at 12.8°C (55°F) for 18 hours. Following this preliminary incubation, a conventional plate count was done. This method was thought to identify milk that had been subjected to less than desirable sanitary conditions at the farm level. Maxcy and Liewen (1989) found that preliminary incubation at the recommended temperature (12.8°C) did not have a selective effect for specific groups of microorganisms. Thus, apparently the PI-SPC procedure is not extremely reliable as a means of evaluating raw milk quality. Certainly, the time involved for this procedure minimizes its effectiveness in screening raw milk supplies.

5. Coliforms—According to *Standard Methods for the Examination of Dairy Products* (Marshall, 1992), coliforms are a group of bacteria that comprise all aerobic and facultatively anaerobic, gram-negative, non–spore-forming rods able to ferment lactose and produce acid and gas at 32°C or 35°C within 48 hours. Typically, coliforms are used as a measure of sanitary conditions in the processing and packaging of pasteurized dairy products. Coliforms are destroyed by pasteurization, hence, any coliforms found in the pasteurized product indicate postpasteurization contamination.

Coliforms may also be of value in checking raw milk. There is no legal standard for the numbers of coliforms that might be present in raw dairy ingredients. It is suggested that a value of 100 coliforms per milliliter be used as an initial screening tool for raw milk. The procedure used would be the same as that outlined in *Standard Methods* (Marshall, 1992). As with pasteurized milk, coliforms are "indicator organisms." This simply means that if coliforms are present, conditions may be suitable for the presence of enteric pathogens, such as *Salmonella*.

6. Adenosine triphosphate bioluminescence assays—In an excellent overview of how ATP bioluminescence can be used in the food industry, Griffiths (1996) agrees with other researchers (Bautista et al., 1992; Griffiths et al., 1991; Reybroeck and Schram, 1995; Sutherland et al., 1994) that these assays may be used successfully for determination of microbial loads in raw milk within 10 minutes. Griffiths (1996) described that the milk is incubated in the presence of a somatic cell-lysing agent and then filtered through a bacteria-retaining membrane. The microorganisms retained on the filter are then lysed with the lysate being assayed for ATP activity. He stressed that microbial populations down to 10^4 cfu/mL can be detected with a greater precision than with the SPC.

Griffiths (1996) described the work of Pahuski et al. (1991), which involved a "concentrating" reagent, Enliten, that clarifies milk and allows removal of microorganisms by centrifugation. These workers indicated that a combination of this treatment along with an ATP assay enabled detection of microbial levels down to 2×10^4 cfu/mL within 6 to 7 minutes.

B. Dairy Ingredients

Many dairy ingredients other than raw milk are received by dairy and food processing plants. Some of these products include nonfat dry milk, whey powder, whey protein concentrates and isolates, condensed skim milk, condensed whole milk, sweetened condensed skim milk and whole milk, cream, and butter. These ingredients must also be tested to ensure their overall quality and that they meet established microbiological criteria. The SPC and the coliform count using violet red bile agar (VRBA) are outlined in *Standard Methods for the Examination of Dairy Products* (Marshall, 1992). This compilation of accepted methods is descriptive with regard to sampling and the quantity of ingredient required for appropriate analysis. Representative samples of each incoming batch should be tested to ensure acceptability. When receiving dried products, a statistically valid number of samples should be obtained. Various sampling procedures have been used by companies with the military standard MIL-STD-105D being a well accepted method for determining the number of samples to be taken. A rough approximation for sampling is based on the following formula (does not take into account degree of severity):

$$\text{Number of samples} = \frac{\sqrt{\text{Batch size}}}{10}$$

The number of samples should be randomly drawn to ensure representative sampling and testing of the entire batch.

C. Nondairy Ingredients

Many ingredients other than dairy products are brought into dairy processing plants. Examples of such products include fruits, nuts, stabilizers, emulsifiers, fat replacers, sucrose and other sweeteners, and spices. The key to ensuring the quality of all ingredients, especially nondairy ingredients, lies with the requirement of a product specification sheet. Each supplier that provides products to a dairy processing plant should provide an individual product specification sheet for each item sold to that company. The specification sheet, which should be updated annually, should contain a description of the product as well as guidelines that the product must meet. Microbiological testing should be outlined on the product specification sheet. This includes the type of tests to be done and either the method outlined or a reference to the procedure to be followed. The specification should ensure that ingredients have been tested for specified pathogens and are known to be "pathogen free." Again, the SPC and the coliform count are commonly used procedures in evaluating the quality of many of these ingredients. Counts are typically related to the grade of product being received. Samples must be obtained as soon as the products arrive so accurate and prompt microbial analysis can be accomplished.

The following is an outline of microbiological testing that should be done on incoming raw dairy ingredients and nondairy ingredients, as recommended by myself and Randolph (personal communication, 1996).

Microbiological Testing of Raw Milk	
Test	Suggested Standard
1. Direct microscopic count—Every tanker (before unloading)	200,000 cfu/mL
2. Coliform (violet red bile agar)—Every tanker (backtrack to individual producer if necessary)	100 cfu/mL
3. Standard Plate Count (PCA)—Silos daily	100,000 cfu/mL
4. PI-SPC (18 h at 12.8°C)—Silos daily (backtrack if necessary)	300,000 cfu/mL

The PPC or the PI-SPC is especially critical for cheese operations because presence of psychrotrophic bacterial proteases can adversely affect yield as well as quality of these concentrated products.

IV. LINE SAMPLING/TESTING

One of the most important aspects of microbiological testing of milk and milk products is line sampling. If only the finished product is tested, then it is only known whether the finished product is "good" or "bad"; however, if the shelf-life of the product is less than desirable, it is not known where the postpasteurization contamination occurred. To gain such information is the purpose of line sampling. In a fluid milk operation, line samples should be obtained at the following locations:

1. At or immediately after the high-temperature, short-time pasteurizer— This is done to ensure that neither the regenerative plates nor cooling plates have pinhole leaks.
2. Preceding pasteurized milk storage tanks—This verifies the cleanliness of the pasteurized milk lines leading from the pasteurizer to storage tanks.
3. Line sample leading from the pasteurized milk storage tanks—This is done to ensure cleanliness of the storage tank itself.
4. Immediately preceding entry of the milk into the separate fillers.

By checking each of these locations, postpasteurization contamination can be pinpointed.

Because most dairy processing plants have welded pipelines and do not disassemble all of their piping, the method for obtaining aseptic line samples becomes critical. One very efficient way of obtaining good samples is by use of the QMI Aseptic Sampler (Food and Dairy Quality Management, Inc.—QMI, St. Paul, MN). The aseptic samplers are inserted into stainless steel elbows for ease of sample extraction. Even though virtually any size sample can be taken, a minimum of 50 mL and preferably 60 mL should be used. There is a greater chance of detecting microorganisms that could be detrimental to product shelf-life from a larger sample.

Regular grommets can also be inserted and then a syringe and needle can be used to extract samples of similar size. Samples in the syringes can be used for any number of microbiological evaluations. The primary bacterial types of concern in these samples are coliforms and psychrotrophic bacteria. To enhance enumeration of psychrotrophic bacteria, a step commonly used is to incubate the sample (in the syringe) at 21°C for 18 hours. Following this preincubation, the sample can either be plated for SPC or for coliforms (VRBA). The preliminary incubation is not absolutely necessary but it does enhance enumeration of any psychrotrophs or heat-injured coliforms that might be present. A SPC on the fresh milk is virtually meaningless. Thus, the different options to consider with regard to microbiological evaluation of line samples are: (a) fresh milk coliform count— VRBA, (b) PI-VRBA, (c) PI-SPC, and (d) PI plus any other selective media designed to enumerate psychrotrophic bacteria, such as PI + CVT (crystal violet tetrazolium agar).

After counts are obtained (counts should be viewed as the same as for any finished fluid product), Gram stains of preparations from colonies on plates can be made to determine whether the microorganisms appearing in "spoiled" products are similar to those observed in line sampling. This can be a direct indication of the presence of bacteria that are reducing shelf-life.

V. SHELF-LIFE PREDICTING TESTS FOR FLUID MILK-TYPE PRODUCTS AND ESTIMATION OF ACTUAL PRODUCT SHELF-LIFE

The term "shelf-life" can be used interchangeably with the term "keeping quality," which is defined as the time a product remains acceptable in flavor after packaging. The question then becomes, what is an acceptable shelf-life for fluid milk products? Before answering this question, the temperature at which the product is held when shelf-life testing is done must be specified. The temperature most commonly used is 7°C (45°F), which is chosen because it approximates the temperature of dairy cases in supermarkets and the home refrigerator (Bishop and White, 1985; White, 1991). Also, as has previously been pointed out (Bishop and White, 1985), in all shelf-life prediction studies, the "potential" shelf-life is actually what is being measured because the experimental sample stored in a cooler in the laboratory is not subjected to the rigors of distribution and transportation.

Almost all tests that are designed to predict the shelf-life of dairy products are based on detection of gram-negative psychrotrophic bacteria (especially the pseudomonads). These microorganisms cause most shelf-life problems, especially in fluid milk and cottage cheese. Regardless of the method, the key (White, 1991) to predicting the shelf-life of milk and milk products is that the method must be rapid—reliable and meaningful results must be obtained within 72 hours and ideally within 24 hours.

In addition, results of tests to predict shelf-life must be compared or correlated with the actual product shelf-life. Thus, to determine whether or not a particular test to predict shelf-life is effective, the actual product shelf-life must be assessed. The actual product shelf-life is determined by holding the samples at 7°C and testing them every day until an off-flavor develops. The shelf-life is then estimated as the day the off-flavor developed minus one. To minimize the number of times the container is opened and closed, the products do not need to be tasted until after day 10 (assuming that the product had a shelf-life of 10 days or more). It is important in determining basic product shelf-life to use the same container because each filler head (on a gallon filler) can yield significantly different results. In selecting samples from a filler, it is good to rotate the samples obtained so that, over a given period, all filler heads can be sampled.

Correlation between the results of shelf-life prediction and actual product shelf-life at 7°C can be ranked using the following scale: excellent = >0.90; good = 0.80 to 0.89; fair = 0.70 to 0.79 (Bishop, 1988, 1993; White, 1991). Because of low initial numbers of bacteria in freshly pasteurized milk, most shelf-life testing consists of preincubating the product (in its original container) at 21°C for 18 hours followed by some rapid bacterial detection method (White, 1991, 1993, 1996).

The Moseley Keeping Quality Test consists of incubating the finished product in its original carton at 7°C for 5 to 7 days followed by doing the SPC. This test has been used for many years by dairy processors as a way of evaluating the "staying power" of their products. The big drawback is the length of time required for results, that is, 7 to 9 days before actual counts are obtained. As newer tests to predict shelf-life are developed, the tendency is for dairy processors using the Moseley Keeping Quality Test to correlate results of the new test with those of their regular test. This is not the way to evaluate a new test. The results of any test to predict shelf-life should be correlated with actual product shelf-life not with the results of another test. Erroneous conclusions may be drawn. Thus, the best testing protocol is a preliminary incubation of the product so any psychrotrophs present can be enumerated rapidly. Many time and temperature combinations have been evaluated but the one set of conditions that seems to optimize outgrowth and enumeration of the psychrotrophs is incubation for 18 hours at 21°C. Therefore, the preliminary incubation (PI) mentioned in the remainder of this chapter represents 18 hours at 21°C.

Some of the proven methods to predict shelf-life are as follows:

1. Moseley Keeping Quality Test.
2. PI plus various plating methods: PI + SPC (incubation of plates at 32°C for 48 hours); PI + mPBC (incubation of plates at 21°C for 25 hours) (mPBC = modified psychrotrophic bacteria count on PCA); PI + CVT (One liter of PCA containing 1 mL of a 0.1% crystal violet solution followed by sterilization, cooling, and addition of 2, 3, 5-triphenyl tetrazolium chloride [TTC]) (Plates are incubated a 21°C for 48 hours.) (Marshall, 1992); PI + VRBA (incubation of plate at 32°C for 24 hours).
3. Bioluminescence.
4. Catalase detection.
5. Limulus amoebocyte lysate (LAL) assay—This procedure involves detection of endotoxins produced specifically by gram-negative bacteria (White, 1993).
6. Impedance microbiology.

7. Dye reduction (HR1, HR2) (Randolph, Personal Communication, 1996).
8. Reflectance colorimetry (the LABSMART System [Gary H. Richardson, Logan, UT])—This is a tristimulus reflectance colorimeter that monitors dye pigment changes caused by microbial activity.

These methods reflect the most current information about the basics of shelf-life prediction techniques. However, no one procedure is ideally suited for every plant application.

Bishop and White (1985) used PI + impedance detection time (IDT) to successfully predict the shelf-life of fluid milk. For fluid milk products, the PI + IDT yielded the highest correlation (r = 0.94) between test result and actual product shelf-life at 7°C. By comparison, the correlation obtained for the Moseley Keeping Quality Test was r = 0.75. Because of the 7 to 9 days required before results are available from the Mosely test and because fluid milk products have a shelf-life of approximately 14 to 21 days at 7°C, there is no question which test would be of more value to the processor. Any of the tests discussed that can give results within 72 hours are of more value not only in predicting shelf-life but also in controlling the sanitary operation of the plant. Fung (1994), in an excellent overview of rapid detection methods, described 10 attributes of an ideal rapid or automated microbiological assay system for food:

1. Accuracy—Especially sensitive for false-negative results
2. Speed—Accurate results within 4 hours
3. Cost—Designed for each application
4. Acceptability—Must be "official"
5. Simplicity—Ideally "dip-stick" technology
6. Training—Adequate for test or kit
7. Reagents and supplies—Stability, consistency, availability
8. Company reputation—Performance of product is critical
9. Technical service—Rapid and thorough
10. Space requirements—Should not take up a whole laboratory

Most of the tests discussed meet most of these criteria.

Another method is described by Bishop (1988) as the Virginia Tech Shelf-Life Method (VTSLM), which involves a preliminary incubation (21°C) followed by simple plating. He describes this method as being reliable, accurate, relatively rapid, economical, and familiar to laboratory personnel. He advocates aseptically transferring 10 mL of a pasteurized fluid milk product into a sterile test tube and incubating the tube and its contents at 21°C for 18 hours. The sample is then mixed well and diluted 1:1000 with the diluted sample being plated on PCA and incubated at 21°C for 25 to 48 hours. He indicated that this method provides an estimate of the growth potential of psychrotrophic bacteria that may be present in the sample. The time variation for the plate incubation indicates the difference between agar and 3M-Petrifilm methods. If PCA is used, add 50 ppm of a filter sterilized solution of 2, 3, 5 triphenyl tetrazolium chloride (TTC) to the melted and cooled (44°C–46°C) agar before pouring plates. Only the red colonies should be counted. Counts can then be extrapolated to indicate estimated shelf-life. Shelf-life categorization by VTSLM (Bishop, 1988, 1993) follows:

Petrifilm/Agar		
Count (cfu/plate)	Total Count (cfu/mL)	Estimated Shelf-Life (Days)
≤1	≤1,000	≥14
1–200	1,000–200,000	10–14
≥200	≥200,000	≤10

By continuing to do the test to predict shelf-life on a regular basis and reacting to the results, confidence can be instilled from quality assurance and production standpoints. Most spoilage of fluid milk-type products occurs from presence of pseudomonads and related gram-negative bacteria. The tests discussed tend to emphasize detection and enumeration of gram-negative rods.

Gutíerrez et al. (1997) reported on generating monoclonal antibodies against live cells of *Pseudomonas fluorescens*, which were used in an indirect ELISA format to detect *Pseudomonas* spp. and related psychrotrophic bacteria in refrigerated milk. The researchers indicated that development of an ELISA technique using these specific antibodies would facilitate rapid screening of refrigerated milk for detection of high concentrations of bacterial cells. They reported a good correlation ($r = 0.96$) between the colony numbers of psychrotrophic bacteria from commercial milk samples maintained at 4°C by the SPC method and the ELISA technique. These authors stressed the advantages of the indirect ELISA technique as being its versatility, simplicity, and speed.

There is still somewhat of an art in predicting the shelf-life of dairy products. Because there is no one perfect test for all needs, processors must carefully select the one or two tests that best fit into their overall quality assurance program. The key points (White, 1991, 1996) regarding prediction of shelf-life are as follows:

1. Know the actual potential shelf-life of the products as measured at 7°C (45°C).
2. Select the test to predict shelf-life that best fits the total program.
3. Routinely do the tests and develop a history, categorizing the results.
4. Ensure top management commitment to define a course of action in case product failure is projected by the tests.

VI. MICROBIOLOGICAL TESTING OF MILK AND NONCULTURED PRODUCTS

A. Fluid Milk Products

Shelf-life becomes critical for fluid milk products. Shelf-life of pasteurized milk has been defined (White, 1991) as the time between packaging and when the milk becomes unacceptable to consumers. Because the actual product shelf-life is between 10 and 21 days at 7°C, rapid shelf-life prediction tests, as discussed previously, become critical. Dairy processors generally do such a good job of cleaning and sanitizing that the number of contaminating bacteria (psychrotrophs) is so small that some form of preincubation is required to obtain numbers large enough for rapid detection tests to enumerate them.

For the reasons just stated, the following are recommended for microbiological testing of fluid milk-type products:

1. Estimation of coliforms—At a minimum, a coliform (VRBA) test should be done on representative samples of all fresh products. Randolph (personal communication, 1996) and I agree that a better test would be a "stress" coliform test wherein the product is incubated at 21°C for 18 hours followed by coliform estimation on VRBA. According to *Standard Methods for Examination of Dairy Products* (Marshall, 1992), plates are incubated at 32°C and counted after 24 hours of incubation. The PI-VRBA allows for outgrowth of heat-injured coliforms, which might not show up on a coliform count made directly on fresh products. Petrifilm or VRBA agar in regular petri dishes may be used. Whereas VRBA agar is normally used for detection of coliforms, PI allows for detection of some psychrotrophic types that may be present.

2. Shelf-life prediction tests—Any of the shelf-life predicting tests discussed previously may be used. Specifically, it is recommended to use one of the following: PI + SPC (18 hours at 21°C plus 48 hours of plate incubation at 32°C); PI + CVT (product incubation for 18 hours at 21°C followed by incubation of crystal violet tetrazolium agar for 48 hours at 21°C); PI + other rapid detection methods, for example, PI + bioluminescence, PI + impedance detection, and PI + reflectance colorimetry.

These other systems can be very effective and accurate in predicting shelf-life. Because of the cost of some of the systems, it may be necessary to use them for more than one test, such as for raw milk evaluation, equipment cleanliness, and culture viability in addition to shelf-life prediction. Smithwell and Kailasapathy (1995) described a rapid test for detection of psychrotrophs wherein milk is mixed with a selective agent (benzalkonium chloride) and a bacterial indicator (tetrazolium salt) and incubated at 30°C. The researchers indicated that gram-positive bacterial growth is suppressed by the benzalkonium chloride, and they stipulated that, if gram-negative bacteria are present, they grow and multiply.

Once the numbers reached approximately 10^7/mL, the tetrazolium salt is reduced and the color of the milk changes from white to red. This is similar to the HR1-HR2 test described by Randolph (personal communication, 1996). The authors caution that the time required for this color reaction to occur depends on the amount of milk examined and the level and activity of bacteria present. These reduction-type tests lack the sophistication of some of the other test methods, but they do have the major benefit of being visible so shelf-life tests can be observed by plant employees. This increases the interest by plant personnel in the sanitary processing and packaging of their fluid milk products.

3. Sensory evaluation of representative samples of fresh product—Milk from all fillers and all labels should be tasted fresh. Samples can be combined to minimize the total number of samples that need to be discarded.

4. Sensory evaluation at end of shelf-life—Samples need to be tasted at some point at or beyond the code date. This time can be extended as shelf-life of the product improves. Many dairies express this type of evaluation in terms of a certain percentage of products that are good (or bad) after a certain number of days at 7°C (45°F). Ideally, 100% of the products would be good when evaluated at day 21. As a rule, the number of days in which 90% or more of the products remain good can be used. Thus, a dairy may start out testing after 10 days at 7°C until success is achieved on a continual basis in 90% or more products being good. Subsequently, the sensory evaluation may be gradually moved to anywhere from 14 to 21 days until continual success is noted. If new shelf-life problems occur, evaluations may have to revert to a shorter time to achieve satisfactory results.

The quality of the raw milk, as discussed previously, is still a very important issue. Celestino et al. (1996) indicated that storage of bulk raw milk resulted in increased numbers of lipolytic and proteolytic bacteria. They found that, on the average, the number of psychrotrophs as a proportion of the total plate count increased from 47% to 80% after 2 days of storage at 4°C. Thus, finished product quality can definitely be affected if raw milk is stored for too long a time (legally no more than 72 hours, ideally no more than 48 hours) (see Chap. 2).

B. Cottage Cheese—Noncultured Dressing

In evaluating the microbiological quality of cottage cheese, the places where cottage cheese could become contaminated (from a keeping quality standpoint) must be considered. There are only three things that consistently cause shelf-life problems to the cottage cheese industry. They are

1. Wash water—The wash water must have proper pH and chlorine level.

2. Dressing—The cream dressing, whether for full fat, low fat, or nonfat cottage cheese, affects the quality of the finished product. If the dressing contains

many psychrotrophic bacteria, the desired shelf-life will not be obtained. This is especially true in dressings to which no culture has been added.

3. The packaging operation—The blending of curds and dressing and filling of cottage cheese cartons constitute excellent opportunities during which psychrotrophic bacteria can gain entry into the finished product. Great care must be exercised to ensure that only cleaned and sanitized food contact surfaces are being used.

These three areas hold true whether the cottage cheese operation is very small with all operations other than packaging being handled within the cheese vat or whether the operation is large with separate washer coolers, blenders, and packaging machines. Thus, samples should routinely be taken to ensure the microbial quality of each of these areas. First, daily samples of the wash water should be obtained and plated for coliforms and psychrotrophic bacteria (by any of the methods discussed previously). Second, daily samples should be obtained of the dressing and tested to ensure microbiological quality. Again, both coliform and psychrotrophic testing should be done. Third, a statistically valid number of samples should be used to evaluate finished product quality. (See previous discussion of this subject in this chapter; see also Chap. 9.)

In other words, cottage cheese with noncultured dressing should be handled very similarly to fluid milk products. If cultured dressing is used, the primary test to use is a coliform (VRBA) test on the fresh product.

With regard to how cottage cheese should be sampled, *Standard Methods for Examination of Dairy Products* (Marshall, 1992) prescribes the use of a sterile blender-container on a balance and tared to which 11 g of cottage cheese are added aseptically along with 99 mL of warmed (40°–45°C), sterile 2% sodium citrate solution. The sample is blended for 2 minutes, after which the product is diluted (if needed) and plated. Also, a Stomacher might be used (11 g of sample and 99 mL of diluent) to blend the cheese sample.

Another method used by some dairies for microbiological examination of cottage cheese is simply plating the dressing found in the container of finished product. This works for some freshly dressed cheeses, but many cheeses do not have enough dressing from which separate extractions can be made. In these instances, blending the cheese either in a sterile blender or in a Stomacher is necessary.

C. Frozen Dairy Desserts

The microbiological evaluation of frozen dairy desserts consists of two basic parts: (a) ingredients and mix samples and (b) finished product. Some of the ingredients used in ice cream that should be tested microbiologically include fluid dairy products, dry dairy products (especially nonfat dry milk and whey powder), fruits, nuts, colors, flavors, stabilizers, and emulsifiers.

Fruits and nuts may be weighed (Marshall, 1992) into wide-mouth containers (11-g portions should be used) to which 99 mL of dilution water are added. The mixture is soaked for 5 minutes, shaken vigorously, and plated. The recommended tests to be used for these type products are:

1. Coliform count on fresh samples.
2. Yeast and mold counts (see Chap. 8 and *Standard Methods for Examination of Dairy Products* [Marshall, 1992]). Probably the most commonly used medium for yeast and mold counts is acidified potato dextrose agar. These plates must be incubated at 25°C for 5 days with counted plates having between 15 and 150 colonies.
3. SPC.

All fluid milk products, including fluid milk, cream, and condensed skim and whole milk are plated as described previously.

Stabilizers and emulsifiers should be plated using 1 g in 99 mL of dilution water (Marshall, 1992). The sample is shaken vigorously for 15 seconds and allowed to hydrate at 20°C to 40°C for up to 20 minutes. The product is then plated for SPC and coliform count (VRBA).

For finished products, a statistically valid number of samples representing each type of product and each label change should be obtained. Finished product samples should be thawed at a temperature of up to 40°C for no more than 15 minutes (Marshall, 1992). A coliform count on fresh product is a good indication of whether sanitary methods were used in processing and handling the mix and finished product. Psychrotrophs can also be a problem. White and Marshall (1973) indicated that heat-stable enzymes produced by typical psychrotrophs could cause a measurable effect on ice cream mix that approached significance from a sensory evaluation standpoint. (See Chap. 4.)

D. Butter

By definition, butter must contain at least 80% milk fat. It seems, then, that the microbiological quality of butter is not as critical as it is with other dairy products, yet microorganisms can and do survive and grow quite well in butter and related products. White and Marshall (1973) evaluated the effect of heat-stable proteases on several dairy products, including butter. They did not find any significant effect of the proteases; this is not surprising because butter contains only about 1% protein. Microorganisms with high lipolytic activity would be expected to have a greater effect on high-fat products. *Standard Methods for Examination of Dairy Products* (Marshall, 1992) lists the following tests that can be done on butter or margarine-type products: SPC, coliform count (VRBA), proteolytic count, psychrotrophic count, lipolytic count, *Enterococcus* count, and yeast and mold count. Furthermore, I and other authors (Baer, personal communication, 1997; Richter, personal communication, 1997) recommend the following tests be done routinely in a creamery operation (all counts should be reported on a per gram of butter basis):

1. SPC (1:1000 dilution as recommended by *Standard Methods for Examination of Dairy Products* [SM])
2. Coliform count (VRBA—1:2–1:10 dilution—SM)
3. Lipolytic count (1:100 dilution—SM)
4. Yeast and mold count (1:2–1:10 dilution—SM). Wilster (1957) recommended a standard of 50 yeast and molds per gram of melted butter. This seems high for present-day circumstances. (See Chap. 5.)

E. Dry Milk and Whey Products

Dry dairy ingredients are used in a wide variety of products, including other food products as well as dairy products. The quality of the finished products can be affected by the quality of these milk ingredients. Nonfat dry milk adds many desirable properties to dairy foods; however, these desirable properties are minimized when inferior powders are used. The same may be said of the use of sweet whey powder and especially the newer whey protein concentrates and whey protein isolates. These ingredients may be purchased in various amounts but typically the product arrives in 40- to 50-pound bags or even in totes.

With regard to microbiological analyses, most dairies are performing the SPC and coliform count (VRBA). Three to 5 mL of agar overlay may be used on surfaces of solidified plates before incubation if spreading of colonies is a problem when these products are tested (Marshall, 1992). With a dried product that has obviously been exposed to some heat treatment, presence of spore-forming bacilli can be common. Also, the DMC may be used to evaluate incoming samples of nonfat dry milk and whey products. Typically, this analysis is done by making a 1:10 dilution (11 g of product in 99 mL dilution water) of the sample before it is examined microscopically. *Standard Methods for Examination of Dairy Products* (Marshall, 1992) recommends use of 2% sodium citrate solution in making these 1:10 dilutions if certain samples dissolve less readily. The reports would show as DMC/g of NDM or whey powder. (See Chap. 3.)

F. Ultra-High Temperature Products

With a commercially sterile product, the presence of any microorganisms able to grow under conditions of product storage is considered detrimental to the shelf-life of the product. Also, because the product normally is held at ambient temperatures, any slight contamination during the aseptic packaging process will damage the product.

Bockelmann (1989) indicated that, under current circumstances, the reject rate for ultra-high temperature (UHT)-type products is approximately 1 defective (unsterile) unit per 100,000 produced packages. To improve beyond this point, for example to achieve a reject rate of 1 per 100,000,000, would be impossible because of construction limits of the equipment (Bockelmann, 1989). He stated that for UHT plants in use at that time, sterilization effects of between 10 and 12 could be assumed. He said that of 10^{10} to 10^{12} bacteria spores fed into the process, one spore would survive and that the microbiological end result of such a process was dependent on (a) the sterilization effect of the UHT process and (b) the bacterial spore count in the raw product.

Thus, the number of bacterial spores present in raw milk is of definite importance when dealing with a "sterile" finished product. According to *Standard Methods for Examination of Dairy Products* (Marshall, 1992), 200 mL of raw milk should be placed in a sterile Erlenmeyer flask with a screwcap lid. The milk should be heated to 80°C for 12 minutes and then cooled immediately in an ice bath and plated on PCA with added starch and plates incubated at 32°C for 48 hours. Even though the plates could be incubated at 7°C for 10 days for psychrotrophic spore counts, the mesophilic spore count as just outlined should provide more meaningful information on UHT-type products.

For finished product testing, *Standard Methods for Examination of Dairy Products* (Marshall, 1992) recommends swabbing the outside surface of the finished product container with 70% alcohol. The needle of a sterile, single-service hypodermic syringe should then be inserted through the package wall and the appropriate amount of sample removed. Because the product is thought to be sterile, precise measurements are normally not needed—because any contamination is considered bad.

Bockelmann (1989) used the sterilization effect and the maximum acceptable defect rate as means of establishing the following proposed standard for spore counts in raw materials such as raw milk:

Standard Spore Count in Raw Materials (UHT Sterilization Effect: Approximately 11; Maximum Acceptable Defect Rate: 1:1000)

Number Surviving per Milliliter	Aim	Action	Limit
10 min, 80°C	<100	~1,000	~10,000
10 min, 100°C	<10	~100	~1,000

With regard to packaging material for UHT products, Bockelmann (1989) indicated the infection rate resulting from the manufacturing process of these packaging materials to be insignificant (i.e., 0.5 microorganism per 100 cm^2), about 3% to 5% of the bacteria were identified as *Bacillus* spores.

Bernard (1983) made several observations on some of the other microbiological considerations for testing aseptic processing and packaging systems. He indicated that, before establishing appropriate times, temperatures, and exposure concentrations to provide for commercial sterility, appropriate test organisms must be determined for each particular sterilization medium. Some of the test organisms for the different sterilization media are as follows:

Sterilization Medium	Organism
Superheated system	*Bacillus stearothermophilus* (1518)
	Bacillus polymyxa (PSO)
H$_2$O$_2$ and heat	*Bacillus subtilis* strain A
H$_2$O$_2$ and UV	*Bacillus subtilis* strain A

In addition to the sensory and physical/chemical testing done on UHT finished products, microbiological testing is also critical. Edwards (1983) indicated that SPCs and coliform counts, among other tests, of aseptically processed products done immediately after packaging are ineffective as quality control procedures because of the extremely low number of viable organisms present in an unsterile container and to the very low numbers of unsterile containers. He said that, to provide a more effective and more rapid method of detecting these low numbers of viable organisms, samples are typically incubated at an elevated temperature (e.g., 35°C [95°F]) to allow for rapid growth of most microorganisms that might be present. He stressed that incubation time may vary depending on product characteristics and types of tests to be used to detect nonsterility. It is necessary to incubate UHT samples at an elevated temperature (e.g., 35°C) for approximately 1 to 3 days. Even if bacteria have been substantially heat injured, this time-temperature combination allows for outgrowth of any survivors. Also, this combination facilitates early detection of enzymes, especially the proteases.

Edwards (1983) indicated that there are two types of samples that should be obtained: (a) aimed samples and (b) random or timed samples. Aimed samples are obtained when the risk of contamination is greater than during normal operations such as during start-up and splices. Evaluation consists of container integrity tests and product incubation. Random or timed samples are obtained during normal operation. Evaluation of these samples consists of container integrity tests, product incubation, and shelf-life monitoring.

These samples should be obtained at different locations such as after the packaging machine, after the downstream equipment, and from the warehouse (Edwards, 1983). If nonsterility is observed, resampling of the product should be done from that period, with evaluation by container integrity tests and product incubation. The defect rate in the aseptic processing and packaging systems, which Edwards (1983) said was the most common, was 1 in 10,000, and some of the sources of nonsterility were inadequate heat treatments of the product, inadequate equipment sterilization, contamination of equipment after sterilization, inadequate package sterilization, contamination of package after sterilization, faulty package material, nonhermetical seal, improper machine adjustment, damage from downstream equipment, and damage from handling and shipping.

VII. MICROBIOLOGICAL TESTING OF CULTURED DAIRY PRODUCTS

In this discussion, cultured dairy products include cultured milk (buttermilk), cultured or acidified cottage cheese, cultured or acidified sour cream, and yogurt. A total count or SPC is not suitable for measuring the microbiological quality of these products because a viable bacterial culture has been added to each of them. Even for noncultured cottage cheese dressings, an SPC on fresh product is meaningless because of the low numbers of microorganisms present after pasteurization. Thus, the coliform count (VRBA) is the primary microbiological test that is used in evaluating cultured dairy products.

For any of the cultured milks (e.g., whole, low-fat, or skim buttermilk), the coliform count may be determined by plating 1:1 on VRBA. With regard to cottage cheese, ideally, the product should be blended in a sterile blender. *Standard Methods for Examination f Dairy Products* (Marshall, 1992) recommends use of a sterile spatula to aseptically transfer 11 g of cottage cheese into the sterile blender, which had been preweighed. Then, 99 mL of warmed (40°C–45°C), sterile 2% sodium citrate solution are added. The product is then thoroughly mixed for 2 minutes. The product is then plated with 1 mL of the blended 1:10 dilution being transferred to a VRBA plate (or Petrifilm).

As discussed previously, an alternative method used by some dairy plant laboratories to test for the presence of coliforms in cottage cheese is to simply plate the dressing directly out of the cottage cheese carton. This can be somewhat difficult, especially if the cottage cheese is relatively dry. Blending yields more consistent results.

Goel et al. (1971) evaluated the duration that coliforms would survive in yogurt, buttermilk, sour cream, and cottage cheese during refrigerated storage.

They noted a marked decrease in numbers of most coliforms tested in yogurt, buttermilk, and sour cream after 24 hours of storage at 7.2°C. Hence, there is a definite need to test for the presence of coliforms in these type products within 24 hours of manufacture and packaging. With cottage cheese, there was less of a decrease in numbers of coliforms than for the other cultured dairy products. Barber and Fram (1955) cautioned that coliformlike colonies on VRBA should be confirmed for yogurt and other products containing fruit or added sweetener.

Also, yeast and mold counts are done by some dairies on some of the cultured dairy products. These counts could be done on buttermilk, cottage cheese, or yogurt. Many times, yogurt develops a yeast or mold problem as opposed to any bacterially related shelf-life–ending problems. *Standard Methods for Examination of Dairy Products* (Marshall, 1992) lists the following media to be used for yeast and mold enumerations: (a) acidified potato dextrose agar, (b) yeast extract-dextrose-chloramphenicol agar, and (c) dichloran–rose bengal–chloramphenicol (DRBC) agar. In addition, Petrifilm makes a yeast and mold agar that is used by many dairy laboratories.

The most common flavor criticism of cottage cheese, sour cream, and buttermilk-type products is that they "lack flavor" or are "flat." Because the incubation time or temperature has not allowed the culture of bacteria to produce sufficient flavor, the resulting product tends to have a flat flavor. Because of this, presence of any contaminating microorganisms, especially coliform or psychrotrophic-type bacteria, or yeast and molds, can cause relatively slight off-flavors to become more pronounced because of the absence of competing desirable flavor notes. Extreme effort should be made to enhance bacterial starter (e.g., acid and diacetyl) activity to the point where desirable flavors may be noted in products such as sour cream and buttermilk.

VIII. MICROBIOLOGICAL TESTING OF RIPENED CHEESES

Natural cheeses, regardless of variety, readily support growth of many microorganisms, even though moisture content, salt content, pH, and other compositional factors vary from cheese to cheese. Cheeses may contain pathogenic bacteria (e.g., *Listeria monocytogenes, Staphylococcus aureus, Salmonella*) (see Chap. 11.) This is the exception and not the rule because cheese is a concentrated dairy product, and if all "make" procedures have been followed and good manufacturing practices adhered to, the probability of foodborne pathogens being present is remote. This is true of Cheddar cheese, for example, as long as the pH in the finished product is controlled (<5.3).

Standard Methods for Examination of Dairy Products (Marshall, 1992) recommends any one of three procedures to mix a cheese sample for subsequent microbiological analysis:

1. Transfer 11 g of cheese into 99 mL of sterile aqueous 2% sodium citrate at 40°C to 45°C. The cheese is then blended for 2 minutes and plated either direct (1:10) or with further dilutions.

2. Weigh 1 ± 0.01 g into a presterilized 177-mL Whirl-Pak bag (NASCO, Inc., 901 Janesville Ave., Fort Atkinson, WI 58538). The cheese is then macerated, after which 9 mL of 2% sodium citrate at 40°C are added. The bag is closed with the contents rolled and then plated.

3. Eleven grams of cheese and 99 mL of diluent are mixed in a Stomacher 400 (Dynatech Laboratories, Inc., 900 Slaters, Alexandria, VA 22314). The cheese is blended for 2 minutes then plated.

The microbiological tests that are done on hard cheese may vary from one processor to another; however, the coliform count and the *Staphylococcus* count should be done. *Staphylococcus* counts are especially critical when there is an abnormally high pH value. It is recommended that a *Staphylococcus* count be automatically done on any Cheddar-type cheese with a pH greater than 5.2.

Interpretation of the coliform count is the same as for any dairy product, that is, a high count indicates unsanitary conditions involved in processing and packaging the product. As discussed previously, coliforms are "indicator organisms." This means that occurrence of coliforms indicates that conditions are suitable for the presence of enteric pathogens. This does not mean that pathogens are definitely present but that the cheese was handled in a manner to allow enteric pathogens to be present. Coliforms are important indicators and hence this test should not be ignored.

IX. FUTURE OF MICROBIOLOGICAL TESTING OF DAIRY PRODUCTS

There is a tremendous amount of work being done regarding development of rapid detection methods for total numbers of both bacteria and specific organisms, primarily pathogens. Karwoski (1996) and Fung (1994, 1995) discussed different areas of research in food microbiology, and a summary follows of what these two authors have reported:

1. Sample preparation—Two useful instruments in this area are the Stomacher and the Gravimetric Diluter (Spiral Biotech, Bethesda, MD).
2. Total viable cell count—Various alternatives include the following (White, 1996): Automated spiral plating method (Spiral Biotech, Bethesda, MD), Isogrid System (QA Laboratories, Ltd., San Diego, CA) (all colonies have

square shape, reported to be easier to count, Petrifilm (3M Co., St. Paul, MN), Redigel System (RCR Scientific, Inc., Goshen, IN), and Direct epifluorescent filter technique (DEFT) slides read by systems such as the Bio-Fos Automated Microbiology System (FOS Electric, Denmark).
3. Differential cell count.
4. Pathogenic organisms.
5. Enzymes and toxins.
6. Metabolites and biomass.

In an article dealing with microbiological testing in the dairy industry, White (1966) summarized some of the methods that Fung had reviewed. Some of these methods are as follows:

1. Microbial ATP detection—Bioluminescence as a screening tool for accepting raw milk shows great promise. Reybroeck and Schram (1995) outlined a test that took less than 6 minutes. They described this method as being very useful as a sensitive and rapid semiautomatic method for fast microbiological screening of raw milk on arrival at a dairy plant.

2. Impedance detection in foods (Bactometer, bioMérieux, Vitek, Inc., Hazelwood, MO, and the Malthus System, Crawley, UK).

3. Omnispec Bioactivity Monitor System (Wescor, Inc., Logan, UT)—A tristimulus reflectance colorimeter monitors dye pigment changes caused by microbial activity. The LABSMART System highlights this use of reflectance colorimetry.

4. Catalase test—This test is very useful in detecting strongly catalase-positive bacteria, such as pseudomonads.

5. Many miniaturized diagnostic kits for identification of microorganisms (e.g., API, Enterotube, R/B, Minitek, MicroID, and IDS).

6. Genetic techniques (Fung, 1995)—DNA/RNA probes are a sensitive method for detection of pathogens (e.g., *Listeria* and *Salmonella* detection using The Gene-Trak Assay System [Gene-Trak, Systems, Framingham, MA]). Sensitivity 1×10^5 organisms per milliliter broth (Giese, 1995). Wolcott (1991) indicated that PCR has become the preferred method for amplifying DNA. This enables detection of target microorganisms in hours rather than days. This procedure has tremendous potential in all areas of food microbiology, including dairy microbiology. The BAX System (Dupont Experimental Station, Wilmington, DE) for screening *Salmonella* is one example.

7. Enzyme-linked immunosorbent assay (ELISA)—systems produced in the United States by Organon Teknika (Durham, NC) use monoclonal antibodies as a diagnostic test, especially for foodborne pathogens. Development of the ELISA technique using monoclonal antibodies specific to *Pseudomonas* and related psychrotrophic bacteria as outlined by Gutíerrez et al. (1997) shows great promise.

8. Vitek Immuno Diagnostic Assay System (VIDAS)—A multiparametric immunoanalysis system uses the enzyme-linked fluorescent immunoassay (ELFA) method. All intermediate steps are automated (Fung, 1994).

There continues to be a need for methods that can rapidly detect the presence of certain types of bacteria. Personnel at a dairy plant must be able to determine whether equipment is clean, to rapidly screen all incoming raw ingredients, and to rapidly (<24 hours) predict the shelf-life of finished products. By monitoring raw ingredients, monitoring the processing and packaging environment, and providing a more limited testing of finished products, a dairy processor becomes much more proactive in eliminating safety and quality hazards.

Other innovations such as addition of carbon dioxide to milk and other dairy products such as cottage cheese serve to extend the shelf-life of the products (Hotchkiss and Chen, 1996; Sierra et al., 1996). Certain questions have been raised that relate to packaging for such products (e.g., high barrier films being required to retain the CO_2).

Thus, much has changed in the testing of milk and milk products by dairy processors. Environmental samples for pathogens are commonly being sent to commercial testing laboratories, more sophisticated equipment is being found in the laboratories, and many of the laboratories are becoming larger because of consolidation and takeovers of smaller operations. However, one significant fact cannot be forgotten: for the dairy industry to continually provide safe, long-lasting products to the American consumer, rapid, accurate, and reliable testing must be done. It is extremely important for management to react to the data provided by this testing. As confidence is gained by quality assurance personnel and production management, the American consumer will continue to receive dairy products that are as good and safe as products produced anywhere in the world.

REFERENCES

Adams DM, Barach JT, Speck ML. Heat resistant protease produced in milk by psychrotrophic bacteria of dairy origin. J Dairy Sci 58:828, 1975.

Adams DM, Barach JT, Speck ML. Effect of psychrotrophic bacteria from raw milk on milk proteins and stability of milk proteins to ultrahigh temperature treatment. J Dairy Sci 59:823, 1976.

Babel FJ. Activity of bacteria and enzymes in raw milk held at 4.0°C (abstr). J Dairy Sci 36:562, 1953.

Barach JT, Adams DM, Speck ML. Low temperature inactivation in milk of heat-resistant proteases from psychrotrophic bacteria. J Dairy Sci 59:391, 1976.

Barber FW, Fram H. The problem of false coliform counts on fruit ice cream. J Milk Food Technol 18:88–90, 1955.

Bautista DA, McIntyre L, Laleye L, Griffiths MW. The application of ATP bioluminescence for the assessment of milk quality and factory hygiene. J Rapid Methods Automat Microbiol 1:179–193, 1992.

Bernard D. Microbiological considerations of testing aseptic processing and packaging systems. In: Proceedings: Capitalizing on Aseptic. NFPA Conference, Washington, DC, 1983.

Bishop JR. Establish a dairy product quality assessment program. 3M-Oct. 1988.

Bishop JR. Estimating shelf-life—A basic part of total dairy quality assurance. UW Dairy Pipeline. 5(2):1, 1993.

Bishop JR, White CH. Estimation of potential shelf-life of pasteurized fluid milk utilizing bacterial numbers and metabolites. J Food Prot 48:663, 1985.

Bishop JR, White CH. Assessment of dairy product quality and potential shelf-life— a review. J Food Prot 49:739, 1986.

Bockelmann BV. Quality control of aseptically packaged food products. In Reuter H, ed. Aseptic Packaging of Food. Lancaster, PA: Technomic Publishing, 1989.

Byrne RD Jr, Bishop JR, Boling JW. Estimation of potential shelf-life of pasteurized fluid milk utilizing a selective preliminary incubation. J Food Prot 52:805, 1989.

Celestino EL, Lyer M, Roginski H. The effects of refrigerated storage on the quality of raw milk. Aust J Dairy Technol 51:59, 1996.

Champagne CP, Laing RR, Roy D, Mafu AA, Griffiths MW. Psychrotrophs in dairy products: their effects and their control. Crit Rev Food Sci Nutr 34:1, 1994.

Edwards J. Q. A. programs for processors and packages—low acid foods. In: Proceedings: Capitalizing on Aseptic. NFPA Conference, Washington, DC, 1983.

Flickinger B. Plant sanitation comes to life—an evaluation of ATP-bioluminescence systems for hygiene monitoring. Food Quality 2(Jun/Jul):22, 1996.

Fung DYC. Rapid methods and automation in food microbiology: a review. Food Rev Int 10:357, 1994.

Fung DYC. What's needed in rapid detection of foodborne pathogens. Food Technol 49(6):64, 1995.

Giese J. Rapid microbiological testing kits and instruments. Food Technol 49(7):64, 1995.

Goel MC, Kulshrestha DC, Marth EH, Francis DW, Bradshaw JG, Read RB Jr. Fate of coliforms in yogurt, buttermilk, sour cream, and cottage cheese during refrigerated storage. J Milk Food Technol 34:54, 1971.

Griffiths MW. The role of ATP-bioluminescence in the food industry: new light on old problems. Food Technol 50:62, 1996.

Griffiths MW, McIntyre L, Sully M, Johnson I. Enumeration of bacteria in raw milk. In Stanley PE, Kricka LJ, eds. Bioluminescence and Chemiluminescence, Current Concepts. Chirchester, UK: John Wiley and Sons, 1991, pp. 479–482.

Griffiths MW, Phillips JD. The application of the pre-incubation test in commercial dairies. Aust J Dairy Technol 41:71–79, 1986.

Guinot-Thomas P, Al-Ammoury M, LeRoux Y, Laurent F. Study of proteolysis during storage of raw milk at 4°C: Effect of plasmin and microbial proteinases. Int Dairy J 5:685, 1995.

Gutíerrez R, González I, García T, Carrera E, Sanz B, Hernandes PE, Martin R. Monoclonal antibodies and an indirect ELISA for detection of psychrotrophic bacteria in refrigerated milk. J Food Prot 60:23, 1997.

Hotchkiss J, Chen JH. Microbiological effects of the direct addition of CO_2 to pasteurized milk (abstr). J Dairy Sci 79(suppl. 1):87, 1996.

Johns CK. Preliminary incubation for raw milk samples. J Milk Food Technol 23:137, 1960.

Karwoski M. Automated direct and indirect methods in food microbiology: A literature review. Food Rev Int 12:155, 1996.

Marshall RT, ed. Standard Methods for the Examination of Dairy Products. 16th ed. Washington, DC: American Public Health Association, 1992.

Martin JH. Significance of bacterial spores in milk. J Milk Food Technol 37:94, 1974.

Martin JH. Heat resistant mesophilic microorganisms. J Dairy Sci 64:149, 1981.

Maxcy RB, Liewen MB. Function of preliminary incubation of raw milk samples for quality control. J Dairy Sci 72:1443, 1989.

Pahauski E, Martin L, Stebritz K, Priest J, Dimond R. Rapid concentration procedure for microorganisms in raw milk. J Food Prot 54:813, 1991.

Prosser JI, Killham K, Glover LA, Rattray EAS. Luminescence-based systems for detection of bacteria in the environment. Crit Rev Biotechnol 16:157, 1996.

Reybroeck W, Schram E. Improved filtration method to assess bacteriological quality of raw milk based on bioluminescence of adenosine triphosphate. Neth Milk Dairy J 49:1, 1995.

Rollema HS, McKellar RC, Sorhaug T, Suhren G, Zadow JG, Law BA, Poll JK, Stepaniak L, Vagias G. Comparison of different methods for the detection of bacterial proteolytic enzymes in milk. Milchwissenschaft 44:491, 1989.

Sierra I, Prodanoy M, Calvo M, Olano A, Vidal-Valverde C. Vitamin stability and growth of psychrotrophic bacteria in refrigerated raw milk acidified with carbon dioxide. J Food Prot 59:1305–1310, 1996.

Smithwell N, Kailasapathy K. Psychrotrophic bacteria in pasteurized milk. Problems with shelf-life. Aust J Dairy Technol 50:28–31, 1995.

Sutherland AD, Bell C, Limond A, Deakin J, Hunter EA. The Biotrace method for estimating bacterial numbers in milk by bioluminescence. J Soc Dairy Technol 47:117, 1994.

White CH. The art and science of predicting shelf-life. Dairy Field 4:24–25, 1991.

White CH. Yogurt and fermented milk drinks. Presented at International Symposium on Dairy Science and Technology, Canilec, Mexico, 1992.

White CH. Rapid methods for estimation and prediction of shelf-life for milk and dairy products. J Dairy Sci 76:3126, 1993.

White CH. Microbiological testing in the dairy industry. Food Testing Anal 2(Aug/Sept): 22, 1996.

White CH, Marshall RT. Reduction of shelf-life of dairy products by a heat-stable protease from *Pseudomonas fluorescens* P26. J Dairy Sci 56:849, 1973.

Wilster GH. Practical Buttermaking. 8th ed. Corvalis, OR: OSC Cooperative Association, 1957, pp 9, 230.

Witter LD. Psychrophilic bacteria—A review. J Dairy Sci 44:983, 1961.

Wolcott MJ. DNA-based rapid detection methods for the detection of foodborne pathogens. J Food Prot 54:387, 1991.

Zall RR, Chen JH, Murphy SC. Effects of heating and storing milk in milk plants before pasteurization. Cult Dairy Prod J 17(4):5–9, 1982.

14
Treatment of Dairy Wastes

WILLIAM L. WENDORFF
University of Wisconsin—Madison, Madison, Wisconsin

I. INTRODUCTION

Dairy plants process a wide variety of products including milk, cheese, butter, ice cream, yogurt, nonfat dry milk, whey, and lactose. The volume and composition of dairy wastes from each plant depends on the types of products produced, waste minimization practices, types of cleaners used, and water management in the plant. Because most dairy plants process a number of milk products, waste streams may vary widely from day to day. The main source of dairy effluents are those arising from the following:

1. Spills and leaks of products or by-products
2. Residual milk or milk products in piping and equipment before cleaning
3. Wash solutions from equipment and floors
4. Condensate from evaporation processes
5. Pressings and brines from cheese manufacture

Dairy plant operators may choose from a wide variety of methods for treating dairy wastes from their plants. This may range from land application for small plants to operation of biological wastewater treatment systems for larger plants.

Some dairy plants may pretreat the effluents and discharge them to a municipal wastewater treatment plant. Dairy wastes are segregated and treated separately from sanitary wastes generated in employee facilities. The objectives of treating dairy wastes are to (a) reduce the organic content of the wastewater, (b) remove or reduce nutrients that could cause pollution of receiving surface waters or groundwater, and (c) remove or inactivate potential pathogenic microorganisms or parasites.

The level of treatment needed for dairy wastewater for each plant is dictated by the environmental regulations applicable to the location of the dairy plant. The Environmental Protection Agency (EPA) establishes general regulations concerning discharges to surface waters and ground water. Each state environmental regulatory agency is responsible for ensuring compliance to those regulations. Each plant must have a discharge permit for each outfall discharging to surface waters. The limits within that permit depend on the flow and type of surface water into which the treated wastewater is discharged. If a plant discharges wastewater to municipal sewers for treatment, the municipal treatment system may require pretreatment of high-strength wastes to bring the waste load down to that of domestic sewage strength. This allows for proper treatment of the wastewater before it is discharged to the surface water. For land applications, state regulatory agencies dictate hydraulic loadings and maximum levels of toxic substances that can be landspread on each unit of land.

II. DAIRY PLANT EFFLUENTS

A. Quantity of Dairy Wastes

Wastes from manufacture of milk products contain milk solids in varying concentrations. Up to 5% of the milk received by a dairy plant may be lost in waste discharges from the plant (Carawan et al., 1979; Harper and Blaisdell, 1971; Harper et al., 1985). Typical product losses for fluid milk and ice cream plants are listed in Table 1. With increased environmental restrictions, dairy plants have instituted waste minimization procedures to reduce the loss of milk solids and improve utilization of milk by-products (Harper and Carawan, 1978; Harper et al., 1985; Wendorff, 1995). These water and waste management programs also emphasize water conservation practices in plants to reduce the overall volume of dairy wastes that need to be treated.

B. Composition of Dairy Wastes

Because more than 95% of the waste load from dairy plants comes from milk or milk products, it is of value to know the average composition of these products (Table 2). Milk solids are primarily composed of fats, proteins, and

TABLE 1 Product Losses for Fluid Milk and Ice Cream Processing Plant

Process	Product Losses	
	Fluid milk (%)	Ice cream (%)
Receiving	.23	.20
Separation	.75	—
Clarification	.08	—
Milk storage	.44	.28
Standardizing	—	.08
Blending	—	.10
Pasteurization	.58	1.00
Pasteurized storage	.25	.40
Flavoring and fruit	—	.30
Freezing	—	.50
Filling	.50	.75
Conveying	.10	.40
Hardening	—	.04
Storage	.10	.04
Miscellaneous	.30	.04
Total	3.33	4.13

Source: Harper et al. (1985).

TABLE 2 Average Composition of Milk and Milk Products (100 g)

Product	Fat (g)	Protein (g)	Lactose (g)	Total Organic Solids (g)
Skim milk	0.08	3.5	5.0	8.56
2% Milk	2.0	4.2	6.0	12.2
Whole milk	3.5	3.5	4.9	11.1
Half and half	11.7	3.2	4.6	19.5
Heavy cream	40.0	2.2	3.1	45.3
Chocolate milk	3.5	3.4	5.0	18.5
Churned buttermilk	0.3	3.0	4.6	8.0
Cultured buttermilk	0.1	3.6	4.3	10.0
Sour cream	18.0	3.0	3.6	24.6
Yogurt	3.0	3.5	4.0	10.8
Evaporated milk	8.0	7.0	9.7	27.0
Ice cream	10.0	4.5	6.8	41.3
Whey	0.3	0.9	4.9	6.3

Source: Harper and Blaisdell (1971).

carbohydrates. Other constituents in dairy wastewater may include sweeteners, gums, flavorings, salt, cleaners, and sanitizers.

Biochemical oxygen demand (BOD) is the amount of dissolved oxygen (DO) consumed by microorganisms for the biochemical oxidation of organic solids in wastewater. The analytical procedure for determining BOD measures the dissolved oxygen consumed by a seeded, diluted wastewater sample incubated at 20°C for 5 days (American Public Health Association, 1992). One gram of milk fat has a BOD of 0.89 g, whereas milk protein, lactose, and lactic acid have BOD values of 1.03, 0.65, and 0.63 g, respectively. The range of BOD values for various milk and milk products is given in Table 3. Roughly, 1 kg of BOD in dairy wastewater represents 9 kg of whole milk. Chemical oxygen demand (COD) is the amount of oxygen necessary to oxidize the organic carbon completely to CO_2, H_2O, and ammonia. The COD is measured colorimetrically after refluxing a sample of wastewater in a mixture of chromic and sulfuric acid (American Public Health Association, 1992). If the BOD/COD ratio of wastewater is less than 0.5, then the organic solids in the waste are not easily biodegraded. The BOD/COD ratio for dairy wastes has been reported to range from 0.55 to 0.70 (Harper et al., 1985; Marshall, 1978).

Some minor constituents, such as phosphorus and chloride, are also very important in the treatment of dairy wastes. Phosphorus is the element that limits the plant and algal growth in surface waters. Discharge of any significant levels of phosphorus in waste effluents to surface waters can lead to decreased water quality in lakes and streams. Milk and milk by-products can contribute significant quantities of phosphorus to dairy wastes. The phosphorus content of milk is

TABLE 3 Reported BOD_5 Values and Percentage Contribution of Milk Components to Product BOD_5

| | | Contribution to BOD_5 by | | |
| | | Milk Fat | Milk Protein | Lactose |
Product	BOD_5 (mg/L)	(%)	(%)	(%)
Skim milk	67,000	6.3	49.3	44.5
Whole milk	104,000	17.8	43.3	39.0
Half and half	156,000	62.4	19.7	17.9
Heavy cream	399,000	89.2	5.7	5.0
Churned buttermilk	68,000	4.2	48.2	46.7
Evaporated milk	208,000	34.6	35.0	30.6
Ice cream	292,000	30.7	15.9	15.2
Whey	34,000	5.9	20.6	70.8

Source: Harper and Blaisdell (1971).

approximately 1000 mg/L, whereas whey contains 450 to 575 mg/L (Wendorff, 1991, Wendorff and Matzke, 1993). Salty whey and brines can contribute significant levels of chloride to dairy wastewater. Chloride concentrations in excess of 400 mg/L in effluents discharged to streams can result in chronic toxicity to sensitive water insects such as *Daphnia magna*. Because chloride cannot be removed with biological or chemical treatments, waste minimization is the only method for reducing chloride in dairy wastes (Wendorff, 1995).

III. TREATMENT OF MILK WASTE

Wastes from processing milk products are almost entirely composed of organic material in solution or colloidal suspension although some larger suspended solids may be present in wastewater from cheese or casein manufacturing plants. Sand and other foreign material is present in limited amounts as a result of floor or truck washes. Because milk waste contains very little suspended matter, preliminary settling of solids does not result in any appreciable reduction of BOD. However, a screen and grit chamber with 0.95-cm mesh wire screen is recommended to remove large particles to prevent clogging of pipes and pumps in the treatment system. This is especially important if the waste is to be pumped with high-pressure pumps, as in spray irrigation. After preliminary treatment in the screen and grit chamber, the waste should be pumped to an equalization tank. With wide variations in wastewater flow, strength, temperature, and pH, some reaction time is required to allow neutralization of acid and alkaline cleaning compounds and to allow for complete reaction of residual oxidants from cleaning solutions with organic solids of dairy waste. Ideally, a minimum of 6 to 12 hours of equalization should be provided to allow for waste stabilization. The equilibrated waste can then be treated with one of the following systems or a combination of treatment systems: (a) land application, (b) treatment ponds or lagoons, (c) activated sludge, (d) biological filtration, or (e) anaerobic digestion.

A. Land Application

Because many dairy plants are located in rural areas, land application of process wastewater and waste by-products may be the simplest and most economical means of treating dairy wastes. Wastewater may be applied in a ridge and furrow system or by spray irrigation. Pollutants in the dairy wastewater are removed by a combination of physical and biological processes. The soil serves as an effective filter to physically remove particulate and colloidal material from process wastes. The upper 12 to 15 cm of soil can remove as much as 30% to 40% of the BOD and COD (Law et al., 1969).

Soluble organic compounds in dairy wastewater and particulate material filtered by the soil are degraded by the heterotrophic microorganisms in the soil. Table 4 lists the range of numbers for major groups of microorganisms in a fertile agricultural soil of midwest United States. Major genera of bacteria in soils include *Arthrobacter, Bacillus, Achromobacter, Flavobacterium,* and *Pseudomonas* (Goodfellow, 1968). Soil microorganisms are contained within biofilms absorbed to colloids or soil particles (Metting, 1993).

1. Biochemical Oxygen Demand Removal

Under aerobic conditions, soil microorganisms degrade the organic pollutants completely to CO_2 and BOD removal should be more than 99%. If the concentration of BOD or the volume of wastewater is too great for the soil capacity, anaerobic conditions may result. Spyridakis and Welch (1976) reported that anaerobic conditions in the soil surface result in a low rate of biological activity and, thus, a tendency for sludge accumulation, production of ferrous sulfide, or accumulation of polysaccharides. Allison (1947) demonstrated that soil clogging was a result of biochemical activity by microorganisms within the soil and not the result of filling soil spaces with sludge from the wastewater. Lactose and milk proteins are easily decomposed by anaerobic soil bacteria. However, fats and oils are more resistant to decomposition and tend to accumulate in the soil under anaerobic conditions. By providing periods of rest between applications to allow

TABLE 4 Relative Number of Flora and Fauna Commonly Found in Surface Soils[a]

	Number	
Organisms	per square meter	per gram
Microflora		
Bacteria	$10^{13}-10^{14}$	10^8-10^9
Actinomycetes	$10^{12}-10^{13}$	10^7-10^8
Fungi	$10^{10}-10^{11}$	10^5-10^6
Algae	10^9-10^{10}	10^4-10^5
Microfauna		
Protozoa	10^9-10^{10}	10^4-10^5
Nematoda	10^6-10^7	$10-10^2$
Other fauna	10^3-10^5	
Earthworms	30–300	

[a]Generally considered 15 cm deep, but, in some instances (e.g., earthworms), a greater depth is used.
Source: Brady (1990).

the soil to dry, clogging problems disappear and aerobic conditions return to the soil surface. Treatability of a large volume of low BOD waste may be limited by the percolation capacity of the soil, whereas a small volume of waste with high BOD is more apt to be limited by the oxidative capacity of the microorganisms and sorptive capacity of the organic matter in the soil (Spyridakis and Welch, 1976).

Parkin and Marshall (1976) reported application rates for dairy plant wastewater of 130 m³/ha to 1500 m³/ha (1 hectare = 2.49 acres) on New Zealand pasture land. A rest period of 10 to 60 days was used to allow the soil bacteria to decompose the effluent and the soil to dry out. Guichet et al. (1991) reported application rates of 45 mm of liquid per month for wastewater sludge from a butter and cheese processing plant in France. They observed rapid decomposition of the lactose in the sludge, but reported a gradual accumulation of lipids in the treated soil. A regular application of dairy wastewater sludge to soil for 25 years resulted in a two-fold increase in the level of organic matter in the soil. The additional moisture and added buffering of pH from the dairy waste greatly improved mineralization of organic matter in the soil (Guichet et al., 1991).

2. Nitrogen and Phosphorus Removal

Nitrogen from dairy wastes is removed by sedimentation of protein adsorption to soil and volatilization of ammonia, uptake by crops, and biological denitrification (Lance, 1972). Milk proteins may be degraded by proteolytic soil bacteria of microflora present in the milk waste from the dairy plant. The ammonia from protein breakdown is biologically oxidized to nitrate by a process known as nitrification.

Nitrification is a two-step process whereby ammonia is first converted to nitrite and then to nitrate. Conversion of ammonia to nitrite is accomplished by *Nitrosomonas* sp. whereas conversion of nitrite to nitrate is completed by *Nitrobacter* sp. Nitrification can also be brought about by certain heterotrophs, including fungi such as *Aspergillus flavus*, some species of *Penicillium*, and bacteria (e.g., *Arthrobacter*) (Hattori, 1973). Other heterotrophic bacteria, such as *Achromobacter, Corynebacterium, Agrobacterium*, and *Alcaligenes*, can convert ammonia to nitrite. Nitrification requires aerobic conditions because gaseous oxygen is involved in the reaction. Nitrification of ammonia releases hydrogen ions, resulting in acidification of the soil.

Nitrogen uptake by plants generally does not exceed 60% to 70% of added inorganic fertilizers. Only with careful management of organic nitrogen sources, such as milk proteins, can increased nitrogen uptake by crops be experienced. The remaining nitrate in the soil may be lost by leaching to groundwater or by biological denitrification. Denitrification occurs when nitrate is reduced to nitrogen gas under anaerobic conditions in the soil. Heavy clay soils tend to favor denitrification because of its effect on soil structure (Law et al., 1970).

Phosphorus in dairy wastes is removed by absorption to soil particles, chemical precipitation, and uptake by crops. Generally phosphorus is effectively removed in the upper 0.3 to 0.6 m of soil (Spyridakis and Welch, 1976). Soils have very reactive surfaces containing iron, aluminum, and calcium, which readily form insoluble phosphates. Normally, the content of organic phosphorus in a soil is higher than that of inorganic phosphorus. Mineralization of phosphorus in organic matter results through the action of bacteria, actinomycetes, and fungi. Up to 50% of the phosphorus from organic fertilizers can be effectively removed by crops (Fried and Broeshart, 1967).

High-strength dairy wastes such as whey, whey permeate, and antibiotic-contaminated milk can effectively be used as sources of plant nutrients for agricultural crops (Kelling and Peterson, 1981; Peterson et al., 1979; Wendorff, 1989). Sharrat et al. (1962) pointed out that the nitrification of organic nitrogen from whey proteins was controlled by the carbon to nitrogen (C:N) ratio of the whey. If the C:N ratio of the whey is too great, nitrogen is incorporated into the cells of the microbes and so is unavailable to plants for some time. Conversely, if the C:N ratio is small, the microbes, through nitrification, convert much of the nitrogen in whey to nitrate within several weeks. Nitrate determinations on soil receiving whey indicated that the organic nitrogen in whey was readily converted to nitrates during the first and second seasons after application. Sharratt et al. (1959) reported increased growth of bluegrass the second year after application of whey. They credited this extra production to the slow breakdown of nitrogen compounds in the whey. Whey and whey permeate also contain high concentrations of phosphorus, which can be used for plant growth (Peterson et al., 1979; Wendorff and Matzke, 1993). Most phosphorus in whey is inorganic phosphorus, which is readily available for plant uptake.

High levels of soluble salts in whey and whey permeate may limit the application rates to certain soils and crops. Some salt-sensitive crops such as soybeans, green bans, and red clover are susceptible to leaf burn and application rates should not exceed 13 mm/yr (Kelling and Peterson, 1981). Chloride levels in whey greatly exceed the drinking water standard of 250 mg/L and application rates should be restricted to no more than 26,000 L/ha/yr to avoid leaching of significant levels of chloride to the groundwater (Matzke and Wendorff, 1993; Wendorff, 1993).

3. Removal of Microorganisms

Unlike domestic wastewater, dairy plant wastewater and dairy wastes do not contain significant levels of human pathogens that may be of concern for irrigating processed food crops. Extensive field observations indicate that bacteria and viruses are efficiently removed from wastewater as it percolates through the soil. Removal of bacteria by soils is inversely proportional to the particle size of the soils. Viruses may be transported to greater depths in the soil than bacteria because of their smaller size (Drewry and Eliassen, 1968). However, percolation through even the coarsest soil will remove bacteria and viruses within 1 to 2 m (McGauhey, 1968). The potential leaching of bacteria or viruses from sludge or wastewater to the groundwater is minimal (Bitton, 1994).

B. Treatment Ponds or Lagoons

Dairy plants in rural areas with insufficient farmland available for land application may be able to use ponds or lagoons for economical treatment of dairy wastes. A pond or lagoon normally consists of a shallow basin designed for treatment of dairy wastewater without extensive equipment and controls. The three types of ponds used are aerobic, facultative, and anaerobic.

1. Aerobic Ponds

Aerobic ponds are generally 0.5 to 2.0 m deep, and the contents are mechanically mixed and aerated to allow penetration of sunlight necessary for growth of algae. The algae produce oxygen through photosynthesis and use waste products from the bacteria involved in the biological breakdown of milk wastes. At 20°C, a BOD removal of 85% can be experienced with an aeration period of 5 days (Bitton, 1994). Pickett (1988) reports retention times of up to 90 days for wastewater from a cheese plant.

2. Facultative Ponds

Facultative ponds are the most common type of treatment ponds for high-strength dairy wastes. Treatment is achieved by action of aerobic, anaerobic, and facultative microorganisms as outlined in Figure 1. In the upper zone, oxygen is supplied by photosynthetic green and blue-green algae. The algae also take up some of the nitrogen and phosphorus from the dairy wastes. In the aerobic zone, heterotrophic bacteria degrade organic matter in dairy wastes and produce CO_2 and micronutrients needed by the algae. Some of the typical bacteria involved in this process include genera such as *Pseudomonas, Achromobacter,* and *Flavobacterium* (Sterritt and Lester, 1988). Dead bacteria and algae settle to the bottom of the pond and are degraded by anaerobic microorganisms. During anaerobic decomposition, methane, hydrogen sulfide, carbon dioxide, and nitrogen may be released to the atmosphere. Although some carbon is lost with the escape of CO_2 or CH_4, most of it is converted to microbial biomass. Zooplankton (rotifera, cladocera, and copepoda) feed on bacterial and algal cells (Bitton, 1994). However, unless sludge is periodically removed from the base of the pond, little

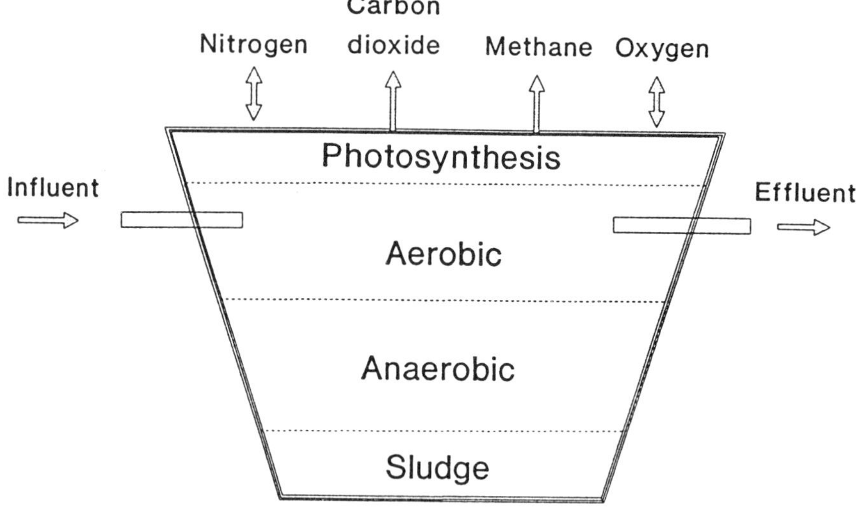

FIGURE 1   Microbial activities in a waste treatment pond.

carbon reduction is obtained with facultative ponds. BOD removals of up to 90% can be obtained in facultative ponds, depending on climatic conditions. Although the aerobic phase of treatment is fairly tolerant of temperature variations, the anaerobic phase is very sensitive, with activity almost ceasing at or below 17°C (Sterritt and Lester, 1988). Retention time in facultative ponds range from 5 to 30 days.

3. Anaerobic Ponds

Anaerobic ponds are generally used for pretreating dairy wastes with high protein and fat levels or for stabilizing settled solids. Organic matter is biodegraded and gases such as CH_4, CO_2, and H_2S are produced. To effectively reduce the BOD in the anaerobic effluent, an aerobic process must follow to allow aerobic micro-organisms to use up the residual breakdown products. The typical retention time for anaerobic treatment ponds ranges from 20 to 50 days (Metcalf and Eddy, Inc., 1991).

C. Activated Sludge

Activated sludge is one of the most popular methods for treating dairy wastes. The process consists of aerobic oxidation of organic matter to CO_2, H_2O, NH_3 and cell biomass, followed by sedimentation of the activated sludge. A portion of the activated sludge is returned to the aeration tank to continue the treatment cycle (Fig. 2).

1. Activated Sludge Microorganisms

Activated sludge contains a large mass of various microorganisms plus organic and inorganic particles. The concentration of biomass in the aeration or contact tank is normally called the mixed liquor suspended solids (MLSS). Bacteria make up the largest portion of activated sludge in the aeration process. Bitton (1994) noted that more than 300 strains of bacteria thrive in activated sludge. Bacteria are primarily responsible for oxidation of organic matter and formation of polysaccharides and other polymeric materials that aid in flocculation of the microbial biomass. Table 5 lists some of the bacterial genera found in activated sludge. Estimates of the aerobic bacterial counts in activated sludge are approximately 10^{10}/g of MLSS or 10^7 to 10^8/mL (Sterritt and Lester, 1988). Hanel (1988) stated that the active fraction of bacteria in activated sludge flocs represents only 1% to 3% of total bacteria present. This indicates that the major portion of activated sludge is actually dead cells and extracellular material. Activated sludge does not normally favor growth of yeast, algae, or fungi.

Protoza may represent up to 5% of the MLSS. Protoza are predators of bacteria in activated sludge; they help reduce effluent suspended solids and soluble BOD. Sterritt and Lester (1988) estimated approximately 5×10^4 protoza in typical activated sludge. Most protoza present in activated sludge are ciliates, although amoeba and flagellates may also be present under certain conditions. The predominant genera of ciliates in activated sludge are *Opercularia, Vorticella, Aspidisca, Carchesium*, and *Chilodonella*. Protoza are also responsible for significant reduction of pathogenic bacteria and viruses in activated sludge.

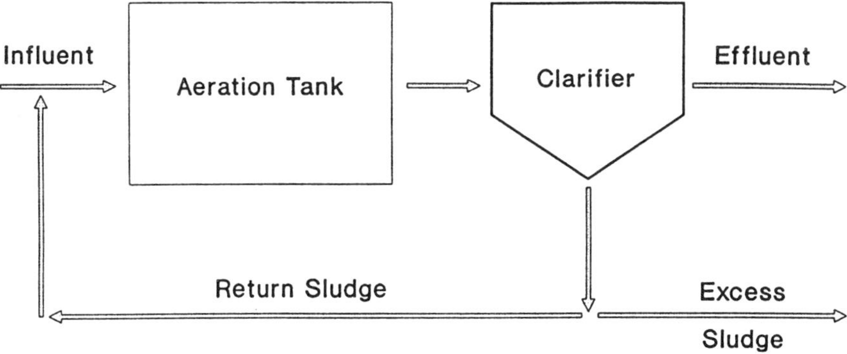

FIGURE 2 The conventional activated sludge treatment system.

TABLE 5 Bacterial Genera Found in
Activated Sludge

Major Genera	Minor Genera
Zoogloea	*Aeromonas*
Pseudomonas	*Aerobacter*
Comomonas	*Micrococcus*
Flavobacterium	*Spirillum*
Alcaligenes	*Acinetobacter*
Brevibacterium	*Gluconobacter*
Bacillus	*Cytophaga*
Achromobacter	*Hyphomicrobium*
Corynebacterium	
Sphaerotilus	

Source: Sterritt and Lester (1988).

Reductions of *Escherichia coli* and coxsackie and polio virus in excess of 90% have been reported (Sterritt and Lester, 1988).

Rotifers are multicellular organisms that are present in activated sludge that is aging. Their role includes removal of freely suspended bacteria and aiding in floc formation by producing fecal pellets surrounded by mucus (Curds and Hawkes, 1975). The four most common genera of rotifers present in activated sludge include *Philodina, Habrotrocha, Notommata,* and *Lecane*.

2. Conventional Process

In the conventional activated sludge process, dairy wastewater is introduced into the aeration tank along with a portion of activated sludge from the clarifier. Air is incorporated into the waste mixture with diffusers or mechanical aerators. The air serves two purposes in the aeration tank: first, to supply oxygen to the aerobic microorganisms and, second, to keep the activated sludge floc thoroughly mixed with the incoming wastewater to allow maximal efficiency in oxidation of the organic matter. The key parameters controlling the operation of the activated sludge process are rate of (a) aeration in the tank, (b) return of activated sludge to the aeration tank, and (c) waste or excess sludge discharged from the treatment system. Normal detention time for conventional activated sludge treatment of municipal or low-strength wastewater is 4 to 8 hours (Bitton, 1994). However, dairy wastewaters may require longer detention times, 15 to 40 hours, to reduce BODs to an acceptable level (Jones, 1974). This type of process is called an extended aeration system. Jones (1974) reported that BOD removal efficiencies in excess of 90% are attainable for dairy wastewaters with extended aeration treatment. Bangsbo-Hansen (1978) also reported that effluent standards of 20 mg of BOD/L could be met if the BOD of incoming dairy wastewater was between 700 and 1200 mg/L.

3. Contact Stabilization Process

Another modification of the activated sludge treatment is a three-step process known as the contact stabilization process (Fig. 3). This process allows for a 30-minute detention time in the contact tank in which the microorganisms obtain their food. The sludge containing the organisms and their food is separated in the clarifier. The sludge that is to be returned to the contact tank is first sent to an aerated stabilization tank for 4 to 8 hours during which time the microorganisms finish digesting their food. By aerating only the sludge that is being returned to the initial contact tank, less tank space and less air are required. This system produces less sludge and is better suited for shock loading. Fang (1991) reported that the BOD of dairy wastewater could be reduced by 99% and total Kjeldahl nitrogen by 91% after a total detention time of 19.8 hours in this type of system.

4. Sequencing Batch Reactor Process

The most popular activated sludge treatment uses sequencing batch reactors (SBRs). As shown in Figure 4, the SBR process is a single tank fill-and-draw system that provides for activated sludge aeration, settling, effluent withdrawal, and sludge recycling. Usually, a plant has two or more SBR tanks, thus allowing for the filling of one tank while the other is going through the reaction sequence.

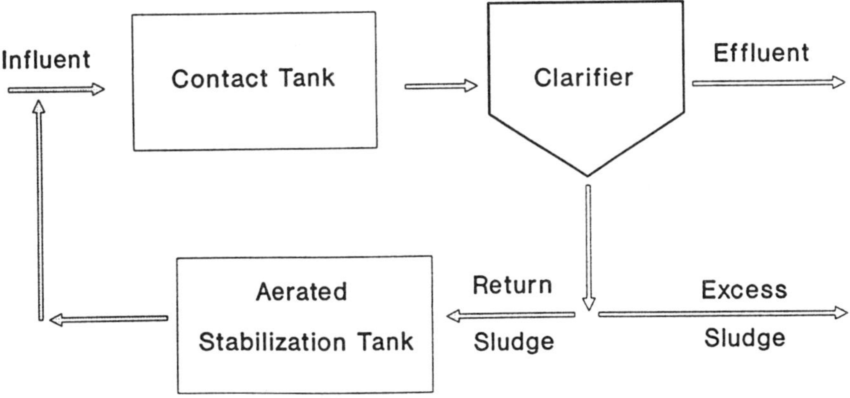

FIGURE 3 Activated sludge system with contact stabilization.

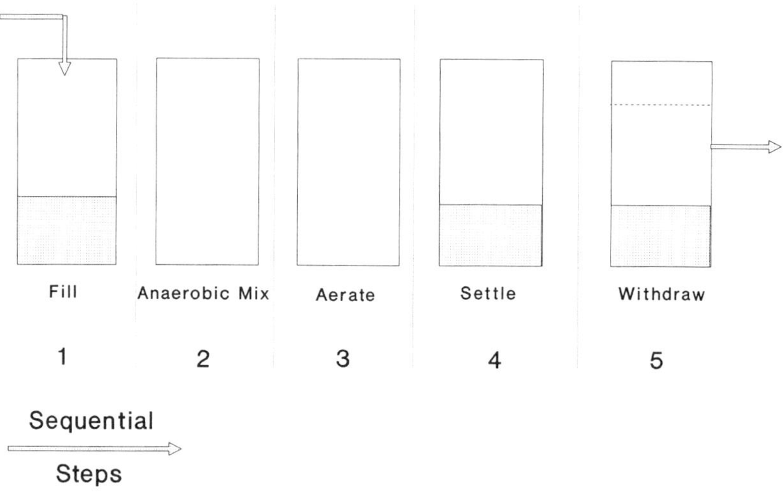

FIGURE 4 Treatment steps using a sequencing batch reactor.

Once the tank is filled, wastewater is mixed, without aeration, to allow uptake of soluble fermentation products. The aeration step provides for oxidation of organic matter in the wastewater. The activated sludge is then settled and treated effluent is drawn off to complete the cycle. This process normally operates over longer detention times than conventional activated sludge systems and allows for wide variations in strength of waste. Schulte (1988) reported that elimination of clarifiers and sludge pump stations, along with flexibility and adaptability to automated process control, made the SBR process more cost effective on creamery wastewater than the other activated sludge processes.

5. Nitrogen and Phosphorus Removal

In removing nitrogen from dairy wastewater with activated sludge processes, the nitrogen must first be removed by nitrification (see Sec. III.A.2) followed by denitrification. The growth of *Nitrosomonas* and *Nitrobacter* sp. in activated sludge depends on BOD of the mixed liquor and retention time of the sludge. The growth rate of nitrifiers is slower than that of heterotrophs in activated sludge, so an aged sludge is needed for conversion of ammonia to nitrate. Hawkes (1983) reports that nitrification is expected at a sludge age of more than 4 days. Nitrification proceeds well in a two-stage activated sludge system in which BOD is removed in the first stage and nitrifiers complete the nitrification in the second stage (U.S. Environmental Protection Agency, 1977). To remove nitrate from the waste effluent, denitrification must occur under anaerobic or anoxic conditions before the treated effluent is discharged to surface waters. This can be accomplished by using a three-stage activated sludge system in which BOD reduction takes place in the first, nitrification in the second, and denitrification in the third stage. Methanol or settled sewage serves as the carbon source for the denitrifiers (Curds and Hawkes, 1983).

Phosphorus removal in activated sludge systems usually requires a combination of anaerobic and aerobic stages in the process. Facultative organisms in the initial anaerobic zone produce acetate and fermentation products from the soluble BOD of the waste. Microorganisms able to remove high levels of phosphorus use these fermentation products and store them with the aid of energy from the hydrolysis of stored polyphosphates during the anaerobic period. During the aerobic stage of the process, stored products are depleted and soluble phosphorus is taken up, with excess amounts stored as polyphosphates (Buchan, 1981). Fuhs and Chen (1975) identified bacteria of the *Acinetobacter* genus as high phosphorus-storing microbes active in activated sludge systems. The high phosphorus-containing sludge must be completely removed from the effluent to ensure compliance with effluent phosphorus limitations. In some instances, alum or ferric chloride may be added to the effluent before the secondary clarifier to remove additional soluble phosphorus before final discharge of the treated effluent (U.S. Environmental Protection Agency, 1987).

6. Flocculation

Settling of sludge in the clarifier usually proceeds best when the microbial growth rate is slow and nutrient concentrations are very low. Extracellular polysaccharides and slimes produced by *Zooglea ramigera* and other activated sludge organisms play a leading role in bacterial flocculation and floc formation (Norberg and Enfors, 1982). Good sludge settling and BOD removal occurs at high MLSS concentrations. Microbial flocculation can be enhanced with addition of polyelectrolytes, alum, or iron salts as coagulants (Bitton, 1994).

Poor settling of sludges may be observed if excess production of exopolysaccharides by bacteria occurs in the activated sludge. This nonfilamentous bulking may be corrected with chlorination (Chudoba, 1989). Filamentous bulking is caused by excessive growth of filamentous bacteria such as *Sphaerotilus* sp. (Sterritt and Lester, 1988). Low level of dissolved oxygen in the aeration tank is the primary factor contributing to growth of this filamentous bacterium in activated sludge (Lau et al., 1984).

D. Biological Filtration

1. Trickling Filter

Biological filters, such as trickling or percolating filters, are one of the earliest types of biological waste treatment. In a biological filter, the biofilm is attached to a support substance such as gravel, stones, or plastic materials. As wastewater is pumped over the biofilm, it oxidizes the organic matter and removes nutrients such as nitrogen and phosphorus.

A basic trickling filter is composed of a tank containing a filter medium to a depth of 1.0 to 2.5 m, a wastewater distributor that applies the waste solution evenly over the medium bed, and a final clarifying tank to remove sludge and solids sloughing off the filter medium (Fig. 5). In some instances, wastewater is recirculated through the system to provide for added dissolved oxygen to the primary influent and greater removal of BOD (U.S. Environmental Protection Agency, 1977). The two most important factors affecting microbial growth on the support medium are the flow rate of wastewater and the size and geometric configuration of the support material. In the initial start up of the filter, the

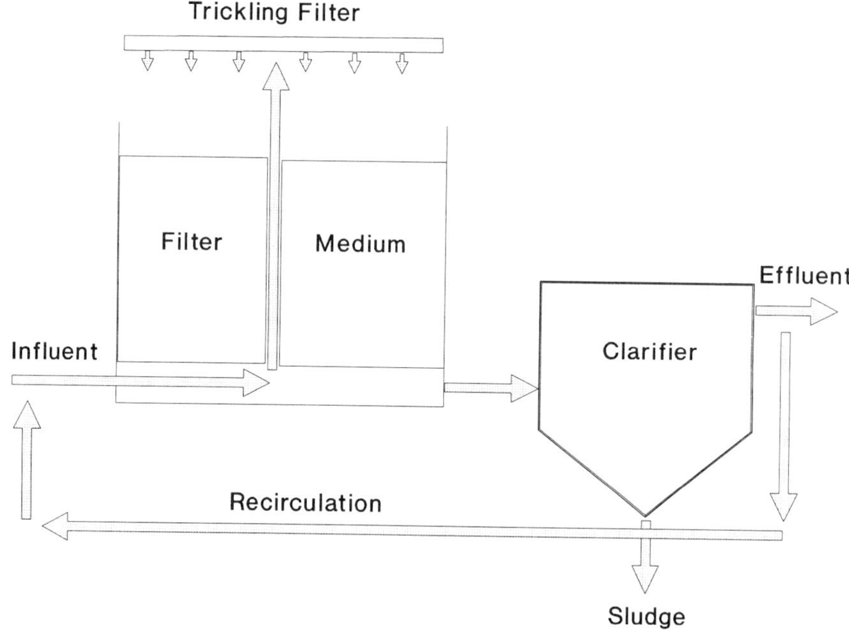

Figure 5 Trickling filter waste treatment system.

medium surface is colonized by gram-negative bacteria followed by filamentous bacteria. The biofilm formed on the support material is called a zoogleal film and is composed of bacteria, fungi, algae, protozoa, and other life forms such as rotifers, nematodes, snails, and insect larvae (Bitton, 1994). Some of the bacterial genera active in trickling filters are *Flavobacterium, Pseudomonas, Achromobacter*, and filamentous bacteria such as *Sphaerotilus* (Sterritt and Lester, 1988). Growth conditions on the outer surface of the biofilm are aerobic but the inner portion of the biofilm next to the support material tends to be anaerobic.

Trickling filters are categorized by the loading rate to the filter medium. Low-rate trickling filters (<40 kg BOD/100 m^3/day) allow for nitrification and more complete removal of nutrients from wastewater. High-rate filters (60–160 kg BOD/100 m^3/day) rarely have nitrification take place and have lower treatment efficiencies (U.S. Environmental Protection Agency, 1975). BOD removal by trickling filters is approximately 85% for low-rate filters and 65% to 75% for high-rate filters (U.S. Environmental Protection Agency, 1977).

2. Rotating Biological Contactor

One biofilm reactor that operates much as the trickling filter is the rotating biological contactor (RBC). The RBC unit consists of a horizontal shaft with discs of medium that are rotated through the primary effluent. Because only about 40% of the medium is submerged, the biofilm growing on the medium obtains its food from the effluent and oxygen from the air above the solution. Increased rotation of the discs improves oxygen transfer and enhances the contact between the biofilm and wastewater (March et al., 1981). The biofilm on RBC is composed of a diverse mixture of eubacteria, filamentous bacteria, protozoa, and metazoa. Alleman et al. (1982) have identified *Beggiatoa* sp. as the primary bacteria in the outer aerobic layer of the biofilm and *Desulfovibrio*, a sulfate-reducing bacterium, in the inner anaerobic layer. The advantages of the RBC are shorter treatment times, lower cost of operation, and production of a readily dewatered sludge that settles easily (Weng and Molof, 1974). For high-strength dairy wastes, Surampalli and Baumann (1992) reported that the first section of the RBC must be enlarged to provide sufficient dissolved oxygen for adequate reduction of the BOD. For effective nitrification, the second stage of the RBC must have an increased rotational speed to promote growth of nitrifying bacteria.

E. Anaerobic Digestion

Anaerobic digestion has been used to stabilize waste treatment sludges for many years. However, in recent years it has also been designed to treat high-strength dairy wastes. In anaerobic breakdown of dairy wastewater, lactose is first fermented to lactic acid and fats and proteins are hydrolyzed to organic acids, amino acids, aldehydes, and alcohols. Second, the intermediate organic compounds are converted to methane and CO_2. Because anaerobic digestion does not require oxygen for decomposition of organic material, operating costs for treatment are greatly reduced from that of aerobic treatments. However, it is a much slower treatment process that is more susceptible to toxic upsets (Bitton, 1994).

1. Conventional Process

A typical anaerobic digester is shown in Figure 6. The anaerobic digester is a large fermentation tank in which fermentation, sludge settling, sludge digestion, and gas collection take place simultaneously. Many dairy plants use a two-stage system in which the first stage is the complete mixing of the contents of a fermentation tank and the second stage is in a digester in which the contents are allowed to stratify. The two-stage anaerobic process allows for higher loading rates and shorter hydraulic retention times (Ghosh et al., 1985). In anaerobic treatment of wastewater, fermentation of sugars, amino acids, and fatty acids is primarily carried out by strict and facultative anaerobic bacteria such as *Bacteroides, Bifidobacterium, Clostridium, Lactobacillus*, and *Streptococcus* (Sterritt and Lester 1988). The production of methane from the fermentation intermediate compounds is accomplished by methanogenic bacteria, which are strict anaerobes. Approximately two-thirds of the methane is derived from acetate

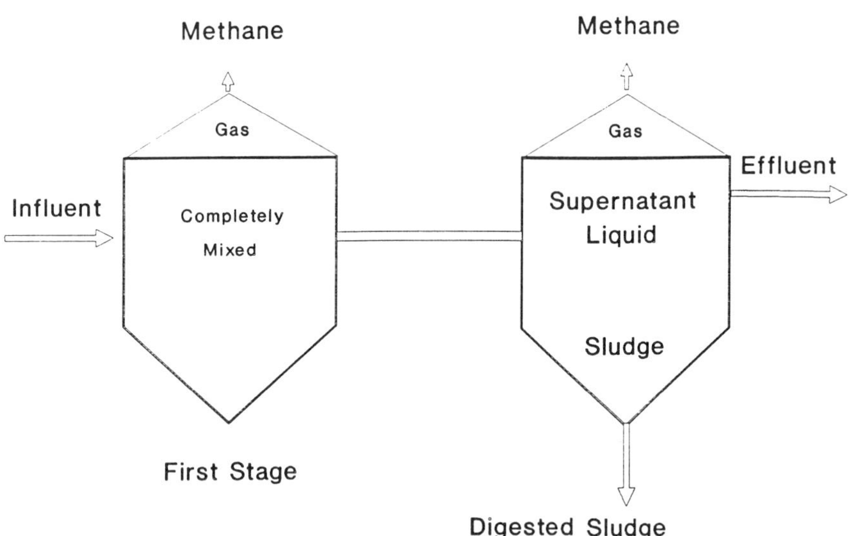

FIGURE 6 Two-stage anaerobic digester for dairy wastewater.

conversion by acetotrophic methanogens and the other one-third is the result of carbon dioxide reduction by hydrogen (Mackie and Bryant, 1981). Methanogens are difficult to grow in pure culture, but Balch et al. (1979) developed a classification scheme for some of the species involved in this process. Perle et al. (1995) reported that milk fat was inhibitory to methanogenic bacteria, and dairy effluents should be treated by anaerobic digestion only after the milk fat concentration was less than 100 mg/L. They also indicated that anaerobic cultures at the startup of anaerobic digestion should be acclimatized to casein to ensure proper degradation of casein in the process. Methanogenic bacteria are also sensitive to acidic conditions with complete inhibition below pH 6.0 (Britton, 1994). With efficient operation of one- or two-stage anaerobic digesters, dairy plants should experience BOD reductions of 78% to 95% (Fang, 1991; Guiot et al., 1995). Biogas from the digester contains up to 67% methane (Lebrato et al., 1990).

2. Upflow Anaerobic Filter

Upflow anaerobic filters operate similarly to trickling filters, but growth conditions for microbes are anaerobic. The primary effluent is pumped into the base of the reactor containing a support medium for growth of the biofilm. The upward flow of the wastewater keeps the suspended solids in solution. In some instances, the support material is replaced with sand to form a fluidized-bed reactor. This type of reactor is effective for low-strength wastes (COD of <600 mg/L) (Speece, 1983). Anderson et al. (1994) indicated that, to obtain high organic loading rates on anaerobic upflow filters, a porous medium must be used in the column to allow for sufficient biomass development. Temperature of the effluent is important for proper fermentation and production of biogas. Viraraghavan and Kikkeri (1990) reported average COD removals in three anaerobic filters were 92%, 85%, and 78% at 30°C, 21°C, and 12°C, respectively. The volume of biogas generated was lower at lower temperatures but the percentage of methane in biogas was higher at lower temperatures. Under efficient operation, anaerobic filters reduce dairy wastewater BODs by 90% to 97% and produce biogas with 54% to 75% methane (Kaiser and Dague, 1994; Sammaiah et al., 1991).

3. Upflow Anaerobic Sludge Blanket

The upflow anaerobic sludge blanket (UASB) digester consists of a tank with a bottom layer of packed sludge, a sludge blanket, and an upper liquid layer. Wastewater flows up through the sludge blanket of active biomass. Settler screens separate the sludge from treated effluent and biogas is collected at the top of the digester (Lettinga et al., 1980). The granular sludge aggregates that form contain three layers of bacteria (MacLoed et al., 1990). The inner layer contains *Methanothrix*-like cells that act as nucleation centers. The middle layer contains bacterial rods that include both H_2-producing acetogens and H_2-consuming organisms. The outermost layer contains a mixture of fermentative and H_2-producing bacteria. Dairy wastes function well in the UASB process because granulation of sludge is favored by soluble carbohydrates (Wu et al., 1987). Kato et al. (1994) reported that for dairy wastes with COD less than 2000 mg/L, acidification instead of methanogenesis was the rate-limiting step in COD reduction. However, Elliott et al. (1991) found that when treating a high-strength waste such as whey permeate, the rate of acid production was too rapid and acetate and propionates accumulated to concentrations that were inhibitory to the methanogenic bacteria. Rico-Gutierrez et al. (1991) indicated that some addition of alkali may be necessary in the startup of the reactor to maintain buffering capacity until a mature bacterial population is established and methanogenesis is proceeding in a uniform manner.

IV. TREATED DAIRY EFFLUENTS

Effluents from waste treatment systems must be sufficiently reduced in BOD and biological nutrients (e.g., P, NH_3) that discharge to surface waters does not significantly affect aquatic life. Environmental regulatory agencies specify limits for composition of effluents discharged to each type of stream or watershed. To reduce the volume of dairy wastewater to be treated and reduce treatment costs, careful attention must be given to minimizing losses of milk and milk products in the dairy plant. With good product conservation and selection of an effective waste treatment process, dairy plant operators should be able to operate profitably and meet environmental requirements.

REFERENCES

Alleman JE, Veil JA, Canaday JT. Scanning electron microscope evaluation of rotating biological contactor biofilm. Water Res 16:543, 1982.

Allison LW. Effect of microorganisms on permeability of soil under prolonged submergence. Soil Sci 63:439, 1947.

American Public Health Association. In: Greenberg AE, Clesceri LS, Eaton AD, eds. Standard Methods for the Examination of Water and Wastewater. 18th ed. Washington, DC: American Public Health Association, 1992.

Anderson GK, Kasapgil B, Ince O. Comparison of porous and non-porous media in upflow anaerobic filters when treating dairy wastewater. Water Res 28:1619, 1994.

Balch WE, Fox GE, Magnum LJ, Woese CR, Wolfe RS. Methanogenesis: reevaluation of a unique biological group. Microbiol Rev 143:260, 1979.

Bangsbo-Hansen D. Dairy wastewater. Nordeuropaeisk-Mejeri-Tidsskrift 44:101, 1978.

Bitton G. Wastewater Microbiology. New York: John Wiley & Sons, 1994.

Brady NC. The Nature and Properties of Soils, New York: Macmillan, 1990.

Buchan L. The location and nature of accumulated phosphorus in seven sludges from activated sludge plants which exhibited enhanced phosphorus removal. Water SA 7:1, 1981.

Carawan RE, Jones VA, Hansen AP. Wastewater characterization in a multiproduct dairy. J Dairy Sci 62:1243, 1979.

Chudoba J. Activated sludge-bulking control. In: Cheremisinoff PN, ed. Encyclopedia of Environmental Control Technology. Vol. 3. Wastewater Treatment Technology. Houston, TX: Gulf Publishing, 1989, p 171.

Curds CR, Hawkes HA. Ecological Aspects of Used-Water Treatment. Vol. 2. London: Academic Press, 1983.

Drewry WA, Eliassen R. Virus movement in ground water. J Water Poll Control Fed 40:R257, 1968.

Elliott JM, Parkin KL, Johnson EA. Biomethanation of cheese whey permeate by anaerobic digestion using a sewage sludge inoculum. Final Res. Report, Department of Food Science, Madison, WI: University of Wisconsin, 1991.

Fang HHP. Pretreatment of wastewater from a whey processing plant using activated sludge and anaerobic processes. J Dairy Sci 74:2015, 1991.

Fried M, Broeshart H. The Soil-Plant System. New York: Academic Press, 1967.

Fuhs GW, Chen M. Microbial basis for phosphate removal in the activated sludge process for the treatment of wastewater. Microbiol Ecol 2:119, 1975.

Ghosh S, Ombregt JP, Pipyn P. Methane production from industrial wastes by two-phase anaerobic digestion. Water Res 19:1083, 1985.

Goodfellow M. Properties and composition of the bacterial flora of a pine forest soil. J Soil Sci 19:154, 1968.

Guichet J, Jambu P, Dinel H. Changes in the organic matter of a Rendzina soil caused by wastewater sludge disposals from dairy processing plants. Pedologie 41:149, 1991.

Guiot SR, Safi B, Frigon JC, Mercier P, Mulligan C, Tremblay R, Samson R. Performance of a full-scale novel multiplate anaerobic reactor treating cheese whey effluent. Biotechnol Bioeng 45:398, 1995.

Hanel L. Biological Treatment of Sewage by the Activated Sludge Process, Chichester, UK: Ellis Horwood, 1988.

Harper WJ, Blaisdell JL. Proceedings of Second National Symposium on Food Processing Wastes, Water Pollution Control Res. Series 12060. Doc NO EP 1.16:12060, pp 509–545. Washington, DC: Environmental Protection Agency, 1971.

Harper WJ, Carawan RE. Approaches to minimizing water usage and wastes in dairy plants. Int Dairy Fed Bull Doc 104. Brussels, Belgium: International Dairy Federation, 1978, p 53.

Harper WJ, Delaney RAM, Igbeka IA, Parkin ME, Schiffermiller WE, Ross TE, Williams RA. Strategies for water and waste reduction in dairy food plants. EPA/600/2-85/076. Washington, DC: Environmental Protection Agency, 1985.

Hattori T. Microbial Life in the Soil. New York: Marcel Dekker, 1973.

Jones HR. Pollution Control in the Dairy Industry. Park Ridge, NJ: Noyes Data Corp, 1974.

Kaiser SK, Dague RR. The temperature-phased anaerobic biofilter process. Water Sci Technol 29:213, 1994.

Kato MT, Field JA, Kleerebezem R, Lettinga G. Treatment of low strength soluble waste-waters in UASB reactors. J Ferment Bioeng 77:679, 1994.

Kelling KA, Peterson AE. Using whey on agricultural land—A disposal alternative. UW Extension Bull A3098. Madison, WI: University of Wisconsin, 1981.

Lance JC. Nitrogen removal by soil mechanisms. Water Poll Control Fed 44:1352, 1972.

Lau AO, Strom PF, Jenkins D. Growth kinetics of *Sphaereotilus natans* and a floc former in pure and dual continuous culture. J Water Pollut Control Fed 56:41, 1984.

Law JP Jr, Thomas RE, Myers LH. Nutrient removal from cannery wastes by spray irrigation of grassland. FWPCA, 16080-11/69. Washington, DC: U.S. Department of the Interior, 1969.

Law JP Jr, Thomas RE, Myers LH. Cannery wastewater treatment by high rate spray on grassland. J Water Poll Control Fed 42:1621, 1970.

Lebrato J, Perez-Rodriguez JL, Maqueda C, Morillo E. Cheese factory wastewater treatment by anaerobic semicontinuous digestion. Resources Conserv Recycling 3:193, 1990.

Lettinga G, van Velsen AFM, Hobma SW, de Zeeuw W, Klapwijk A. Use of upflow sludge blanket (USB) reactor concept for biological wastewater treatment, especially for anaerobic treatment. Biotechnol Bioeng 22:699, 1980.

Mackie DM, Bryant MP. Metabolic activity of fatty acid-oxidizing bacteria and the contribution of acetate, propionate, butyrate and CO_2 to methanogenesis in cattle waste at 40 and 60°C. App Environ Microbiol 41:1363, 1981.

MacLeod FA, Guiot SR, Costerton JW. Layered structure of bacterial aggregates in an upflow anaerobic sludge bed and filter reactor. Appl Environ Microbiol 56:1598, 1968

March D, Benefield L, Bennett E, Linstedt D, Hartman R. Coupled trickling filter-rotating biological contactor nitrification process. J Water Pollut Control Fed 53:1469, 1981.

Marshall KR. The characteristics of effluents from New Zealand dairy factories. Int Dairy Fed Bull Doc 104. Brussels, Belgium: International Dairy Federation, 1978, p 123.

Matzke S, Wendorff WL. Chloride in cheese manufacturing wastes to be landspread on agricultural land. Bioresource Technol 46:251, 1993.

McGauhey PH. Manmade contamination hazards. Ground Water 6(3):10, 1968.

Metcalf and Eddy, Inc. Wastewater Engineering: Treatment, Disposal and Reuse. New York: McGraw-Hill, 1991.

Metting JB Jr. Structure and physiological ecology of soil microbial communities. In: Metting FB Jr, Soil Microbial Ecology. New York: Marcel Dekker, 1993, p 3.

Norberg AB, Enfors SO. Production of extracellular polysaccharide by *Zooglea ramigera*. Appl Environ Microbiol 44:1231, 1982.

Parkin MF, Marshall KR. Land disposal of dairy factory effluent by spray irrigation. Int Dairy Fed Bull Doc 104. Brussels, Belgium: International Dairy Federation, 1978, p 151.

Perle M, Kimchie S, Shelef G. Some biochemical aspects of the anaerobic degradation of dairy wastewater. Water Res 29:1549, 1995.

Pickett M. F & A Dairy Products finds efficient wastewater system. Dairy Field 171(5):26, 1988.

Peterson AE, Walker WG, Watson KS. Effect of whey application on chemical properties of soils and crops. J Agric Food Chem 27:654, 1979.

Rico-Gutierrez JL, Garcia-Encina PA, Fernandez-Polanco F. Anaerobic treatment of cheese-production wastewater using a UASB reactor. Bioresource Technol 37:271, 1991.

Sammaiah P, Sastry CA, Murty DVS. Dairy wastewater treatment using an anaerobic contact filter. Indian J Environ Prot 11:418, 1991.

Schulte SR. SBR treatment of food process wastewater. Proceedings of 1988 Food Processing Waste Conference, Georgia Institute of Technology, Atlanta, GA, 1988.

Sharratt WJ, Peterson AE, Calbert HE. Whey as a source of plant nutrients and its effect on the soil. J Dairy Sci 42:1126, 1959.

Sharratt WJ, Peterson AE, Calbert HE. Effect of whey on soil and plant growth. Agron J 54:359, 1962.

Speece RE. Anaerobic biotechnology for industrial waste treatment. Environ Sci Technol 17:416A, 1983.

Spyridakis DE, Welch EB. Treatment processes and environmental impacts on waste effluent disposal on land. In: Sanks RL, Asano T, ed. Land Treatment and Disposal of Municipal and Industrial Wastewater. Ann Arbor, MI: Ann Arbor Science Publishers Inc., 1976, p 45.

Sterritt RM, Lester JN. Microbiology for Environmental and Public Health Engineers. London: E & FN Spon, 1988.

Surampalli RY, Baumann ER. RBC kinetics in treating domestic and industrial dairy wastewater under low and high organic loading conditions. Water Pollution Res J Canada 27:665, 1992.

U.S. Environmental Protection Agency. Process Design Manual for Nitrogen Control. Washington, DC: Office of Technology Transfer, 1975.

U.S. Environmental Protection Agency. Wastewater Treatment Facilities for Sewered Small Communities. Report EPA/625/1-77/009, 1977.

U.S. Environmental Protection Agency. Phosphorus Removal. Design Manual EPA/625/1-87/001, 1987.

Viraraghavan T, Kikkeri SR. Effect of temperature on anaerobic filter treatment of dairy wastewater. Water Sci Technol 22:191, 1990.

Wendorff WL. Landspreading whey permeate: watch the parameters. UW Dairy Pipeline 1(1):1, 1989.

Wendorff WL. Phosphorus in dairy wastes—a real challenge. UW Dairy Alert, Nov. 15, 1991. Madison, WI: University of Wisconsin, 1991.

Wendorff WL. Revised guidelines for landspreading whey and whey permeate. UW Dairy Alert, June 1, 1993. Madison, WI: University of Wisconsin, 1993.

Wendorff WL. Managing chloride and phosphorus by minimizing wastes. UW Dairy Pipeline 7(4):1, 1995.

Wendorff WL, Matzke S. Phosphorus in major varieties of whey produced in Wisconsin. Dairy Food Environ Sanit 13:166, 1993.

Weng CN, Molof AH. Nitrification in the biological fixed film RBC. J Water Pollut Control Fed 46:1674, 1974.

Wu W, Hu J, Gu X, Zhao Y, Zhang H. Cultivation of anaerobic granular sludge with aerobic activated sludge as seed. Water Res 21:789, 1987.

Index